REDD+ Crossroads Post Paris:
Politics, Lessons and Interplays

Special Issue Editors

Esteve Corbera
Heike Schroeder

MDPI • Basel • Beijing • Wuhan • Barcelona • Belgrade

MDPI

Special Issue Editors
Esteve Corbera
Universitat Autònoma de Barcelona
Spain

Heike Schroeder
University of East Anglia
UK

Editorial Office
MDPI AG
St. Alban-Anlage 66
Basel, Switzerland

This edition is a reprint of the Special Issue published online in the open access journal *Forests* (ISSN 1999-4907) from 2016–2017 (available at: http://www.mdpi.com/journal/forests/special_issues/redd_impact).

For citation purposes, cite each article independently as indicated on the article page online and as indicated below:

Lastname, F.M.; Lastname, F.M. Article title. *Journal Name*. **Year**. *Article number*, page range.

First Edition 2018

ISBN 978-3-03842-707-0 (Pbk)
ISBN 978-3-03842-708-7 (PDF)

Cover photo courtesy of Esteve Corbera

Table of Contents

About the Special Issue Editors v

Preface to "REDD+ Crossroads Post Paris: Politics, Lessons and Interplays " vii

Erlend A. T. Hermansen, Desmond McNeill, Sjur Kasa and Raoni Rajão
    Co-Operation or Co-Optation? NGOs Roles in Norways International Climate and Forest
    Initiative
    doi:10.3390/f8030064 1

Jovanka Spirić, Esteve Corbera, Victoria Reyes-García and Luciana Porter-Bolland
    A Dominant Voice amidst Not Enough People: Analysing the Legitimacy of Mexicos REDD+
    Readiness Process
    doi:10.3390/f7120313 28

Sam Airey and Torsten Krause
    Georgetown aint got a tree.   We got the trees—Amerindian Power & Participation
    in Guyanas Low Carbon Development Strategy
    doi:10.3390/f8030051 50

Thu Ba Huynh and Rodney J. Keenan
    Revitalizing REDD+ Policy Processes in Vietnam: The Roles of State and Non-State Actors
    doi:10.3390/f8030053 74

Irmeli Mustalahti, Mathias Cramm, Sabaheta Ramcilovic-Suominen and Yitagesu T. Tegegne
    Resources and Rules of the Game: Participation of Civil Society in REDD+ and FLEGT-VPA
    Processes in Lao PDR
    doi:10.3390/f8020050 91

Adeniyi P. Asiyanbi, Albert A. Arhin and Usman Isyaku
    REDD+ in West Africa: Politics of Design and Implementation in Ghana and Nigeria
    doi:10.3390/f8030078 112

Dara Y. Thompson, Brent M. Swallow and Martin K. Luckert
    Costs of Lost opportunities: Applying Non-Market Valuation Techniques to Potential
    REDD+ Participants in Cameroon
    doi:10.3390/f8030069 136

Kimberly R. Marion Suiseeya
    Transforming Justice in REDD+ through a Politics of Difference Approach
    doi:10.3390/f7120300 151

Maria Fernanda Gebara and Arun Agrawal
    Beyond Rewards and Punishments in the Brazilian Amazon: Practical Implications of the
    REDD+ Discourse
    doi:10.3390/f8030066 165

Leif Tore Trdal and Pal Olav Vedeld
    Livelihoods and Land Uses in Environmental Policy Approaches: The Case of PES and
    REDD+ in the Lam Dong Province of Vietnam
    doi:10.3390/f8020039 192

Timothy Cadman, Tek Maraseni, Hugh Breakey, Federico López-Casero and Hwan Ok Ma
Governance Values in the Climate Change Regime: Stakeholder Perceptions of REDD+
Legitimacy at the National Level
doi:10.3390/f7100212 . . . . **212**

Ignacia Holmes, Catherine Potvin and Oliver T. Coomes
Early REDD+ Implementation: The Journey of an Indigenous Community in Eastern
Panama
doi:10.3390/f8030067 . . . . **229**

Courtney Work
Forest Islands and Castaway Communities: REDD+ and Forest Restoration in Prey Lang
Forest
doi:10.3390/f8020047 . . . . **247**

Alexander Pfaff, Francisco Santiago-Ávila and Lucas Joppa
Evolving Protected-Area Impacts in Mexico: Political Shifts as Suggested by Impact
Evaluations
doi:10.3390/f8010017 . . . . **268**

Taiji Fujisaki, Kimihiko Hyakumura, Henry Scheyvens and Tim Cadman
Does REDD+ Ensure Sectoral Coordination and Stakeholder Participation? A Comparative
Analysis of REDD+ National Governance Structures in Countries of Asia-Pacific Region
doi:10.3390/f7090195 . . . . **282**

Mareike Blum and Sabine Reinecke
Towards a Role-Oriented Governance Approach: Insights from Eight Forest Climate
Initiatives
doi:10.3390/f8030065 . . . . **299**

**Pamela McElwee, Van Hai Thi Nguyen, Dung Viet Nguyen, Nghi Huu Tran, Hue Van Thi Le,
Tuyen Phuong Nghiem and Huong Dieu Thi Vu**
Using REDD+ Policy to Facilitate Climate Adaptation at the Local Level: Synergies and
Challenges in Vietnam
doi:10.3390/f8010011 . . . . **323**

Mari Mulyani and Paul Jepson
Does the One Map Initiative Represent a New Path for Forest Mapping in Indonesia?
Assessing the Contribution of the REDD+ Initiative in Effecting Forest Governance Reform
doi:10.3390/f8010014 . . . . **347**

Johanne Pelletier, Nancy Gélinas and Margaret Skutsch
The Place of Community Forest Management in the REDD+ Landscape
doi:10.3390/f7080170 . . . . **368**

About the Special Issue Editors

Esteve Corbera (Ph.D., MSc) is an environmental social scientist who studies how climate change programs and biodiversity conservation policies impact livelihoods, resource management institutions and peoples behaviour. He works at the Institute of Environmental Science and Technology (ICTA), Universitat Autnoma de Barcelona, where he co-leads the Laboratory for the Analysis of Socio-Ecological Systems in a Globalised World (LASEG). Previously, he has worked at the University of East Anglia, the University of Cambridge, and The Tyndall Centre for Climate Change Research. He is an editor of the journal Geoforum, and was an author in the 5th Assessment Report of the Intergovernmental Panel on Climate Change. He holds a PhD in Development Studies from the University of East Anglia. He can be contacted at esteve.corbera@uab.cat and followed at www.estevecorbera.com.

Heike Schroeder is a senior lecturer (associate professor) in climate change and international development at the School of International Development, University of East Anglia. Her work focuses on the UNFCCC process, forest governance and REDD+ and urban climate governance. She is a member of the Tyndall Centre for Climate Change Research, the Earth System Governance project under Future Earth and IDDRIs Scientific Council. Previously, Heike has held positions at the Environmental Change Institute, University of Oxford and the Bren School of Environmental Science and Management, University of California, Santa Barbara. She holds a Ph.D. in political science from the Free University of Berlin.

Preface to "REDD+ Crossroads Post Paris: Politics, Lessons and Interplays "

Abstract: This article introduces the special issue REDD+ crossroads post Paris: politics, lessons and interplays. The contributions to the special issue suggest, first, that REDD+ design in the studied countries has generally lacked social legitimacy and sidelined key actors who can considerably influence land-use sector dynamics. Second, they show that REDD+ early actions have tended to oversimplify local realities and have been misaligned and local needs. Third, REDD+ efforts have remained constrained to the forestry or climate mitigation policy sectors and have thus suffered from a lack of policy harmonization. As REDD+ moves from its preparedness to its implementation phase, more research efforts should be aimed at analysing the power relations that underpin and determine the design and implementation of REDD+ policies and actions, the potential for and limits to the vertical and horizontal coordination of land-use policies and management, and the processes of resistance to or accommodation of REDD+ practices on the ground. In doing so, we advocate for multi- and transdisciplinary research that does not take for granted the benefits of REDD+ and which critically scrutinizes the multiple goals of this ambitious international policy framework, and where it sits within the broader Paris Agreement implementation agenda.

Keywords: REDD+; environmental governance; politics; conflict; climate change

1. Introduction

Reducing Emissions from Deforestation and Forest Degradation, conserving and enhancing forest carbon stocks, and sustainably managing forests (REDD+), has become a reference framework for national forest governance across many tropical and subtropical forest countries [1,2] These countries have used funds from multilateral and bilateral aid platforms to re-organise forest management and conservation policy around the idea of mitigating climate change, including the development of national REDD+ strategies and both carbon accounting and benefit-sharing protocols [3]. In parallel, international conservation organizations have mobilised REDD+ principles and assumptions to promote small-scale pilot activities across the tropics, in order to capture the economic value of any resulting land-use emission reductions, mostly and rather limitedly through voluntary carbon markets [4].

REDD+ emerged at a time when an alternative path to the arguably dead-end of the Kyoto Protocol needed to be carved out that would allow all countries to participate in long-term cooperative action. The prospect of avoided deforestation being a cheap mitigation option [5] convinced many countries to endorse the adoption of the REDD+ mechanism as part of the 2007 Bali Action Plan that contained the building blocks of what has ultimately become the 2015 Paris Agreement. Thus, the idea of payment for standing forests was born and is now a core element of many forest nations in their Nationally Determined Contributions (NDCs) under the Paris Agreement [6]. International programmes, in particular the World Banks Forest Carbon Partnership Facility (FCPF) and UN-REDD, have been instrumental in paving the way for host country capacity building and readiness activities to enable implementation on the ground [7].

The political traction that REDD+ has sparked in the international climate change negotiations and multilateral funding frameworks has translated into considerable academic attention. A search in Scopus, an international repository of scientific articles, reveals that the number of articles with REDD in the title alone has exponentially increased from two articles in 2007 to over 550 articles in 2017. The search ignores of course the articles that might have studied REDD+ policy and projects using other terminology, which implies that REDD+ analyses are probably more numerous. Generally speaking, research on REDD+ can be classified across two very broad schools of thought. On the one hand, there are those who have looked skeptically at the framework, departing from the premise that REDD+ might be, or might become, yet another top-down, blueprint approach to environmental management that might reify inequities in access to forest and land resources while limiting the right to develop of host countries (e.g., [8,9]). On the other hand, there are those who have approached REDD+ more optimistically, looking at how the framework could support both conservation and development aspirations through better design and implementation [10,11].

It is commonly agreed across these two schools that REDD+ policy preparedness and pilot project implementation have been characterized by numerous challenges, some related to broader UNFCCC and aid policy features and others related to in-country lack of capacities, policy misalignments and an unfavourable political economy [12,13]. There is also consensus that REDD+ programs and actions can fall short in terms of their legitimacy [14,15], as well as result in social conflicts, unexpected and counterproductive tenure arrangements [16] or limited wellbeing gains [17,18]. However, it is also true that the short shelf life of REDD+, coupled with the fact that forest and land-use governance reform require long-term thinking and a profound change in national and international political economies [19], increases the risk that both scholars and the international policy community might throw the baby out with the bathwater.

This special issue is an attempt to contribute to this burgeoning research on REDD+, providing theoretically grounded and empirical evidence on three analytical domains: the politics of REDD+ design; the lessons from REDD+ early actions; and REDD+ policy interplays. The contributions are a response to a question that the two guest editors considered timely to pose in 2015, i.e., how is REDD+ unfolding and with which consequences? With many countries developing their national REDD+ strategies or hosting pilot projects over the last few years, it was important to ask about the political nature of these endeavours and to gather specific evidence on how various governments have designed and rolled out their REDD+ strategies, and how and why a range of public and private actors have or have not become involved in such processes. It was important to explore which ratio-nales, techniques, views and values are being contested and constructed in the design of REDD+ national strategies and pilot projects, which conflicts have emerged and why, or how coordination across competing actors and interests has been pursued. Relatedly, it was timely to look for more evidence on the impacts of policies and projects on specific environments, peoples, and ecosystems, since in some cases these interventions might have been in place for a few years by then. Finally, if one assumed that reducing deforestation and improving forest governance required harmonizing land-use policies and eliminating both public and private incentives that result in the unsustainable management of forests, then it was necessary to collect evidence on the ability of REDD+ processes to improve forest policy regulations and influence both development and land-use policies.

The contributions below shed light on the above-mentioned issues, and we think that, combined with other emerging evidence, they provide a solid knowledge basis upon which to build the research agenda of REDD+ in its post-preparedness phase. As developing countries move toward and into implementation of their national REDD+ strategies, in the broader context of their NDCs, some of the red flags identified to date in terms of REDD+ policy design and on-the-ground implementation will require further attention as new questions will also become relevant. In the closing of this article, we highlight what we think these questions might be and in doing, so we hope to be drawing a sensible agenda for the near future.

2. Politics of REDD+ Design

The collection starts with a batch of eight articles that analyse the politics of REDD+ design, as well as of related funding frameworks. The fact that this is the topic of the special issue deserving more coverage can be explained by the fact that REDD+ in its early years mostly consisted of developing national policy strategies and supporting pilot, often short-lived projects, while the translation of these policy processes and the impacts of pilot projects are yet to be seen or realized. An underlying concern across these contributions relates to the lack of legitimacy that these processes and projects have suffered from, and the problematic absence of social actors that play a key role in development planning and land-use change.

The first contribution explores the role that NGOs have played in crafting and facilitating Norways International Climate and Forest Initiative (NICFI), the worlds largest single country attempt to support the development of climate and REDD+ related policies and projects over the last decade in developing countries. Hermansen et al. [1] use qualitative process-tracing to analyse how NGOs in Norway, Brazil, Indonesia, and Tanzania have cooperated with the Norwegian and their own country governments to facilitate the early start of REDD+ policy design and implementation. The authors argue that cooperation should be distinguished from co-optation to the extent that NGOs remain critical of certain aspects of REDD+ and have found forums and ways to exert an influence on and to change NICFIs operating procedures. In Norway, NGOs have opted to use private forums to voice their concerns directly to government and have remained strategically quiet in the public arena about the challenges of REDD+ implementation on the ground, albeit with some exceptions from development-focused NGOs. The authors show that public criticism of REDD+ by NGOs has been minimal in Brazil and Tanzania where NGOs have been mostly cooperative, driven by the fact that they became core recipients of aid and thus key implementation actors.

The articles by piric et al. [15] and Airey and Krause [20] explore the process of REDD+ policy design in Mexico and Guyana, respectively. In these two Latin American countries, the authors observe a lack of inclusion and recognition of key actors in the policy process, although their analytical and scalar focus is different. Spiric and colleagues explore the legitimacy of the political processes leading to Mexicos REDD+ national strategy over the period 2011–2014, grounded in stakeholder and discourse analysis. The authors highlight that the government agencies have retained most decision-making power, while representatives of relevant land-use sectors and local communities have been absent from formal decision-making forums. They also show the existence of three main narratives about REDD+ in Mexico and demonstrate that the narrative predominantly held by government agencies, multilateral and international conservation organizations has found its way into the strategys final draft much more than other narratives supported by NGOs and academia. Therefore, Spiric and colleagues argue that Mexicos REDD+ process should become more inclusive, decentralised, and better coordinated to allow for the deliberation and institutionalisation of different actors ideas in REDD+ design (ibid.).

Airey and Krause, in turn, use in-depth interviews with the Amerindian community of Chenapou to analyse the extent to which they had been informed about and participated in the design of Norway-funded Guyanas Low Carbon Development Strategy (LCDS), in which any ensuing REDD+

policy and project developments should be grounded. Their analysis suggests that the government has neither dedicated sufficient resources nor used appropriate participatory tools to inform and involve the people of Chenapou in the design of the LCDS. Additionally, no attempt was made to address the fact that Amerindian communities are not properly represented in the countrys political system, with their traditional village council institutions disconnected from the state administration and thus from decision-making powers. For the authors, this is problematic on political and ethical grounds, but also conducive to risks in future REDD+ implementation. In their own words, by not engaging with the community, the actors in power (i.e., the governments of Guyana and Norway) miss out on the value of local knowledge and input, which may hold significant benefit for projects (ibid.).

Another two articles dig deeper into issues related to participation and representation in South-East Asia. Huynh and Keenan [21] explore how Vietnamese policy actors view REDD+ policy development and their influence on these processes. In line with Spiric et al., the authors examine the influence of state and non-state actors on the 2012 National REDD+ Action Program (NRAP) processes, using a combination of document analysis and semi-structured interviews. They find that most non-state actors in REDD+ exerted a minor influence on NRAP developments, and demonstrate that national NGOs were much more entangled and thus satisfied with the states rather authoritarian decision-making procedures than international NGOs, which felt more excluded and less able to influence the policy process. For the authors, in the coming years, non-state actors will be able to take advantage of the fact that the countrys ruling party is increasingly eager to receive information on best REDD+ practice and, in this context, they should find the right tactics to improve the participatory character of REDD+ policy and to expand and build extra-sectoral partnerships across multiple organisational boundaries, including those beyond the forestry sector (ibid.).

In Lao PDR, Mustalahti et al. [22] analyse the importance and involvement of civil society organisations (CSOs) in the design and implementation of REDD+ and the European Union Action Plan on Forest Law Enforcement, Governance and Trade (FLEGT), since these are two related policy frameworks that aim to address deforestation, forest degradation and promote sustainable forest management. Informed by both surveys and interviews, the authors show that CSOs have been key players in the roll-out of both frameworks to date, but they argue that such involvement has not resulted in significant changes in the governments objectives and decisions over both frameworks. In line with Huynh and Keenan above, Mustalahti et al. distinguish between international and national CSOs but suggest that, while both experience powerlessness in front of the state, the latter are further hampered by their lack of knowledge, capacities and resources to follow-up, participate and possibly benefit from the two policy processes. Furthermore, the authors point out that CSOs should better coordinate among themselves, since doing so might increase their technical capacity and maximize the resources available to mobilize other CSOs working in environment and development issues more broadly, which could ultimately increase their decision-making power over REDD+ and FLEGT policy processes.

The first batch of articles focused on the politics of REDD+ design ends with another empirically grounded article exploring REDD+ policy in two African countries, and with a contingent valuation study of REDD+ costs and a theoretically informed review of REDD+ research through a justice lens. Asiyanbi et al. [23] investigate the early development of REDD+ in Ghana and Nigeria and they distinguish between the processes and principles that have guided the writing of REDD+ national strategies from the early steps conducted to roll-out the strategies, namely, institutional and capacity building, carbon accounting and property rights over land and carbon. Regarding the former, and in contrast with other contributions in the collection, they show that policy design included the diversity of all relevant actors through multiple meetings and forums over time, and captured their contrasting interests through broad, ambitious and possibly contradictory aims and objectives in policy documents. Regarding the latter, they demonstrate that in both countries, there have been intra-governmental struggles over who should gain institutional control of the REDD+ process, and favoritism over which social actors should be involved and benefit from capacity building. Simultaneously, state administrations have used REDD+ to assert control over land, forest and carbon at the expense of community rights and public access to timber and non-timber forest products.

Thompson et al. [24] investigate farmers Willingness To Accept (WTA) compensation for REDD+ in southern Cameroon. Their analysis allows understanding the costs or foregone benefits that farmers face when engaging in REDD+ activities and the appeal of such activities when they land in complex social-ecological landscapes characterized by land-use change and social heterogeneity. The study demonstrates that farmers WTA is much higher than the previously calculated opportunity costs for Forests 2017, 8, 508 5 of 11 conserving forests in Cameroon and elsewhere, and they also show that such WTA varies according to expectations over agricultural markets and tenure security, as well as according to farmers age and household land-use patterns. The findings challenge the assumption that REDD+ is a cheap mitigation option and they shed light on the importance of delivering conservation incentives paying attention to the diversity of farming strategies and livelihood needs. Finally, Marion Suiseeyas [25] offers a comprehensive review of justice-related REDD+ debates and demonstrates how these have been generally grounded in simplistic visions of either distributive or procedural justice, ignoring the importance of recognition as the third pillar of justice. According to her, embracing the recognition principle would make of REDD+ a less technocratic endeavour, more sensitive to the values, identities, lifeways, and voices that might be displaced as a result of REDD+ activities.

3. Lessons from REDD+ Early Actions

When it comes to understanding the early lessons of REDD+ early actions, there are several themes that the six articles below point to repeatedly. They include the peril of oversimplification and generalisation at the local level, thereby missing local dynamics and complexities and leading to adverse outcomes; the peril of not aligning project goals with local needs; the relatively low levels of deforestation in the poorest segments of society; and the potential of conflict if tenure rights are unclear or ignored.

To begin, Gebara and Agrawal [26], apply Foucaults concepts of governmentality and technologies of governance to argue that local complexities, heterogeneities and multi-dimensional capacities are often forsaken for linear and rational simplicity. Their case study focuses on reactions to rewards and punishments by smallholder farmers in Brazil. They find that the techniques of remuneration and coercion on which a rewards and punishments approach rests are only supporting limited behavioural changes on the ground, achieving negative adaptations and deforestation practices, reducing positive feedbacks and, above all, only producing short-term outcomes at the expense of positive long-term land use changes. They conclude by emphasising the importance of looking at local heterogeneities and capacities and the need to promote trust, altruism and responsibility towards others and future generations (ibid.) as one key lesson from REDD+ early actions. Conceptually, the authors contribute to the governmentality literature by adding to its focus on rationalities of governance a regard for social practices within which such rationalities are embedded.

Next, Trdal et al. [27] apply a livelihoods framework to discuss the link between livelihoods and land use amongst smallholder farmers. Their findings, based on work in Vietnam, similarly indicate that one-dimensional perspectives of the drivers behind deforestation do not lead to appropriate solutions to address the underlying dynamics at stake. Instead, the authors suggest addressing issues of land tenure and the scarcity of productive lands, and generating viable off-farm income alternatives through engaging with multiple stakeholders, which could include business-oriented households in control of a commodity trade (coffee in this case) and of land transactions. Another finding of this study is that the poorest segments of the population are found to deforest the least. Their data showed that middle to better-off households cultivate most land (including for coffee production) and have cleared more forest land for agriculture over the years than poorer households. This is true especially when taking into account uncertified, or illegal, clearing. The authors therefore conclude that the focus on the linkages between poverty and coffee-related forest encroachment seems to be overemphasised in PES and REDD+ policies and discourses. Their main argument is therefore that, in order to enable a more comprehensive understanding of land-use change and its management, the focus should be expanded beyond the poverty-environment nexus (ibid.).

Cadman et al. [28] use an online survey and two workshops targeting a number of REDD+ stakeholder groups to investigate the quality of REDD+ governance in Nepal and Papua New Guinea (PNG), in order to infer lessons on the frameworks legitimacy and to identify what matters most for its future effectiveness, given distinct national priorities and contexts. The authors demonstrate that the governance principles of inclusiveness, resources, accountability, and transparency are ranked Forests 2017, 8, 508 6 of 11 very highly by most participants in both countries, even if they are distinctively prioritised in different ways, while the principles of agreement, dispute settlement, and problem solving were the least prioritised and again in different orders. A key finding of the article is not so much the existence of certain consensus on which principles should matter most in designing and implementing REDD+, but the fact that in both countries the lack of resources and inclusiveness are two related and important caveats of REDD+ implementation, which taken together compromise the transparency and accountability of REDD+ governance. In other words, a limited participation of key stakeholders given the lack of funding to facilitate such participation might lead to unfair decisions and suspicion over who made such decisions and why. For the authors, their findings also challenge the idea that REDD+ governance will be on safe grounds if focused narrowly on FPIC or safeguards, rather than on broader governance principles that merit attention across potentially diverging socio-political and cultural contexts.

Next, Holmes et al. [29] also find that success of REDD+ requires more nuance and alignment with local needs. The authors used semi-structured interviews and participatory methods over the span of 11 years in an Embera community in Panama to examine how local communities sought to reduce emissions from deforestation and benefit from carbon offset trading while improving local livelihoods. Their findings lead the authors to suggest the development of a support system for implementation through bridging institutions and the use of REDD+ for conflict resolution around tenure issues, in addition to economic incentives, as ways to respond to the needs of local communities and improve their livelihoods. In conclusion, they propose that successful implementation of REDD+ projects in small rural or indigenous communities requires a shift in paradigm, away from evidence-based payments and towards an integrated development approach (ibid.). This sums up a key lesson derived from this work: REDD+ takes much longer to be implemented because it requires redefining livelihood strategies, i.e., transformative change, and this cannot simply be achieved through evidence-based payments based on the current market price for carbon. Rather, evidence-based payments need to be combined with development and poverty reduction.

This issue of conflict resolution is also picked up byWork [30]. Taking a landscape approach to analysing forest-based climate change mitigation policies and land grabs in Cambodia, Work argues that forest-based climate change mitigation policies can reduce conflict through attention to tenure rights, responsibilities and authority of citizens, and by improving project landscapes. However, given the need for resource capture to generate economic growth, conflict is intrinsic to climate change policies. This exposes, according to the author, deep contradictions at the heart of climate change policies that require attention if nothing but forest islands and castaway communities are to be avoided (ibid.).

Finally, Pfaff et al. [31], in a study on shifts in impacts of Protected Areas (PAs) in Mexico over time, comparing the 1990s with the period 2000–2005 when conservation politics had allegedly shifted, find that PAs have indeed shifted in impacts from being paper parks to achieving a reduction of 3.2conservation of forests, and that the strictness and proximity to cities have a positive impact on their success. With this research, they contribute to understanding variation in forest impacts over time, which is much less understood compared to impacts across space.

4. REDD+ Interplays

The special issue now turns to five articles that shed light on the extent to which REDD+ policy or early actions have been able to influence other policy frameworks that might contradict REDD+ objectives, or to what extent REDD+ can be an opportunity to coordinate different policy sectors. The contributions suggest overall the existence of tension between what has been subsumed under REDD+ as additional to deforestation, namely forest degradation, sustainable forest management, conservation and enhancement of forest carbon stock, as well as non-carbon benefits and safeguards, and what significantly impacts deforestation rates but remains outside Forests 2017, 8, 508 7 of 11 of the scope of the REDD+ mechanism. Additionally, as also highlighted by the first batch of articles, the contributions demonstrate that the agricultural and extractive industries have been problematically sidelined, purposely or not, from REDD+ policy discussions and policy harmonization efforts, which have remained constrained to the forestry or climate mitigation policy sectors. Equally, it is shown that REDD+ governance is characterised by a lack of integration of local, national and international concerns.

Fujisaki et al. [32], based on a comparative study of five countries, conclude that, despite structural differences across countries, REDD+ can potentially encourage new forms of environmental governance that promote a cross-sectoralism and stakeholder participation. However, their research identifies a lack of operationalisation of cohesiveness and inclusiveness in environmental policymaking processes in most developing countries, including in REDD+. They argue that cohesiveness within a broader governance system is key to defining the capacity of REDD+ governance, i.e., harmonising different policy sectors and interests that have impacts on forests. Furthermore, they highlight the need for inclusiveness and policy support for those affected by REDD+ to ensure their voices are heard in decision-making processes to overcome structural inequalities. In examining whether emerging governance arrangements help REDD+ development by delivering participatory mechanisms for policy coordination, they find that REDD+ potentially encourages a new form of environmental governance that promotes a cross-sectoral approach and stakeholder participation.

Blum and Reinecke [33] draw on social roles to examine eight different REDD+ governance processes in terms of the variety of practical authoritative roles enacted in administration, finance, decision making and knowledge production. By systematically identifying the ways in which different roles are filled, the authors develop a typology of structural role practices and underlying rationales. This can help assess legitimacy and empirically operationalise REDD+ governance performance. Their study illustrates that if new governance practices are created, the actors involved often bring in their understandings about roles (including working culture, values and administrative routines), which may not correspond with what was envisioned in the first place. Therefore, to avoid conflict during later phases, and for more clarity, expectations and conceptualizations of roles need to be discussed in the initial phase.

McElwee et al. [34] examine if REDD+ policies and projects on the ground acknowledge that climate change is likely to impact forests and forest users, if this knowledge is built into REDD+ policy and activities, how households in forested areas subjected to REDD+ policy are vulnerable to climate change, and how REDD+ can help or hinder adaptation to climatic changes. Using stakeholder interviews, focus groups and household surveys in three provinces in Vietnam, the authors find a lack of coordination between mitigation and adaptation policy with regard to REDD+ at the sub-national level. Policies for forest-based mitigation at national, subnational and project levels have paid little attention to the adaptation needs of local communities, where climate changes are already being experienced. They also suggest that there is an untapped potential for understanding how REDD+ activities could facilitate increased resilience. Most studied projects and policies did not explicitly target their activities to focus on adaptation or resilience, and in one case at least, negative livelihood impacts increased household vulnerability to climate change. Key barriers to integration include sectoral specialisation, a lack of attention in REDD+ projects to livelihoods and inadequate support for ecosystem-based adaptation.

In discussing the perils and merits of REDD+, or the question of whether REDD+ invites a skeptical or an optimistic outlook, Mulyani and Jepson [35] find that in the case of Indonesia REDD+ and institutional interplay have led to notable transformations in its forest institutions. Based on path dependence theory and eighty semi-structured interviews with REDD+ policy actors, the authors seek to determine the extent to which the REDD+ initiative has created a critical juncture for institutional change, the authors examine Indonesias one map initiative as the governments response to a call for greater transparency in implementing REDD+ and find that it has indeed led to improved transparency, public participation and coordination among ministries. It has shown an ability to break the old Forests 2017, 8, 508 8 of 11 path-dependence of map making but the research also uncovered several historical events that preceded the REDD+ initiative as contributing factors to the relative success of REDD+ in effecting forest governance reform.

Finally, Pelletier et al. [36] examine the interplay between REDD+ and community forest management (CFM) to determine the latters potential role of in achieving forest carbon benefits and social co-benefits for forest communities. Their systematic review of CFM case studies uncovers that there is strong evidence of CFMs role in reducing degradation and stabilising forested landscapes. However, the review also shows less evidence about its role in reducing deforestation. For social benefits, they find that CFM contributes to livelihoods, but its effect on poverty reduction may be limited. This might be because CFM may not deal adequately with the distribution of benefits within communities or user groups. The authors, like others in this special issue, argue that the recognition of rights for forest communities is one first step in promoting CFP-based REDD+ interventions and achieving both, positive biophysical and positive social outcomes.

5. Conclusions: A Renewed Research Agenda for REDD+

Back in 2011, we sketched out a research agenda for REDD+ which was built around three pillars: (i) the politics of REDD+ national strategy design, specifically how governments are designing REDD+ strategies, what degree of coordination and reform across policies and sectors is being sought and achieved and how different government and non-governmental actors are being involved in such discussions, and why [37]; (ii) the interplay between REDD+ and other policies and market processes in the land-use sector, and in particular whether REDD+ is able to transcend forest sector regulations, based on cross-sectoral and coordinated policy bodies and, . . . , how REDD+ policies and actions unfold in local contexts, through existing commercial networks, extension services and both legal and illegal markets for natural resources (ibid.); and (iii) the impacts of REDD+ policies and measures, including how these might transform institutions and livelihoods related to forest conservation and land-use, and how resource and carbon monitoring systems might change such institutions and be harmonized across scales.

This research agenda has been thoroughly pursued over the last few years, as demonstrated by the articles in this special issue and others beforehand (e.g., [38]). The contributions to this special issue generally demonstrate that the development of REDD+ national strategies have been characterized by participation deficits and coordination challenges, including the reproduction of illegitimate forms of top-down decision making, or the lack of coordination across land-use related policy sectors, which can ultimately undermine the legitimacy and effectiveness of REDD+, now and in the future. However, the development of REDD+ strategies has contributed in some countries to open up policy-making processes, and has improved land-use and forest governance, for example by unearthing and subsequently resolving existing conflicts over forest access and resource use. In turn, research focusing on the role that specific land-use lobbies have played in influencing the contents of national REDD+ strategies or in jeopardizing policy coordination processes remains rare. In other words, detailed analyses of the power relations underpinning REDD+ decision-making and implementation are lacking.

Nuanced understandings of how forest and rural communities have responded to or become engaged in REDD+ pilot projects have been published more recently [39–41] and, as REDD+ moves from the preparedness to its early implementation phase-with some countries signing Emission Reduction Purchase Agreements (ERPAs) and delivering quantifiable carbon emission reductions in forthcoming years-, we think that research on REDD+ implementation is likely to flourish. In-depth analyses of social-ecological impacts will thus be required, ideally combining distinct forms of expertise to address the multiple dimensions of REDD+, i.e., from spatial assessments to measure its environmental effectiveness to ethnographic accounts of local institutional and cultural change [42,43]. In line with others [44], we advocate for multi- or transdisciplinary research that takes into account the complexity of the social-ecological systems within which REDD+ actions will unfold, as well as Forests 2017, 8, 508 9 of 11 the latters (often problematic) rationale and multiple goals. Future research should be aware that many of those who have a stake in land-use management and forest conservationincluding the rural poormight not perceive reducing or stabilizing emissions in the land-use sector through external incentives as the right thing to do, and research should acknowledge that such efforts might in many contexts be peoples last concern. Embracing such premises can allow researchers and practitioners to be more sensitive to local peoples needs, interests and concerns regarding REDD+, as well as to become more aware of whose REDD+ it is, and why.

To conclude, and if we ask ourselves the question that underpins this special issue, it is not surprising that we are not able to provide a definite answer. In spite of being set as an international policy framework with well-established procedures and development phases, the special issue contributions show that REDD+ is unfolding differently in host countries, depending on state governance structures, cultural practices, and understandings of who should be ultimately responsible for operationalizing and realizing REDD+ across policy and implementation scales. In all countries, of course, the political economy of land-use and changing market dynamics will be critical in determining the viability and success of REDD+ activities on the ground. We cannot provide a definitive answer either as regards the consequences of REDD+. The special issue contributions, however, suggest that REDD+ runs the risk of replicating the mistakes of other policy approaches to environmental management, such as protected areas, integrated conservation and development projects and more recently payments for ecosystem services. These include the prioritization of technical and environmental goals over social justice considerations, the disconnect of project rationales and activities from local realities, and the insufficient attention to the actual drivers of environmental change and biodiversity loss. We are indeed at a crossroads now between relegating REDD+ to a similar fate as many previous forest conservation attempts on one side and forging ahead with much stronger attention paid to local needs, dynamics and complexities and the actual drivers of deforestation on the other.

Acknowledgments: Esteve Corbera acknowledges the financial support of the UAB-Banco de Santander Talent Retention programme and notes that this work contributes to ICTA-UAB Mara de Maeztu Unit of Excellence (MDM-2015-0552).

Conflicts of Interest: The authors declare no conflict of interest.

Esteve Corbera, Heike Schroeder

Special Issue Editors

References

1. Hermansen, E.A.T.; McNeill, D.; Kasa, S.; Rajo, R. Co-Operation or Co-Optation? NGOs Roles in Norways International Climate and Forest Initiative. Forests 2017, 8.
2. Vijge, M.J.; Brockhaus, M.; di Gregorio, M.; Muharrom, E. Framing National REDD+ Benefits, Monitoring, Governance and Finance: A Comparative Analysis of Seven Countries. Glob. Environ. Chang. 2016, 39, 57–68.
3. Dunlop, T.; Corbera, E. Incentivizing REDD+: How Developing Countries Are Laying the Groundwork for Benefit-Sharing. Environ. Sci. Policy 2016, 63, 44–54.
4. Sunderlin,W.D.; Sills, E.O.; Duchelle, A.E.; Ekaputri, A.D.; Kweka, D.; Toniolo, M.A.; Ball, S.; Doggart, N.; Pratama, C.D.; Padilla, J.T.; et al. REDD+ at a Critical Juncture: Assessing the Limits of Polycentric Governance for Achieving Climate Change Mitigation. Int. For. Rev. 2015, 17, 400–413.

5. Stern, H.N.; Britain, G. The Economics of Climate Change: The Stern Review; Cambridge University Press: Cambridge, UK, 2007.

6. Savaresi, A. A Glimpse into the Future of the Climate Regime: Lessons from the REDD+ Architecture. Rev. Eur. Comp. Int. Environ. Law 2016, 25, 186–196.

7. Angelsen, A.; Brockhaus, M.; Sunderlin, W.D.; Verchot, L.V. Analysing REDD+: Challenges and Choices; Center for International Forestry Research (CIFOR): Bogor, Indonesia, 2012.

8. DeShazo, J.L.; Pandey, C.L.; Smith, Z.A. Why REDD Will Fail, 1st ed.; Routledge: New York, NY, USA, 2016.

9. Lund, J.F.; Sungusia, E.; Mabele, M.B.; Scheba, A. Promising Change, Delivering Continuity: REDD+ as Conservation Fad. World Dev. 2017, 89, 124–139.

10. Brown, M.I. Redeeming REDD: Policies, Incentives and Social Feasibility for Avoided Deforestation, 1st ed.; Routledge: New York, NY, USA; London, UK, 2013.

11. Turnhout, E.; Gupta, A.;Weatherley-Singh, J.; Vijge, M.J.; de Koning, J.; Visseren-Hamakers, I.J.; Herold, M.; Lederer, M. Envisioning REDD+ in a Post-Paris Era: Between Evolving Expectations and Current Practice. InWiley Interdisciplinary Reviews: Climate Change; JohnWiley & Sons, Inc.: New York, NY, USA, 2017; Volume 8.

12. Angelsen, A. REDD+ as Result-Based Aid: General Lessons and Bilateral Agreements of Norway. Rev. Dev. Econ. 2017, 21, 237–264.

13. Burgess, N.D.; Bahane, B.; Clairs, T.; Danielsen, F.; Dalsgaard, S.; Funder, M.; Hagelberg, N.; Haule, C.; Kabalimu, K.; Kilahama, F.; et al. Getting Ready for REDD+ in Tanzania: A Case Study of Progress and Challenges. Oryx 2010, 44, 339–351.

14. Glover, A.; Schroeder, H. Legitimacy in REDD+ Governance in Indonesia. Int. Environ. Agreem. Politics Law Econ. 2017, 17, 695–708.

15. piric, J.; Corbera, E.; Reyes-Garca, V.; Porter-Bolland, L. A Dominant Voice amidst Not Enough People: Analysing the Legitimacy of Mexicos REDD+ Readiness Process. Forests 2016, 7.

16. To, P.; Dressler,W.; Mahanty, S. REDD+ for Red Books? Negotiating Rights to Land and Livelihoods through Carbon Governance in the Central Highlands of Vietnam. Geoforum 2017, 81, 163–173.

17. Shrestha, S.; Shrestha, U.B.; Bawa, K.S. Contribution of REDD+ Payments to the Economy of Rural Households in Nepal. Appl. Geogr. 2017, 88, 151–160.

18. Vatn, A.; Kajembe, G.; Mosi, E.; Nantongo, M.; Silayo, D.S. What Does It Take to Institute REDD+? An Analysis of the Kilosa REDD+ Pilot, Tanzania. For. Policy Econ. 2017, 83, 1–9.

19. Minang, P.A.; van Noordwijk, M. The Political Economy of Readiness for REDD+. Clim. Policy 2014, 14, 677–684.

20. Airey, S.; Krause, T. Georgetown Aint Got a Tree. We Got the Trees-Amerindian Power & Participation in Guyanas Low Carbon Development Strategy. Forests 2017, 8.

21. Huynh, T.B.; Keenan, R.J. Revitalizing REDD+ Policy Processes in Vietnam: The Roles of State and Non-State Actors. Forests 2017, 8.

22. Mustalahti, I.; Cramm, M.; Ramcilovic-Suominen, S.; Tegegne, Y.T. Resources and Rules of the Game: Participation of Civil Society in REDD+ and FLEGT-VPA Processes in Lao PDR. Forests 2017, 8.

23. Asiyanbi, A.P.; Arhin, A.A.; Isyaku, U. REDD+ in West Africa: Politics of Design and Implementation in Ghana and Nigeria. Forests 2017, 8.

24. Thompson, D.Y.; Swallow, B.M.; Luckert, M.K. Costs of Lost Opportunities: Applying Non-Market Valuation Techniques to Potential REDD+ Participants in Cameroon. Forests 2017, 8.

25. Suiseeya, K.R.M. Transforming Justice in REDD+ through a Politics of Difference Approach. Forests 2016, 7.

26. Gebara, M.F.; Agrawal, A. Beyond Rewards and Punishments in the Brazilian Amazon: Practical Implications of the REDD+ Discourse. Forests 2017, 8.

27. Trdal, L.T.; Vedeld, P.O. Livelihoods and Land Uses in Environmental Policy Approaches: The Case of PES and REDD+ in the Lam Dong Province of Vietnam. Forests 2017, 8.

28. Cadman, T.; Maraseni, T.; Breakey, H.; Lpez-Casero, F.; Ma, H.O. Governance Values in the Climate Change Regime: Stakeholder Perceptions of REDD+ Legitimacy at the National Level. Forests 2016, 7.

29. Holmes, I.; Potvin, C.; Coomes, O.T. Early REDD+ Implementation: The Journey of an Indigenous Community in Eastern Panama. Forests 2017, 8.

30. Work, C. Forest Islands and Castaway Communities: REDD+ and Forest Restoration in Prey Lang Forest. Forests 2017, 8.

31. Pfaff, A.; Santiago-vila, F.; Joppa, L. Evolving Protected-Area Impacts in Mexico: Political Shifts as Suggested by Impact Evaluations. Forests 2017, 8.

32. Fujisaki, T.; Hyakumura, K.; Scheyvens, H.; Cadman, T. Does REDD+ Ensure Sectoral Coordination and Stakeholder Participation? A Comparative Analysis of REDD+ National Governance Structures in Countries of Asia-Pacific Region. Forests 2016, 7.

33. Blum, M.; Reinecke, S. Towards a Role-Oriented Governance Approach: Insights from Eight Forest Climate Initiatives. Forests 2017, 8.

34. McElwee, P.; Nguyen, V.H.T.; Nguyen, D.V.; Tran, N.H.; Le, H.V.T.; Nghiem, T.P.; Vu, H.D.T. Using REDD+ Policy to Facilitate Climate Adaptation at the Local Level: Synergies and Challenges in Vietnam. Forests 2017, 8.

35. Mulyani, M.; Jepson, P. Does the one Map Initiative Represent a New Path for Forest Mapping in Indonesia? Assessing the Contribution of the REDD+ Initiative in Effecting Forest Governance Reform. Forests 2017, 8.

36. Pelletier, J.; Glinas, N.; Skutsch, M. The Place of Community Forest Management in the REDD+ Landscape. Forests 2016, 7, 170.

37. Corbera, E.; Schroeder, H. Governing and Implementing REDD+. Environ. Sci. Policy 2011, 14, 89–99.

38. Schroeder, H.; McDermott, C. Beyond carbon: Enabling justice and equity in REDD+ across levels of governance. Ecol. Soc. 2014, 19, 31.

39. Corbera, E.; Martin, A.; Springate-Baginski, O.; Villaseor, A. Sowing the Seeds of Sustainable Rural Livelihoods? An Assessment of Participatory Forest Management through REDD+ in Tanzania. Land Use Policy 2017.

40. Nantongo, M.G. Legitimacy of Local REDD+ Processes. A Comparative Analysis of Pilot Projects in Brazil and Tanzania. Environ. Sci. Policy 2017, 78, 81–88.

41. Poudyal, M.; Ramamonjisoa, B.S.; Hockley, N.; Rakotonarivo, O.S.; Gibbons, J.M.; Mandimbiniaina, R.; Rasoamanana, A.; Jones, J.P.G. Can REDD+ Social Safeguards Reach the right People? Lessons from Madagascar. Glob. Environ. Chang. 2016, 37, 31–42.

42. Bos, A.B.; Duchelle, A.E.; Angelsen, A.; Avitabile, V.; de Sy, V.; Herold, M.; Joseph, S.; de Sassi, C.; Sills, E.O.; Sunderlin,W.D.; et al. Comparing Methods for Assessing the Effectiveness of Subnational REDD+ Initiatives. Environ. Res. Lett. 2017, 12.

43. Sunderlin,W.D.; de Sassi, C.; Ekaputri, A.D.; Light, M.; Pratama, C.D. REDD+ Contribution Towell-Being and Income Is Marginal: The Perspective of Local Stakeholders. Forests 2017, 8.
44. Visseren-Hamakers, I.J.; Gupta, A.; Herold, M.; Pea-Claros, M.; Vijge, M.J.Will REDD+Work? The Need for Interdisciplinary Research to Address Key Challenges. Curr. Opin. Environ. Sustain. 2012, 4, 590–596.

Article

Co-Operation or Co-Optation? NGOs' Roles in Norway's International Climate and Forest Initiative

Erlend A. T. Hermansen [1,*], Desmond McNeill [2], Sjur Kasa [3] and Raoni Rajão [4]

[1] CICERO Center for International Climate Research, P.O. Box 1129 Blindern, 0318 Oslo, Norway
[2] Centre for Development and Environment, University of Oslo, P.O. Box 1116 Blindern, 0317 Oslo, Norway; desmond.mcneill@sum.uio.no
[3] Inland Norway University of Applied Sciences, Postbox 400, 2418 Elverum, Norway; sjur.kasa@hihm.no
[4] Department of Production Engineering, Universidade Federal de Minas Gerais, Av. Antônio Carlos, 6627 Pampulha, Belo Horizonte, Brazil; rajao@ufmg.br
* Correspondence: erlend.hermansen@cicero.oslo.no; Tel.: +47-220-047-68

Academic Editor: Timothy A. Martin
Received: 31 October 2016; Accepted: 22 February 2017; Published: 28 February 2017

Abstract: This paper investigates non-governmental organisation (NGO) involvement in policy processes related to Norway's International Climate and Forest Initiative (NICFI) comparing four countries: Norway, Brazil, Indonesia, and Tanzania. Based on documents and interviews, NGO involvement is mapped using a conceptual framework to categorise and compare different roles and modes of engagement. NGOs have co-operated with government in policy design and implementation, albeit to varying degrees, in all four countries, but expressed relatively little public criticism. Funding seems to have an influence on NGOs' choices regarding whether, what, when, and how to criticise. However, limited public criticism does not necessarily mean that the NGOs are co-opted. They are reflexive regarding their possible operating space, and act strategically and pragmatically to pursue their goals in an entrepreneurial manner. The interests of NGOs and NICFI are to a large extent congruent. Instead of publicly criticising a global initiative that they largely support, and thus put the initiative as a whole at risk, NGOs may use other, more informal, channels to voice points of disagreement. While NGOs do indeed run the risk of being co-opted, their opportunity to resist this fate is probably greater in this instance than is usually the case because NICFI are so reliant on their services.

Keywords: REDD+; NGOs; policymaking; co-operation; co-optation; policy entrepreneurship

---

1. Introduction: NGOs' Roles in REDD+ Policymaking Processes

Since REDD+ (Reducing Emissions from Deforestation and forest Degradation) was introduced on the international climate policymaking scene in 2003 [1–3], and officially entered the United Nations Framework Convention on Climate Change (UNFCCC) agenda in 2007 [4], this policy instrument for curbing tropical deforestation and degradation has evolved considerably [5]. According to Angelsen and McNeill [6], the most important changes to REDD+ are the following: "(i) the focus has moved from carbon only to multiple objectives; (ii) the policies adopted so far are not only, or even primarily, directed at achieving result-based payments; (iii) the subnational and project, rather than national, levels are receiving a large share of resources; and (iv) the funding to date is mainly from international aid and the national budgets of REDD+ countries, and not from carbon markets" [6] (p. 31). All these factors have, to varying extents, combined to increase the role of civil society and non-governmental organisations (NGOs) in REDD+ policy and practice. Social safeguards, where local participation plays a major part, have become established as important for REDD+ [7], and it has been argued that NGOs can play important roles in mediating between different actors and interests with diverse motives [8].

However, scholars have also expressed some scepticism with regard to concerns such as participation: "it is unclear to what extent these are mere rhetoric or whether they represent genuine motivation to address such issues [in] the context of REDD+" [9] (p. 25).

While there is only a limited literature on the relationship between NGOs and REDD+ there is, however, a more substantial body of social science literature studying NGOs roles and modes of involvement in environmental policymaking and implementation more generally [3,10–15]. While this literature has examined a number of cases in different countries [3,12], and sometimes compared similar cases across countries [15], most of the comparative work has been rather general, comparing environmentalism across countries as a whole [11,16,17]. Little attention has been given to comparative studies of NGOs' role in designing and implementing one and the same policy initiative simultaneously in different countries [18]. This is why O'Neill [19] (p. 135) asserts that there is a need for further comparative study of how environmental movements and organisations address political challenges and opportunities within and across borders.

This paper will make a contribution to filling that gap, by comparing the roles of NGOs in Norway's International Climate and Forest Initiative (hereafter NICFI). Norway is the world's largest contributor to the global REDD+ initiative, accounting for about 73% of pledged funds globally [20]. In addition to exploring the interface between NGOs and the Norwegian government domestically, we will use as case studies the REDD+ schemes supported in Brazil and Indonesia, being the two largest bilateral initiatives in the NICFI portfolio, and Tanzania, the first country to host NICFI initiatives as well as being the African country outside the Congo Basin with most subnational REDD+ initiatives [21] (p. 219).

Environmental NGOs (ENGOs) cannot be treated as one homogeneous mass, because organisations differ substantially on a number of variables [19], e.g., with regard to organisational form (e.g., grassroots, professional foundation), financing (members, public, private sector, project-based), relationship to government and private sector (close vs. distant) as well as tactics and choice of levers for action (confrontation vs. co-operation). NGOs are generally associated with civil society and are often expressions of broader social movements. However, environmental NGOs may only represent the views of small interest groups comprised mostly by experts and/or scientists. For this reason, Yearley [12] argues that neither a definition that focuses on organisational characteristics nor a definition that focuses on the organisations' issues and purposes (e.g., working for societal transformations of different kinds) is adequate. Rather, he asserts that only an essentially descriptive definition is acceptable [12] (p. 25). In other words, Yearley directs attention to the organisations' actual practices and suggests that these should be studied on a case-by-case basis.

In this paper we will follow Yearley's maxim and conduct an empirically driven descriptive comparative analysis, studying NGOs' involvement in NICFI-related policy processes in Norway, Brazil, Indonesia, and Tanzania. A central idea in the paper is to "follow the funding", and see how the size and relative share of NICFI funding contributes to shaping the organisations' roles and practices in the respective countries. We focus on the organisations most involved in Norway and Brazil, Indonesia, and Tanzania, and discuss the choices and implications of organisations taking on different roles ('insider' as policy designers, policy advisors and implementers; 'outsider' as critics; or in combinations) in different stages of the REDD+ policy processes. To compare across countries, we use a conceptual framework to map the roles played by organisations in different stages of the policy processes.

As NICFI perceives 'civil society organisations' (to use NICFIs own terms) to possess knowledge, expertise, and networks vital to REDD+, broad involvement of such organisations has been an integral part of their strategy from the very start. A vast variety of organisations receive civil society funding from NICFI, from broad grassroots social movements working on civil and indigenous people's rights, to highly specialised organisations working on e.g., forestry and biodiversity, to high-profile international organisations performing professional advocacy and lobbying. In this paper, beneficiaries of NICFI civil society funding will all be treated under the broad common category of non-governmental organisations (NGOs). Further distinctions, for instance between environmental and development NGOs, for instance, will be made when relevant.

NGOs can be seen as an expression of public concern over issues, serving to give civil society a voice in public policymaking [3]. Numerous scholars in the social sciences have often encouraged public involvement in policymaking in order to arrive at sound, democratic, and legitimate decisions, especially in controversial issues [22–27]. However, commentators have also warned that public involvement strategies are sometimes designed to reduce or circumvent public resistance, rather than taking public concerns seriously [28–30]. In her highly influential paper on public involvement in policy issues, Marres maintains that "empirical case studies in STS make clear that we cannot expect that 'public involvement' will be easy to distinguish from less authentic forms of lobbying or 'public window dressing'" [31] (p. 771). We agree with Marres on this point, and see this as a strong argument for more empirical studies that can improve our theoretical understanding by looking into the intersections of public involvement, lobbying, and 'public window dressing'. A pertinent question is how many different roles NGOs can take on, and whether different roles are complementary or conflicting. The case of NICFI is particularly suited for such inquiry since we are here dealing with rather different publics relating to different issues (for instance general nature conservation, bio-diversity, indigenous peoples, and human rights), though in this case co-operating closely in relation to one issue, namely REDD+.

In summary, we will describe the nature and extent of NGOs involvement in four countries (Norway, Brazil, Indonesia, and Tanzania), analyse their roles and relations to NICFI in each case (ranging from co-operative to confrontational), and draw some explanatory conclusions.

In the remainder of the paper, we first review relevant literature and operationalise this into an analytical framework. Next, we present the empirical material from the four countries. After that, we map the cases into the analytical framework and draw out some comparative observations. We conclude that NGO involvement in REDD+ policymaking in all countries has been distinctly entrepreneurial. Although the roles adopted by NGOs in the four countries have varied considerably, a common finding across the cases is that choices seem to have been guided by a pragmatic focus on results and policy impact. Norway stands out as the case where a mixed strategy of criticism and co-operation has been most clearly apparent. Giving up some public visibility in exchange for continuing to have direct contact with policymakers (both bureaucrats and elected officials) and "a place at the table" has been the price that NGOs have paid to achieve policy impact—which in this instance has been very substantial.

2. Study Design, Methods and Materials

In the latest report from Norad (the Norwegian Agency for Development Cooperation) about NICFIs support scheme for civil society, it states: "The non-governmental organisations play different roles. They serve as advocates, watchdogs, and independent verifiers, as well as knowledge and service providers" [32] (p. 6). Thus, official Norwegian policy documents portray these roles as complementary, exemplifying what by many is seen as "the Norwegian government/NGO model". However, as will be set out in the subsequent section, different strands of literature disagree whether, how, and to what degree NGOs actually can combine different roles; for instance close co-operation (including financially) with government, and public confrontation. Here, "the Norwegian government/NGO model" (elaborated below) is particularly interesting. A central objective of this paper is to compare and analyse how the NICFI funding scheme—where "the Norwegian government/NGO model" is implicitly embedded—plays out in the different countries.

The study is designed as a 'least likely' comparative case study [33] (p. 121) as Norway and the other countries we compare are very different across a range of dimensions. Put differently, we hypothesise that we will not find patterns resembling "the Norwegian government/NGO model" in the other countries. The countries do, however, have in common one important independent variable, namely being recipients of substantial NICFI funding. This makes it relevant to trace and seek to explain variations on the dependent variables, i.e., the roles the NGOs take in different countries, and feed that back to theory testing [33] (pp. 115–119).

The study uses process-tracing, a method which draws on a number of empirical sources to trace the (intervening) independent variables and possible variations on the dependent variable (i.e., what roles NGOs take), as well as uncovering and discussing explanations for that possible variance [33] (p. 206). The empirical basis consists of written documents, media coverage, and interviews (both formal and informal) as well as participant observation in meetings with NGOs and government officials. The time period studied extends from 2005 until 2016, and data collection was conducted from 2008 until 2016. Political and administrative documents have been analysed together with evaluations and other reports from both the public sector and NGOs. Moreover, a systematic analysis of Norwegian media coverage of NICFI has been carried out. This analysis consisted of a search combining the keywords "rainforest" and "climate" in the database "Retriever", which contains all published articles in the Norwegian media. A similar approach was also used in Brazil to follow the media coverage of three of the country's main outlets. The justification for focusing especially on Norway is twofold: first, it is by far the largest REDD+ donor globally (73% of the funds [20]); second, as our references to the literature show, Norway has been identified as a rather unusual case with regard to the relationship between NGOs and policymakers.

In Norway, the formal interviews consisted of 11 semi-structured interviews and one telephone interview conducted with NGOs, bureaucrats, politicians, and academics. In the case of Brazil 19 interviews were conducted with representatives from the Ministries of Environment, Science and Technology, and Foreign Affairs and three of the main NGOs involved in the creation and implementation of the Amazon Fund. Informants were asked about the background for REDD+ generally and NICFI specifically, how policies had been designed and evolved in the different countries, funding schemes for NGOs, NGO involvement, criticism expressed by NGOs—both public and private—as well as prospects for the future. All interviewees gave their informed consent before the interview.

In September 2012, as part of this research, a workshop was organised in Norway with senior representatives from NGOs, public administration and academia, in total 10 persons. The purpose of the workshop was to discuss the background, status and future of REDD+ generally and NICFI specifically, including the relations between the different actors in Norway. As a follow-up to the workshop, the authors in October 2013 organized a conference in Norway (attended by about 100 people) about deforestation in Brazil, attended by politicians (at the level of state secretary in Brazil and Norway), senior officials, leading academics, NGOs, and civil society. The authors had all followed the progress of Norway's involvement in REDD+: participating in and arranging seminars and conferences, advising policymakers (mainly bureaucrats), and supervising and commenting on students' doctoral and masters theses on the topic. In addition, they had had more than two dozen informal conversations in the course of their active research involvement in REDD+ in all four countries and collected in the context of participant observations in meetings between NGOs and officials from the Brazilian and Norwegian governments. The study was conducted in accordance with the guidelines set out by the Norwegian Centre for Research Data (Norsk Samfunnsvitenskapelig Datatjeneste—NSD) [34].

3. Analytical Framework: Inside, Outside—or Both?

In this section we review different accounts of the "the Norwegian government/NGO model" and construct a conceptual framework to guide the analysis. Dryzek et al. [11] compared the relationship between the state and environmental NGOs in four countries (the United States, the United Kingdom, Germany, and Norway) (A weakness of the analysis of Dryzek et al. [11] is that they analyse environmentalism within states, while many environmental NGOs in fact operate across borders, as exemplified by RFN. This study contributes to the growing literature on environmental NGOs and globalization, see Rootes [35]). In the case of Norway, the authors found that the relationship between the state and NGOs was characterised by so-called active inclusion, meaning that the NGOs receive substantial financial support from the state and are actively invited to contribute to policymaking processes in a consensus-oriented manner. However, the authors also warn that NGOs run the risk of being co-opted by the state in this way: "States such as Norway cultivate groups that moderate their demands in exchange for state funding and guaranteed participation in policy making, to the extent they can hardly be called NGOs" [11] (p. 171). The authors conclude that a close co-operation between NGOs and the state might increase NGO influence on state policy, as least apparently, but might also come at the expense of co-optation, pre-empting what otherwise could have been effective outsider criticism. Basically, it comes down to a trade-off between inside vs. outside strategies.

Three years after the Dryzek et al. [11] study was released, Grendstad et al. [13] published a study where they criticise the conclusions of Dryzek et al. [11] regarding the case of Norway. According to Grendstad et al. [13], Norwegian society is much more open and inclusive than the study of Dryzek et al. [11] indicates. The Norwegian societal model, characterised by the *"inclusive polity and the state-friendly society"* [13] (p. 2, italics in original), has resulted in a tradition of Norwegian ENGOs being more co-operative, pragmatic, and less openly confrontational than their counterparts abroad. Norwegian NGO's have not seen the same need to distance themselves from the state and oppose it, since they do not see the state as a "threat" as such, in contrast to the situation in many other countries. Rather, the NGOs consider that they can achieve more by co-operating with the state, instead of constantly opposing it: "Rather than seeing the organizations being pressed to become moderate, we see them as being moderate from the start. It is this symmetry that makes cooperation come so naturally for both parts" [13] (p. 39). In other words, Grendstad et al. [13] maintain that it is fully possible for Norwegian NGOs to co-operate closely with the state and receive funds without being co-opted by the state.

Dryzek et al. [11] and Grendstad et al. [13] study environmental NGOs (ENGOs), but we find it relevant to refer also to the work of scholars who focus on development NGOs. This is for two reasons. One is that this gives further insight into the 'Norwegian model'. The second is that REDD+ relates, in practice, to a mix of environment, climate and development issues, and NICFI is funded from the development aid budget.

In Norway, according to Steen [36], also in the field of development aid "relations between government and NGOs can be characterised as 'near' or 'close'" [36] (p. 157). Similar to that which Grendstad et al. find in the environmental field, Steen concludes that "although Norad is based on a well-regulated bureaucratic system, there are no indications that this has created any serious conflict with the NGOs" [36] (p. 157). However, Tvedt [37], who for decades has been studying Norwegian NGOs working with development aid, has a rather different view. Tvedt's [37] basic claim is that development NGOs have gradually aligned with state interests in a corporatist manner, not least because of huge financial dependency on the state. According to Tvedt, "institutional relationships and patterns of interaction tendentiously eliminate distinctions between different sub-systems in society that contribute to the pluralism of the community—e.g., organisations, research and the state" [37] (p. 622). The result is a consensual "aid system" where the state has the upper hand, although the corporatist system has also given opportunities for NGOs and researchers to shape policy: "The state has gradually invaded the organisations and the research institutes, while the leaders of the organisations and the researchers have broken the state's monopoly of foreign policy" [37] (p. 630).

In brief, Tvedt's [37] argument is that generous funding for civil society has dampened what might have been highly relevant criticism; in other words co-opting civil society voices [37].

Summing up, we see three main positions along a continuum of relative proximity to the state. At one end of the scale we find Steen [36] and Grendstad et al. [13], who argue that Norway is a special case where a hybrid, dual strategy of co-operation and confrontation is feasible without the NGOs having to compromise. At the opposite end of the scale we find Tvedt [37], who claims that Norwegian civil society development organisations already are co-opted/silenced by the state through generous funding systems. Dryzek et al. [11] occupy a middle position, acknowledging how a close co-operation between NGOs and the state may increase NGO influence on state policy, but perhaps at the expense of co-optation, preventing what otherwise could have been effective outsider criticism.

The literature we have reviewed here relates to Norwegian NGOs. A pertinent question with regard to Brazil, Indonesia, and Tanzania is thus what roles are adopted by NGOs in these countries: how, if at all, is "the Norwegian government/NGO model" (regardless of whether one initially agrees with the position of Steen [36] and Grendstad et al. [13], Dryzek et al. [11] or Tvedt [37])—embedded in the NICFI funding schemes—"exported" and adapted to these country contexts.

In this paper, we will operationalise the positions of Steen [36] and Grendstad et al. [13], Dryzek et al. [11] and Tvedt [37], by placing them on a continuum between two contrasting strategies: co-operation and confrontation. The category of co-operation is divided into two sub-categories, relating to the two roles that NGOs may play: *policy design* in the initial phase and *implementation* on the ground. The second strategy, confrontation, is manifested by *public criticism* of REDD+ policies. In brief, we will classify NGO involvement according to three different roles; inside as policy designers (co-operation), inside as implementers (co-operation), and outside as public critics (confrontation).

The *policy design* role implies that NGOs have been active in promoting and advocating REDD+ on the international, national, and/or local level, and assisted and/or advised government in designing policies and/or institutions for governing REDD+ (for instance commenting on draft documents, participating in hearings etc.) The *implementation* role comprises both providing policy advice in the implementation phase, for instance regarding portfolio development, as well as actual implementation "on the ground". The last role, *public criticism*, refers to explicit criticism by NGOs of REDD+ in general, or NICFI in particular, in the media or in public events, meetings, and workshops. Lack of public criticism in combination with a high degree of co-operation (either in policy design and/or implementation) is seen as an indicator of co-optation. To the degree in which we have data, we will also map and discuss private criticism.

We will now present relevant information about the four cases to enable us to complete the analysis.

4. Results: Country Case Studies

In this section, the empirics from the country cases will be set out in detail. Before delving into the details, we have in Table 1 compared the countries across some key dimensions relevant to REDD+ to give the reader a quick and comparable overview of some important factors relevant to the analysis.

Table 1. A comparison of Norway, Brazil, Indonesia, and Tanzania across key dimensions related to Reducing Emissions from Deforestation and Forest Degradation (REDD+) and Norway's International Climate and Forest Initiative (NICFI). Forest data from Global Forest Watch. Data on land area, population and governance from Wikipedia.

Dimension	Norway	Brazil	Indonesia	Tanzania
Land area	0.4 million km^2	8.5 million km^2	1.9 million km^2	1 million km^2
Population (2016)	5.2 million	206.4 million	255.5 million	51.8 million

Table 1. *Cont.*

Dimension	Norway	Brazil	Indonesia	Tanzania
Governance	Unitary parliamentary constitutional monarchy	Federal presidential constitutional republic	Unitary presidential constitutional republic	Unitary presidential republic
Tree cover loss 2001–2014	418,456 ha	38,336,733 ha	18,507,771 ha	1,699,305 ha
Tree cover gain 2001–2012	172,935 ha	7,586,752 ha	6,970,546 ha	304,101 ha
Percent tree cover 2000	38%	62%	86%	30%
Tree cover 2000	12 Mha	519 Mha	161 Mha	26 Mha
Main drivers of deforestation	Not applicable	Land grabbing, agriculture (mainly beef and soy), logging, mining	Agriculture (mainly palm oil), logging, pulp, mining	Fuelwood and charcoal, small-scale agriculture, logging
Main NGOs involved in REDD+	Rainforest Foundation Norway, World Wide Fund for Nature (WWF) Norway, Friends of the Earth Norway	Instituto de Pesquisa Ambiental da Amazônia (IPAM), Instituto Socioambiental (ISA), The Nature Consevancy (TNC), Instituto do Homem e Meio Ambiente da Amazônia (IMAZON), World Wide Fund for Nature (WWF) Brazil	Wahana Lingkungan Hidup Indonesia (WALHI), Komunitas Konservasi Indonesia Warung Informasi Konservasi (KKI WARSI), Aliansi Masyarakat Adat Nusantara (AMAN)	Tanzania Forest Conservation Group (TFCG), Mpingo Conservation, Development Initiative (MCDI)

4.1. Norway

In the case of Norway, ENGOs have been heavily involved from the very start. In fact, NICFI originates from a proposal from the leaders of two ENGOs: Lars Løvold from the Rainforest Foundation Norway (RFN) and Lars Haltbrekken from Norwegian Society for the Conservation of Nature/Friends of the Earth Norway (FOEN) (see [3,38,39] for thorough analyses, including timelines of major events (particularly [39] (pp. 26–29) and [38] (p. 41)). This heavy NGO influence on NICFIs inception has continued to be a long-lasting characteristic of the initiative and we will therefore devote some space to explain the emergence of the initiative.

In September 2007, the leaders of RFN and FOEN sent a letter to leading Norwegian politicians. The authors—"Lars & Lars"—here proposed that Norway should allocate 6 billion NOK (approximately 1 billion USD) (The NOK/USD exchange rate has varied considerably in recent years. For convenience we have used a figure of 6:1 throughout this paper, and will in subsequent cases omit the word 'approximately') annually over five years to protect rainforest as a climate mitigation measure in a payment-for-performance mechanism, very similar to what has become the REDD+ mechanism. The two ENGOs then went on to lobby their proposal, both to the sitting 'red-green' coalition government (consisting of the Labour Party, the Socialist Left Party and the Centre Party) and the centre/right opposition, arranging meetings with political parties and attending hearings in Parliament. They even flew in the prominent REDD+ proponent Marcio Santilli from the Brazilian Instituto Socioambiental (The Socioenvironmental Institute—ISA) to back their case [3]. Also Brazilian Minister of the Environment, Marina Silva, attended an event organised by RFN. In order to gain broad support for their proposal, they also induced other Norwegian ENGOs to back the rainforest proposal.

Parallel to this, under public pressure, the political opposition challenged the cabinet to enter into a 'climate settlement' (a cross-party Parliamentary agreement on climate policy). The discussions in Parliament soon heated up and turned into a political bidding war over climate policy integrity [3]. Then, just a couple of months after the letter was sent, Prime Minister Stoltenberg announced at the

UN Framework Convention on Climate Change (UNFCCC) Conference of the Parties (COP) 13 in Bali that Norway would grant 3 billion NOK (500 million USD) annually for rainforest protection. The decision came as a surprise to the general Norwegian public and most of the public administration, and was strikingly similar to the proposal from the two ENGOs apart from the sum, which was cut by half from the original proposal—but still extremely large in relative terms [3]. The ENGOs had argued that the funding should not be allocated from the aid budget, but as a political compromise, NICFI was funded by an *increase* in the Norwegian aid budget, thus simultaneously achieving Norway's political target of allocating 1% of Gross National Income (GNI) to Official Development Assistance (ODA) [38]. Approximately one month after the Bali meeting NICFI was established as a part of the cross-party climate settlement. In 2012, a new climate settlement was agreed, prolonging NICFI until 2020. At COP 21 in Paris in December 2015, Norway announced that NICFI would be extended to 2030.

When NICFI was approved by Parliament in January 2008, Norway did not have the governmental apparatus in place for implementing it. The Ministry of Foreign Affairs (hereafter MFA) and Norad (the Norwegian Agency for Development Cooperation, a Directorate under the Norwegian Ministry of Foreign Affairs) were both considered as possible hosts for the initiative, as NICFI was funded from the ODA-budget. However, NICFI was primarily thought of as a climate initiative, not a development initiative, despite being funded as aid. Erik Solhem (Erik Solheim in May 2016 became Executive Director of the United Nations Environment Programme, UNEP), who at that point had the double minister-post as Minister of the Environment and International Development (international development is a branch under MFA), agreed with Prime Minister Stoltenberg that the Ministry of the Environment should host NICFI. Part of the reason for this decision was that Norad was considered too risk-averse and bureaucratic by Solheim. Here is Solheim's own account from our interview:

> *"They had to have the activist attitude, they had to have daily direct contact with the political leadership. If not it would just be bogged down in bureaucracy, it would be killed by all the bureaucratic objections that the system produces every day and hour. The point is that this is a political project to drive the climate fight. To locate it within Norad was never an alternative as I see it, because the leaders of the program needed daily access to me as the Minister. This was a signature Norwegian effort to change global behaviour and could only succeed at the highest level of politics, meaning at the centre of the Ministry."* (our translation)

Instead, a small fast-working task force called "the climate and forest secretariat" (in Norwegian 'klima-og skogsekretariatet'—hereafter KOS) consisting of just a few persons was established in the Ministry of the Environment to manage NICFI. An experienced diplomat, Hans Brattskar, was handpicked by Solheim to head the secretariat. Other staffers had background from NGOs, bureaucracy, and management consulting. However, there were very few people with development experience, at least when first established. Norad was given an advisory role and tasked with managing the vast pot of civil society funds. (Although under the Ministry of Foreign Affairs, in matters regarding NICFI Norad reports to the Ministry of Climate and Environment). KOS is still the primary policy and implementing body and has now grown to 17 staff in Norway [40]. In, addition, KOS has a network of eight staff located in six different Norwegian embassies. But the number of staff in KOS is still relatively small compared to the scale of the funds they manage.

The competence on tropical forests in the Norwegian state bureaucracy was virtually non-existent at the time when NICFI was established. Although Norad and the MFA had experience in large international development projects, and advised KOS on these issues from a general perspective, they too had very little competence on tropical forests. The first director of NICFI and manager of KOS, Hans Brattskar, in fact had no experience with or competence on rainforests at all. RFN, on the other hand, has a very substantial knowledge base due to almost three decades of work on rainforest projects. Consequently, competence from RFN played a decisive role in the initial setup of NICFI. RFN even educated the KOS staff through courses on rainforest issues [41]. In the early phase, KOS also commissioned the high-profile professional not-for-profit organisation "Meridian Institute" to facilitate the production of a number of influential reports [42] drawing on input from a range of different actors. Over the years KOS has somewhat expanded its own competence base, while also making substantial use of international consultants. Nevertheless, RFN has remained an important partner for KOS and NICFI more generally (for instance through a close co-operation with Norwegian embassies) regarding policy design and portfolio development, both in Norway and abroad.

In 2014, 125 million NOK (21 million USD) of RFN's total revenue of 151 million NOK (25 million USD) stemmed from Norad and the Ministry of Foreign Affairs [43]. RFN received sizeable funding from MFA and Norad before NICFI was established, most notably through MFA's Amazon programme, but their total budget increased substantially thanks to REDD+. The foundation has been given a central role in the implementation phase of NICFI and greatly increased its staffing. RFN is by now the largest rainforest conservation organisation in Europe. The clearest indicator of RFNs central role in advising KOS—and NICFI more broadly—is the organisation's share of the NICFI funds for civil society administered by Norad: RFN is the second largest beneficiary, and in the period 2009–2012 received 7.5% of the total funds, next after CIFOR (Center for International Forestry Research), which received 12.3% [44] (p. 113). Third on the list was WWF International, with 6.8%. The funding period 2013–2015 shows a similar pattern: in the total period 2009–2015 CIFOR is on top of the list, RFN second and WWF International third. Both RFN and WWF Norway, and particularly the former, have frequent meetings with KOS and are in continuous dialogue with them.

Thus, RFN has certainly played a significant role in advising on policy design and implementation, i.e., co-operation. However, what about confrontation? There are a few instances where RFN has voiced its concerns in public. In the early phase, RFN engaged in the question of what organisations Norway should co-operate with (e.g., favouring the UN over the World Bank), and criticised inadequate rainforest funds in the state budget. RFN and FOEN have also on a few occasions voiced public criticism towards the official Norwegian position on REDD+ in the international climate negotiations under the UNFCCC. In 2010 [45] and 2012, letters were sent signed by Lars Løvold (RFN) and Lars Haltbrekken (FOEN) to the Ministers in charge of NICFI (The 2010 letter was addressed to Minister of the Environment and International Development (Solheim) 2010), while the 2012 letter was addressed to the Minister of the Environment (Solhjell) and the Minister of International Development, Holmås). Both letters addressed a recurring point of disagreement between RFN and the Norwegian government regarding the official Norwegian position on REDD+: whether Norway should pursue a narrow, carbon-focused approach to REDD+, or take a broader view on the issue. RFN and FOEN have repeatedly criticised the Norwegian official position for being too narrow and carbon-focused, and instead argued for a broader approach where safeguards such as human rights, biodiversity, good governance, and inclusive processes are given a more central position. In addition, in both letters Løvold and Haltbrekken argue for their original position, that REDD+ should not be a carbon offsetting mechanism similar to the Clean Development Mechanism (CDM), but come in addition to other mitigation efforts. RFN and FOEN have also made an effort to generate media attention to their position on these issues, but have found it quite difficult to catch journalists' attention, as the issues may seem rather technical and beyond the scope of public interest. Norwegian public authorities have on their side largely avoided debating the broad/narrow REDD+ issue in public.

Another example of public criticism emerging from the media analysis relates to RFN's criticism of the Norges Bank Investment Management (NBIM) investment practices. NBIM is the Norwegian state's sovereign wealth fund; the world's largest of its kind. On several occasions, RFN has both publicly and privately criticised NBIM's investments for having negative consequences for rainforests, through NBIM's investment in petroleum extraction, mining, palm oil, and pulp. RFN argues that it is paradoxical that Norway, as the world leader in REDD+ finance, also (through NBIM) invests up to 40 times more in commercial activities that destroy rainforests. After RFN—and others—had pushed this case for several years, NBIM finally took action in 2013 and included "severe environmental damage" among the criteria for excluding companies from their investment portfolio. In 2014, "climate damage" was included in the exclusion criteria, which also include rainforest destruction. RFN advised

NBIM in the process of developing the exclusion criteria. Although NBIM have several times pulled out of companies associated with rainforest destruction, RFN is still not satisfied. The last time they expressed strong public criticism on the NBIM issue was 22 May 2015 [46]. Approximately one year later, on 9 March 2016, RFN publicly acknowledged that NBIM was on the right track, having reduced the portfolio of rainforest destroying companies by 20 per cent in one year [47]. However, RFN also stated that NBIM still had a way to go. It is highly likely that RFN will continue to pursue this case in the Norwegian public sphere. In summary, RFN publicly criticises Norwegian authorities, such as the Ministry of Environment, and NBIM, and indirectly the Ministry of Finance and Norges Bank (the Norwegian national bank)—but not to any similar extent NICFI, KOS, Norad, or the Ministry of Foreign Affairs, i.e., RFN's biggest funders.

In fact, KOS has unofficially backed RFN efforts on the NBIM issue, as they saw that RFN could (and actually did) achieve things that KOS was not able to achieve themselves. RFN, on their side, allied with MFA for pushing KOS towards including development issues into the REDD+ strategy, rather than taking a narrow carbon-focused approach. These points also remind us that we should be careful seeing the state as representing a single position articulating one coherent voice. Rather, it has been documented that the Norwegian state operates with polyphonic voices on environmental issues [48]. That certainly applies to REDD+ and NICFI as well, as there have been tensions between KOS/NICFI/The Ministry of Environment, MFA, and Norad from the very start regarding NICFI policy and practice. RFN has been aware of these tensions and, when appropriate, attempted to take advantage of them to pursue their agenda. Finally, whenever Norwegian politicians have discussed possible cutbacks in NICFI funding, for instance in the fall of 2013, as well as 2015 and 2016, RFN and FOEN have mobilised heavily in the media against politicians (mainly from the Conservative Party and the Progress Party, the latter not being part of the cross-party parliamentary climate settlement) to prevent budget cuts. At one point RFN even called for a broader political debate about NICFI by publishing a story on their website titled "Let the forest into the Parliament" [49] (our translation).

RFN (together with FOEN) have also been critical of NICFI's engagement in Indonesia in Norwegian media [50]. While Solheim praised the 2011 moratorium on new forest concessions as an "important step forward" [39] (p. 21) the two ENGOs criticised the moratorium for having loopholes and limitations, and suggested that financial transfers to Indonesia should be reduced [39,51]. On the one-year anniversary of the logging moratorium, the two ENGOs together with Greenpeace Norway arranged a public event where they handed Solhjell (newly appointed Minister of the Environment) a letter from several Indonesian NGOs, expressing concern that deforestation in Indonesia was not decreasing, because a strong lobby had been able to weaken the moratorium [39,52]. The Norwegian ENGOs argued that Norwegian authorities should push Indonesia to prolong the moratorium, and stand up against the pressure from the lobby. At the same time, Løvold from RFN recognised that the moratorium had led to more openness around forest issues, which opened up opportunities for Norway to exert pressure on the Indonesian government [39,52].

WWF International is the third largest recipient of NICFIs/Norad's civil society funds. WWF Norway, by contrast, only received 0.2% of the NICFI civil society funds in 2009 [44] (p. 114). WWF Norway has several projects funded by Norad and the Ministry of Foreign Affairs, but a significant part of WWF Norway's project portfolio is not directly related to REDD+, and many projects were agreed before NICFI came into play. In 2014, 85 million NOK (14 million USD) of their total budget of 117 million NOK (20 million USD) came from Norad, the Ministry of Foreign Affairs and the Ministry of Climate and Environment [53]. Despite being a significant player, WWF Norway has not been involved to the same degree as RFN in NICFI, but like RFN, WWF Norway also has meetings with KOS and co-operates with them. WWF Norway also has an advisory role in WWF International's REDD+ projects, which are more focused on implementation "on the ground" compared to RFN's engagement. WWF Norway does not have the same public voice on rainforest issues in Norway as RFN, although it is generally rather vocal in criticising the Norwegian government, primarily on domestic policies, including climate policy.

Ironically, WWF has itself been subject to public criticism in Norway regarding implementation of a forest project in Tanzania funded by Norway [39] (p. 18). In November 2011, three Norwegian researchers wrote an opinion piece where they accused WWF of being too close to Tanzanian authorities and thus indirectly involved in activities depriving local population of their livelihoods [54]. The chairman of WWF Norway responded and rejected the allegations [55]. In January 2012, many of the same claims were repeated in a news article based on a published study [56,57]. In the wake of these incidents, a fierce debate broke out between the researchers and WWF in Norway, culminating in a public seminar about the issue. The controversial project was, it later turned out, not connected to REDD+ as such [58]. A few months later, two other Tanzanian REDD+ pilot projects were stopped [59]. What is interesting for this paper is that these events did not attract a lot of media attention [39] (p. 18). NICFI as such was not questioned.

FOEN, although instrumental in initiating and advocating for what became NICFI [3,38], has since then taken a rather limited role. After succeeding with their proposal back in 2007/2008, RFN and FOEN have arrived at a sort of division of labour: RFN co-operates closely with government authorities and is actively involved in advising and assisting KOS, occasionally voicing rainforest issues in the public arena. FOEN—which primarily focuses on environment rather than development issues—steps in when there are discussions about more overarching issues, such as cuts in NICFI funding, or whether REDD+ should be part of an international carbon trading mechanism (of which they are critical). Put differently, FOEN mainly concentrates on the domestic domain, and does not co-operate directly with or advise KOS. Neither does FOEN confront KOS or criticise NICFI in public to any large extent. FOEN has received only a small share of NICFI's funding for civil society, a grant amounting to 0.7% of that budget—which was shared with RFN [44] (p. 114).

That Norway's REDD+ funding is classified as aid is particularly anomalous with regard to Brazil, as this middle-income "BRIC" country has consequently become the largest recipient of Norwegian ODA funds [39] (p. 16). The decision to finance NICFI with aid money (which did not please Brazil) was taken for political reasons; it enabled Norway in achieving the aid target of 1% of gross national income (GNI) [38] (p. 42). There are good arguments against treating global public goods such as REDD+ as development aid [60] (p. 119). And some NGOs working with poverty and development, as well as the Christian Democratic Party and the Socialist Left Party, have from the very start been critical of the possible negative effects on poverty reduction by redirecting aid money to rainforest projects in relatively rich countries like Brazil [39] (p. 16) [61,62]. But ENGOs have been comparatively silent on this point, seeking to sidestep the issue by emphasising the possible poverty amelioration effects of REDD+ [39,61].

Perhaps this points to a bigger issue: that the ENGOs have been vocal with public criticism of overarching NICFI policy design issues (e.g., Norway's position on REDD+ in the UNFCCC negotiations), but relatively silent on implementation issues, at least in the public sphere. This observation also supports the point made above, that RFN publicly criticises NBIM, and indirectly the Ministry of Finance and Norges Bank (the Norwegian national bank)—but not NICFI/KOS, Norad, the Ministry of Climate and Environment or Ministry of Foreign Affairs, i.e., RFN's biggest funding sources. Development NGOs, however, have publicly criticised these agencies, directly and/or indirectly [61,62]. Part of the reason why ENGOs have been silent on implementation issues may be that they have had good internal channels to KOS for critique, feedback and advice.

Summing up, we observe that ENGOs have co-operated extensively with the state and been instrumental in policy design of the Norwegian rainforest initiative, as well as giving policy implementation advice regarding, for example, which agreements to enter into and on what conditions. Including NGOs and drawing on their competence has been Norwegian government policy from the start of NICFI [44] (p. 7). Regarding confrontation, public criticism of NICFI as such, or KOS and related agencies, has from NGOs been limited, but not entirely absent. Development NGOs have arguably been more vocal on implementation issues "on the ground", while ENGOs have been more vocal on overarching policy design issues. Although some researchers have publicly voiced criticism of various kinds, this criticism has not resonated loudly in the Norwegian public sphere. Overall, there is surprisingly little debate about NICFI in the Norwegian public domain [39,63], which is interesting given that a core aim of the Norad civil society funding is to "enhance the engagement of civil society at national level to generate an open, inclusive, and comprehensive debate on REDD+" [44] (p. 7).

4.2. Brazil

The development of REDD+ in Brazil is marked by an internal conflict between different sectors of the government and a strong participation of NGOs. When the concept of REDD+ was proposed by a group of Brazilian and North-American scientists and activists in the mid-2000s [1,2], it was well received by the Ministry of Environment under Marina Silva (2003–2008) and Carlos Minc (2008–2010). In addition to the favourable political context, Brazilian NGOs have developed over the years the capacity to exert national and international pressure to reduce deforestation in the Amazon and in this way influence environmental policies in the country [64]. One of the advocacy efforts of civil society during this period was a zero-deforestation pact carried forward by a group of Brazilian researchers from the Federal University of Minas Gerais and a group of NGOs led by IPAM that included cost calculations for ending Amazon deforestation by 2015 through a mechanism similar to REDD+ [3,65]. At the same time, more conservative sectors of the government such as the Ministry of Foreign Affairs openly opposed carbon markets with credits from REDD+ because it would allow developed countries to maintain their high emission levels. In private conversations, some diplomats and scientists representing Brazil in the climate negotiations also recognised that REDD+ may be problematic, since the creation of legal obligations for Brazil regarding conservation of the Amazon would represent a threat to its sovereignty [66,67]. These concerns were also made public during the initial stages of the creation of international fund for the Amazon that later became the Amazon Fund. When asked by reporters about what he thought about the creation of the fund, President Luis Inácio Lula da Silva said that he was worried about the possibility of foreign intervention in the Amazon:

> *"I will talk with the Minister [of Environment . . .] it is a concern that if someone donates US$ 10 he will soon think that he owns the Amazon".* (our translation)

Brazilian and Norwegian NGOs played a key role in circumventing the political obstacles for the establishment of the Amazon Fund. On the Norwegian side Lars Løvold from the RFN was able to communicate to NICFI/KOS the scepticism voiced by some Brazilians against channelling funding through the World Bank [39] (p. 15), a solution adopted in a previous large-scale scheme funded by the G7 countries (known as PPG7), that would restrict Brazilian control over the fund. As an alternative it was proposed that BNDES (the Brazilian Development Bank) would receive and distribute the resources to projects in the Amazon. This solution not only safeguarded Brazilian sovereignty, but was also well received by NICFI/KOS, which recognised that it would be difficult to manage such a fund from a distance. On the Brazilian side, key NGOs, such as ISA (Instituto Socioambiental, Socio-Environmental Institute) and IPAM (Instituto de Pesquisa Ambiental da Amazônia, Amazon Environmental Research Institute), followed closely the negotiation process. This enabled the emergence of a governance structure for the Amazon Fund in which representatives from the Federal government, Amazonian states, and NGOs would have equal voice in the fund's steering committee (COFA). Many of the Brazilian officials involved in the negotiations on behalf of the Ministry of Environment also had strong ties to the third sector having worked in NGOs for many years prior to joining the government. Therefore, people with NGO-background were involved in governmental decision-making in both Brazil and Norway, and were important for the outcome of the negotiations [68]. The Minister Carlos Minc would later recognise in an interview that this innovative arrangement was crucial for the approval of the Amazon Fund:

> *"Lula almost did not sign [. . .] but I explained that the Fund is much more autonomous and sovereign than the PPG7. The resources will be managed by the Brazilian National Development*

*Bank and the people from the federal government, scientific community, NGOs and the state
governments of the Amazon."* [69,70] (our translation)

The Memorandum of Understanding between Brazil and Norway was signed in Brasília on
16 September 2008. According to this agreement, Norway would until 2015 donate up to 1 billion USD
to support actions to further reduce deforestation in the Amazon. In order to obtain this donation Brazil
would need to continue reducing its emissions from deforestation in relation to their historical levels.
The amount receivable was calculated by considering a reference level corresponding to the average
emissions taking place due to deforestation in the Amazon between 1996 and 2005. By measuring
the difference between this reference level and the actual emissions taking place between 2006 and
2010, the Amazon Fund would be able to receive a donation of 5 USD for every ton of CO_2 (tCO_2) of
avoided emissions from deforestation.

While the fund's steering committee was responsible for establishing the main directives and
priorities of the Amazon Fund, it was up to BNDES technical team to decide which projects were eligible
for funding. These approvals, however, took place very slowly, mostly due to the complex bureaucratic
requirements from BNDES. Many NGOs were already used to dealing with large bureaucracies, such
as Norad and USAID, to obtain funding. However, according to members of NGOs that submitted
projects to the fund, BNDES was more stringent as it adopted the same procedure and criteria used to
evaluate large-scale investments such as the building of dams and airports. With these procedures it is
expected that the funds disbursed by BNDES become more traceable, and the likelihood of corruption
reduced (The Brazilian government is currently facing major corruption scandals involving the
country's main political parties and large public and private companies but to our knowledge BNDES
staff have not been the target of corruption accusations). Thus, many NGOs had to make substantial
investments in order to engage in lengthy application processes and to adapt their organisational
structure and procedures. Consequently, by the end of 2012 the fund disbursed only 71 million Reais
(approximately 23 million USD), threatening Norway's' 1% ODA target since the funds remained in
Norwegian bank accounts [39,71]. Thus, the Amazon Fund's slow implementation became a point
of criticism both in Brazil and in Norway even in a context where deforestation rates continued to
decline [72]. The situation started to change in 2013 when the maximum value allowed to individual
projects was raised, and state governments in the Amazon applied for large projects. At the same time,
the BNDES staff learned rapidly and were able to start evaluating the applications to the fund more
effectively, while NGOs adapted to the requirements set by the fund.

With the exception of the criticism in 2012 concerning the high level of bureaucracy and slow
pace of project evaluation, the Amazon Fund has been largely shielded from public criticism since its
inception. In closed meetings, however, NGOs have voiced their concerns about different aspects of the
fund's management to representatives from the Norwegian and Brazilian governments. In particular,
concerns were raised on various occasions that the projects supported by the fund may not be leading
to additional reductions in deforestation as initially expected. It should be noted that the disconnect
between many projects of the Amazon Fund and the reduction in deforestation can be traced to the
very design of the fund. As mentioned above, the fund receives donations in proportion to results
already achieved; thus in principle the projects it supports are not required to show evidence of further
reductions in order to be justified. Furthermore, it was decided in 2013 that the fund was to support
mainly "structuring" projects that aim at improving environmental governance in the long-term rather
than produce direct short-term results.

Nevertheless, there has been mounting private criticism by NGOs and researchers seeking to
increase the effectiveness of the fund in different ways. First, the fund was criticised for taking
a passive stance whereby it waits for project proposals rather than actively looking for situations
that may require resources to slow deforestation. Second, some NGOs and researchers pointed
out that the fund has invested a sizable share of its resources into the implementation of the rural
environmental registry (CAR), despite evidence that so far this policy tool did not contribute to reduce
deforestation [73,74]. Third, contradictions have been pointed out between projects supported by

the fund. In one case, a representative of an influential NGO criticised a large grant to equip the National Force. While this elite police force plays an important role in escorting forest rangers from IBAMA (the federal environmental agency), the National Force has also used violence against native Indians opposing the construction of Belo Monte Dam in the state of Pará—a project at odds with the socioenvironmental aims of the fund [75]. Even though these are important matters, and members from NGOs were very concerned, these issues did not find a place in the traditional spaces of public activism such as the press, reports, protests, and social media.

The contradiction between closed-door criticism and public silence (and thus apparent approval) of the Amazon Fund by large NGOs can be explained by several factors. First, the Amazon Fund has become an important financial source for the third sector as a whole and especially for NGOs, as illustrated in Figure 1 below. By the end of 2015, 31 of the 69 projects accepted by the Amazon Fund were proposed by third sector organizations. This corresponds to 30% of the R$ 1 billion (approximately US$ 300 million) approved by the fund. It should be noted that FUNBIO and FBB, the two largest recipients from the third sector, are foundations managed indirectly by the Brazilian government. Furthermore, NGOs supporting indigenous groups and small farmers such as the Centro de Trabalho Indigenista, have also received substantial resources. Nevertheless, almost all large environmental NGOs with a long history of engagement in the region have projects approved by the fund, corresponding to nearly half of the grants to the third sector (see Figure 1). The increasing reliance on government funds implies that NGOs could lose part of their ability to promote the environmental agenda independently. In this regard, a director of a large environmental NGO reported in a private conversation that his organisation strives to keep the contribution from the Amazon Fund below one third of its total budget in order to avoid being co-opted by the government. This implies that NGOs perceive that the Amazon Fund provides not only an opportunity for the expansion of conservation projects on the ground but also poses a risk for environmental policy advocacy.

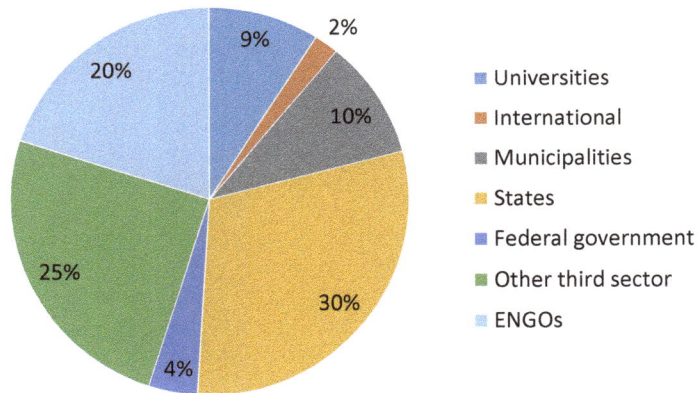

Figure 1. Projects funded by the Amazon Fund categorised according to the nature of the party responsible between 2008 and 2015. Data from BNDES (Brazilian Development Bank).

Second, the central role of NGOs in the governance of the Amazon Fund implies sharing responsibility for its performance. As in the case of WWF in Norway, NGOs end up having their reputation tied to the fund by implementing projects. For instance, some of the same NGOs that privately criticised the excessive emphasis of the fund on the implementation of CAR were also members of the steering committee that approved in 2013 the change to the fund that gave priority to structuring projects. Furthermore, the Amazon Fund granted to environmental NGOs substantial resources for the implementation of large projects, such as "Sustainable Settlements in Amazonia" from IPAM and the "Green Turn" by TNC with total investments of approximately US$ 8 and 5 million, respectively. Therefore, by criticising the fund, they would indirectly become more vulnerable as well.

Third, NGOs fear that by criticising the Amazon Fund they might hinder an initiative that, despite its shortcomings, has qualities that other governmental initiatives lack. The virtues of the Amazon Fund become particularly clear when it is compared to the Brazilian national REDD+ strategy: the mechanism establishing the rules for the distribution of financial resources from UNFCCC's Green Fund. The discussion of the national strategy started in 2008 in parallel with the implementation of the Amazon Fund. But in contrast to the latter, the Ministry of Environment has been very reluctant to create a governance structure that gives equal voting rights to NGOs and state governments. The Ministry of Environment finally published the national REDD+ strategy and the decree detailing its governance structure in November 2015 following a process that, according to the members of the NGOs and state governments, was largely undemocratic (This is also clear in a letter sent by the coalition of state governments of the Amazon to the Federal Government in 2014 complaining that no joint meeting to develop the national REDD+ strategy was held since 2012). As expected, BNDES was nominated as the manager of the resources, but in contrast to the balanced structure of the Amazon Fund the board of REDD+'s national committee is composed of eight representatives from the Federal Government and two representatives from the states, one from the municipalities and two from civil society. Seeing this decision as unfair, a coalition that includes most large NGOs in Brazil requested, in an open letter, that the Ministry of Environment adopt for the REDD+ national commission the same governance structure used in the Amazon Fund [76]. Even though the governance structure of the fund was perceived by NGOs as being overwhelmingly influenced by the Ministry of Environment, many saw the Amazon Fund as a very important first step in the direction of the establishment of larger and better mechanisms in the future since the REDD+ strategy opened the possibility of allowing states to receive resources directly from donors, thus bypassing BNDES.

Fourth, NGOs fear that criticism of the Amazon Fund may disturb the delicate political balance that made possible the collaboration with Norway. If the fund is publicly criticised, Norwegian society may call into question the investment of public money in the Amazon Fund and request that this rather be allocated to more urgent matters such as the refugee crisis in Europe. While some members of the NGOs we interviewed expressed their willingness to open a public debate about the Amazon Fund, they fear that if not well managed this process may backfire and end an influx of resources that is crucial for the continuation of many environmental policies in Brazil. Furthermore, even if in the unlikely event that Norwegian diplomacy changed its "light approach" [60] and became more demanding following an outburst of public criticism of the fund, this would be risky since more conservative sectors of the Brazilian government could accuse Norway of interfering in the country's sovereignty. This suggests that NGOs see the Amazon Fund as an initiative that could be improved from the inside, rather than attacked from the outside by adopting traditional social activism tactics.

Finally, since the creation of the Amazon Fund in 2012, Brazilian NGOs have been engaged in many important public disputes with the Brazilian government. The most important one was the revision of the Forest Code approved in 2012 that provided an amnesty for 58% of the area illegally deforested before 2008 [77]. Even though many sectors of the Brazilian society, including not only NGOs but also researchers and artists, were publicly against the revision of the code, the government successfully used the strong reductions in deforestation rates occurring in that period as part of the argument for the creation of a lighter yet "enforceable" law. Similarly, NGOs have also lost the fight against the government in relation to the construction of the Belo Monte and other hydroelectric dams in the Amazon (paradoxically also funded by BNDES) that have been causing major social and environmental impacts in the region [78]. At the same time, there has been an increasing distance between the Ministry of the Environment and NGOs in Brazil after Marina Silva left the Lula government in 2008, and Dilma Rousseff took office in 2011. In this process many of the more progressive senior officials working on climate have been gradually replaced by officials that work more in line with the agenda set by the conservative Ministry of Foreign Affairs. In this context, and similar to the Norwegian case, many NGOs may regard it as inappropriate to express strong public criticism of the Amazon Fund since it is seen overall as a positive mechanism, especially in comparison to other governmental policies.

The above suggests that ENGOs find it difficult to publicly criticise the Amazon Fund, given the relatively open governance structure of the fund and its history of co-operation. ENGOs are aware of, and privately point out, the many shortcomings and contradictions of the Amazon Fund. But if the fund were to cease to exist because of withdrawal by Norway, ENGOs and the environmental cause would lose more than they win—in a context in which it has been increasingly difficult for them to influence key policies such as the Forest Code.

4.3. Indonesia

As the second largest deforester worldwide, Indonesia stood out as a country which NICFI/KOS almost had to engage with, despite substantial challenges, including internal conflict between different sectors of the government, a powerful private sector, and high levels of corruption. In contrast to Brazil, Indonesia was included in NICFI's portfolio with a much more measured approach emphasising a logical sequence of phases from readiness to full implementation [79]. However, some have nevertheless suggested that Norway applied an unduly 'light touch' in their relationship with the Indonesian government [60]. Knowledge about the challenges of governance in the Indonesian forest sector is widespread in Norway, and such concerns were often aired in public meetings about REDD+. Adding to this cautiousness, Norwegian authorities had some negative experience with an environmental agreement with Indonesia initiated by the Brundtland government (1990–1996) as part of its strategy to strengthen economic ties with Indonesia. This agreement also included rainforest conservation projects in Sumatra. However, these were harshly criticised for insufficient implementation and inadequate concerns for local populations by a Norwegian consultancy charged with evaluating the agreements [80].

Angelsen [79] (p. 17) notes that NICFI's agreement with Indonesia was delayed for several reasons, among them a low willingness to reform on the Indonesian side. After lengthy discussions, Norway and Indonesia signed a Letter of Intent (LoI) in late May 2010, at the same time as the large "Oslo Climate and Forest Conference" was arranged in Norway's capital. The bilateral agreement pledges up to USD 1 billion in performance-based payments before 2020, in a sequential approach starting with preparations, capacity-building and pilot project identification [81]. President Yudhoyono was very supportive of REDD+, and shortly after the LoI was signed he created a REDD+ Task Force to implement it. Indonesia presented a REDD+ strategy in June 2012 (one of the conditions of the agreement), and established a REDD+ agency in 2013. A moratorium on new forest concessions was established in 2011, extended to 2017.

According to a 2014 report [82], Indonesian stakeholders diverged in their views on REDD+, and the institutional situation "on the ground" in Indonesia was perceived as cluttered [39] (p. 20). This observation led the authors to ask: "[A] major question for NICFI is how to support the necessary changes, and whether and how to do so within the context of the pay-for-performance approach." [82] (p. 10). Implementation of REDD+ has indeed proved difficult, but it is widely agreed that one of its effects has been to enhance the role of civil society. Based in part on their earlier experience in Indonesia, and, more generally, on the need for local participation for REDD+ to be successful, Norway included as one of seven 'principles' in the LoI the following requirement:

"b. Give all relevant stakeholders, including indigenous peoples, local communities and civil society, subject to national legislation, and, where applicable, international instruments, the opportunity of full and effective participation in REDD+ planning and implementation." (While NGOs have often been critical of conditionality, they do sometimes in effect recommend it—to promote what they regard as valid ends.)

Promoting 'participation' has been sought mainly by involving NGOs, to try and ensure that the people affected have the opportunity to influence the decisions taken. RFN has been actively involved in Indonesia for many years, initially mainly through their collaboration with KKI WARSI (Komunitas Konservasi Indonesia Warung Informasi Konservasi, Indonesian Conservation Community, an alliance of twenty NGOs in Southern Sumatra), starting in 1996 with the project mentioned above. Their strategy has been to partner with local groups, such as KKI WARSI, WALHI (Wahana Lingkungan Hidup Indonesia, The Indonesian Forum for Environment) which is part of the Friends of the Earth network) and AMAN (Aliansi Masyarakat Adat Nusantara, Indigenous Peoples Alliance of the Archipelago), building capacity and seeking to influence government policy. In Indonesia, a large number of NGOs—both national and local—have been heavily involved in REDD+, often with indirect financial support from NICFI. In the most recent allocation WALHI has been granted NOK 13 million for the period 2016–2018, directly—not via RFN [83]. Norway has also provided funding to Indonesian NGOs via UN-REDD, which has been a major recipient of Norwegian REDD+ money. The NGOs have been engaged to undertake a range of different tasks: to assist in selecting pilot sites; provide socio-economic background information; activate communities by communicating about REDD+; to assist in seeking 'free, prior, informed consent' (FPIC); and to prepare action plans. In future, when further progress has been made, they may also, for example, be engaged to oversee implementation, and monitor progress. It should nevertheless be noted that the total funding involved is far less than the amount allocated to Brazilian NGOs.

The attitudes—and responses—of Indonesian NGOs towards REDD+ vary widely. At one extreme, WALHI (Friends of the Earth, Indonesia) have a slogan: "REDD Wrong Path—Pathetic Ecobusiness" which sums up their attitude, echoing the views of some international NGOs which have been sceptical to REDD+ for a number of reasons, for example: rejection of the market and commodification of nature; scepticism towards private property rights; objection to the poor solving the problems of the rich; and concern about diversion of attention from the real villains, who are not the poor but large-scale commercial operators [84]. More supportive of REDD+ is KKI WARSI, in South Sumatra, with their primary focus on the forest rather than the indigenous people [85]. An intermediate position is taken by AMAN (Indonesia's Indigenous Peoples' Organisation). Their slogan "No Rights, No REDD" indicates conditional support, and reflects their primary focus on the people, and an emphasis on land tenure as the key issue [84]. WALHI's open, public criticism of REDD was manifested very clearly by their criticism of the huge Australia-supported Kalimantan Forests and Climate Partnership (KFCP) project whose aim was to repair the damage done to the forest by the 'mega-rice' project. So great was the criticism that the project was abruptly terminated.

A study of Indonesian media coverage by CIFOR also reveals "a broad range of stances by NGOs" [86] (p. 17), from advocates to adversaries; with WALHI identified as a particularly strong critic. Such a varied response—as evidenced by WALHI, AMAN and KKI WARSI—is not an uncommon situation where NGOs are faced with public policies that may be controversial. Some work 'inside'—co-operating with (some would say co-opted by) the government—while others work outside, as critics. A sort of 'division of labour' is established. What is novel in Indonesia is the sheer extent of public participation—often critical—which would not have been possible under the Suharto regime. Whatever else, Norway's support to REDD+ in Indonesia has strengthened the hand of NGOs. While it might be claimed that some have moderated criticism of REDD+ because of the benefits of Norwegian funding, others certainly have not.

NICFI/KOS did not have (and chose not to establish) the capacity to become involved in detail in the planning and implementation of REDD+ in Indonesia [60]. The Norwegian Embassy in Jakarta had until recently only 1–2 staff members employed to deal with REDD+; one of whom, at a crucial period, was a former RFN staff member. NICFI has thus been very dependent on RFN, both for advice and also by virtue of their contacts with Indonesian NGOs. RFN have, on their side, had to walk a difficult middle path: supportive of REDD+ while aware of, and when appropriate pointing out, its potential shortcomings; collaborating with, and supporting, local NGOs some of which are very critical. Their policy has been to pursue a "dual goal of protecting the rainforest *and* securing the rights of its inhabitants" [87] (p. 12, italics in original). Strengthening the rights of local people over the forest may be seen as both an end in itself and the most effective way of limiting deforestation. Largely for this reason, certainly in Indonesia, there has been a notable shift of focus in REDD+, from the trees to the people [88]. Additionally, the organisations which have been involved in REDD+ have been not only environmental NGOs but also those concerned with human rights, such as AMAN and HuMa (Perkumpulan untuk Permbaharuan Hukum Berbasis Masyarakat dan Ekologis—Association for Community Based and Ecological Law Reform) (which played a central role in bringing about a crucial ruling of the Indonesia Constitutional Court regarding the Revised Forestry Law).

RFN, together with AMAN, has defended NICFI against criticism in REDD-Monitor, the very influential website based in Indonesia. In a joint opinion piece in Development Today (26 March 2014), reposted in REDD-Monitor (27 March 2014) two staff members from RFN and AMAN write: "We would argue that more has been achieved in forest protection and the rights of indigenous peoples in Indonesia since the signing of the Indonesia–Norway agreement than over the course of the previous 15 years".

In summary, Indonesian NGOs have been involved with NICFI as both collaborators and critics, and Norwegian NGOs, and most particularly RFN, have played a number of different roles: as policy design advisers, implementers (if only indirectly through their collaboration with local NGOs) and critics. NICFI has been extremely reliant on their expertise, and has provided very substantial funding. The challenging task of combining these very different roles has been alleviated by two main factors: one, a close relationship of mutual trust between RFN and NICFI; second the policy summarised as 'no rights, no REDD' which has made it possible to support REDD+ on a provisional basis—the same slogan, and conditional support, adopted by AMAN, the major Indonesian NGO.

4.4. Tanzania

The agreement with Tanzania was Norway's first REDD+ bilateral agreement. A Letter of Intent (LoI) was signed by PM Stoltenberg and Solheim on April 21 2008, during their visit to Tanzania. By comparison with the agreements with Brazil and Indonesia, the (LoI) on a Climate Change Partnership on April 21 2008, places more emphasis on "REDD-Readiness", though also aiming at the further purpose to "implement programmes on adaptation and mitigation of climate change" [89] (p. xiii) (By comparison, Brazil was the most 'ready to go' with Indonesia as an intermediate case). The Norwegian government promised 500 million NOK (83 million USD) over five years and the first contract, for the development of a national REDD+-strategy, was signed in 2009.

The project idea was actually developed at the Norwegian Embassy in Tanzania in early 2008 before NICFI was established, following discussions between one of the members of staff and a local NGO. It continued to be mainly managed through the Norwegian Embassy in Dar es Salaam, the main driver of the project. REDD+ activities in Tanzania have been hugely dependent on Norwegian funds. As noted in the mid-term review [89] REDD+ policy development was entirely financed by NICFI, as well as all activities of the REDD+ task force and Tanzania's REDD+ secretariat. This funding was used mainly for three purposes: establishment of pilot projects, by NGOs; support to policy and planning, mainly through the local Institute of Resource Assessment, based at the University of Dar es Salaam; and a large number of research projects, mainly based at Sokoine University of Agriculture (SUA) in Morogoro (NICFI is also the biggest financier of UN-REDD in Tanzania with a budget allocation shared between UNDP, UNEP and FAO). However, as the subsequent review found [90], there has been a lack of 'national ownership' which has hindered progress; whether this is because of, or despite, Norway's substantial funding is open to debate. The evaluation also states: "NICFI support has been the driving force behind the national readiness process, but there is no evidence yet that the Tanzanian government is able to commit to building sustainable and permanent institutions without continued external support, for example from Norway" [91] (p. 318).

The decision to prioritise pilot projects is largely explained by the pressures on NICFI to demonstrate early results [91] (p. 313). And the main reason for giving the task of implementation to NGOs, effectively by-passing government to a very large extent, was the issue of corruption. It was particularly problematic for the Norwegian Embassy that at the time of the REDD+ negotiations they were involved in a dispute concerning the relevant government body, the Ministry of Natural Resources and Tourism, over a major environmental programme that the Embassy had supported over many years (After a long and thorough audit, the Ministry repaid Tsh 2,802,480,600 (2 million USD) [92].

In contrast to the Brazilian and Indonesian projects, neither RFN nor other Norwegian NGOs were involved in the Tanzanian REDD+ programme. But NGOs with a base in Tanzania have been very important in its implementation. NGOs in Tanzania were invited to submit proposals, and the nine pilot projects were selected from 46 applications, evaluated by the National REDD+ Task Force and the Embassy. The NGOs were primarily Tanzanian, often with international links: involved in conservation, for example Tanzania Forest Conservation Group (TFCG), and Mpingo Conservation and Development Initiative (MCDI); several were wildlife organisations, but also included were, for example, Tanzania Traditional Energy Development and Environment Organisation (TATEDO), and CARE.

The relationship of these NGOs to Norway was, in this case, simply that of a sub-contractor: they were hired to implement a set of pilot projects [93,94]. They were not engaged as advisers—as in the case of RFN in Brazil for example—although the contract with one of the NGOs, which may be regarded as typical, did include, in addition to implementing the pilot project, 'advocacy at national and international levels' [91] (p. 326). Also, the NGOs were quite active in giving comments on the draft national REDD+ plan (National Strategy for REDD+: draft December 2010, final version March 2013). While not critical of the initiative, they did express strong views about certain issues, most notably advocating a project-based, 'nested' approach over a centrally controlled national fund.

As noted, Norwegian support included a major funding of applied research, primarily involving Sokoine University of Agriculture in Tanzania and the Norwegian University of Life Sciences in Norway (This Section 4.4 is based in part on this programme, some of the results of which are forthcoming and some recently published in [95]). Whereas Norwegian NGOs have not been much involved in REDD+ in Tanzania, staff from these universities—in Tanzania and Norway—have to some extent provided advice. Progress with REDD+ has proved disappointing, and now that the nine pilot projects have come to an end, the scale of Norwegian funding has been reduced.

In summary, the Tanzanian case differs considerably from those of Norway, Brazil, and Indonesia. NGOs in Tanzania, with substantial funding from Norway, have played a modest advisory role, but have been the key players in implementation of the pilot projects, which have now come to an end. They have not expressed public criticism, but they have themselves (in one case) been criticised in Norwegian media, as described in the Norway section above.

5. Analysis and Discussion

Overall, the evidence suggests that NGO/government relations in Norway constitute a rather special case—not least because NICFI originates with an NGO proposal, and KOS and NICFI more generally has been very dependent on policy advice. Norwegian NGOs have co-operated closely with the government authorities from the very start—including policy design—and regularly advised on overarching policy issues, such as safeguards and portfolio development. Confrontation—in the form of public criticism—has to a relatively small extent occurred in Norway, and then towards certain actors (mainly political leadership), rather than towards NICFI/KOS, Norad or the MFA, i.e., their biggest funders. Furthermore, to the degree that there actually has been public criticism from NGOs, this has mainly been about the "big issues", such as cutbacks in the NICFI budget and Norway's official REDD+ position in the UNFCCC negotiations, and not implementation issues "on the ground" in the recipient countries. But there are also examples (most recently on 19 February 2017) that RFN (lately together with Greenpeace Norway) has voiced harsh public criticism towards NICFI-funded activities in the Democratic Republic of the Congo [96]. RFN has also expressed public criticism against NICFI-sponsored activities in Guyana [97].

However, regular meetings with government authorities have provided the NGOs with ample opportunities to voice points of contention, closed off from the public arena. Precisely because

NGOs have had this opportunity, they have not had the same need to go public when disagreeing with NICFI/KOS. Issues of contention—for instance on how strictly "carbon-focused" NICFI should be—have, with a few exceptions, been discussed mainly in private, which has been a very conscious choice by the NGOs. In fact, the NGOs seem to perform a sort of "stage management" [98], carefully considering what issues to sort out backstage (in the form of private criticism) or frontstage (public criticism). Put differently, what is de-politicised ("cooled down") and sorted out backstage vs. what is politicised ("heated up") [30] on the public scene is by no means coincidental, rather the result of careful choices by the NGO concerned. For example, Norwegian NGOs have experienced that private criticism, including letters to one or several ministers, may be more effective than harsh public criticism in the media. The following quote summarises RFNs position: "Løvold says that RFN's main task is to protect the rainforest and indigenous peoples' rights, and therefore it is much more important for the organisation to have real influence than to, as he puts it: 'In periods of powerlessness, yell out in the press'. Løvold believes that Norway has done a good job in preventing deforestation, and therefore it would be wrong and inappropriate to come forth as a harsh critic in the media" [41] (p. 37). Thus, NGOs' roles, or more precisely choices regarding what roles they take, have clear entrepreneurial aspects to them. By this we mean that the NGOs are highly reflexive and aware of their possible operating space—including gains and losses by taking on different roles. Guided by their core objectives, they carefully choose their strategy, including how they frame and draw attention to issues ("issue entrepreneurship", see [3] for a more thorough description), and proposing policy solutions (i.e., policy entrepreneurship [99]).

It is interesting to compare our account of REDD+ and how Norwegian NGOs have dealt with the issue of carbon capture and storage (CCS). In the latter case, implementation of Norway's CCS policy has been strongly criticised in public by certain ENGOs, although the same ENGOs initially were instrumental in establishing the Norwegian CCS policy [100]. This fierce criticism stands in stark contrast to NGOs relationship to NICFI and REDD+. However, unlike in the NICFI policy process, ENGOs have gradually become more marginalised in the CCS implementation process, as they have become less important in supplying technical knowhow throughout the policy process. This suggests that inclusion in implementation processes may dampen NGOs' public criticism, as inclusion provides them with an effective means to influence the policy process. Conversely, less inclusion may leave public criticism the only viable route to influence and if NGOs are gradually excluded, public criticism may soar, as exemplified by the CCS case.

In view of the massive scale of the REDD+ budget, it is noteworthy how minimal the public debate in Norway has been [63]. Aid is often criticised in the media—but mainly for not yielding projected results [101]. Criticisms could have been levelled at REDD+, not only on these grounds, but across a wide range of issues—such as adopting market-based incentives, inadequate attention to the views and interests of indigenous people, and hasty and risky decisions without sufficient knowledge bases. One explanation for the lack of debate may indeed be because it is usually NGOs that are active in promoting such debate.

The weight of the evidence suggests that a dual strategy of co-operation/confrontation has been possible in Norway, although the NGOs have favoured co-operation more than public confrontation. Norwegian NGOs have experienced that sustained dialogue with the government—including private criticism—yields effective results. The reasons why Norwegian NGOs have comparatively large scope for combining different roles is a combination of the long-established 'corporatist' "Norwegian government/NGO model" (see Grendstad et al. [13], Steen [36]), and the special case of REDD+ where government has been so dependent on NGO knowledge and networks.

It is worth noting, however, that some development NGOs have been more critical in public than ENGOs, for instance regarding issues related to implementation "on the ground", and questioning the "crowding out" effect that the NICFI scheme has on poverty reduction [39] (p. 16). ENGOs have been comparatively silent on these issues, and instead focused their public criticism on more overarching policy issues. The fact that ENGOs have received notably more NICFI funding than development and humanitarian NGOs suggests that funding may guide the choice of whether or not to adopt a public confrontation strategy on implementation issues. The same pattern can be observed in Brazil.

Moving to the other countries, the nature and extent of NGO involvement has varied in part because of the different approaches taken by NICFI/KOS in each case. In Brazil, where the state was viewed by Norway as a competent and trustworthy partner, REDD+ was designed as a national programme—for which responsibility was largely delegated to the Brazilian state; and very considerable funds have been allocated for use by NGOs. By contrast, in Tanzania, Norway was much more sceptical to the government apparatus and chose to render its support mainly in the form of pilot projects. For this reason local NGOs benefited greatly since they were given the task of implementing these pilots. Indonesia constitutes an intermediate case, where support is given to a combination of national activities and pilot projects at regional level. Although the amount of funds allocated to Indonesian NGOs was relatively less, compared with Brazil and Tanzania, the support to NGOs has played a very important part in increasing their political influence.

In Brazil, Indonesia and Tanzania non-Norwegian NGOs have not been very confrontational. However, there is some variation both between and within countries. In Brazil, some NGOs have identified important issues with the design and implementation of the Amazon Fund, especially in relation to the lack of strategic vision and the apparent low efficacy of the fund in promoting short-term deforestation reductions. Yet, Brazilian NGOs were facing much more pressing issues such as the dismantling of key aspects of the Forest Code in 2012 and the construction of Belo Monte and other large hydroelectric dams. Therefore, it is understandable why many members from NGOs perceived the Amazon Fund as an overall positive initiative that could be improved more by engagement from the inside instead of subjecting the fund to public criticism that could risk the withdrawal of donor support. In Indonesia, even some of those that have received funding have been critical. Some major NGOs—both environmental and human rights—have indeed been very critical of REDD+ as an initiative. Additionally, they have also been critical of specific projects and programmes dependent on financial support from NICFI. Criticism has been expressed both in the media and also in public meetings organised to inform about and promote REDD+. In Tanzania, NGOs have only to a minimal degree expressed criticism.

Against this background, we may briefly draw some comparisons.

Co-operation in policy design: In all four countries NGOs have, to varying extents, co-operated with government in the policy design of the initiative. (In Tanzania, the Norway-supported REDD+ initiative consisted primarily of project-based activities, though one NGO gave input to the initial project idea).

Co-operation in implementation: Here, Norway is the exception, for the simple reason that REDD+ is not implemented in Norway. In all the other three countries, NGOs have certainly co-operated. This has in the cases of Brazil and Indonesia sometimes been contracted via local NGOs, sometimes in partnership/co-operation with Norwegian NGOs; and sometimes contracted by the respective government, using Norwegian funds. In Tanzania, NGOs functioned as direct sub-contractors to Norway

Confrontation: Here the picture becomes more complex. In Norway, ENGOs have expressed some public criticism, but then mainly regarding overarching policy issues, and not towards their main funders. On the other hand, Norwegian NGOs have expressed substantial private criticism. The case of Brazil is similar: it seems clear that the NGOs have prioritized co-operation above confrontation, and instead expressed private criticism. The experience in Indonesia is mixed. The simplest case is Tanzania, where criticism by NGOs has been minimal.

The discussion in this section is summarised in Table 2 below.

Table 2. A schematic comparison of different roles NGOs have taken regarding NICFI in Norway, Brazil, Indonesia and Tanzania.

Country	Co-Operation		Confrontation
	Policy design	Implementation	Criticism
Norway	Heavily involved, particularly in the start-up phase	Not applicable	Some targeted public criticism, substantial private criticism
Brazil	Heavily involved from the very start, but less as the Amazon Fund's work has evolved	National and local NGOs heavily involved, Norwegian NGOs work in partnership with selected Brazilian NGOs	Limited public criticism, more private criticism
Indonesia	Considerably involved, particularly providing policy advice	National, and local NGOs involved; Norwegian NGOs work in partnership with selected Indonesian NGOs	Some strong public criticism, some private criticism
Tanzania	To a limited degree	NGOs (not Norwegian) as sub-contractors	Minimal criticism

6. Conclusions

Overall, the evidence across the cases suggests that NGOs have prioritised co-operation above confrontation. However, although public criticism has been limited, it has not been entirely absent and in all cases, perhaps with Tanzania as the exception, NGOs have expressed private criticism. In sum, this seems to suggest that "the Norwegian government/NGO model", as presented by Grendstad et al. [13] and Steen [36], has to some extent been exported to the other countries, particularly through the NICFI funding scheme for civil society. The point of Dryzek et al. [11] about trade-offs between inside and outside strategies seems highly relevant in all the cases. There are also good reasons to extend the argument by adding that NGOs are highly aware of and reflexive regarding their possible operating space, and perform trade-offs in an entrepreneurial manner, focused on results.

Close co-operation seems to correlate with limited public confrontation. Largely in line with the position of Dryzek et al. [11]—the evidence suggests that the relative levels of funding and co-operation do influence NGO's choices regarding levels of and channels for criticism (public vs. private). For instance, Friends of the Earth International and Climate Action Network (CAN) are less involved in REDD+ policymaking processes than their Norwegian counterparts, but voice more public criticism towards REDD+.

Limited public criticism does not, however, necessarily mean that criticism has ceased; it may simply mean that criticism instead is expressed in settings away from the public arena. Particularly the evidence from Norway and Brazil suggests that instead of publicly criticising a global initiative that they largely support and have huge stakes in—and thus put the initiative as a whole at risk—NGOs to a larger degree have used other more informal channels to voice points of disagreement. NGOs' opportunity to resist being co-opted is probably greater in this instance than is usually the case—because the interests of NGOs and NICFI/KOS are to a very large extent congruent, and NICFI/KOS is so dependent on NGO services.

Choosing private channels for criticism may be distinguished from co-optation. A clear pattern is that NGOs knowingly trade away some public visibility in exchange for having direct contact with policymakers (both bureaucrats and elected officials) and "a place at the table"; this has been the price to pay for achieving policy impact—the ultimate goal for most NGOs—which in this case has been very substantial. If, however, NICFI initiates policies conflicting with NGOs views, there are examples that NGOs go public, such as in the case of Norway's position on REDD+ in the UNFCCC, indicating they are not co-opted, at least not silenced.

Close co-operation and financial dependence does indeed increase the risk of co-optation. However, we find that the NGOs are very much aware of their possible operating space and pitfalls, and navigate strategically and pragmatically to pursue their goals in an entrepreneurial manner. This point becomes clear when we observe what issues the NGOs publicly criticise—or not. Acting in an entrepreneurial manner still implies making tough choices to navigate the fine line between co-operation and co-optation; deciding when and how to play a critical role—in public or in private.

Returning to the title of the paper—whether the relationship between NICFI and NGOs is best described as an instance of co-operation or co-optation—this analysis has revealed a much more nuanced picture. Instead—drawing on experience from several cases relating to one specific government initiative—we believe that our case is revealing of an entrepreneurial, pragmatic and largely effective NGO approach which may be summarised in the phrase *result-oriented pragmatism* [102].

Thus, we support Soneryd's [103] call to take an *agnostic* scholarly approach when analysing public involvement in policymaking, including NGOs. NGO practices and choices clearly have entrepreneurial aspects to them, which are best understood by employing the sort of empirically driven study that Yearley [12] advocates.

Acknowledgments: The research was funded by The Research Council of Norway, grant 207622/E10 'Dissemination of Scientific Knowledge as a Policy Instrument in Climate Policy' and NORAD/IPAM grant to UFMG. CICERO Centre for International Climate and Environmental Research—Oslo covers the Article Processing Charge (APC) for publishing in this open access journal. We are grateful to Arild Angelsen, Bård Lahn, Göran Sundqvist, Solveig Aamodt, Jennifer Joy West and panellists at the 2015 NFU conference for valuable comments on earlier drafts. Hermansen would also like to thank Irene Øvstebø Tvedten and the TIK reading group (Hilde Reinertsen, Ann-Sofie Kall, Sylvia Irene Lysgård, Eirik Frøhaug Swensen) for interesting discussions, and for coining the concept 'result-oriented pragmatism'. Thanks also to all the informants for sharing their knowledge, perspectives, and valuable time. This paper is dedicated to the memory of Sjur Kasa.

Author Contributions: Erlend A. T. Hermansen (main author), Desmond McNeill and Sjur Kasa conceived and designed the study; Erlend A. T. Hermansen, Desmond McNeill, Sjur Kasa, and Raoni Rajão collected the data; Erlend A. T. Hermansen, Desmond McNeill, and Raoni Rajão analysed the data; Erlend A. T. Hermansen, Desmond McNeill, Sjur Kasa, and Raoni Rajão contributed with materials/analysis tools; Erlend A. T. Hermansen, Desmond McNeill, and Raoni Rajão wrote the paper.

Conflicts of Interest: The authors declare no conflict of interest. The funding sponsors had no role in the design of the study; in the collection, analyses, or interpretation of data; in the writing of the manuscript, or in the decision to publish the results.

References

1. Santilli, M.; Moutinho, P.; Schwartzman, S.; Nepstad, D.; Curran, L.; Nobre, C. Tropical Deforestation and the Kyoto Protocol. *Clim. Chang.* **2005**, *71*, 267–276. [CrossRef]
2. Moutinho, P.; Santilli, M.; Schwartzman, S.; Rodrigues, L. Why ignore tropical deforestation? A proposal for including forest conservation in the Kyoto Protocol. *Unasylva* **2005**, *56*, 27–40.
3. Hermansen, E.A.T. Policy window entrepreneurship: The backstage of the world's largest REDD+ initiative. *Environ. Politics* **2015**, *24*, 932–950. [CrossRef]
4. Angelsen, A. REDD+ as Result-based Aid: General Lessons and Bilateral Agreements of Norway. *Rev. Dev. Econ.* **2016**. [CrossRef]
5. Seymour, F.; Busch, J. *Why Forests? Why Now? The Science, Economics, and Politics of Tropical Forests and Climate Change*; Brookings Institution Press: Washington, DC, USA, 2016.
6. Angelsen, A.; McNeill, D. The Evolution of REDD+. In *Analysing REDD+: Challenges and Choices*; Angelsen, A., Brockhaus, M., Sunderlin, W.D., Verchot, L.V., Eds.; Center for International Forestry Research (CIFOR): Bogor, Indonesia, 2012.
7. McDermott, C.L.; Coad, L.; Helfgott, A.; Schroeder, H. Operationalizing social safeguards in REDD+: Actors, interests and ideas. *Environ. Sci. Policy* **2012**, *21*, 63–72. [CrossRef]
8. Hayes, T.; Persha, L. Nesting local forestry initiatives: Revisiting community forest management in a REDD+ world. *For. Policy Econ.* **2010**, *12*, 545–553. [CrossRef]

9.   Peskett, L.; Brockhaus, M. When REDD+ goes national: A review of realities, opportunities and challenges. In *Realising REDD+: National Strategy and Policy Options*; Angelsen, A., Brockhaus, M., Kanninen, M., Sills, E., Sunderlin, W.D., Wertz-Kanounnikoff, S., Eds.; Center for International Forestry Research (CIFOR): Bogor, Indonesia, 2009; pp. 25–44.

10. Bortne, Ø.; Selle, P.; Strømsnes, K. *Miljøvern Uten Grenser?* 1st ed.; Gyldendal Akademisk: Oslo, Norway, 2002.

11. Dryzek, J.S.; Downes, D.; Hunold, C.; Schlosberg, D.; Hernes, H.K. *Green States and Social Movements: Environmentalism in the United States, United Kingdom, Germany and Norway*, 1st ed.; Oxford University Press: New York, NY, USA, 2003.

12. Yearley, S. *Cultures of Environmentalism: Empirical Studies in Environmental Sociology*, 1st ed.; Palgrave Macmillan: Basingstoke, UK, 2005.

13. Grendstad, G.; Selle, P.; Strømsnes, K.; Bortne, Ø. *Unique Environmentalism: A Comparative Perspective*, 1st ed.; Springer: New York, NY, USA, 2006.

14. Pralle, S.B. Agenda-setting and climate change. *Environ. Politics* **2009**, *18*, 781–799. [CrossRef]

15. Pralle, S.B. *Branching Out, Digging In: Environmental Advocacy and Agenda Setting*, 1st ed.; Georgetown University Press: Washington, DC, USA, 2006.

16. Jamison, A.; Eyerman, R.; Cramer, J. *The Making of the New Environmental Consciousness: A Comparative Study of the Environmental Movements in Sweden, Denmark and the Netherlands*, 1st ed.; Edinburgh University Press: Edinburgh, UK, 1990.

17. Rootes, C. *Environmental Protest in Western Europe*, 1st ed.; Oxford University Press: New York, NY, USA, 2003.

18. Jamison, A. Climate change knowledge and social movement theory. *Wiley Interdiscip. Rev. Clim. Chang.* **2010**, *1*, 811–823. [CrossRef]

19. O'Neill, K. The comparative study of environmental movements. In *Comparative Environmental Politics: Theory, Practice, and Prospects*; Steinberg, P.F., VanDeveer, S.D., Eds.; MIT Press: Cambridge, MA, USA, 2012; pp. 115–142.

20. Climate Funds Update. Available online: http://www.climatefundsupdate.org/data (accessed on 26 October 2016).

21. Sills, E.O.; Atmadja, S.S.; de Sassi, C.; Duchelle, A.E.; Kweka, D.L.; Resosudarmo, I.A.P.; Sunderlin, W.D. *REDD+ on the Ground: A Case Book of Subnational Initiatives across the Globe*; 6021504550; Center for International Forestry Research (CIFOR): Bogor, Indonesia, 2014.

22. Wynne, B. Misunderstood misunderstanding: Social identities and public uptake of science. *Public Underst. Sci.* **1992**, *1*, 281–304. [CrossRef]

23. Wynne, B. Uncertainty and environmental learning: Reconceiving science and policy in the preventive paradigm. *Glob. Environ. Chang.* **1992**, *2*, 111–127. [CrossRef]

24. Wynne, B. Public Understanding of Science. In *Handbook of Science and Technology Studies*, 1st ed.; Jasanoff, S., Markle, G.E., Petersen, J.C., Pinch, T., Eds.; Sage: Thousand Oaks, CA, USA, 1995; pp. 361–388.

25. Wynne, B. May the sheep safely graze? A reflexive view of the expert-lay knowledge divide. In *Risk, Environment & Modernity: Towards a New Ecology*; Lash, S., Szerszynski, B., Wynne, B., Eds.; Sage: Thousand Oaks, CA, USA, 1996; pp. 27–43.

26. Irwin, A.; Michael, M. *Science, Social Theory and Public Knowledge*, 1st ed.; Open University Press: Maidenhead, UK, 2003.

27. Callon, M.; Lascoumes, P.; Barthe, Y. *Acting in an Uncertain World: An Essay on Technical Democracy*, 1st ed.; MIT Press: Cambridge, MA, USA, 2009.

28. Irwin, A. The Politics of Talk. *Soc. Stud. Sci.* **2006**, *36*, 299–320. [CrossRef]

29. Sundqvist, G.; Elam, M. Public Involvement Designed to Circumvent Public Concern? The "Participatory Turn" in European Nuclear Activities. *Risk Hazards Crisis Public Policy* **2010**, *1*, 203–229. [CrossRef]

30. Sundqvist, G. 'Heating up' or 'Cooling Down'? Analysing and Performing Broadened Participation in Technoscientific Conflicts. *Environ. Plan. A* **2014**, *46*, 2065–2079. [CrossRef]

31. Marres, N. The Issues Deserve More Credit Pragmatist Contributions to the Study of Public Involvement in Controversy. *Soc. Stud. Sci.* **2007**, *37*, 759–780. [CrossRef]

32. Norad. *The Norwegian Climate and Forest Funding to Civil Society. Key Results 2013–2015*; Norad—Norwegian Agency for Development Cooperation: Oslo, Norway, 2016.

33. George, A.L.; Bennett, A. *Case Studies and Theory Development in the Social Sciences*, 1st ed.; MIT Press: Cambridge, MA, USA, 2005.

34. Norwegian Centre for Research Data. Available online: http://www.nsd.uib.no/personvern/en/index.html (accessed on 3 January 2017).
35. Rootes, C. Facing south? British environmental movement organisations and the challenge of globalisation. *Environ. Politics* **2006**, *15*, 768–786. [CrossRef]
36. Steen, O.I. Autonomy or dependency? Relations between non-governmental international aid organisations and government. *Voluntas* **1996**, *7*, 147–159. [CrossRef]
37. Tvedt, T. International Development Aid and Its Impact on a Donor Country: A Case Study of Norway. *Eur. J. Dev. Res.* **2007**, *19*, 614–635. [CrossRef]
38. Hermansen, E.A.T. I Will Write a Letter and Change the World: The Knowledge Base Kick-Starting Norway's Rainforest Initiative. *Nord. J. Sci. Technol. Stud.* **2015**, *3*, 34–46. [CrossRef]
39. Hermansen, E.A.T.; Kasa, S. *Climate Policy Constraints and NGO Entrepreneurship: The Story of Norway's Leadership in REDD+ Financing*; Center for Global Development: Washington, DC, USA, 2014.
40. Ministry of Climate and Environment. Available online: https://www.regjeringen.no/en/topics/climate-and-environment/climate/climate-and-forest-initiative/kos-innsikt/theteam/id734275/ (accessed on 2 January 2017).
41. Tvedten, I.Ø. *Silent Struggles. The Depoliticization of Norway's International Climate and Forest Initiative*; University of Oslo: Oslo, Norway, 2011.
42. Meridian Institute. Available online: http://www.redd-oar.org/index.html (accessed on 23 October 2016).
43. Rainforest Foundation Norway. Available online: http://d5i6is0eze552.cloudfront.net/documents/Publikasjoner/Aarsmeldinger/2014.pdf?mtime=20150701144010 (accessed on 27 September 2016).
44. Norad. *Real-Time Evaluation of Norway's International Climate and Forest Initiative. Lessons Learned from Support to Civil Society Organisations*; Norad—Norwegian Agency for Development Cooperation: Oslo, Norway, 2012.
45. Friends of the Earth Norway. Available online: http://naturvernforbundet.no/getfile.php/Dokumenter/h%C3%B8ringsuttalelser%20og%20brev/2010/Klima/rf-nnv-redd-cancun.pdf (accessed on 27 September 2016).
46. Aftenposten. Available online: http://www.aftenposten.no/okonomi/--Oljefondet-ma-bruke-makt-mot-regnskogverting-39865b.html (accessed on 24 February 2017).
47. Norsk Rikskringkasting (NRK). Available online: https://www.nrk.no/norge/oljefondet-far-skryt-fra-regnskogfondet-1.12844081 (accessed on 24 February 2017).
48. Fløttum, K.; Espeland, T.J. Norske klimanarrativer-hvor mange "fortellinger"? En lingvistisk og diskursiv analyse av to norske stortingsmeldinger. *Sakprosa* **2014**, *6*, 4.
49. Rainforest Foundation Norway. Available online: http://www.regnskog.no/no/nyheter/nyhetsarkiv/regnskogfondet/slipp-skogen-inn-p%C3%A5-stortinget (accessed on 31 October 2014).
50. Nationen. Available online: http://www.nationen.no/tunmedia/naturvernforbundet-krever-kutt-i-skogavtalen-med-indonesia/ (accessed on 24 February 2017).
51. Ministry of Climate and Environment. Available online: https://www.regjeringen.no/no/aktuelt/lanseringen-av-moratoriet-er-et-viktig-s/id643916/ (accessed on 27 September 2016).
52. Greenpeace Norway. Available online: http://www.greenpeace.org/norway/no/press/releases/Regnskogfondet-Greenpeace-Naturvernforbundet-og-12-orangutanger-serverer-kake/ (accessed on 27 September 2016).
53. WWF Norway. Available online: http://awsassets.wwf.no/downloads/wwf_arsmelding_2014_web.pdf (accessed on 29 October 2016).
54. Aftenposten. Available online: http://www.aftenposten.no/meninger/kronikk/Naturvern_-bistand-som-ikke-hjelper-174389b.html (accessed on 24 February 2017).
55. Aftenposten. Available online: http://www.aftenposten.no/meninger/Uriktig-om-miljovern-174230b.html (accessed on 24 February 2017).
56. Beymer-Farris, B.A.; Bassett, T.J. The REDD menace: Resurgent protectionism in Tanzania's mangrove forests. *Glob. Environ. Chang.* **2012**, *22*, 332–341. [CrossRef]
57. Aftenposten. Available online: http://www.aftenposten.no/norge/Forskere-Norske-klimapenger-kan-bidra-til-tvangsflytting-168306b.html (accessed on 24 February 2017).
58. Burgess, N.D.; Mwakalila, S.; Munishi, P.; Pfeifer, M.; Willcock, S.; Shirima, D.; Hamidu, S.; Bulenga, G.B.; Rubens, J.; Machano, H.; et al. REDD herrings or REDD menace: Response to Beymer-Farris and Bassett. *Glob. Environ. Chang.* **2013**, *23*, 1349–1354. [CrossRef]

59. Aftenposten. Available online: http://www.aftenposten.no/norge/To-norske-regnskogprosjekter-har-havarert-i-Tanzania-113984b.html (accessed on 24 February 2017).

60. McNeill, D. Norway and REDD+ in Indonesia: The Art of Not Governing? *Forum Dev. Stud.* **2015**, *42*, 113–132. [CrossRef]

61. Aftenposten. Available online: http://www.aftenposten.no/norge/Redder-skogen--men-svikter-de-fattigste-175551b.html (accessed on 24 February 2017).

62. Morgenbladet. Available online: http://test.morgenbladet.no/aktuelt/2013/04/brasil-er-norges-bistandsyndling-igjen (accessed on 24 February 2017).

63. Borge, L. *Trær Vokser Ikke på Penger-og andre Medie-og Publikumsperspektiver på Regnskogmilliardene*; University of Oslo: Oslo, Norway, 2014.

64. Hochstetler, K.; Keck, M.E. *Greening Brazil: Environmental Activism in State and Society*, 1st ed.; Duke University Press: Durham, NC, USA, 2007.

65. Nepstad, D.; Soares-Filho, B.S.; Merry, F.; Lima, A.; Moutinho, P.; Carter, J.; Bowman, M.; Cattaneo, A.; Rodrigues, H.; Schwartzman, S. The end of deforestation in the Brazilian Amazon. *Science* **2009**, *326*, 1350–1351. [CrossRef] [PubMed]

66. Van der Hoff, R.; Rajão, R.; Leroy, P.; Boezeman, D. The parallel materialization of REDD+ implementation discourses in Brazil. *For. Policy Econ.* **2015**, *55*, 37–45. [CrossRef]

67. Hochstetler, K.; Viola, E. Brazil and the politics of climate change: Beyond the global commons. *Environ. Politics* **2012**, *21*, 753–771. [CrossRef]

68. Kasa, S. The Second-Image Reversed and Climate Policy: How International Influences Helped Changing Brazil's Positions on Climate Change. *Sustainability* **2013**, *5*, 1049–1066. [CrossRef]

69. Folha de São Paulo. Available online: http://www1.folha.uol.com.br/fsp/brasil/fc3105200810.htm (accessed on 24 February 2017).

70. Folha de São Paulo. Available online: http://www1.folha.uol.com.br/fsp/brasil/fc2406200820.htm (accessed on 24 February 2017).

71. Dagbladet. Available online: http://www.dagbladet.no/nyheter/norske-regnskogmilliarder-star-ubrukt-pa-konto/63561106 (accessed on 24 February 2017).

72. Globo.com. Available online: http://g1.globo.com/natureza/noticia/2012/01/fracasso-do-fundo-amazonia-causa-desconforto-entre-paises-doadores.html (accessed on 24 February 2017).

73. Rajão, R.; Azevedo, A.; Stabile, M.C.C. Institutional subversion and deforestation: Learning lessons from the system for the environmental licencing of rural properties in Mato Grosso. *Public Adm. Dev.* **2012**, *32*, 229–244. [CrossRef]

74. Azevedo, A.; Rajão, R.; Costa, M.; Stabile, M.; Alencar, A.; Moutinho, P. Cadastro Ambiental Rural e sua influência na dinâmica do desmatamento na Amazônia Legal. *Boletim Amazônia em Pauta* **2014**, *3*, 1–16.

75. Folha de São Paulo. Available online: http://www1.folha.uol.com.br/mercado/2014/05/1460527-manifestantes-bloqueiam-acesso-a-belo-monte-ha-seis-dias-no-para.shtml (accessed on 24 February 2017).

76. Instituto Envolverde. Available online: http://www.envolverde.com.br/1-1-canais/carta-aberta-sobre-a-recem-criada-comissao-nacional-de-redd/ (accessed on 24 February 2017).

77. Soares-Filho, B.; Rajão, R.; Macedo, M.; Carneiro, A.; Costa, W.; Coe, M.; Rodrigues, H.; Alencar, A. Cracking Brazil's Forest Code. *Science* **2014**, *344*, 363–364. [CrossRef] [PubMed]

78. Bratman, E.Z. Contradictions of Green Development: Human Rights and Environmental Norms in Light of Belo Monte Dam Activism. *J. Lat. Am. Stud.* **2014**, *46*, 261–289. [CrossRef]

79. Angelsen, A. *REDD+ as Performance-Based Aid. General Lessons and Bilateral Agreements of Norway*; No. 2013/135; UNU World Institute for Development Economics Research (UNU-WIDER): Helsinki, Finland, 2013.

80. Ny Tid. Available online: https://www.nytid.no/pa_stubbene_los/ (accessed on 24 February 2017).

81. Statsministerens Kontor. Available online: https://www.regjeringen.no/globalassets/upload/SMK/Vedlegg/2010/Indonesia_avtale.pdf (accessed on 24 February 2017).

82. Lash, J.; Dyer, G. *Norway's International Climate and Forest Initiative: A Strategic Evaluation*; Norwegian Ministry of Climate and Environment: Oslo, Norway, 2014.

83. Norad. Available online: https://www.norad.no/en/front/funding/climate-and-forest-initiative-support-scheme/grants-2013-2015/projects/just-governance-to-address-underlying-causes-of-deforestation/ (accessed on 24 February 2017).

84. Lenes, J.T. *Understanding the Making of REDD and the Kalimantan Forest and Climate Partnership (KFCP) in Central Kalimantan through Different Modes of Engagement*; University of Oslo: Oslo, Norway, 2014.
85. Sari, I.M. *Community Forests at a Crossroads: Lessons Learned from Lubuk Beringin Village Forest and Guguk Customary Forest in Jambi Province-Sumatra, Indonesia*; University of Oslo: Oslo, Norway, 2013.
86. Cronin, T.; Santoso, L. *REDD+ Politics in the Media: A Case Study from Indonesia*; Center for International Forestry Research (CIFOR): Bogor, Indonesia, 2010.
87. Rainforest Foundation Norway. *Strategy 2008–17*; Rainforest Foundation Norway: Oslo, Norway, 2012.
88. Howell, S. 'No RIGHTS–No REDD': Some Implications of a Turn Towards Co-Benefits. *Forum Dev. Stud.* **2014**, *41*, 253–272. [CrossRef]
89. Norad. *Real-Time Evaluation of Norway's International Climate and Forest Initiative*; Contributions to National REDD+ Processes 2007–2010. Country Report: Tanzania; Norad—Norwegian Agency for Development Cooperation: Oslo, Norway, 2011.
90. Norad. *Real-Time Evaluation of Norway's International Climate and Forest Initiative*; Synthesising Report 2007–2013; Norad—Norwegian Agency for Development Cooperation: Oslo, Norway, 2014.
91. Norad. *Real-Time Evaluation of Norway's International Climate and Forest Initiative*; Synthesising Report 2007–2013. Annexes 3–19; Norad—Norwegian Agency for Development Cooperation: Oslo, Norway, 2014.
92. JamiiForums. Available online: http://www.jamiiforums.com/threads/tanzania-forced-to-refund-embezzled-funds-to-norway.92290/ (accessed on 24 February 2017).
93. Resset, H. *To Fly a Plane While Building it: NGO's Role in the Development of REDD+ in Tanzania*; University of Oslo: Oslo, Norway, 2012.
94. Furuly, M. *Who should be the Main Actor in Governing REDD+ Projects on the Ground? A Comparative Study of the Relationship between NGOs and the Local Government in the Implementation of Three REDD+ Pilot Projects in Tanzania*; Norwegian University of Life Sciences: Ås, Akershus, Norway, 2016.
95. Kulindwa, K.A.; Silayo, D.S.; Zahabu, E.; Lokina, R.; Hella, J.; Hepelwa, A.; Shirima, D.; Macrice, S.; Kalonga, S. *Lessons and Implications from REDD+ Implementation: Experiences from Tanzania*; Sokoine University of Agriculture: Morogoro, Tanzania, 2016.
96. Dagbladet. Available online: http://www.envolverde.com.br/1-1-canais/carta-aberta-sobre-a-recem-criada-comissao-nacional-de-redd/ (accessed on 24 February 2017).
97. Bergens Tidende. Available online: http://www.bt.no/nyheter/lokalt/Klimaprosjekt-kan-ende-med-regnskog-rasering-284132b.html (accessed on 24 February 2017).
98. Hilgartner, S. *Science on Stage: Expert Advice as Public Drama*, 1st ed.; Stanford University Press: Stanford, CA, USA, 2000.
99. Kingdon, J.W. *Agendas, Alternatives, and Public Policies*, 2nd ed.; Longman: New York, NY, USA, 2003.
100. Tjernshaugen, A. The growth of political support for CO_2 capture and storage in Norway. *Environ. Politics* **2011**, *20*, 227–245. [CrossRef]
101. Reinertsen, H. *Optics of Evaluation. Making Norwegian Foreign aid an Evaluable Object, 1980–1992*; University of Oslo: Oslo, Norway, 2016.
102. Lysgård, S.I.; Swensen, E.F. Bevegelser i Miljøfeltet: Kjetting og Kapital – Ja Takk – Begge Deler? (unpublished).
103. Soneryd, L. What is at stake? Practices of linking actors, issues and scales in environmental politics. *Nord. J. Sci. Technol. Stud.* **2015**, *3*, 18–23. [CrossRef]

Article

A Dominant Voice amidst Not Enough People: Analysing the Legitimacy of Mexico's REDD+ Readiness Process

Jovanka Spirić [1,*], Esteve Corbera [1], Victoria Reyes-García [1,2] and Luciana Porter-Bolland [3]

[1]   Institute of Environmental Science and Technology, Universitat Autònoma de Barcelona,
      08193 Bellaterra, Spain; esteve.corbera@uab.cat (E.C.); victoria.reyes@uab.cat (V.R.-G.)
[2]   Institució Catalana de Recerca i Estudis Avançats (ICREA), Pg. Lluís Companys 23, 08010 Barcelona, Spain
[3]   Instituto de Ecología, A. C., Red de Ecología Funcional, Carretera Antigua a Coatepec 351, El Haya,
      91070 Xalapa, Mexico; luciana.porter@inecol.mx
*   Correspondence: vankajo@yahoo.com; Tel.: +381-638275058

Academic Editors: Damian C. Adams and Timothy A. Martin
Received: 28 September 2016; Accepted: 30 November 2016; Published: 10 December 2016

Abstract: In the development of national governance systems for Reducing Emissions from Deforestation and forest Degradation (REDD+), countries struggle with ensuring that decision-making processes include a variety of actors (i.e., input legitimacy) and represent their diverse views in REDD+ policy documents (i.e., output legitimacy). We examine these two dimensions of legitimacy using Mexico's REDD+ readiness process during a four-year period (2011–2014) as a case study. To identify REDD+ actors and how they participate in decision-making we used a stakeholder analysis; to assess actors' views and the extent to which these views are included in the country's official REDD+ documents we conducted a discourse analysis. We found low level of input legitimacy in so far as that the federal government environment agencies concentrate most decision-making power and key land-use sectors and local people's representatives are absent in decision-making forums. We also observed that the REDD+ discourse held by government agencies and both multilateral and international conservation organisations is dominant in policy documents, while the other two identified discourses, predominantly supported by national and civil society organisations and the academia, are partly, or not at all, reflected in such documents. We argue that Mexico's REDD+ readiness process should become more inclusive, decentralised, and better coordinated to allow for the deliberation and institutionalisation of different actors' ideas in REDD+ design. Our analysis and recommendations are relevant to other countries in the global South embarking on REDD+ design and implementation.

Keywords: discourses; legitimacy; Mexico; REDD+; stakeholder analysis

1. Introduction

Reducing Emissions from Deforestation and forest Degradation, plus the conservation, sustainable management of forests, and enhancement of forest carbon stocks (REDD+) is an emergent global forest governance regime under the auspices of the United Nations Framework Convention on Climate Change (UNFCCC). REDD+ is intended to align the views of a variety of actors who are active across different social and political jurisdictions on how to frame and address the problem of deforestation and forest degradation in developing countries [1–6]. During the last decade, REDD+ has become increasingly relevant in the environmental and land-use policy agendas of many developing countries. To date, around 50 countries in the global South are involved in the so-called *readiness* phase—the first of three phases in REDD+ design and implementation [7,8]. These countries have been building institutional capacity and developing national strategies, which contain guidelines for the design of REDD+ policies and activities that will be operationalised during the *implementation* phase, while any potential resulting emission reductions will be accounted for and rewarded in the *performance* phase [9,10].

The "Warsaw Framework for REDD+" provides the rules and guidance that developing countries should follow to ensure an effective and sustained REDD+ implementation [11], and several studies have investigated the legitimacy of such international REDD+ governance negotiations [12–14]. However, as REDD+ will ultimately be implemented at the national level [15] and jurisdictional or sub-national approaches to REDD+ have been officially accepted as interim measures towards a full national approach [11], the legitimacy of REDD+ national governance (i.e., how non-state actors are involved in REDD+ decision-making and to which extent their ideas and views permeate the design and implementation of REDD+ policies and activities [16,17]) will be determined by host countries' governance processes, which deserve detailed scrutiny (e.g., [18] this special issue).

The legitimacy of environmental governance encompasses the legitimacy of the process as well as its outcomes, the so-called *input* and *output* legitimacy [19,20]. Input legitimacy concerns the extent to which actors are recognised, included, and represented in negotiations. It also refers to whether negotiations are transparent and all participants are accountable to one another and engage in discussions and decision-making on a voluntary and equal basis [20–22]. Output legitimacy refers to the level of actors' acceptance of adopted decisions and the outcomes of their implementation [20,23–25]. In the context of REDD+ readiness, input legitimacy concerns the design of national REDD+ decision-making processes through which governments should share their power in addressing deforestation and forest degradation with other actors. In turn, output legitimacy concerns the extent to which the general idea of REDD+ is either contested or accepted by these actors and how their different views on REDD+ key issues become acknowledged in government approved national REDD+ documents.

Despite the growing importance of REDD+ social safeguards, which call for a full and effective participation of all relevant stakeholders [26], most REDD+ pursuing countries still face the challenge of designing legitimate decision-making processes [27–37]. Using Mexico as a case study, this article contributes to these debates by investigating the legitimacy of the country's REDD+ readiness process over a four-year period. To evaluate the level of input legitimacy, we identify the key REDD+ actors and analyse their relevance, influence, and interest in the readiness process, as well as their perceptions on the legitimacy of national REDD+ discussions. To evaluate the level of output legitimacy, we identify the principal discourses that have emerged around REDD+ and calculate the degree to which the ideas embedded in actors' discourses are reflected in the two most advanced REDD+ policy documents at the time of this research: *Mexico's Emission Reductions Initiative Idea Note* (ER-PIN, August 2013) and the fifth draft of the *National REDD+ Strategy* (ENAREDD+, November 2014; with ENAREDD+ representing its Spanish acronym—similar naming/referencing conventions will appear throughout this document for the names of other forums and organisations that were originally in Spanish).

2. Case Study and Methods

2.1. REDD+ Governance in Mexico

Mexico showed an early interest in REDD+ and was the first country to submit the *Readiness Plan Information Note* (R-PIN) in 2008 and the *Readiness Preparation Proposal* (R-PP) in 2010 to the Forest Carbon Partnership Facility (FCPF)—the World Bank's multilateral readiness platform that provides technical and financial support to REDD+ developing countries [38–40]. Furthermore, aiming to explore and test different REDD+ institutional and financial arrangements, Mexico's government has been implementing *REDD+ early actions* since 2010 through the existing National Forest Commission's (CONAFOR) programmes and the newly launched "special programmes" [41]. Several local REDD+ pilot projects promoted by non-governmental organisations (NGOs) have also been implemented [42–44].

In 2009, the Mexican federal government created a cross-sectorial body to coordinate government sectors relevant to REDD+, known as the Working Group for REDD+ (GT-REDD+) [45]. A year later, in 2010, the government promoted a multi-stakeholder consultative forum, known as the national

Technical Advisory Committee for REDD+ (CTC-REDD+, hereafter CTC) to involve non-state actors in the REDD+ readiness phase. Several sub-national CTCs were also established in 2011 in states selected for early REDD+ implementation (i.e., Oaxaca, Chiapas, Yucatán, Campeche, Quintana Roo, and Jalisco) [46]. In addition, the government reformed the General Law on Ecological Equilibrium and Environmental Protection (LGEEPA) and the General Law on Sustainable Forest Development (LGDFS) to facilitate REDD+ design and implementation in the same year [47,48]. In 2013, another national level consultative multi-stakeholder forum, the Working Group on ENAREDD+ (GT-ENAREDD+), was established under the National Forest Council (CONAF) [49].

The REDD+ readiness process in Mexico had a first intermediate product in 2010: *Mexico's REDD+ Vision*. This document identified sustainable rural development as the key governing principle for REDD+ implementation and defined five strategic lines for the development of the national REDD+ strategy (ENAREDD+): (a) institutional arrangements and public policies; (b) financing mechanisms; (c) monitoring, reporting, and verification (MRV) systems—in place to measure the country's performance in terms of forest related emissions and removals [50]; (d) communication, participation, and transparency; and (e) environmental and social safeguards [51]. From 2011 to 2014, six ENAREDD+ drafts were produced providing further details on specific activities to be implemented within each strategic line [45,51–54]. In 2013, CONAFOR elaborated *Mexico's Emission Reductions Initiative Idea Note* (ER-PIN), a document required by the FCPF Carbon Fund, which expands on key REDD+ design issues such as benefit-sharing, scope of activities (only conservation or also productive activities), and expected co-benefits (non-carbon benefits, including social, environmental, and governance benefits [11]) [55,56]. As Figure 1 suggests, although CONAFOR endorses the official national REDD+ documents, they should theoretically include the views and ideas of the actors involved in the CTC and GT-ENAREDD+ forums.

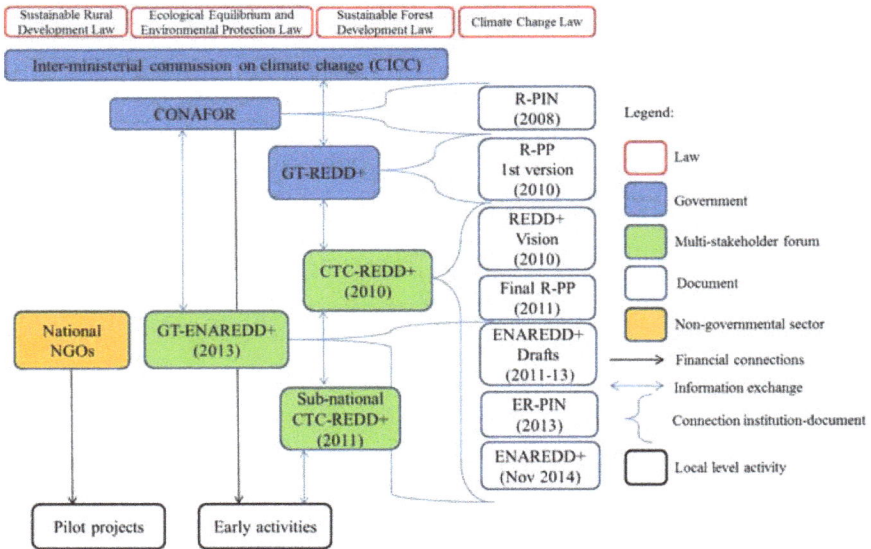

Figure 1. Main laws, documents, and activities in Mexico's REDD+ readiness governance system and their authoring and/or implementing actors and consultative forums. Source: own elaboration. REDD+, Reducing Emissions from Deforestation and forest Degradation; ENAREDD+, Mexico's national REDD+ strategy in Spanish; CONAFOR, National Forest Commission; R-PIN, Readiness Plan Information Note; R-PP, Readiness Preparation Proposal; ER-PIN, Mexico's Emission Reductions Initiative Idea Note; CTC-REDD+, Technical Advisory Committee for REDD+; GT-REDD+, Working Group for REDD+; GT-ENAREDD+, Working Group on ENAREDD+; NGOs, non-governmental organisations.

2.2. Data Collection

Data collected for this article includes secondary and primary sources. Secondary sources include information collected from Mexico's REDD+ actors' publications and reports, key official documents, open letters and other media communications, and minutes from the above-described multi-stakeholders REDD+ forums, both in English and in Spanish, and published up to December 2014. These documents helped us in understanding the evolution of REDD+ in Mexico because they included written statements that captured actors' general views on REDD+ design and implementation. We acknowledge that actors can change their position in regards to REDD+ issues, but for the purpose of the analysis presented here, we only consider the predominant position of the actor for a given theme over the research period. The first author also collected primary data on participant actors' discussions and decision-making procedures during her participation in REDD+ related forums and events. These events included: The Commission for State Development Planning—Quintana Roo, 17 and 28 June 2011; *U'yool'che's* REDD+ workshop with local communities, 9 and 10 July 2011; The Nature Conservancy's deforestation workshop, 10 and 11 July 2011; *U'yool'che's* workshop on the community's protected area, 12 July 2011; *Universidad Nacional Autónoma de México's* roundtable on the Law on Ecological Equilibrium and Environmental Protection, 20 July 2011; *Consejo Mexicano de Silvicultura Sostenible's* REDD+ workshop, 9 August 2011; *Aliance Sian Ka'an-Calakmul's* REDD+ workshop, 16 August 2011; CTC-Quintana Roo session, 14 October 2013; *El Consejo Regional Indígena y Popular de Xpujil's* meeting, 29 November 2013; and the state of Campeche environmental agency's working meeting on REDD+ with CONAFOR, 14 February 2014.

The information from the documents and meetings helped us evaluate the actors' interest in and power to influence the REDD+ readiness process, and strengthened the analysis of their views on REDD+ design and implementation. They helped in developing a list of potential interviewees with a stake in Mexico's forestry and land-use sectors as well as of actors that could be affected by future REDD+ implementation in the country. This non-comprehensive list of interviewees included representatives from all sectors relevant to REDD+ design and implementation at federal level. At the local level, it included actors from two early action regions of the Yucatán peninsula and the state of Chiapas, which were prioritised to investigate the legitimacy of the readiness phase at sub-national level. Additional REDD+ actors were identified through snowball sampling in initial interviews [57,58].

The first author conducted 41 face-to-face semi-structured interviews during two periods of fieldwork (June–August 2011 and September 2013–February 2014). Specifically, 16 interviewees belonged to national and three to international NGOs, ten to federal and three to state governments, six to academic institutions, and three involved local representatives from the communities of *Felipe Carrillo Puerto* in Quintana Roo, and *La Mancolona* and *Xmaben* in Campeche. These communities were chosen because they are located in REDD+ development priority regions and were part of a larger research project in which the authors were involved. Local communities' views about REDD+ were also captured through the secondary sources highlighted above, such as newspapers and official letters, as well as in REDD+ forums and events. The interview guide included three sections addressing (i) actors' roles in REDD+ readiness and their external reasons and internal motivations to participate (or not) in the process; (ii) actors' perceptions of the REDD+ decision-making process; and (iii) actors' general opinions on REDD+ and its design and operational issues. The interviews lasted about one and a half hours, were conducted in Spanish, recorded (with consent), and subsequently transcribed for analysis.

2.3. Data Analysis

2.3.1. Qualitative Content Analysis

We used a qualitative content analysis software programme (MaxQDA) to assign codes related to eight key issues in REDD+ design and implementation to paragraphs or sentences in the interview and event transcripts, as well as in the analysed publications and reports. The eight key issues selected were deforestation drivers, REDD+ definitions, implementation scale, scope of activities, carbon rights, land tenure, participation, and decision-making procedures. The segments containing the same codes were grouped in documents and translated into English. The results of an interpretative analysis of these documents informed the stakeholder and the discourse analyses.

2.3.2. Stakeholder Analysis

We used stakeholder analysis to evaluate the input legitimacy of Mexico's REDD+ readiness phase. Specifically, we used the interview transcripts and the published documents to evaluate an actor's *relevance, power to influence,* and *interest* in REDD+. For these three characteristics, we used a three-grade system (high, moderate, and low). *Relevance* was determined based on the potential impact that a given actor's activities might have on REDD+ effectiveness (i.e., in reducing land-use emissions and increasing removals) [59]. *Influence* was determined based on the extent to which an actor was likely to persuade or coerce other actors into following certain courses of action in REDD+ design and implementation [60]. Since the level of persuasion and coercion was not specifically assessed in the interviews, we used information about the actor's REDD+ financial resources and position in formal social hierarchies, specifically in REDD+ decision-making bodies to assess the actor's *influence* [61]. Finally, *interest* was measured based on (i) the actor's role as financial investor in REDD+; (ii) the frequency of actor's participation in both governmental and alternative REDD+ readiness events; and, (iii) the amount of documents that contribute to REDD+ discussions produced by the actor [60,62] (Table 1).

Table 1. Description of actor's relevance, influence, and interest in REDD+. Source: own elaboration.

Level / Attributes	High	Moderate	Low
Relevance	Actors who design or implement public policies and activities that directly contribute to land-use change, either increasing or decreasing carbon stocks	Actors who provide financial resources and/or information for the development of specific land-use change activities that either increase or decrease carbon stocks	Actors whose activities do not have an impact (or whose impact would be hard to prove) on land-use change
Influence	Actors who are primary recipients of relevant financial resources for REDD+ or who can directly influence policies given their position in formal social hierarchies	Actors receiving REDD+ financial resources from government, thus steering REDD+ design in ways that meet their expectations	Actors who do not hold significant financial REDD+ resources and who are not present in formal REDD+ decision-making forums
Interest	Actors who financially invest in REDD+ and/or regularly participate in either the governmental or alternative REDD+ forums, contributing to discussions in oral and/or written forms	Actors with all preconditions to participate (e.g., financial resources, invited to meetings) but who only intermittently participate in the governmental or alternative REDD+ oral or written discussions	Actors who were formally invited to participate in either the governmental or alternative REDD+ forums, but neither participated nor communicated their views on REDD+

2.3.3. Discourse Analysis

Stakeholders articulate their views of reality through discourses [63], which are defined as the shared ways of understanding and perceiving the world [64,65]. Environmental discourses frame how we conceive environmental problems (e.g., deforestation) and related policies (e.g., REDD+) [66]. Given that by constructing and reproducing REDD+ discourses, actors legitimise or delegitimise REDD+ governance [65–71], we used discourse analysis to assess the output legitimacy of REDD+ governance in Mexico's readiness phase as well as the actors' perceptions on the legitimacy of REDD+ readiness discussions. The analytical framework we developed to identify and examine REDD+ discourses in

Mexico combines three elements suggested by Dryzek [65]: key storylines, main discursive agents, and key metaphors.

Key storylines are a collection of actors' stances on ten central REDD+ dimensions including (i) three *REDD+ conceptual dimensions* (i.e., REDD+ goals and its role within climate change governance; drivers of deforestation; and the role of local communities in deforestation) and (ii) seven *REDD+ strategic dimensions* (i.e., REDD+ implementation scale; benefit-sharing strategy; scope of activities; co-benefits; safeguards; impact on land and tenure; and carbon rights). The discursive agents represent the actors who, through the storylines, are characterised as the archetypes of "heroes" and "culprits" (i.e., those who positively or negatively contribute to forest conservation and REDD+ effectiveness), or as "winners" and "losers" (i.e., those who will benefit the most or become worse off from REDD+). Finally, key metaphors are two or three keywords or phrases used in storylines to symbolise the discourse (e.g., 'win-win-win').

We used the results of qualitative content analysis to identify ten key storylines that each REDD+ actor in Mexico put forward for each conceptual and strategic dimension highlighted above. Specifically, we grouped into *discourse coalitions* the actors who share the usage of a particular set of storylines [64]. We further explored overlaps and conflicts between different discourse coalitions by identifying whether they promote the same or opposed storylines. Hajer [72] suggests that a discourse becomes *hegemonic* if policy actors are required to use its vocabulary and concepts to appear credible (*discourse structuration*), and if such discourse fully permeates into policy decisions and institutional arrangements and practices (*discourse institutionalisation*). In reality, discourse structuration and institutionalisation occur only to a certain degree, resulting in a weaker form of hegemony or *discursive domination* [72].

As we analysed discourses about REDD+, and all policy actors used the REDD+ terminology in their storylines, we evaluated the degree of *discourse structuration* by calculating the proportion to which the storylines on conceptual REDD+ dimensions were represented, explicitly or implicitly, in either of the two most advanced REDD+ policy documents by the time we concluded data collection (i.e., the fifth ENAREDD+ draft and the ER-PIN). We evaluated *discourse institutionalisation* by calculating the extent to which the storylines on strategic REDD+ dimensions were represented, explicitly or implicitly, in any of the named documents. Therefore, if a discourse had all its storylines on REDD+ conceptual issues (in our case three) and half or more (in our case between four and seven) of those focused on strategic dimensions represented in the documents, we considered that discourse as 'dominant'. If a discourse had less than three of the storylines related to REDD+ conceptual issues and less than four of the storylines concerned with strategic REDD+ issues represented in the documents, we considered it as 'marginalised' (see Table 2).

Table 2. Rules to determine the degree of discourse influence according to the representation of storylines discussing REDD+ conceptual and strategic dimensions as they appear in selected policy documents. The number of storylines is in brackets. Source: own elaboration.

Conceptual Dimensions / Strategic Dimensions	All Storylines (3)	Less Than Three Storylines (2 or less)
Four or more storylines (4–7)	Dominant	Marginalised
Less than four storylines (0–3)	Marginalised	Marginalised

Finally, to assess the overall legitimacy of REDD+ readiness in Mexico, we plotted the actors' perceptions of input and output legitimacy using a two-grade system (granted and not granted) (Table 3). We used the actors' perceptions on the legitimacy of the REDD+ decision-making process organised through the CTC to represent the level of input legitimacy. In the case of output legitimacy, and since the actors' perceptions on the content of REDD+ documents were not specifically assessed in the interviews, we used the degree of discourse influence as an indicator of the likely actors' levels of acceptance and endorsement of REDD+ policy documents.

Table 3. Rules to evaluate the overall legitimacy of REDD+ governance in Mexico. Source: own elaboration.

Input Legitimacy / Output Legitimacy	Granted	Not Granted
Granted	Actors who perceive the CTC as a legitimate forum and share a 'dominant' discourse	Actors who perceive the CTC as an illegitimate forum and share a 'dominant' discourse
Not granted	Actors who perceive the CTC as a legitimate forum and share a 'marginalised' discourse	Actors who perceive the CTC as an illegitimate forum and share a 'marginalised' discourse

3. Results

3.1. REDD+ Stakeholders

The identified actors were categorised into eight stakeholder groups based on their *relevance* and ability to *influence* REDD+ readiness. Out of the nine theoretically possible stakeholder groups, there were none with low relevance but high ability to influence REDD+ design. We named the identified groups as: *top holders, followers, frontliners, money patrons, intermediaries, on the ground, information providers*, and *observers*, deriving labels from the most common role or position that group members have in REDD+ readiness (Figure 2).

Top holders include federal environmental agencies that are very relevant for REDD+ effectiveness as they design environmental and forestry policy and legal frameworks that should be conducive to forest conservation and reduce rates of land-use change. Among the *top holders*, the Environment Ministry (SEMARNAT) holds the highest position in the formal social hierarchy, which makes it very influential in REDD+. SEMARNAT's high level of influence and interest in REDD+ emanates from the joint participation in readiness discussions and activities of its decentralised federal agencies for protected areas (CONANP), environmental protection (PROFEPA), and ecology and climate change (INECC), in addition to a permanent interdepartmental commission for knowledge and use of biodiversity (CONABIO). Despite being one of SEMARNAT's decentralised agencies, CONAFOR is the most influential REDD+ actor in Mexico, as it is the national focal point under the UNFCCC negotiations and controls the largest sum of REDD+ readiness financial resources. CONAFOR is also the CTC convener and the lead author of the policy documents analysed.

Followers include the federal ministry of agriculture (SAGARPA), which is responsible for the design and implementation of agricultural public policies, a major driver of deforestation. Despite SAGARPA's high position in the formal hierarchy, this agency does not manage REDD+ funds and has had a secondary role in the readiness phase. SAGARPA's frequency of participation in CTC meetings has declined over the years, currently showing a moderate level of interest in REDD+. *Followers* also include the state environmental agencies in early action regions (e.g., SMAAS-Campeche; SEMA-Quintana Roo; SEMAHN-Chiapas), who have a moderate influence because the sub-national authorities manage only a limited share of the federal budget and are financially subordinated to the central government [73]. These sub-national agencies, however, have high interest in REDD+, being leaders of state CTCs and lead authors of sub-national REDD+ strategies.

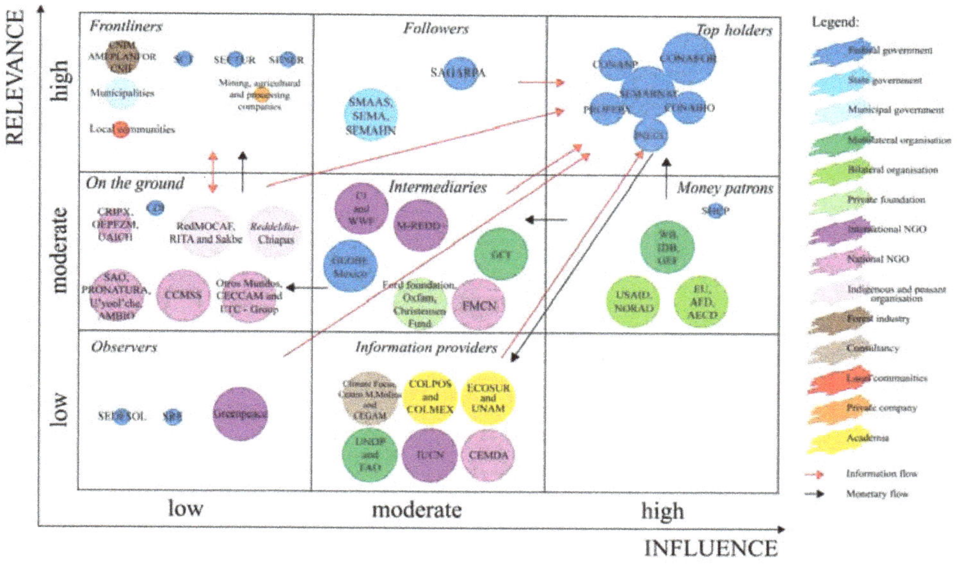

Figure 2. REDD+ stakeholder's groups according to their relevance and influence. Colour corresponds to actor typology (figure right column legend), and size of the circle to the actor's level of interest in REDD+ implementation. Source: own elaboration. SEMARNAT, Ministry of Environment and Natural Resources in Spanish; CONANP, National Commission of Protected Areas; PROFEPA, Office of the Federal Attorney for Environmental Protection; INECC, National Institute of Ecology and Climate Change; CONABIO, National Commission for Knowledge and Use of Biodiversity; SAGARPA, Ministry of Agriculture, Livestock, Rural Development, Fisheries and Food; SMAAS, Ministry of Environment and Sustainable Use, Campeche; SEMA, Ministry of Ecology and Environment, Quintana Roo; SEMAHN, Ministry of Environment and Natural History, Chiapas; SENER, Ministry of Energy; SECTUR, Ministry of Tourism; SCT, Ministry of Communications and Transport; AMEPLANFOR, Mexican Association of Forest Planters; CNIM, National Chamber of Wood Industry; CNIF, National Chamber of Forest Industry; SHCP, Ministry of Finance and Public Credit; WB, World Bank; IDB, Inter-American Development Bank; GEF, Global Environment Facility; EU, European Union; AFD, French Development Agency; AECD, Spanish Agency for International Development Cooperation; USAID, United States Agency for International Development; NORAD, Norwegian Agency for Development Cooperation; GCF, Governors' Climate and Forests Task Force; FMCN, Mexican Fund for the Conservation of Nature; M-REDD+, Mexico REDD+ Alliance; CI, Conservation International; WWF, World Wildlife Fund for Nature; CECCAM, Center for Studies for Change in the Mexican Countryside; ETC-Group, Action Group on Erosion, Technology and Concentration; RedMocaf, Mexican Campesino Forest Producers Network; RITA, Indigenous Network of Environmental Tourisms; CCMSS, Mexican Civil Council for Sustainable Forestry; CRIPX, Popular Regional Indigenous Council of Xpujil; UAICH, Union of Indigenous Beekeepers from Chenes region; OEPFZM, Organization of Forest Ejido Producers of the Maya Zone; SAO, Environmental Services of Oaxaca; CDI, National Commission for Indigenous Development; ECOSUR, College of the South Border; UNAM, National Autonomous University of Mexico; COLPOS, College of Postgraduates; COLMEX, College of Mexico; CEGAM, Centre of Specialists in Environmental Management; CEMDA, Mexican Centre for Environmental Law; IUCN, International Union for Conservation of Nature; UNDP, United Nations Program for Development; FAO, Food and Agriculture Organisation; SEDESOL, Ministry of Social Development; SRE, Ministry of Foreign Affairs.

Frontliners include a range of actors from different social sectors operating at different geographical and jurisdictional levels, including the Ministries of Tourism (SECTUR), Communications and Transport (SCT), and Energy (SENER); state agricultural agencies; municipal authorities; and some private mining, agricultural, and processing companies and commercial forest plantations (e.g., *Asociación Mexicana de Plantadores Forestales* (AMEPLANFOR), *Cámara Nacional de la Industria Maderera* (CNIM), and *Cámara Nacional de la Industria Forestal* (CNIF)). These actors design or implement policies and activities that influence the country's deforestation rate, which may influence REDD+ success in the future. The group also includes local communities, which own approximately 70% of the forests [74] and 52% of the agricultural land in the country [75] and who implement land-use policies and activities on their lands. *Frontliners* have low influence on REDD+ policy-making due to their limited access to REDD+ financial resources and their lack of information and clear roles in the readiness phase, which has resulted in little involvement and sense of ownership of policy documents. Only municipal authorities and private forest plantations owners participate in REDD+ related events occasionally and are therefore characterised as moderately interested.

Money patrons include moderately relevant actors, such as the World Bank (WB), the Inter-American Development Bank (IDB), the Global Environment Facility (GEF), and the development agencies of Norway (NORAD), the United States of America (USAID), France (AFD), Spain (AECD), and the federal Ministry of Finance (SHCP). These actors have provided financial resources for the development of REDD+ in Mexico and, in doing so, they have enforced their procedures and timetables. For example, the Mexican government has to follow the FCPF's Strategic Environmental and Social Assessment (SESA) process to meet environmental and social safeguards, as well as it has to respond to the FCPF's timetable in order to be eligible for REDD+ implementation funding under the Carbon Fund in the future. For the same reason, multilateral agencies are considered to be very interested in REDD+. Only the SHCP showed limited interest in participating in REDD+ events.

Intermediaries include the federal agency in charge of legislative power (GLOBE Mexico), a consortium of national and international NGOs (the M-REDD+ Alliance—leader of the "Mexico's REDD+ project" (M-REDD+) including The Nature Conservancy (TNC), Rainforest Alliance (RA), Woods Hole Research Center (WHRC), and Natural Areas and Sustainable Development (ENDESU)), two large international NGOs (World Wildlife Fund (WWF) and Conservation International (CI)), voluntary carbon market developers (e.g., The Californian Governors' Climate and Forests Task Force (GCF)), and private foundations and funds (e.g., Moore foundation, The Conservation, Food and Health (CFH) foundation, Climate Works). *Intermediaries* are moderately relevant as they provide knowledge and/or financial resources to the government and/or to actors on the ground for the implementation of pilot REDD+ activities. Such financial resources vest them with a moderate level of influence on REDD+, but given their frequent participation in the readiness phase they have been considered highly interested actors.

On the ground actors include moderately relevant actors who have a role in promoting or facilitating certain land-use and land-change activities at local level. They encompass the National Commission for Indigenous Development (CDI) and numerous civil society organisations (CSOs), such as NGOs developing carbon forestry and/or REDD+ pilot projects (*Consejo Mexicano de Silvicultura Sostenible* (CCMSS), *Servicios Ambientales de Oaxaca* (SAO), *Cooperativa AMBIO, PRONATURA*, and *U'yool'che*), peasant and indigenous peoples' organisations (*Red Mexicana de Organizaciones Campesinas Forestales* (RedMocaf), *Red Indígena de Turismo* (RITA), *Sakbe-Comunicación y Defensa*), CSOs partners in the *Reddeldia*-Chiapas movement (e.g., *Otros Mundos Chiapas, Centro de Estudios para el Cambio en el Campo Mexicano* (CECCAM), Action Group on Erosion, Technology and Concentration (ETC-Group)), and many local CSOs (e.g., *Consejo Regional Indígena y Popular de Xpujil* (CRIPX), *Unión de Apicultores Indígenas de los Chenes* (UAICH), *Organización de Ejidos Productores Forestales en la Zona Maya* (OEPFZM)). These actors neither have important roles in formal decision-making, nor hold significant REDD+ financial resources, therefore have little impact on REDD+ design. However, most members of this group participate frequently in REDD+ events, with the exception of CDI and some local CSOs who demonstrate low and moderate interest in REDD+, respectively.

Information providers include national research institutions (*Universidad Nacional Autónoma de México* (UNAM), *El Colegio de la Frontera Sur* (ECOSUR), *El Colegio de Postgraduados* (COLPOS), and *El Colegio de México* (COLMEX)), two UN agencies (The United Nations Program for Development (UNDP), Food and Agriculture Organisation (FAO)), a large international NGO (International Union for Conservation of Nature (IUCN)), a national NGO (*Centro Mexicano de Derecho Ambiental* (CEMDA)), and independent expert advisories (Climate Focus, *Centro Mario Molina, Centro de Especialistas en Gestión Ambiental* (CEGAM)). These actors do not influence land-use change activities and thus will not influence REDD+ effectiveness. They, however, have a role as facilitators and observers of REDD+ readiness and often provide CONAFOR with information on REDD+ technical and governance issues, which makes them moderately influential actors. All *information providers* are frequent participants in REDD+ readiness and have produced relevant literature on the topic.

Finally, *observers* include the Ministry of Foreign Affairs (SRE), the Ministry of Social Development (SEDESOL), and the international NGO Greenpeace. The activities of these actors do not affect land-use, so they are considered irrelevant in regards to REDD+ effectiveness. Moreover, these actors have no influence on REDD+ design, as they do not have a specific role in the formal REDD+ decision-making processes nor manage REDD+ financial resources. However, SRE and SEDESOL are considered REDD+ actors because they have helped CONAFOR in negotiating foreign investments for REDD+ and could be important in ensuring that REDD+ activities translate into social co-benefits. Greenpeace is considered a REDD+ actor for its demonstrated high interest in the readiness phase, for participating in the CTC, and for publishing REDD+ related documents.

3.2. REDD+ Discourses

We classified REDD+ actors into three discourse coalitions according to their views regarding REDD+'s conceptual and strategic dimensions. Given that a shared degree of relevance, interest, and influence by several actors does not necessarily mean they hold a shared discourse, members of the same stakeholder group might belong to different discourse coalitions. We named each coalition according to its members' general attitude towards REDD+, namely: the *opposition, advocacy*, and *reform* coalitions. The *reformists'* discourse shares some storylines with the *advocates* and some with the *opponents*, but the later have no storyline in common and can be considered antagonistic (Figure 3).

The *opposition* coalition includes numerous CSOs, peasant and indigenous peoples' organisations from Chiapas, and some local communities of the Lacandon rainforest implementing REDD+ pilot projects, members of *frontliners*, and *on the ground* stakeholder groups. The shared discourse of this coalition is that REDD+ is not the solution to climate change because deforestation is not its main driver and because REDD+ potentially allows the global North to offset its greenhouse gas emissions cheaply in the global South, instead of reducing domestic emissions or paying off a climate debt. *Opponents* warn that REDD+ may introduce changes in forest tenure rights that could constrain local people's access to forest resources and lead to the loss of local knowledge and forest conservation behaviours. Overall, in the *opposition* coalition storylines, local communities are characterised as both heroes and losers in REDD+, while the government, private companies, financial institutions, and large international NGOs are seen as culprits and winners. *Opponents* argue that the process of consulting with local people in the REDD+ readiness phase has been subject to moral manipulation:

> "They [the government] say to the communities: . . . we are fighting climate change, and we will pay you to help. Are you with us? The expected answer is: Of course we are." (representative of *Otros Mundos Chiapas* in [76]).

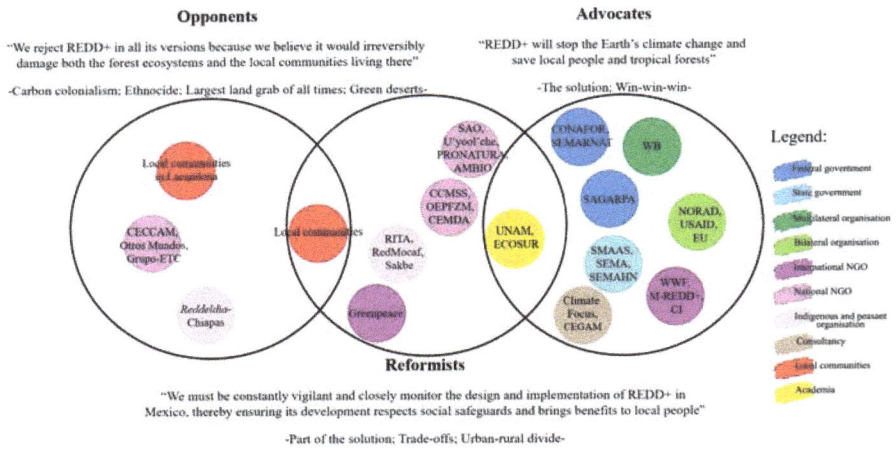

Figure 3. Composition (non-exhaustive) of the three REDD+ discourse coalitions in Mexico. The overlaps represent storylines on specific REDD+ issues shared between discourse coalitions. Each coalition is described by key metaphors and a quote paraphrased from interviews and documents to summarise the main storylines. Source: own elaboration.

Opponents do not participate in official REDD+ forums, which they consider illegitimate. However, they use informal mechanisms such as alternative forums, protests, campaigns, official letters, and printed and video material to call for structural changes to address deforestation and forest degradation (e.g., [77–81]). They maintain that community-based forest management founded upon democratic consultation is a viable alternative to REDD+.

Advocates include representatives of federal and sub-national government ministries and agencies, as well as multilateral and bilateral organisations and carbon market developers, and international NGOs and consultancies. These are in most part members of *top holders, followers, money patrons,* and *intermediaries* stakeholder groups. *Advocates* consider REDD+ as an opportunity to bring economic, social, and environmental co-benefits to rural communities that might help solve deforestation and forest degradation in developing countries. In the *advocates'* view, deforestation drivers in Mexico include unsound land-use policies, uncontrolled urban and tourism infrastructure development, and illegal logging, all activities in which local communities participate.

Most *advocates* generally prefer a national approach to REDD+, as it allows a centralised control of REDD+ funding. However, under the current slow and uncertain development of UNFCCC's negotiations, *advocates* would also favour a jurisdictional approach that could facilitate the development of REDD+ activities at the scale of administrative units or eco-regions. *Advocates* support a landscape-based approach to REDD+ implementation that could bring benefits to individuals or groups of forest owners, such as payment for ecosystem services (PES) initiatives coordinated with productive activities under the umbrella of sustainable rural development.

In the *advocates'* view, any potential benefit accruing from the enhancement of forests' carbon stocks should be directed to forest owners, while benefits from emission reductions from avoided deforestation should be attributed to the government. This is because it is not legally possible to assign a property right over something that has never existed (i.e., emissions reductions are essentially the product of performance against a hypothetical reference level baseline) and because deforestation is illegal unless government authorised. As this interviewee notes:

"Under the current legal framework, any change in land-use has to be authorised. Therefore if a person says: I will deforest; it is the same as if he says: I will kill three people, but if I instead kill only one, you have to compensate me." (INECC officer, 05 February 2014).

Advocates suggest that the voluntary nature of REDD+ and the widespread legitimacy of Mexico's land tenure regime should guarantee the alignment of REDD+ activities with the UNFCCC social safeguards. *Advocates* understand safeguards as rights and duties that REDD+ actors should respect. They also perceive the CTC as a legitimate forum that has included all key actors and has been transparent in providing information. *Advocates* argue that, given the high costs of involving local peoples in federal and state decision-making processes, the lack of local communities' participation in the CTC is an intrinsic problem of environmental decision-making in Mexico and beyond.

The *reformists'* coalition involves national NGOs, peasant and indigenous peoples' organisations, and local communities' members of *on the ground* and *frontliners* stakeholder groups, as well as representatives of academia from *information providers* and one international NGO from the *observers* group. *Reformists* support REDD+ as an important element of national climate change policy, but do not consider deforestation as the largest source of greenhouse gas emissions. According to the *reformists*, REDD+ has the potential to provide positive benefits to local communities, but only if it promotes sustainable rural development that combines conservation and productive activities. *Reformists* argue that over-consumption of natural resources by urban populations, contradictory agricultural subsidies, and urban and tourism infrastructure development should be simultaneously tackled to stop deforestation.

In the early readiness phase, all *reformists*, except the academic sector, preferred a jurisdictional over a national REDD+ approach, arguing that the jurisdictional approach would ease the identification and attribution of responsibility to all actors contributing to deforestation in the selected regions, and not only to local communities, mistaken as the main culprits. *Reformists* also consider that the rights over forests' carbon stocks and over the reduction of emissions from avoided deforestation should be recognised as an ecosystem service in the national legal framework and linked to land ownership. For this reason, the peasant and indigenous peoples' organisations in this coalition oppose a landscape approach to REDD+:

" ... if I take care of my forest but my neighbour gives his forest to a mining concession and therefore causes deforestation, the overall [carbon stock and emission reductions] measurements in a given territory will be affected. It is then a question of carbon ownership." (RITA officer, 6 February 2014).

In line with the *advocates*, *reformists* suggest that the legitimacy of the current land tenure regime guarantees that REDD+ does not contribute to alienate land rights in Mexico. However, they emphasise that REDD+ policies and measures should respect international laws and conventions on human and indigenous peoples' rights, and that solid social safeguards should guide REDD+ design and implementation.

There are different positions among *reformists* regarding the legitimacy of REDD+ decision-making. Large national NGOs and academia consider the CTC inclusive and therefore a legitimate forum. Others, mostly peasant and indigenous peoples' organisations, left the CTC to push for the establishment of GT-ENAREDD+ because, in their opinion, the former lacks inclusiveness, transparency, and accountability. The local communities from our sample did not participate in national REDD+ discussions.

3.3. Degree of Discourse Influence

Our analysis suggests that the *advocates'* discourse is 'dominant' as it has all the storylines on both conceptual and structural REDD+ issues explicitly acknowledged in the fifth ENAREDD+ draft and the ER-PIN (i.e., it reaches the highest degree of structuration and institutionalisation). The *reformists'* discourse is 'marginalised' in regards to conceptual issues, with only one of its storylines present in these documents, but partially institutionalised in regards to strategic issues, since four of its related storylines are represented in the named documents. Finally, the *opponents'* discourse is the most 'marginalised' and has the lowest degree of structuration and institutionalisation, because none of its storylines have been identified in the national REDD+ documents (Figure 4; Table 4).

Table 4. Quotes related to the ten key REDD+ issues extracted from the analysed documents and paraphrased for length. Source: own elaboration from the sources indicated in the Table. FCPF, Forest Carbon Partnership Facility; PES, payment for ecosystem services.

Key REDD+ Issues	Official REDD+ Statements
REDD+ goals and role	Deforestation is the third largest source of carbon emissions in the country and worldwide [54] (pp. 11, 20). Any disturbance of tropical forests significantly affects the global carbon cycle and contributes to climate change [54] (p. 12). Mexico has a great potential for REDD+ not only for reducing deforestation and forest degradation, but also for increasing forest carbon stocks [54] (p. 22).
Deforestation drivers	The main drivers of deforestation in the country are unsound agricultural policies followed by tourism, urban and industrial development; lack of coordination across these land-use sectors, and ineffective legislation [54] (p. 20); [55] (p. 28). Forest owners have few incentives to preserve them due to the market demand for specific products (e.g., timber, minerals, food, meat, biofuels, etc.), local needs, and population growth [55] (p. 27).
Local people and deforestation	Greater deforestation occurs in communities without forest management institutions [55] (p. 28). Forest degradation is caused by local forest users' activities (e.g., selective harvesting, overgrazing, extraction of firewood, etc.) [55] (p. 26).
Implementation scale	The FCPF Carbon Fund pays for emission reductions to the National Fund, which transmits the sum (proportionally according to emissions reduction contribution) to a jurisdictional fund (state or interstate fund) [54] (p. 46); [55] (pp. 34, 64).
Benefit-sharing strategy	The scale of activities under REDD+ corresponds to a territory that includes a number of communities and obeys environmental limits (basin, sub-basin, biological corridor) [55] (p. 33).
Scope of activities	Sustainable rural development is the best way to realise REDD+ in Mexico [54] (pp. 26, 33). PES is one of the activities within the special programmes [55] (pp. 31–33). REDD+ offers new opportunities to effective management and expansion of protected areas to contribute to climate change mitigation [54] (p. 25). Community forest management can be more effective in controlling deforestation than protected areas [55] (pp. 31–33). Sustainable agricultural practices could be part of REDD+ [55] (p. 33).
Co-benefits	Co-benefits or collateral benefits refer to the additional benefits to carbon storage resulting from REDD+ implementation, such as poverty reduction, biodiversity conservation and improvement in forest governance [54] (p. 86). REDD+ in Mexico will generate substantial non-carbon benefits because it will be implemented in early action regions [55] (pp. 2, 30, 63).
Social safeguards	Safeguards are designed to prevent and mitigate any direct or indirect negative impact on ecosystems and local people [54] (pp. 67–68). REDD+ is of voluntary nature and a collective consent by community authorities will guarantee respect of social safeguards [55] (p. 80). Safeguard Information System and Safeguard National System should oversee safeguards implementation and respect [54] (p. 70); [55] (p. 14).
Land tenure	Mexico has a sound community land rights system, therefore there is little risk of land rights violations [55] (p. 36).
Carbon rights	The ownership over biomass and carbon stocks is in the hands of forest owners, as sanctioned by national legislation (Art. 134bis, General Law of Sustainable Forest Development). However, it is not legally or technically feasible to attribute emissions reductions that result from avoided deforestation to a particular forest owner within the landscape, who might only hold the rights [not the ownership] to benefit from these emissions reduction [54] (p. 35); [55] (p. 36). Avoided deforestation has not been defined as an environmental service in any legislation to date. If necessary, corresponding law reforms should be promoted [54] (p. 35).

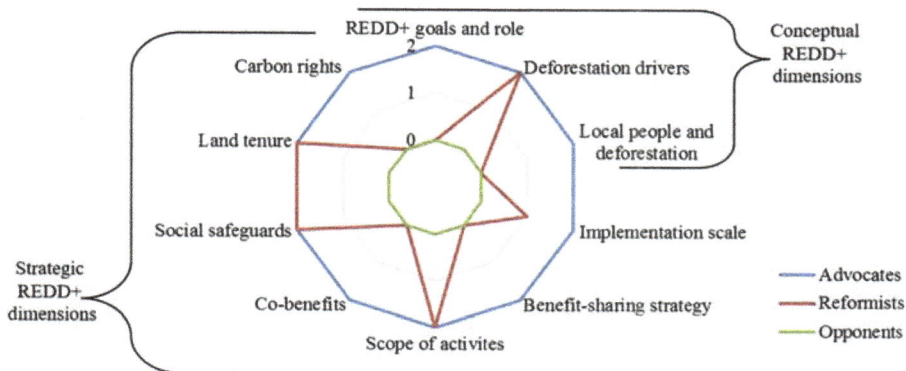

Figure 4. Degree of structuration and institutionalisation of the central storylines on ten key REDD+ dimensions. 0, no storyline included in the REDD+ documents; 1, storyline implicitly included; 2, storyline explicitly included. Source: own elaboration.

In line with the *reformists'* storyline, the REDD+ documents analysed in this article present deforestation as the third largest source of carbon emissions, but in line with the *advocates'* storyline, they also sustain an explicit connection between climate change mitigation and the need to preserve tropical forests. REDD+ is also presented as a key national mitigation measure, the implementation of which will guarantee non-carbon benefits. These documents also suggest that the key underlying drivers of deforestation in Mexico are ineffective legislation and uncoordinated land-use policies, as suggested by both *advocates* and *reformists*. In addition, the documents implicitly consider the critical impact on deforestation of the growing urban population's demand of agricultural goods and forest resources, as noted in the *reformists'* storyline. Nevertheless, the documents implicitly address the *advocates'* storyline on local people's responsibility in causing or halting deforestation by emphasising the importance of forest owners' capacities and incentives to use sustainably their forests, as well as by explicitly identifying local communities as key forest degradation agents.

According to the ENAREDD+ and the ER-PIN documents, REDD+ in Mexico will be operationalised through a set of activities from various land-use sectors with sustainable rural development as a leading principle, as suggested by the *reformists'* storyline. Potential REDD+ activities include PES, the establishment of protected areas, and sustainable agricultural practices, suggested by *advocates*, as well as activities oriented to community forest management, vehemently promoted by the *reformists*. These activities will be operationalised in "REDD+ implementing landscapes" and the ensuing verified carbon emission reductions would be accounted for at a jurisdictional level (state or region) and traded through a national government entity, as supported by the *advocates'* storyline.

Further, in line with the *reformists'* idea, and in addition to the Cancun Agreements' safeguards [26] the fifth ENAREDD+ draft and the ER-PIN include the so called 'national safeguards'. Both documents emphasise the voluntary nature of REDD+ activities and the fact that a collective consent obtained from community authorities should be enough to guarantee respect of social safeguards, as in the *advocates'* storyline. However, by promoting the elaboration of the Safeguard Information System and Safeguard National System that should oversee the implementation and respect of the safeguards, the documents also include *reformists'* concerns on this issue and the importance of respecting communities' land tenure and other customary rights. Yet again, and in line with both the *advocates'* and the *reformists'* discourse, the documents assert that REDD+ implementation should not challenge land ownership nor facilitate land grabbing.

Most importantly, and as included in the *advocacy* coalition storyline, the two analysed documents propose that forest owners hold full ownership over existing carbon stocks and expected carbon gains, while they are only entitled with the rights to benefit from potential emission reductions from avoided deforestation on their lands. This decision is explained by the critical technical difficulties to attribute emission reductions from such activities to one forest owner.

Based on the actors' perceptions of the national REDD+ discussions organised through the CTC and the identified degrees of discourse influence, we suggest that the current REDD+ readiness process in Mexico is considered legitimate, both from an input and output perspective, only by members of the 'dominant' *REDD+ advocacy* coalition, which include national and sub-national government agencies, international NGOs, multilateral and bilateral organisations, and consultancies. Other actors, including some national NGOs and academic institutions, have considered the process legitimate from a decision-making perspective, but not from an output perspective due to the fact that policy documents have been unable to integrate some of their central ideas on key REDD+ issues. More worrisome is the fact that some national NGOs and peasant indigenous peoples' organisations from the *reform* coalition and all members of the *opposition* coalition consider REDD+ illegitimate from an input perspective and refuse to grant REDD+ readiness in Mexico with output legitimacy (Figure 5).

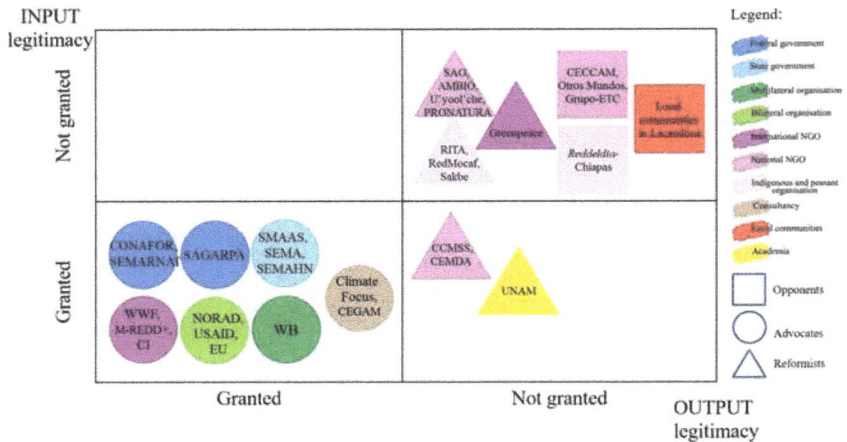

Figure 5. Correlation between the actors' sector, discourse, and perception on the input and output legitimacy of REDD+ governance in Mexico. Source: own elaboration.

4. Discussion

The results above suggest that Mexico's REDD+ readiness process has included to date a variety of actors differing in relevance, influence, and interest. Not surprisingly, the highest capacity to influence REDD+ policy development has lied in the federal government, and particularly in its forestry agency. The latter has justified its control over the REDD+ policy-making process by portraying itself as more capable and reliable authority than sub-national counterparts [82]. Notwithstanding, our research suggests that there has been a limited or almost negligible involvement from other ministries in REDD+ readiness, which could possibly compromise the effectiveness of REDD+ activities in the future.

CONAFOR has shared some decision-making power in REDD+ with academia and international NGOs hired to provide guidance on REDD+ technical and governance issues. However, national NGOs have remained sceptical about the input legitimacy of the REDD+ readiness process [83], considering it a government- and donor-driven process without the meaningful representation of local communities [84,85]. Similarly, non-governmental REDD+ actors in Indonesia have also perceived the national readiness process as externally driven and even detrimental for national sovereignty [36].

Although local communities' low capacity to influence REDD+ readiness has also been found in most developing countries participating in REDD+ [28,35], it is particularly worrisome in a country where rural communities have rights over the majority of forests and agricultural lands [74,75]. Lack of local community involvement can undermine the interest of national NGOs and peasant and indigenous peoples' organisations in the policy process in the future, which would not be a surprising outcome [86]. In the Democratic Republic of the Congo, for example, a group of national NGOs suspended their engagement with the REDD+ coordination process in 2012 [87], and an indigenous peoples' coordinating body withdrew from the Panama's REDD+ readiness process in 2013 [88]. In Mexico, the fact that some organisations left the CTC in disconformity and began to articulate their claims in an alternative forum under CONAF might be regarded both as a complaint to the ongoing process or as a strategic move to open new grounds for broader participation and deliberation [89] (see last paragraph in Section 3.1).

The legitimacy of governance processes is a dynamic state that must be constantly created and recreated among all participants [24,90]. This process implies the need to design REDD+ multi-stakeholder forums flexible enough to guarantee their credibility and legitimacy. In Mexico, the government justified the absence of practical solutions to increase the level of input legitimacy of the readiness process on the grounds that key national and international NGOs already endorsed the CTC-driven process. Yet the fragmentation of the national REDD+ discussion in two national forums revealed the fragility of the decision-making procedures governing the CTC. Subsequently, the lack of clarity regarding how deliberations in both forums had to feed into the national strategy resulted in confusion and ultimately resulted in the government prioritising the views and inputs from those participating in the CTC.

Our results also indicate that Mexico's REDD+ readiness actors have mobilised numerous storylines around which the three identified discourse coalitions have coalesced. The principal differences between the three coalitions reflect divergent perspectives around REDD+ conceptual dimensions and that the *advocacy* and *reform* coalitions differ in their views about strategic REDD+ issues, particularly regarding the attribution of carbon rights (for similar findings see references [91–93]). Not surprisingly, the *advocates'* discourse is 'dominant' in the analysed REDD+ readiness documents, given that this is the discourse promoted by CONAFOR. The dominance of a government-supported discourse has also been found in other REDD+ host countries, including Tanzania [92], Mozambique [93], Indonesia, Brazil, Vietnam, Peru [94], and Nepal [37].

Although it would be tempting to assume that under the 'dominant' coalition there would be little room for alternative arguments, some of the exclusively *reformists'* arguments have found their way into Mexico's REDD+ documents, although the reasons for this seem to be highly contextual. The adoption of safeguards and a jurisdictional approach can be explained by the impact of the global REDD+ debate on the Mexican government's views, whereas the adoption of the sustainable rural development principle and the existence of technical and not only legal difficulties to the attribution of carbon rights at the forest owner's level can be attributed to the government's alignment with the argument promoted by some *reformists*.

A high level of output legitimacy does not imply an even translation of actors' ideas into policy decisions [95], but rather that most actors have a strong sense of ownership over the decisions adopted [96]. In this regard, our analysis reflects that the documents include only a limited number of alternative storylines, while many in the *reformists'* coalition still support the fifth ENAREDD+ draft to be sent for a countrywide public consultation (ongoing at the time of writing). After all, large national NGOs perceive the REDD+ decision-making process as legitimate, while other national NGOs and peasant and indigenous organisations think that involving local people more directly is critical to legitimate REDD+ implementation in the coming future.

The case of the *opposition* coalition is different. It is not surprising that members of this coalition have not endorsed the REDD+ readiness process, principally because they are against some of the REDD+ conceptual dimensions promoted by the other two coalitions, such as the role of forests as carbon sinks, the supposedly secure status of Mexico's tenure regime, or the lack of risk of land grabs by external actors. The informal forums, protests, and campaigns that these actors have organised have proved ineffective to persuade the government to abandon the REDD+ process but, in the future, it seems plausible to think that this coalition will play a critical role in overseeing and scrutinising REDD+ implementation.

5. Conclusions

This paper set out to explore the legitimacy of the REDD+ readiness process in Mexico by analysing REDD+ actors' characteristics and their participation in the related decision-making process (i.e., input legitimacy), and by calculating the degree to which the different actors' ideas are reflected in the resulting official REDD+ policy documents (i.e., output legitimacy). Our results corroborate an existing trend in national REDD+ governance of increasing centralisation of REDD+ policy making and early implementation [18,97] in which the predominant actors are the government, donors, international INGOs, and select CSOs [36,98]. In addition, the design of the correspondent multi-stakeholder forums has been ineffective in bringing forward the perspectives of sub-national and other non-governmental stakeholders [31,98–100]. As other REDD+ pursuing countries, Mexico continues to face the challenge of designing and establishing a legitimate REDD+ readiness process that includes, on an equal basis, a variety of actors in the decision-making process and that acknowledges the actors' diverse views, communicated through different discourses, in REDD+ policy documents.

To improve the legitimacy of REDD+ readiness process, Mexico and other REDD+ host countries should consider designing novel institutional arrangements under a more inclusive and decentralised approach and strengthening the coordination among the existing multi-stakeholder forums. Greater inclusiveness of a more diverse range of public policy sectors and non-state actors, particularly local communities' representative organisations, at both national and sub-national levels, would enrich REDD+ discussions and improve the quality of the national strategy design with alternative storylines, such as those on locally adequate and acceptable REDD+ activities. In this sense, the countrywide public consultation of the ENAREDD+ draft conducted very recently suggests that the Mexican government was well aware of the need to broaden the scope of the participatory process analysed in this article.

We believe, however, that the ensuing REDD+ implementation phase would require additionally efforts to ensure that legitimacy increases. Promoting a greater number of public debates across governance scales to further discuss the advantages and risks of REDD+ activities on the ground, including their possible impact on local tenure systems and ecosystem services management regimes, as well as an equitable benefit-sharing of resulting emission reductions (currently under design) seems a promising way forward. Only by including alternative viewpoints and constructive proposals on these and other issues might Mexico and other REDD+ host countries limit the scope of social contestation and guarantee a higher degree of legitimacy in the operation of such ambitious policy framework.

Acknowledgments: The authors acknowledge two anonymous reviewers, Arild Angelsen, Roser Maneja and Diana Pritchard for constructive comments on an earlier version of this manuscript. Jovanka Špirić acknowledges the financial support received from the FI-AGAUR scholarship of the Catalan government, the Fundació Autònoma Solidària cooperation grant, the Foundation Open Society Institute's Global Supplementary Grants, as well as the additional financial and logistical support from the EU projects: CONSERVCOM funded by Fondo de Cooperación Internacional en Ciencia y Tecnología UE-México (FONCICYT 94395) and Programa de Cooperación Inter-Universitaria e Investigación Científica, Ministerio de Asuntos Exteriores y Cooperación (A/023406/09 and A/030044/10), and COMBIOSERVE (Grant No. 282899). Esteve Corbera acknowledges the financial support of the 'Conflict and Cooperation over REDD+ in Mexico, Nepal and Vietnam' project, supported by the Netherlands Organisation for Scientific Research and the UK Department for International Development (Grant No W07.68.415), as well as of the UAB-Banco de Santander Talent Retention Programme. This work contributes to ICTA-UAB "María de Maeztu Unit of Excellence" (MDM-2015-0552).

Author Contributions: Jovanka Špirić designed the research with inputs from Esteve Corbera, Victoria Reyes-García, and Luciana Porter-Bolland. Jovanka Špirić collected and analysed the data. All authors participated in writing the paper.

Conflicts of Interest: The authors declare no conflict of interest.

References

1. Corbera, E.; Schroeder, H. Governing and implementing REDD+. *Environ. Sci. Policy* **2011**, *14*, 89–99. [CrossRef]
2. Thompson, M.C.; Baruah, M.; Carr, E.R. Seeing REDD+ as a project of environmental governance. *Environ. Sci. Policy* **2011**, *14*, 100–110. [CrossRef]
3. Lederer, M. REDD+ governance. *WIREs Clim. Chang.* **2012**, *3*, 107–113. [CrossRef]
4. Vignola, R.; Guerra, L.; Trevejo, L.; Aymerich, J.P. *REDD+ Governance across Scales in Latin America. Perceptions of the Opportunities and Challenges from the Model Forest Platform*; REDD-Net Publications: London, UK, 2012. Available online: http://repositorio.bibliotecaorton.catie.ac.cr/bitstream/handle/11554/8285/REED_Governance_across_scales_in_Latin_America.pdf (accessed on 14 April 2016).
5. Long, A. REDD+, Adaptation, and sustainable forest management: Toward effective polycentric global forest governance. *Trop. Conserv. Sci.* **2013**, *6*, 384–408. [CrossRef]
6. De la Plaza, C.; Visseren-Hamakers, I.J.; De Jong, W. The legitimacy of certification standards in climate change governance. *Sustain. Dev.* **2014**, *22*, 420–432. [CrossRef]
7. United Nations Collaborative Programme on REDD+ (UN-REDD+). Programme Regions and Partner Countries. 2016. Available online: http://www.un-redd.org/partner-countries (accessed on 12 July 2016).
8. Forest Carbon Partnership Facility (FCPF) REDD+ Countries. 2016. Available online: https://www.forestcarbonpartnership.org/redd-countries-1 (accessed on 12 July 2016).
9. Davis, C.; Nakhooda, S.; Daviet, F. *Getting Ready. A Review of the World Bank Forest Carbon Partnership Facility Readiness Preparation Proposals*; Version 1.3; WRI Working Paper; World Resources Institute: Washington, DC, USA, 2010. Available online: http://www.wri.org/gfi (accessed on 6 August 2016).
10. Bradley, A. *Review of Cambodia's REDD Readiness: Progress and Challenges, Forest and Conservation Project*; Occasional Paper No. 4; Institute for Global Environmental Studies: Hayama, Japan, 2011. Available online: http://redd-database.iges.or.jp/redd/download/link?id=4 (accessed on 6 August 2016).
11. United Nations Framework Convention on Climate Change (UNFCCC). Report of the Conference of the Parties on Its Nineteenth Session, Held in Warsaw from 11 to 23 November 2013. Available online: http://unfccc.int/resource/docs/2013/cop19/eng/10a01.pdf (accessed on 1 August 2016).
12. Cadman, T.; Maraseni, T.N. The governance of REDD+: An institutional analysis in the Asia Pacific region and beyond. *J. Environ. Plan. Manag.* **2011**, *55*, 617–635. [CrossRef]
13. Cadman, T.; Maraseni, T.N. More equal than others? A comparative analysis of state and nonstate perceptions of interest representation and decisionmaking in REDD+ negotiations. *Eur. J. Soc. Sci. Res.* **2013**, *26*, 214–230. [CrossRef]
14. Maraseni, T.N.; Cadman, T. Comparative analysis of global stakeholders' perceptions of the governance quality of the CDM and REDD+. *Int. J. Environ. Stud.* **2015**, *72*, 288–304. [CrossRef]
15. United Nations Framework Convention on Climate Change (UNFCCC). Document FCCC/CP/2010/7/Add.1, Report of the Conference of the Parties Fifteenth Session, Held in Copenhagen from 7 to 19 December 2009. Available online: http://unfccc.int/resource/docs/2010/cop16/eng/07a01.pdf (accessed on 1 August 2016).
16. Lyster, R. REDD+, transparency, participation and resource rights: The role of law. *Environ. Sci. Policy* **2011**, *14*, 118–126. [CrossRef]
17. United Nations Collaborative Programme on REDD+ (UN-REDD+). Legal Analysis of Cross-Cutting Issues for REDD+ Implementation. Lessons Learned from Mexico, Viet Nam and Zambia, 2013. UN-REDD Programme Secretariat International Environment House: Geneva, Switzerland. Available online: http://theredddesk.org/sites/default/files/resources/pdf/2013/legal-analysis-final-web_1.pdf (accessed on 3 September 2016).
18. Cadman, T.; Maraseni, T.; López-Casero, F.; Ok Ma, H. Governance values in the climate change regime: Stakeholder perceptions of REDD+ legitimacy at the national level. *Forests* **2016**, *7*, 212. [CrossRef]
19. Scharpf, F. *Governing in Europe. Effective and Democratic*; Oxford University Press: Oxford, UK, 1999.
20. Bäckstrand, K. Multi-stakeholder partnerships for sustainable development: Rethinking legitimacy, accountability and effectiveness. *Eur. Environ.* **2006**, *16*, 290–306. [CrossRef]
21. Paavola, J. Institutions and environmental governance: A reconceptualization. *Ecol. Econ.* **2007**, *63*, 93–103. [CrossRef]

22. Vatn, A.; Vedeld, P. National governance structures for REDD+. *Glob. Environ. Chang.* **2013**, *23*, 422–432. [CrossRef]

23. Burger, D.; Mayer, C. *Making Sustainable Development a Reality: The Role of Social and Ecological Standards*; Deutsche Gesellschaft für Technische Zusammenarbeit: Kessel, Germany, 2003; p. 50. Available online: http://www.mekonginfo.org/assets/midocs/0002193-economy-making-sustainable-development-a-reality-the-role-of-social-and-ecological-standards.pdf (accessed on 6 August 2016).

24. Parkinson, J. *Deliberating in the Real World: Problems of Legitimacy in Deliberative Democracy*; Oxford University Press: Oxford, UK, 2006.

25. Vatn, A. Environmental Governance—A Conceptualization. In *The Political Economy of Environment and Development in a Globalized World. Exploring the Frontiers*; Kjosavik, D., Vedeld, P., Eds.; Tapir Academic Press: Trondheim, Norway, 2011; pp. 131–152.

26. United Nations Framework Convention on Climate Change (UNFCCC). The Cancun Agreements. 2010. Available online: http://unfccc.int/resource/docs/2010/cop16/eng/07a01.pdf (accessed on 6 August 2016).

27. Angelsen, A. *Moving ahead with REDD: Issues, Options and Implications*; Center for International Forestry Research: Bogor, Indonesia, 2008. Available online: http://www.cifor.org/publications/pdf_files/Books/BAngelsen0801.pdf (accessed on 6 August 2016).

28. Veierland, K. *Inclusive REDD+ in Indonesia? A Study of the Participation of Indigenous People in the Making of the National REDD+ Strategy in Indonesia*; University of Oslo: Oslo, Norway, 2011.

29. Angelsen, A.; Brockhaus, M.; Sunderlin, W.D.; Verchot, L.V. *Analysing REDD+: Challenges and Choices*; Center for International Forestry Research: Bogor, Indonesia, 2012. Available online: http://www.cifor.org/publications/pdf_files/Books/BAngelsen1201.pdf (accessed on 6 August 2016).

30. Aquino, A.; Guay, B. Implementing REDD+ in the Democratic Republic of Congo: An analysis of the emerging national REDD+ governance structure. *For. Policy Econ.* **2013**, *36*, 71–79. [CrossRef]

31. Manyika, K.F.K.; Kajembe, G.C.; Silayo, D.A.; Vatn, A. Strategic power and power struggles in the national REDD+ governance process in Tanzania: Any effect on its legitimacy? *Tanzan. J. For. Nat. Conserv.* **2013**, *83*, 69–82.

32. Mulyani, M.; Jepson, P.R. REDD+ and forest governance in Indonesia: A multistakeholder study of perceived challenges and opportunities. *J. Environ. Dev.* **2013**, *22*, 261–283. [CrossRef]

33. Somorin, O.A.; Visseren-Hamakers, I.J.; Arts, B.; Sonwa, D.J.; Tiani, A.-M. REDD+ policy strategy in Cameroon: Actors, institutions and governance. *Environ. Sci. Policy* **2014**, *35*, 87–97. [CrossRef]

34. Zelli, F.; Erler, D.; Frank, S.; Hein, J.I.; Hotz, H.; Cruz-Melgarejo, A.M.S. *Reducing Emissions from Deforestation and Forest Degradation (REDD) in Peru: A Challenge to Social Inclusion and Multi-Level Governance*; German Development Institute: Bonn, Germany, 2014.

35. Minang, P.A.; Van Noordwijk, M.; Duguma, L.A.; Alemagi, D.; Do, T.H.; Bernard, F.; Agung, P.; Robiglio, V.; Catacutan, D.; Suyanto, S.; et al. REDD+ Readiness progress across countries: Time for reconsideration. *Clim. Policy* **2014**, *14*, 685–708. [CrossRef]

36. Luttrell, C.; Resosudarmo, I.A.P.; Muharrom, E.; Brockhaus, M.; Seymour, F. The political context of REDD+ in Indonesia: Constituencies for change. *Environ. Sci. Policy* **2014**, *35*, 67–75. [CrossRef]

37. Khatri, D.B.; Pham, T.T.; Di Gregorio, M.; Karki, R.; Paudel, N.S.; Brockhaus, M.; Bhushal, R. REDD+ politics in the media: A case from Nepal. *Clim. Chang.* **2016**, *138*, 309–323. [CrossRef]

38. Forest Carbon Partnership Facility (FCPF). Readiness Plan Idea Note (R-PIN)—External Review Form, Mexico. 2008. Available online: https://www.forestcarbonpartnership.org/sites/forestcarbonpartnership.org/files/Mexico_TAP_Consolidated.pdf (accessed on 6 August 2016).

39. Comisió Nacional Forestal (CONAFOR). The Forest Carbon Partnership Facility (FCPF) Readiness Plan Idea Note (R-PIN) Mexico. 2008. Available online: https://www.forestcarbonpartnership.org/sites/forestcarbonpartnership.org/files/Mexico_FCPF_R-PIN.pdf (accessed on 6 August 2016).

40. Comisión Nacional Forestal (CONAFOR). The Forest Carbon Partnership Facility (FCPF). Readiness Preparation Proposal (R-PP) Mexico. 2010. Available online: http://forestcarbonpartnership.org/sites/fcp/files/Documents/tagged/Mexico_120211_R-PP_Template_with_disclaimer.pdf (accessed on 6 August 2016).

41. Comisión Nacional Forestal (CONAFOR). Visión de México Sobre REDD+. Hacia una Estrategia Nacional. 2010. Available online: http://www.conafor.gob.mx:8080/documentos/docs/35/2521Visi%C3%B3n%20de%20M%C3%A9xico%20para%20REDD_.pdf (accessed on 6 August 2016).

42. U'yool'che and Servicios Ecosistémicos de la Selva Maya S.C. *Plan Vivo Project Design Document (PDD), Much Kanan K'aax*; U'yool'che and Servicios Ecosistémicos de la Selva Maya S.C.: Felipe Carrillo Puerto, Mexico, 2011.

43. Consejo Civil Mexicano para la Silvicultura Sostenible (CCMSS). Nota de Idea del Proyecto REDD+ Comunitario en la Zona Maya de José María Morelos, Quintana Roo. 2011. Available online: http://www.ccmss.org.mx/descargas/pin_jmm_140711.pdf (accessed on 6 August 2016).

44. PRONATURA. El Zapotal. 2015. Available online: http://www.pronatura-ppy.org.mx/seccion.php?id=5 (accessed on 6 August 2016).

45. Comisión Nacional Forestal (CONAFOR). Estrategia Nacional para REDD+ (ENAREDD+). Primer borrador. 2011. Available online: http://www.conafor.gob.mx:8080/documentos/docs/35/4859Elementos%20para%20el%20dise%C3%B1o%20de%20la%20Estrategia%20Nacional%20para%20REDD_.pdf (accessed on 6 August 2016).

46. Comisión Nacional Forestal (CONAFOR). Participación. CTC-REDD+ Estatales. 2015. Available online: http://www.conafor.gob.mx/web/temas-forestales/bycc/redd-en-mexico/participacion/ (accessed on 6 August 2016).

47. Ley General de Desarrollo Forestal Sustentable (LGDFS). *El Diario Oficial de la Federación el 25 de Febrero de 2003 (Última Reforma DOF 04-06-2012)*; Cámara de Diputados del H. Congreso de la Unión, Secretaría General, Secretaría de Servicios Parlamentarios: Mexico city, Mexico, 2016.

48. Ley General del Equilibrio Ecológico y la Protección al Ambiente (LGEEPA). *El Diario Oficial de la Federación el 28 de enero de 1988 (Última Reforma DOF 04-06-2012)*; Cámara de Diputados del H. Congreso de la Unión, Secretaría General, Secretaría de Servicios Parlamentarios: Mexico city, Mexico, 2015.

49. Consejo Nacional Forestal (CONAF). Memoria de Gestión de la Renovación del Consejo Nacional Forestal para el Periodo 2013–2014. Available online: https://www.gob.mx/cms/uploads/attachment/file/82979/Memoria_de_Gestion_CONAF_2013_-2014.pdf (accessed on 6 August 2016).

50. Jagger, P.; Brockhaus, M.; Duchelle, A.E.; Gebara, M.F.; Lawlor, K.; Resosudarmo, I.A.P.; Sunderlin, W.D. Multi-level policy dialogues, processes, and actions: Challenges and opportunities for national REDD+ safeguards measurement, reporting, and verification (MRV). *Forests* **2014**, *5*, 2136–2162. [CrossRef]

51. Comisión Nacional Forestal (CONAFOR). Estrategia Nacional Para REDD+ (ENAREDD+). Borrador Octubre de 2012. Available online: http://www.conafor.gob.mx:8080/documentos/docs/35/5303Elementos%20para%20el%20dise%C3%B1o%20de%20la%20Estrategia%20Nacional%20para%20REDD_.pdf (accessed on 6 August 2016).

52. Comisión Nacional Forestal (CONAFOR). Estrategia Nacional Para REDD+ (ENAREDD+). Borrador Julio de 2013. Available online: http://www.conafor.gob.mx:8080/documentos/docs/35/4861Estrategia%20Nacional%20para%20REDD_.pdf (accessed on 6 August 2016).

53. Comisión Nacional Forestal (CONAFOR). Estrategia Nacional Para REDD+ (ENAREDD+). Borrador Abril de 2014. Available online: http://www.conafor.gob.mx:8080/documentos/docs/35/5559Elementos%20para%20el%20dise%C3%B1o%20de%20la%20Estrategia%20Nacional%20para%20REDD_.pdf (accessed on 6 August 2016).

54. Comisión Nacional Forestal (CONAFOR). Estrategia Nacional Para REDD+ (ENAREDD+) (Para Consulta Pública). Available online: http://www.conafor.gob.mx:8080/documentos/docs/35/6462Estrategia%20Nacional%20para%20REDD_%20(para%20consulta%20p%C3%BAblica)%202015.pdf (accessed on 6 August 2016).

55. Comisión Nacional Forestal (CONAFOR). Forest Carbon Partnership Facility (FCPF) Carbon Fund. Emission Reductions Program. Idea Note (ER-PIN) Mexico. Available online: http://www.conafor.gob.mx:8080/documentos/docs/4/6170Propuesta%20de%20Nota%20de%20Idea%20de%20la%20Iniciativa%20de%20Reducci%C3%B3n%20de%20Emisiones%20%28ERPIN%29%20de%20M%C3%A9xico.pdf (accessed on 6 August 2016).

56. Forest Carbon Partnership Facility (FCPF). ER-PIN Template. 2013. Available online: http://www.bankinformationcenter.org/wp-content/uploads/2013/12/FCPF-Carbon-Fund-ER-PIN-v4.pdf (accessed on 6 August 2016).

57. Beardsworth, A.; Keil, T. The vegetarian option: Varieties, conversions, motives, and careers. *Sociol. Rev.* **1992**, *40*, 253–293. [CrossRef]

58. Bernard, H.R. *Research Methods in Anthropology: Qualitative and Quantitative Approaches*, 4th ed.; Altamira Press: Walnut Creek, CA, USA, 2006; pp. 186–210.

59. Angelsen, A.; Bockhaus, M.; Kanninen, M.; Sills, E.; Sunderlin, W.D.; Wertz-Kanounnikoff, S. *Realising REDD+: National Strategy and Policy Options*; Center for International Forestry Research: Bogor, Indonesia, 2009. Available online: http://www.cifor.org/publications/pdf_files/Books/BAngelsen0902.pdf (accessed on 8 November 2016).

60. Mayers, J. *Stakeholder Power Analysis*; International Institute for and Environment Development (IIED): London, UK, 2005.

61. Diefenbach, T.; Sillince, J.A.A. Formal and informal hierarchy in different types of organization. *Organ. Stud.* **2011**, *32*, 1515–1537. [CrossRef]

62. Overseas Development Administration. *Guidance Note on How to Do Stakeholder Analysis of Aid Projects and Programmes*; Overseas Development Department: London, UK, 1995.

63. Parkinson, J. Legitimacy problems in deliberative democracy. *Polit. Stud.* **2003**, *51*, 180–196. [CrossRef]

64. Hajer, M.A. Discourse coalitions and the institutionalisation of practice: The case of acid rain in Great Britain. In *The Argumentative Turn in Policy Analysis and Planning*; Fischer, F., Forester, J., Eds.; Duke University Press: Durham/London, UK, 1993; pp. 43–67.

65. Dryzek, J. *The Politics of the Earth: Environmental Discourses*; Oxford University Press: Oxford, UK, 1997.

66. Adger, W.N.; Benjaminsen, T.A.; Brown, K.; Svarstad, H. Advancing a political ecology of global environmental discourses. *Dev. Chang.* **2001**, *32*, 681–715. [CrossRef]

67. Cashore, B. Legitimacy and the privatization of environmental governance: How Non State Market-Driven (NSMD) governance systems gain rule making authority. *Governance* **2002**, *15*, 503–529. [CrossRef]

68. Steffek, J. The legitimation of international governance: A discourse approach. *Eur. J. Int. Relat.* **2003**, *9*, 249–275. [CrossRef]

69. Steffek, J. Discursive legitimation in environmental governance. *For. Policy Econ.* **2009**, *11*, 313–318. [CrossRef]

70. Steffek, J.; Hahn, K. *Evaluating Transnational NGOs: Legitimacy, Accountability, Representation*; Palgrave Macmillan: Basingstoke, UK, 2011.

71. Buchanan, A.; Keohane, R.O. The legitimacy of global governance institutions. *Ethics Int. Aff.* **2006**, *20*, 405–437. [CrossRef]

72. Hajer, M.A. *The Politics of Environmental Discourse: Ecological Modernization and the Policy Process*; Oxford University Press: New York, NY, USA, 1995; pp. 9–44.

73. Organisation for Economic Co-operation and Development (OECD). *Evaluaciones de la OCDE Sobre el Desempeño Ambiental: México 2013*; OECD Publishing: Paris, France, 2013.

74. Food and Agriculture Organization of the United Nations (FAO). *Evaluación de los Recursos Forestales Mundales, Informe Nacional, México*; FAO: Rome, Italy, 2010. Available online: http://www.fao.org/forestry/20262-1-176.pdf (accessed on 6 August 2016).

75. De Ita, A. Land concentration in Mexico after PROCEDE. In *Promised Land: Competing Visions of Agrarian Reform*; Rosset, P.M., Patel, R., Courville, M., Eds.; Institute for Food and Development Policy: Oakland, CA, USA, 2008.

76. Conant, J. A Broken Bridge to the Jungle: The California-Chiapas Climate Agreement Opens Old Wounds; Global Justice Ecology Project. *Climate Connections*, 7 April 2011. Available online: http://climate-connections.org/2011/04/07/a-broken-bridge-to-the-jungle-the-california-chiapas-climate-agreement-opens-old-wounds/ (accessed on 31 August 2016).

77. Lang, C. People's Forum Against REDD+ in Chiapas, Mexico. *Redd-Monitor*, 24 September 2012. Available online: http://www.redd-monitor.org/2012/09/24/peoples-forum-against-redd-in-chiapas-mexico/ (accessed on 21 September 2016).

78. Papel Revolución. Comunicado REDDeldía Leido en el Foro de Gobernadores pro REDD+. *Papel Revolución*, 27 September 2012. Available online: http://www.papelrevolucion.com/2012/09/comunicado-reddeldia-leido-en-el-foro.html (accessed on 21 September 2016).

79. Reddeldia. Open letter from Chiapas about the Agreement between the States of Chiapas (Mexico), Acre (Brazil) and California (USA). *Reddeldia*, April 2013. Available online: http://reddeldia.blogspot.rs/2013/04/carta-abierta-de-chiapas-sobre-el.html (accessed on 21 September 2016).

80. Otros Mundos AC- Friends of the Earth Mexico. REDD: la codicia por los árboles (El Caso Chiapas: la Selva Lacandona al mejor postor). *Otros Mundos AC, Friends of the Earth Mexico*, 2011. Available online: https://www.youtube.com/watch?v=b0Md6WXj0pM (accessed on 21 September 2016).

81. Centro de Estudios Para el Cambio en el Campo Mexicano (CECCAM). *REDD+ y los territorios indígenas y campesinos*; CECCAM: Mexico city, Mexico, 2012. Available online: http://ceccam.org/sites/default/files/AAA-REDD%2BWeb.pdf (accessed on 21 September 2016).

82. Phelps, J.; Edward, L.W.; Agrawal, A. Does REDD+ threaten to recentralize forest governance? *Policy Forum* **2010**, *328*, 312–313. [CrossRef] [PubMed]

83. Biermann, F.; Betsill, M.M.; Gupta, J.; Kanie, N.; Lebe, L.; Liverman, D.; Schroeder, H.; Siebenhüner, B. *Earth System Governance: People, Places and the Planet. Science and Implementation Plan of the Earth System Governance Project*; Earth System Governance Report 1, IHDP Report 20; The Earth System Governance Project: Bonn, Germany, 2009.

84. Corbera, E.; Estrada, M.; May, P.; Navarro, G.; Pacheco, P. Rights to land, forests and carbon in REDD+: Insights from Mexico, Brazil and Costa Rica. *Forests* **2011**, *2*, 301–342. [CrossRef]

85. Hajjar, R.; Kozlak, A.R.; Inners, J.L. Is decentralization leading to "real" decision-making power for forest dependent communities? Case studies from Mexico and Brazil. *Ecol. Soc.* **2012**, *17*, 12. [CrossRef]

86. Yosie, T.; Herbst, T. *Using Stakeholder Processes in Environmental Decision Making: An Evaluation of Lessons Learned, Key Issues, and Future Challenges*; Ruder Finn Washington: Washington, DC, USA, 1998.

87. Forest Peoples Programme, Civil Society Groups in DRC Suspend Engagement with National REDD Coordination Process. Available online: http://www.forestpeoples.org/topics/redd-and-related-initiatives/news/2012/07/civil-society-groups-drc-suspend-engagement-nationa (accessed on 6 November 2016).

88. Lang, C. COONAPIP, Panama's Indigenous Peoples Coordinating Body, Withdraws from UN-REDD. REDD-Monitor, 2013. Available online: http://www.redd-monitor.org/2013/03/06/coonapip-panamas-indigenous-peoples-coordinating-body-withdraws-from-un-redd/ (accessed on 6 November 2016).

89. Hatanaka, M.; Konefal, F. Legitimacy and standard development in multi-stakeholder initiatives: A case study of the leonardo academy's sustainable agriculture standard initiative. *Int. J. Sociol. Agric. Food* **2012**, *20*, 155–173.

90. Boström, M.; Tamm Hallström, K. Global multi-stakeholder standard setters: How fragile are they? *J. Glob. Ethics* **2013**, *9*, 93–110. [CrossRef]

91. May, P.H.; Millikan, B.; Gebara, M.F. *The Context of REDD+ in Brazil: Drivers, Agents and Institutions*; Occasional Paper 55, revised edition; Center for International Forestry Research: Bogor, Indonesia, 2011.

92. Rantala, S.; Di Gregorio, M. Multistakeholder environmental governance in action: REDD+ discourse coalitions in Tanzania. *Ecol. Soc.* **2014**, *19*, 66. [CrossRef]

93. Quan, J.; Naess, L.O.; Newsham, A.; Sitoe, A.; Fernandez, M.C. *Carbon Forestry and Climate Compatible Development in Mozambique: A Political Economy Analysis*; IDS Working Paper No. 448; Institute of Development Studies: Brighton, UK, 2014.

94. Di Gregorio, M.; Brockhaus, M.; Cronin, T.; Muharrom, E.; Santoso, L.; Mardiah, S.; Büdenbender, M. Equity and REDD+ in the Media: A comparative analysis of policy discourses. *Ecol. Soc.* **2014**, *18*, 39. [CrossRef]

95. Boedeltje, M.; Cornips, J. Input and Output Legitimacy in Interactive Governance. NIG Annual Work Conference, Rotterdam (No. NIG2-01) 2004. Available online: http://hdl.handle.net/1765/1750 (accessed on 6 August 2016).

96. Hemmati, M. *Multi-Stakeholder Processes for Governance and Sustainability beyond Deadlock and Conflict*; Earthscan Publications Ltd.: London, UK; Sterling, VA, USA, 2002.

97. Vijge, M.J.; Brockhaus, M.; Di Gregorio, M.; Muharrom, E. Framing REDD+ in the national political arena: A comparative discourse analysis of Cameroon, Indonesia, Nepal, PNG, Vietnam, Peru and Tanzania. *Glob. Environ. Chang.* **2016**, *39*, 57–68. [CrossRef]

98. Bushley, R.B. REDD+ policy making in Nepal: Toward state-centric, polycentric, or market-oriented governance? *Ecol. Soc.* **2014**, *19*, 34. [CrossRef]

99. Bushley, B.R.; Khatri, D.B. *REDD+: Reversing, Reinforcing or Reconfiguring Decentralized Forest Governance in Nepal*; Discussion Paper Series 11:3; Forest Action Nepal: Patan, Nepal, 2011.

100. Bastakoti, R.R.; Davidsen, C. Nepal's REDD+ readiness preparation and multi-stakeholder consultation challenges. *J. For. Livelihood* **2015**, *13*, 30–43. [CrossRef]

forests

MDPI

Article

"Georgetown ain't got a tree. We got the trees"—Amerindian Power & Participation in Guyana's Low Carbon Development Strategy

Sam Airey [1,*] and Torsten Krause [2]

[1] Uppsala University, Villavägen 16, SE-752 36 Uppsala, Sweden
[2] Lund University Centre for Sustainability Studies, Lund University, Lund, SE-221 00, Sweden; torsten.krause@lucsus.lu.se
* Correspondence: aireysam@gmail.com; Tel.: +44-750-334-9656

Academic Editors: Esteve Corbera and Heike Schroeder
Received: 28 October 2016; Accepted: 14 February 2017; Published: 23 February 2017

Abstract: International bi-lateral agreements to support the conservation of rainforests to reduce greenhouse gas emissions are growing in prevalence. In 2009, the governments of Guyana and Norway established Guyana's Low Carbon Development Strategy (LCDS). We examine the extent to which the participation and inclusion of Guyana's indigenous population within the LCDS is being achieved. We conducted a single site case study, focussing on the experiences and perceptions from the Amerindian community of Chenapou. Based on 30 interviews, we find that a deficit of adequate dialogue and consultation has occurred in the six years since the LCDS was established. Moreover, key indigenous rights, inscribed at both a national and international level, have not been upheld with respect to the community of Chenapou. Our findings identify consistent shortcomings to achieve genuine participation and the distinct and reinforced marginalisation of Amerindian communities within the LCDS. A further critique is the failure of the government to act on previous research, indicating a weakness of not including indigenous groups in the Guyana-Norway bi-lateral agreement. We conclude that, if the government is to uphold the rights of Amerindian communities in Guyana, significant adjustments are needed. A more contextualised governance, decentralising power and offering genuine participation and inclusion, is required to support the engagement of marginal forest-dependent communities in the management of their natural resources.

Keywords: Sustainable development; participation; forest governance; REDD+; indigenous rights; Guyana; REDD+ impacts

1. Introduction

The majority of the world's tropical forests are located in developing countries [1]. Up to a quarter of the total forest area in developing countries could be considered 'community controlled' with many of these forest communities, depending on the region, made up of indigenous groups [2]. These communities are often the most affected by the implementation of forest governance policies [3] with past and present forest management mechanisms frequently acting to dispossess, exclude and marginalise many forest communities [4]. Based on a conservative estimate, at least 20% of the global carbon stored in standing forests is located in indigenous territories [1]. As the value and significance of this stored carbon heightens, there has been a proliferation of novel forest governance mechanisms such as REDD+. The equitable engagement of forest communities and indigenous peoples is seen as pivotal to the success of climate-related forest governance policies [5,6].

The role that forests play in storing carbon has highlighted the need to confront the loss and degradation of forests at international climate change negotiations [7]. In response, REDD+ as a tool has become one of the most prominent examples of a global governance and market mechanism for forest conservation [8]. REDD+ received further support at the UNFCCC conference of the parties meeting in Paris in 2015 [7]. Yet, evidence from on-going pilot projects has been inconclusive [9,10]. In particular, the impacts on people who are affected by REDD+ activities in the design and implementation phases have to be investigated more carefully. Most of the countries that already have, and are likely to receive REDD+ payments in the future, are struggling with poor institutional capacity to meet and ensure social and environmental benefits [10,11].

Increasing pressure on tropical forests and international demands for sustainable forest management and good forest governance in tropical developing countries in general has led to the development of social and environmental safeguards [12]. These safeguards are, however, increasingly scrutinized [13–15]. Social safeguards include, inter alia, the respect for the knowledge, rights and interest of indigenous people (including land tenure rights); free, prior and informed consent; and the equitable sharing of benefits and effective stakeholder participation [16–18]. In particular, citizen participation and involvement in decision-making are considered to be important aspects of good governance [19], specifically with regards to forests and forest resources that are used or owned by local and indigenous people who rely on these for their livelihoods and well-being [20,21].

Yet, participation of local stakeholders can take various forms. Experiences of participation can range from pseudo or 'non-participation' to meaningful participation, determined by who is involved, how they are engaged, and who has control [22–24]. In their discussion of participation in relation to power, Cooke and Kothari [19] understand participation as often representing yet another possible technique to exert control, or a form of neo-colonial oppression, over local people. Participation is a contentious, frequently politically laden process that is also limited by power relations of the wider society [23]. Questions regarding participation depend on the equality of the power relations in the respective setting [25]. In this study, we base our analysis of participation on Arnstein's ladder of citizen participation, which is a frequently used framework [22,24]. Herein, we also understand participation as a major constituent of equity in terms of its procedural and distributional aspects [4].

Indeed, power is at the heart of questions regarding participation and Arnstein's model supports this. Power, and in particular the power devolved to citizen control, offers the metric on which her ladder of participation is based. Each step or rung on the ladder represents an increased balance of power that is in favour of the "have-not" citizens. At its core, "the ladder juxtaposes powerless citizens with the powerful in order to highlight the fundamental divisions between them." [22] (p. 217).

For decades there have been attempts to outline what power is and how it can be defined or studied [26], with limited success [27]. Nevertheless, power is ubiquitous and shapes the way that people interact with, and are subjected to, institutions and policies. With the increasing presence of international actors and donors attempting to influence and change national forest policies, or pushing for stricter conservation to mitigate climate change and foster more sustainable development, power resurfaces as a significant central concept. In line with Eyben et al. [28] we seek to understand the societal and political processes through which power operates, including whose voice is heard and whose is excluded or discounted in these processes. These factors shape the extent to which participation with REDD+ plays out in reality, for instance in the design and implementation of forest conservation policies and mechanisms. When we talk about power in the context of international development and in particular with regards to people, a pivotal concept is empowerment. Here, it is understood as a person or community's ability to mobilize in order to claim their rights and demand responsibilities from the state in which they are citizens.

Aims and Scope of the Paper

In this paper, we focus on the degree of Amerindian participation in the bi-lateral agreement between the governments of Norway and Guyana and the Low Carbon Development Strategy (LCDS), whose stated goal is to offer a model for how to achieve development that does not compromise the natural environment. Little to date has been recorded as to what extent Guyana's indigenous

Amerindian population has been involved in the design and implementation of the LCDS and how they have been affected by it (Notable exceptions include the work of the Amerindian People's Association [29] or Bulkan [30]). Based on an ethnographic case study of an Amerindian community in Guyana's interior region 8, we analyse 1) To what extent the past 6 years (2009–2015) of the Low Carbon Development Strategy's engagements with the community of Chenapou have been 'inclusive', 'broad-based' and participatory and 2) What the potential impacts of the current form of participation and distribution of power are for the LCDS of Guyana and Norway and, more broadly, REDD+.

2. Background

2.1. REDD+ in Guyana: The Guyana-Norway Agreement

In 2006, former president Bharrat Jagdeo recognised the emergent value of stored carbon associated with Guyana's vast tropical rainforests [29] (p. 10). He sought to capitalise on this by presenting Guyana to the international climate fora as an apt site for early Payment for Ecosystem Services (PES) models. Given the right economic impetus, Guyana would commit to protecting its rainforests in support of the global effort to tackle climate change. A commissioned McKinsey report (Full McKinsey report remains unpublished—information can be found at [31]) in 2008 estimated, based on an 'apocalyptic' deforestation rate of 4% per year for 25 years [32], that Guyana was foregoing some US$ 580 million per year by maintaining its natural forests [31] (p. 16). Buoyed by this, Jagdeo proposed an opportunity for international donors to put forward funding to incentivise securing the global ecosystem service provided by Guyana's rainforest whilst supporting the country's economic growth [33,34]. Jagdeo's model drew from the emerging UN REDD+ or 'Reducing Emissions through Deforestation and Forest Degradation and the enhancement of forest carbon stocks' mechanism (The final agreement with Norway does not strictly represent a REDD+ model as the reference levels determine that no absolute reduction in historic deforestation/emissions is required for payments to be made [32,35]).

REDD+ is a global initiative, aimed at incentivising non-Annex 1 nation reductions in deforestation and forest degradation through creating "a financial value for stored forest carbon" [36]. Payments are provided to promote the protection and enhancement of forest carbon stocks, effectively reducing greenhouse gas emissions from developing countries, whilst compensating for their opportunity costs associated with non-exploitation of their forest resources [37].

Globally, Norway has pledged billions of US$ to protect standing forests through a series of bi-lateral agreements and continues to play a leading role in financing forest conservation [38]. Guyana is a tropical developing country with an extensive forested area, i.e., 88% forested land cover [39] and a very low historic rate of deforestation of around 0.03% forest loss per year [32]. Mining, predominantly for gold, diamonds and bauxite, is considered to drive 85% of this deforestation [39], whilst also being a significant income stream for many of the poorest, including many Amerindians, in Guyana [40].

In 2009, Norway and Guyana signed the 'Norway-Guyana Agreement' [41] outlining US$ 250 million of support for Guyana up to 2015 [34]. At the centre of the agreement is the Low Carbon Development Strategy (hereafter LCDS). When it was signed in 2009, a prominent feature was the inclusion and recognition of Guyana's indigenous Amerindian population [34]. Comprising 10.5% of the national population and predominantly located in the forested regions of the country [42], the Amerindian population represents a significant actor in the LCDS functioning. Early reviews of the LCDS commended its articulation as being inclusive and participatory, although questions remained about the implementation of such measures [43]. At its inception, the LCDS was presented as aiming to achieve two overarching goals:

i "transform Guyana's economy to deliver greater economic and social development for the people of Guyana by following a low carbon development path"

ii   "provide a model for the world of how climate change can be addressed through low carbon development in developing countries" [44] (p. 2)

The stated goals were ambitious as the LCDS set out not only to provide a transformation of the Guyanese economy but also to offer a replicable example of low carbon development to the international community. However, during the six years of operation since, independent reports have identified considerable shortcomings in the LCDS facilitating the participation of indigenous groups [35,45,46]. A report by Rainforest Alliance in 2012 found that the government of Guyana failed to respect the rights of Amerindians in the process of setting up and operating the LCDS [46] (p. 6).

2.2. The LCDS and Indigenous Communities

Amerindian titled land represents 14% of Guyana's total forested land (18.5 Mha) [39]. Land title provides partial autonomy under the Amerindian Act of 2006 [47], with control over activities such as large-scale mining still residing with the Government of Guyana (GoG). The indigenous population of Guyana constitutes a considerable landholder and therefore is an important actor within the LCDS process [48]. The importance of the Amerindian population was acknowledged within the details of the Guyana-Norway agreement, which informed the LCDS' articulation, stating:

"The Constitution of Guyana guarantees the rights of indigenous peoples and other Guyanese to participation, engagement and decision making in all matters affecting their well-being. These rights will be respected and protected throughout Guyana's REDD-plus and LCDS efforts. There shall be a mechanism to enable the effective participation of indigenous peoples and other local forest communities in planning and implementation of REDD-plus strategy and activities." [49] (p. 5)

Thus, a Multi-Stakeholder Steering Committee (MSSC) was established to ensure the 'transparency' and 'effective participation' within decisions made regarding the LCDS [34]. Alongside the MSSC, the indigenous communities were directly incorporated into the LCDS through three specific projects: the Amerindian Development Fund, the Amerindian Land Titling project and the Opt-In Mechanism (see Table 1 for a summary of each).

Collectively, these projects effectively represent the LCDS' adherence with the relevant REDD+ indigenous safeguards [18]. Therefore, it is clear that the functioning of these projects is of pivotal importance when determining whether the LCDS is operating with, and in support of, indigenous communities in Guyana.

Table 1. Outline of Amerindian relevant projects within the LCDS mechanism.

	What is it?	Objective(s)	Progress to Date
Multi-stakeholder Steering Committee (MSSC)	An "institutionalized, systematic and transparent process of multi-stakeholder consultation(s)" on the LCDS [49] (p. 5)	To enable the "participation of all potentially affected and interested stakeholders at all stages of the REDD-plus/LCDS process" [49] (p. 4)	IIED report in 2009 noted it to be "credible, transparent and inclusive" [43] (p. 5) but 2012 Rainforest Alliance report found the mechanism "not effectively enabled" [46] (p. 7). Records suggest there have been no documented MSSC meetings since the change of government in 2015 [50]
Amerindian Land Titling (ALT) project	A project "designed to advance the process of titling the outstanding Amerindian lands currently awaiting demarcation and titling" [51] (p.7)	To complete "land titling for all eligible Amerindian communities by 2015" [49] (p. 5)	A number of outstanding title claims, demarcation issues and boundary conflicts persist. The ALT required to establish a second phase [48]

Table 1. *Cont.*

	What is it?	Objective(s)	Progress to Date
Amerindian Development Fund (ADF)	Fund set up by GRIF * to support "socio-economic development of Amerindian communities" by meeting their "own priorities ... and objectives" [44] (p. 9)	To support the 166 recognised Amerindian communities with development plans [44] (p. 24)	Pilot and phase 1 completed: "[A] total of US$ 1,298,577 has been disbursed to ninety (90) communities/villages" [52] (p.2)
Opt-In Mechanism (OIM)	Mechanism intended to allow "indigenous peoples [to] choose [whether] to "Opt-In" to the national REDD+ mechanism and receive a pro rata share of Guyana's REDD+ earnings" or not. [53] (p. 3)	To be operationally piloted by 2015 [34,53]	Extensive delays mean pilot settlement selected but Opt-In pilot process has yet to begin

* Guyana REDD+ Investment Fund (GRIF)—the trust fund established to facilitate funding for LCDS activities.

2.3. Participation in Environmental Governance

The recent Paris climate agreement is testament to the growing role of participation of a wide citizen body in environmental governance. It states that nations " ... should promote, protect, respect, and take into account their respective obligations on all human rights, the right to health, and the rights of indigenous peoples, local communities ... " when "developing policies and taking action to address climate change" [7]. This elevation of participatory methods and acknowledgement of indigenous groups builds on accounts testifying to the potential value accrued by genuine indigenous participation in environmental policy (e.g., [54]). Participation can lead to multiple benefits including (i) increasing the influence of civil society organizations; (ii) ensuring fairness of decisions; (iii) fostering greater voice and equity for underrepresented groups and (iv) enhancing the governments capacity to build consensus and support [54].

In REDD+, the necessity of participation, particularly of local or indigenous groups is addressed through the 2010 'Cancun safeguards'. These acknowledge the importance of the "full and effective participation of relevant stakeholders, in particular indigenous peoples and local communities, in [REDD+] actions" [18]. The safeguards are the basis for a broader architecture of national and international obligations to which Guyana is a signatory, including the United Nations Declaration on the Rights of Indigenous Peoples (UNDRIP), the World Bank's Forest Carbon Partnership Facility (FCPF) and Guyana's own Constitution [46]. Within the articulation of 'effective participation', the principles of free, prior and informed consent (FPIC) are of particular relevance. The guiding principles of FPIC can be summarized as a necessity to provide (i) information about and consultation on any proposed initiative and its likely impacts and (ii) meaningful participation of indigenous peoples and representative institutions [55] (p. 34).

2.4. Case Study Site—Chenapou

Chenapou is a Patamona Amerindian community with a population of approximately 500. It is situated on the banks of the Potaro river, 30 miles upstream from Kaieteur Falls (see Figure 1) within Region 8—the least densely populated region in Guyana [42]. Chenapou is considered relatively remote, with its neighbouring settlement a two-day walk through the North Pakaraima mountains (At the time of writing roads were being cut to neighbouring villages but these are still mud terrain and unreliable in heavy rains, which are common). This geographic seclusion and the lack of infrastructure within Guyana's interior means that Chenapou is also disconnected from most external information. National radio and television signals do not reach the community and any newspapers that do are sporadic and often days, if not weeks, old. As with all recognised Amerindian communities, Chenapou is politically represented by a village council, consisting of an elected *Toshao* or village captain and their associated Councillors. The *Toshao* is selected by a village-wide vote held every 3 years.

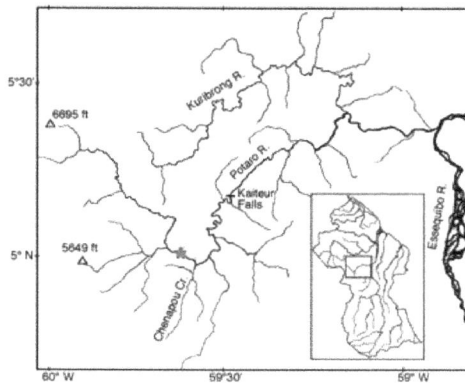

Figure 1. Map showing location of Chenapou (star) on the Potaro river [41]

Chenapou was selected as a case site for a number of reasons. Principally, the first author had an affinity with the community having spent a year living there as a volunteer teacher in 2010–2011. During this year, much time was spent attending village *kayaps* (*Kayaps* are occasions where the community collectively work on a task, often farming related, to support a village member. In return, the host will customarily provide food and drink for the workforce. These are weekly occurrences in Chenapou, usually attended by upwards of 30 people and are very social occasions with the drinking and eating after the *kayap* often going on for many hours whilst playing dominoes, cards or telling stories.), working in the farm and taking fishing and hunting trips with other villagers. These everyday subsistence activities allowed bonds to establish with many in the community, developing the researcher's understanding of local concerns through conversations which often revolved around resource rights and land ownership. From time spent in the village, the researcher learnt basic phrases in the local Patamona language; however, conversations were always held in English. It should be noted that conversational English is spoken by most in the community well, with the exclusion of a small number of the most elderly who only speak Patamona.

In addition, Chenapou offers a valuable site to interpret several ongoing political and environmental processes. It is at the confluence between a series of contested environmental, economic and development dynamics. A push for land title extension, tension over the relationship with the neighbouring national park and a transit site for many illegal gold and diamond operations are all ongoing issues for the community [56].

Chenapou's proximity to sites rich in mineral deposits and a lack of alternative cash-based occupations means that the major income source for most families comes from involvement in artisanal and illegal mining. Whilst subsistence activities such as hunting, farming and fishing are still prevalent, the growing influence and importance of money in transactions and education means that mining has taken on increasing significance over recent decades. Highlighting that mining constitutes the principal driver of deforestation in Guyana [39], has strong ties with corruption and is largely unregulated at the community level [57], underlines the pertinence of the LCDS and REDD+ mechanisms to this community.

Moreover, protracted and ongoing disagreements have revolved around land titling and extension of land rights within Chenapou for at least the past two decades [58]. Since the extension of the neighbouring Kaieteur National Park there have been ongoing, and at times fractious, relations with park management and the government over titling of land. The expansion—from 12.95 km² set in

1973 to 626.8 km² in 1999 [58]—was not only considered to encroach on spiritual as well as traditional hunting, mining and gathering grounds (It should be noted that hunting, fishing and gathering have been permitted in the park for Chenapou residents since 2000; however, mining and logging remain prohibited [59]), but was also imposed rapidly in the absence of adequate consultation [56,58].

This means that both the Amerindian land titling (ALT) project and the 'Opt-in' procedure are of significance to residents of Chenapou. Based on these dynamics, Chenapou represents a pertinent study site in which to ground research, offering illustration of key features associated with LCDS outreach, indigenous participation and the processes that support or undermine both political and social empowerment of the community.

3. Materials and Methods

3.1. Case Study Approach and Selection

We used a single-site qualitative case study based on narrative interviews and ethnographic principles in an Amerindian village. To carry out the research, the first author was granted access to the community and had been given the permission to conduct interviews with its residents. The choice of a single in-depth case study provides certain benefits, for instance, a sensitivity to contextual nuance and complexity which is seen to be "underrepresented" [2] in the more prevalent macro-level research on global environmental change [60]. This is of particular significance when looking at the effects of forest conservation approaches that are global in scope, but local in their impacts.

Based on ethnographic principles, we selected methods which did not impose formal structures—such as narrative interviewing. This was in part motivated by the findings of a 2009 discussion between the neighbouring Kaieteur national park and the community of Chenapou, where community members expressed concern that prior research in the village had been inaccessible and one-sided, excluding the voice of locals [58]. Therefore, we sought a more transparent and accessible research approach, reflecting findings and interpretations iteratively with the community throughout the research. Furthermore, we gave focus to the subtlety of interpretations to reveal people's tacit as well as expressed perceptions. Recorded interviews were supported by observations, note taking and participation in everyday activities within the community.

In order to support the narrative evidence of interviews, we also took account of the village visitor records in Chenapou, which have been documented by the village, as all visitors require official permission to enter an autonomous Amerindian community [47]. Information was available from the period between 1998–2015 with the absence of information for 2010 and 2011. Nevertheless, the upkeep of this record is not always entirely diligent and conclusions taken from it are limited, but still valuable as they give an indication of who visited, for what purpose, when and for how long.

Outside of the community, a broad reading of LCDS policy documents was important to establish an understanding of the development of the policy at a national level. Those documents were not simply regarded as objective texts, but instead were analyzed for their framing, intentions and discourse as is important particularly with 'official' texts [61]. They provide a validation or contrast to findings in community discussions and, through a hermeneutic reading, can provide key insights [61]. Observations were also informed through a wider reading of relevant development, anthropological and political literature, as well as local media sources, around the topic.

3.2. Narrative Interviews

Research was conducted in Guyana for five months in 2015, with two periods of extended stay in Chenapou. During that time, thirty (30) in-depth interviews were conducted and recorded. The interviews were supported by numerous participant observation sessions such as attending village works (*Kayaps*) or community meetings. The majority of these conversations and observations were in English, yet there were also times, for instance when sitting with women whilst preparing cassava (Cassava or manioc is a root vegetable ubiquitous in South American and particularly Amazonian diets. Within indigenous groups in Guyana, it is a staple which is processed in a number of ways to produce both drinks (*cassiri* and *parakari*) and food (e.g., *farine* and cassava bread).), that a more elderly villager would speak in Patamona with a younger relative translating. All recorded interviews, however, were conducted in English.

Selection of participants followed both opportunistic and deliberate sampling as we sought to cast the widest net during first stages of research and narrow interactions as we engaged with more specific members of the community [62]. The more deliberate selections sought out village council members, LCDS workshop attendees or more vocal community members. Effort was made to balance age ranges, gender and status of participants involved. Of the interviews conducted, around 1/3 (11) were with female members of the community, which is likely a reflection of the fact that the researcher is male. It is noted that within Chenapou there is a recognisable degree of gender parity in forums such as village meetings meaning that women, although often fewer in number, are both vocal and supported to be vocal in discussions.

Recorded interviews acted as culminations of a process of discussions, observations and document readings leading up to that point. Within these interviews the dynamic of free discussion and participants' development of answers associated with that of 'narrative research' techniques [62] was emphasised. This meant that interviews often took the form of discussions with other family members or friends present—as is common in village life—wherein we would be eating the local dish of *tuma* (Spicy broth often containing fish or wild meat if there has been a recent hunt) or drinking some *cassiri* (Staple drink made from fermented cassava). We were often sitting by the fire and sometimes the participant(s) would be engaged with activities like weaving crafts or preparing cassava as we spoke. These discussions varied in length but were mostly either around or over an hour.

A relaxed environment of view-sharing and discussion is an important element when considering narratives as co-constructed as well as personal [63]. Allowing for these 'joint narratives' [61], wherein multiple persons could contribute to a discussion, was imperative and allowed us to observe collective perceptions alongside personal views, instead of treating the community as disconnected individuals [64]. This approach was also suited to the existing culture of social interactions in Chenapou as both researcher and interviewee become 'active participants', collectively producing meaning [63] and providing a process that offers sensitivity to the nuanced understanding of how people in Chenapou perceived the LCDS, which has largely been missing in research to date.

3.3. Methods of Analysis—Evaluation of Participation

To evaluate participation, it is necessary first to identify what is meant by participation in this context. Apart from Arnstein's ladder of citizen participation, we draw on Utting's articulation that participation "involves far more than the active and willing involvement of local people in [planning and design]". Instead, "it is also about 'empowerment' or the organized efforts of marginalized groups to transform patterns of resource allocation and increase their control over material resources and resource management decisions." [65] (p. 256). Meaningful participation is understood as guaranteed through "access rights": the rights of public access to information, to public participation in government decision-making, and of access to justice." [54] (p. x).

We set out to assess what kind of participation and engagement has been experienced. Arnstein's typology provides a functional delineation of degrees of citizen participation, offering a frame of reference against which to measure the participants' experience. The ladder presents a spectrum of participation calibrated into a series of sub-sections dependent on the degree of power genuinely placed with the citizenry. It is this understanding of participation, a measure of the degree to which legitimate power is conferred to the citizen, that we adopt when analysing our findings. This offers a useful framing against which we can discern what level of participation people in Chenapou perceive to have experienced with regards to the LCDS. All accounts in the results are anonymised as agreed with the participants, and reference is made only to the recording number (i.e., I:x).

4. Results

4.1. *Effective Participaiton Largely Failed*

Consistently, accounts from community members regarding how they experienced the LCDS or participation in its design and local implementation presented very similar depictions. As this account typifies:

> "They [the government] come in for an hour, mostly just 30 minutes, talk and then leave for their boat to Kaieteur. You can't expect people to understand all that so fast. Then they call that consultation" (I:2)

This sense that the government visit too infrequently, and when they do it is always a rushed and insubstantial session, was evoked by many community members we spoke with. Frustration at this was expressed by numerous participants:

> "These government officials come and spend one day and keep meeting for two hour and beat out and go back to town. They don't got time to stay, they schedule always busy" (I:3)

> "Within two hours we wanted to put forward things they must do. You know, give people a chance to talk, you know?... Only three or four persons got to say something, what about the rest?... And they want you to listen to them and they don't want you to talk" (I:23)

The length of time that government representatives would spend in the community during these visits was a frequent issue in many interviews. There is a sense that the opportunity to actively and meaningfully participate has been curtailed by such short sessions, as these sessions did not adequately allow community members to engage with the topic and offer their input or have their voices heard. Instead, these sessions allowed for the visitors (government) to speak without granting adequate time for community input or response. Furthermore, the reasoning behind such curtailed and infrequent consultation was felt to be unjustified by a small number of community members who were aware of the Guyana-Norway agreement. One participant, who had previously been put forward as the village representative during outreach sessions, put it like this:

> " ... but I know with this LCDS and this FLEGT (EU- FLEGT is a mechanism set up to reduce illegal logging and promote sustainable forestry by certifying sourcing of timber to the EU (EU-FLEGT 2016). It is not a part of the LCDS formally but was confused by many in the community as being associated with the LCDS mechanism.) they are getting a lots of funds to do these [outreach] programmes which they are not really doing. I mean with these workshops that I have been attending so far I have learnt that a lot of money being put into Guyana to do these consultations to do these, what you call it ... outreach programmes they call it—in communities that may be affected by the programme that they are planning now." (I:28)

The village visitor records (It should be noted that visitor records are not considered highly accurate and so conclusions drawn from this source are limited.) supported the claims that the government visits were infrequent. In the 16 years of documented visitors, researchers, representatives from a conservation NGO, and the mining commission have visited Chenapou much more frequently than representatives from the regional or national government (see Appendix A).

4.2. *Knowledge of the LCDS*

> "No one really understand about it ... We ain't getting the understanding. It is only down there [Georgetown] they is getting to know what is happening" (I:14)

Knowledge and understanding of the LCDS is a useful indicator of the success of engagement with people in Chenapou. Across the 30 recorded interviews conducted, more than two-thirds of participants responded that they either had never heard of the LCDS and Guyana-Norway agreement or had heard of it but could give no explanation or detail as to what it was. The ambiguity of understanding is illustrated in the response given when one participant was asked to explain their understanding of the LCDS:

> "Low, something, carbon, something ... I can't remember ... Low carbon something something. We don't really hear nothing about that, them just come and tell we one thing and we don't know ... they just left we in the dark man. They don't like Chenapou people." (I:3)

The general dearth of understanding around the LCDS was often connected to the process through which information has been shared. This account, from the representative who is frequently asked to attend training sessions and inform the entire community, details how the LCDS 'outreach training was conducted and offers explanation as to why knowledge of the LCDS is so limited in Chenapou:

> "I can't remember it was about two years ago ... again as I am saying it was just a ten, fifteen-minute story (explaining the LCDS) so people don't know anything at all in Chenapou about what they are really doing or what LCDS really is. You understand?

> When you see there is this big, big long (LCDS document). I mean they have the book, they have the draft what you call it (LCDS update document) but they don't really explain, I mean one or two people would read it and they may understand it in one or two parts, but then all this different, different things you know. It is difficult, it is difficult." (I:28)

The training session described had occurred two years ago, during a day in which numerous policies had been explained simultaneously with just ten to fifteen minutes dedicated to explaining the LCDS. The only other source of information available to support these outreach sessions are documents such as the LCDS 2010 update mentioned, which is lengthy (100+ pages) and fairly inaccessible [34]. The scant documents that were available in the community were similarly inaccessible, often being highly technical and filled with jargon. These were evidently not designed for a community where English is considered a second language for many and where the majority have only been formally educated to a primary school level [58].

> "The large documents we get (from external programmes) are hard to read and long ... we think that they should come more in our own language [Patamona] that would be easier for us to understand" (I:18)

Short, infrequent consultations, inadequate outreach training sessions and complex information materials have engendered a sentiment amongst many participants that the government have provided a one-way flow of information, effectively curtailing the space for discussion and participation from people in Chenapou.

4.3. 'Lip Service' Participation

When discussing reasons for why outreach sessions were infrequent and, when they did occur, were insubstantial and brief, a consistent interpretation was offered. This quote is illustrative of a sense of dishonesty/instrumentalisation felt by the interviewees in Chenapou regarding consultations:

> "They (politicians) say they are busy and they are making a special time to visit, but really they are just coming, talking and going. Then in town (Georgetown) they can make like they consulted the whole community when they didn't even listen to them!" (I:20)

Here the sentiment of 'lip-service' or 'tick-box' participation behind what are considered superficial and insubstantial consultations is evident. Duplicity between what is conducted within Chenapou and the perception of how that is reported favourably and unrealistically by the government externally is clear. A participant who had recently moved to Georgetown, but had spent most her life in Chenapou, affirmed this sense:

"They say one thing and do another ... they think buck-man (This term is generally considered a derogatory, racial label carrying connotations of a lack of intelligence or development towards an Amerindian.) are stupid when really they are some brilliant people. The government just think they can get away with telling them (Amerindians) what they like and then going back to town and telling everyone they have consulted the Amerindian." (I:2)

The sense of Amerindian voice being willfully discounted in the performance of participation by the government was evident from those accounts and observations made in Chenapou.

Minister of Tourism Visit

Whilst conducting research in the village two prominent government officials—the minister of tourism and a member of the Ministry of Indigenous Peoples Affairs—visited the community. Their visit was explained as an opportunity to respond to, and to understand, the issues that the Chenapou community might have regarding their recent concerns with suicides being committed in the neighbouring national park [66]. Observing this meeting offered evidence as to the nature and experience of interactions between the community and government. In total, the contingent of five government personal and assistants were in Chenapou for a little under two hours having flown directly from Georgetown.

The meeting included speeches by both ministers, taking approximately an hour and a half, followed by a period of questions lasting fifteen minutes. It was clear from the first questions that although the stated intention was to talk about the issues with the park, many in attendance wanted to ask about a broad range of topics. After the second question, however, the minister of tourism prompted the assembled community that " ... we haven't much time" (I:16), attempting to wrap up the session, which was not received well by those waiting to pose further questions. In response to the disquiet produced by this abrupt ending the Minister stated that a final two questions could be asked. After attempting to wrap up after those two questions, more were provoked with a further four questions being posed amongst numerous comments of discontent. At this point, one local in attendance expressed his frustration with the brevity of consultation to the minister:

"Just for future, for the record. Sometimes we are tired with government officials coming to speak and really spending just a small time. Sometimes, you know, some people are very slow and they might have something to say and the time has gone up ...

(interjection from another community member present) That is what happening right now.

... in future we would like that you come to visit often, you would stay with us and listen to all who want to talk and then you could get a good note once everything has taken place and you can have a very balanced view. And I hope that in future our senior minister Allicock (Sidney Allicock—Vice President and Minister of Indigenous Peoples Affairs.) is coming here, we do not want only to have a visit when the time is up (end of administration wherein the government can do no more). That has happened in the last administration and we do not want that to occur in this new administration." (I:16)

The minister responded to this apologetically, was held to another two questions as she made her way to the exit, and whilst attempting to say thanks for the third or fourth time, one vocal member in attendance reasserted the mounting frustration felt: "She going? ... No, no, no man this not no meeting you all can't leave yet. We got enough discussion to have." (I:16)

The small group of government ministers left to a chorus of discontent and grumbling of dissatisfaction at the session. It had been discussed amongst community members for days leading up to it and had lasted a little under two hours. They had wanted to use the opportunity to ask about a broad set of issues. Instead, those in the school building were left frustrated, with many stating that they had many more questions to ask at such a rare occasion as a minister visiting their village and others feeling that this was the same old story of insubstantial government interactions with them.

4.4. Widespread Mistrust

The effects and potential impacts associated with the poor participation are significant. As well as resulting in the lack of understanding of the LCDS, insufficient participation was associated with some broader concerns. These were addressed in many conversations, often after the more specific comments had been made, as accounts reflected participants' thoughts on broader implications.

Perhaps the most common of these responses was one of mistrust in the political system and disenchantment from the political process. The dearth of engagement experienced with the LCDS was manifest in a sense of opposition, with many people in Chenapou feeling the GoG's inactions were deliberate:

> "Really and truly Sir, the government don't like Chenapou. Chenapou has a big voice and will stand firm to government and its rights which the government don't like, so they play politics against us". (I:20)

This illustrates a sense of alienation and opposition common to many we spoke with. The phrasing of politics as something instrumentalised actively against the community emphasises the degree to which some felt a distrust towards the government. The distrust towards the government was palpably expressed by some as outright frustration and anger regarding the general status of Amerindians in Guyana and the mistreatment of their rights. Interviewees expressed concerns about land-grabbing or loss of land sovereignty as well as misappropriation of funding as illustrated by these passages:

> "They (GoG) just grabbing from the Amerindian's all the time. They are destroying our freedom too, with this FLEGT (This participant's description of EU-FLEGT was really a working explanation of REDD+. So when interpreting this comment we infer it more as a reflection on REDD+ (the reference to who should be getting the money shows this) than EU-FLEGT.) thing they gone destroy the freedom, they gone really destroy our freedom, because we accustomed to cutting bushes how we want to but now as they doing this they getting money, they don't want us to do these things. It's affecting us. It is going to start affecting the community, all the communities and furthermore it is we who got the trees, it is we who supposed to get the money, not them." (I:3)

A local teacher echoed these suspicions over where the investment from Norway has gone:

> "And where is it going? In the government pocket . . . Georgetown getting all the money and Georgetown ain't got a tree! We got the trees" (I:22)

These concerns over finances were widespread in the community and also documented in other sources identified [29,35,46]. The Amerindian People's Association noted in its 2014 special report that as of 2012 only US$ 9.2 million out of a total of US$ 69.8 million funding from Norway had been released within two years of the REDD+ programme [29] (p. 70), with a significant proportion going towards the rather ambiguous 'institutional capacity building' [52].

4.5. Respect for Indigenous Rights

The distrust in the political system for many in Chenapou was couched in a distinction between the Georgetown politicians and the Amerindian population. Many felt that at the heart of the issue of LCDS participation was the active lack of respect given to indigenous voice and rights. This was directly addressed by one resident during the minister of tourism's visit to the community:

> "I'm asking the government, or the heads [ministers], to show more respect to what we say as the Patamona people because we born here, we grew here and we know what around us. We know how to live with our mountain, with our rivers and what we say I think this is what should be respected before any other rules and regulations from the government side" (I:16)

Non-inclusion in the LCDS is understood as a denial of the rights given to them as indigenous land owners. It is with a deep sense of injustice that the exclusion has been interpreted from many in Chenapou as exemplified here:

> "Right, so you know these (LCDS) documents. It seems that there is nothing really, nothing to do with the Amerindians communities when you look at it. The FLEGT and the LCDS that is how I see it: nothing really to do with the Amerindian. So, if they could consider us in whatever they are doing in the future that would be so great and then come in to tell us and let us know." (I:18)

Responses to this sense of injustice varied from those of dismay and despondency to more forthright and determined positions. The despondency is illustrated in one participant who, after telling many stories of how the GoG had failed to help develop their community, posed the rhetorical question:

> "Really, how long have we (Chenapou) been behind?" (I:11)

For others, the perception that the LCDS is just another case of the government not engaging with the community adequately, but instead politically ostracizing it, evoked a sense of significant injustice:

> "I study these things and I said look, I think the time is now for us to stand for our rights, it is time. For too long we have been deprived from our rights. We have been deprived in this country for a long, long time. It is now that we know our rights and we try to share with our people that this is our rights, this is what should happen this is not what should happen. Don't let people tell you: 'this here is good for you' when you know it is not good for you. Let the people know that no, that is not good for me this is good for me. That is what we told them . . .

> . . . we decide what is ours, we decide what is good for us, you don't come from the coastland telling us that this should be good for you- we decide together, we decide what is good for us that is what we must tell you. And you must adhere to these things but that is not what is happening with this government, and it has been happening for years, now I am 43 years old . . . " (I:29)

5. Discussion

5.1. Why has Participation been so Poor?

Interviews, supported by observations and document readings, identified clearly that participation of residents from Chenapou in the LCDS has been weak. There has been a failure of adequate outreach techniques. Short and infrequent engagements, often with just one or two representatives from the village attending, are ineffective approaches to ensuring that the whole community is informed.

The governance architecture for Amerindian engagement in the LCDS, outlined in Table 1, has been poorly implemented. The MSSC, although applauded in early assessments [43], has been highly selective [35], ineffectively enabled [46] and appears to have not met since March 2015 [50]. Both the Amerindian Land Titling process and the Amerindian Development Fund have been protracted. The land titling process now requires an extensive second phase after initial projections to complete in 2015 [44], and the development funds have been found to lack transparency [46], with only small sums having filtered through to Amerindian communities whilst most still resides with facilitation bodies such as the UNDP [67]. Moreover, recent research conducted over the past three years has identified numerous Amerindian communities where land titling remains strongly contested and where the ALT has failed to uphold the principles of Free Prior and Informed Consent [68]. The Opt-In procedure, timetabled for completion in 2015 [34,53], has shown almost no progression from conceptual drafts and remains largely stagnant, although a pilot community has been identified [69].

However, to interpret a solely technical reason for this failing risks under-estimating the significant power imbalances present between Guyanese society and the political system. Understanding the lack of participation as a product of an inequality of information and power offers a more nuanced interpretation. This need not amount to a suggestion that the Government of Guyana has deliberately or purposefully not engaged the community of Chenapou, but rather that it is the established norm to not engage them. As Lukes suggests, the "bias of a system" may be held up not only by the deliberate action of an individual or group but also, "by the socially structured and culturally patterned behaviour of groups, and practices of institutions" [26] (p. 26). In other words, an understanding of power as systemically applied, based on a status quo, provides a possible interpretation of the failure of the government to support the participation of a community like Chenapou.

The dynamic of Amerindian communities being omitted from the political process is supported on a national scale by Bulkan's [70] articulation of Guyana's 'racialized geography'. Her concept refers to more than simply the ethnically quantifiable demographics of census data. It serves as a commentary on a deeply embedded social, cultural and political division which is present at the core of Guyanese society. This is encapsulated well by anthropologist Andrew Sander's account:

> "In Guyana the distinction between the Coast and the Interior is more than merely a geographical one. It dominates the Coastal society's conception of its country ... The town ... is a bright, exciting place, full of interesting people. At the other extreme the Bush is a dark, dangerous, uninteresting place, inhabited by fierce animals and backward, furtive Amerindians." [71] (p. 11)

This is an oppositional framing born from the colonial disregard for the indigenous communities in Guyana [70]. Bulkan identifies this cultural divide as persisting in contemporary politics, finding that "although Amerindians constitute almost 10% of the population ... political issues and resource allocations are still dominated by the coastland parties and their concerns" [70] (p. 373). She broadly identifies a restriction on Amerindian autonomy, noting that village councils—the sole political representation of Amerindians—"have no formal link with the regional system of government, which deepens Amerindian isolation from the political process." [70] (p. 375). Power—politically, financially and culturally—has permanently resided in the coast since Guyana's colonial existence, often excluding the Amerindian population. This "racial politics" reflects "the divide-and-rule strategy practiced by small numbers of colonial masters over large numbers of slaves." [40] (p. 249) and appears concurrent today.

Prior research, such as the 2012 Rainforest Alliance report and 2014 Amerindian Peoples Association findings, support the presence of a power imbalance, restricting and subduing Amerindian's political role in the LCDS [29,46]. Donovan et al. identify that Amerindian stakeholders "feel that (the) GoG has not kept them updated and often feel strongly that their voices are not being heard, especially with respect to land titling and traditional land extensions" [46] (p. 20). In the context of Chenapou, the exertion of political power in the form of control over information, dialogue and engagement with the LCDS process has marginalised the community. These people have been subjected to the discernment of the government over their access to information about the LCDS and their capacity to have their voice heard and included in the discussion. For residents of Chenapou, this political exclusion has precedent. Experiences such as insufficient consultation over the expansion of the neighbouring national park [58], and a feeling that government visits are rare and insubstantial (see Appendix A) emphasises this.

This dynamic has also been seen to represent a broader status quo within Guyana where racialized geography is embedded into the political architecture, giving rise to a "system bias" [26] that ostracizes Amerindians from the political process [30,70]. Bharrat Jagdeo and the PPP administration that heralded the LCDS were a pinnacle of this power imbalance, their hegemony prevalent beyond the LCDS [30,35,70]. Chenapou is at the receiving end of this asymmetry, which is highlighted by the non-participation experienced during the past six years of the LCDS process.

5.2. Consequences of Failed Participation: 'Opportunity Costs'

The LCDS and REDD+ can be understood based on the premise of providing financial incentive, on a national level, to replace opportunity costs incurred in reducing deforestation [36]. However, the costs invoked by a failure to adequately engage with indigenous communities may represent the most significant opportunity lost on the part of those establishing the LCDS. By not engaging with the community, the actors in power—both the GoG and Norwegian government—miss out on the value of local knowledge and input, which may hold significant benefit for projects [3]. Moreover, there is a subsequent loss of support from communities due to a politics of disengagement, which has the corrosive implication of translating into a general mistrust of the political system. By ostracising groups from the political process, the pragmatic implementation of political decisions is made increasingly challenging [72]. Broader societal engagement and understanding of political decisions are seen to contribute to the likelihood of successful and higher quality outcomes [25,73].

The range of Amerindian-oriented policies (see Table 1) and the recognition of Amerindians and their indigenous status within both the initial LCDS documents and the Joint Concept Note are commendable. However, in failing to overcome the "most important first step towards legitimate" participation [22] (p. 219), of informing citizens of the political process, the value of these policies is lost. If, as is articulated, the government seeks to support, respect and engage indigenous communities within the LCDS process, then the reality that most of those in Chenapou had no understanding of what it is, does not reflect well. The findings in the national Rainforest Alliance report in 2012 largely support our observations, noting that most Amerindians they engaged with were "still confused about basic principles of the LCDS" [46] (p. 32).

For those in Chenapou who had some awareness of the LCDS, a specific concern was that of the misuse of funding. Many felt that there was an injustice in the fact that Norway had supposedly provided many millions of dollars and yet Chenapou had very little tangible evidence of this funding. There was an incredulity expressed at the notion that the government was unjustly profiting from the LCDS arrangements:

"Georgetown getting all the money and Georgetown ain't got a tree! We got the trees" (I:22)

This sentiment was echoed in the Rainforest Alliance report identifying: "(a) perception that the GoG is receiving LCDS resources whilst beneficiaries in the field are not" [46] (p. 19).

Without an understanding of the LCDS process, Chenapou and other communities are further isolated from the policy, which, at least in the case of Chenapou, gives rise to fear and distrust. In place of the lack of understanding, narratives of land-grabbing, political corruption and malpractice of the government became readily associated with the LCDS. These are built on unclear and imprecise flows of information, proliferating narratives of mistrust and fear which go unchecked by a largely absent government. As accounts from Chenapou reflect, this lack of engagement in politics is not isolated to the LCDS (see I:29). Instead, it is felt that the political exclusion of Amerindian's is consistent across the spectrum of politics and across time, whether for expediency or political calculation [35]. The superficial 'tick-box' outreach, which has operated throughout the LCDS, allows the government to report as though the process has been participatory, whilst acting unimpeded by citizen input. Whilst this may appear to grant the government free reign in political decisions, this is not without consequence.

Distrust from an electorate can be instrumental in whether a national policy will have public support [65]. Moreover, trust endures such that a legacy of distrust can be politically immobilising and requires great effort to overcome [74], meaning that even a political transition with well-meaning intentions may be compromised by pre-existing perceptions and feelings. This is a very real and existing consequence of weak participation that was observable amongst those we spoke with in Chenapou. Moreover, if the LCDS is not understood, it stands little chance of persuading residents in communities like Chenapou to support it and reduce or halt their mining activities. Without influencing those miners in Guyana, the hope of maintaining and preserving the forested environment is unlikely to succeed. Thus, failure to engage communities in the democratic process has clear and lasting implications for the government which should not be underestimated.

5.3. Poor Participation is a Transgression of Fundamental Indigenous Rights

The level of participation observed and expressed from most interviewees in Chenapou falls well below the stated intentions of the LCDS. We found the experiences of those in Chenapou to fall between Arnstein's [22] definition of tokenistic and inconsistent informing and total non-participation. Outreach that did occur was little more than a 'tick-box' exercise, allowing the government to report that it had fulfilled its engagement. No evaluation of the communities' understanding of the key concepts and implications of the LCDS was made, and there is little evidence that input from communities had any effect on the LCDS functioning.

This failure to uphold the founding sentiments of the LCDS is significant as the principles themselves were built upon more fundamental rights. To "participate in decisions that affect their environment" [54] (p. x) is a right that should be given to all people. However, this right is only accessible when information about the impacts is given and the "opportunity to voice opinions and to influence choice(s)" (*ibid.*) is present. Without those there is a restriction of this right, rendering the process a restriction of access to justice, which effectively denies the "democratic legitimacy of environmental governance." [75].

The LCDS and REDD+ programmes fall under the umbrella of a number of safeguards and principles (such as the 'Cancun safeguards' [18]), which often exist to protect the rights of marginal communities. A central tenet of this protection is the principle of FPIC. We have found that the provision of information and consultation, along with the "meaningful participation of indigenous peoples" [55] (p. 34), which constitute FPIC, have not been upheld in the context of Chenapou. Thus, not only has FPIC not been respected, but the "full and effective participation of relevant stakeholders" [18] (p. 26), as called for in the Cancun safeguards, has similarly not been met in the LCDS. By not respecting FPIC and not upholding the Cancun safeguards, the LCDS and GoG have critically failed to respect the rights of the indigenous community of Chenapou.

This sets a worrying precedent when reflected against broader Amerindian discontent in the LCDS [76] and the findings of both the Amerindian People's Association [29] and the 2012 Rainforest Alliance report, which found the LCDS to have failed in protecting "the rights of indigenous peoples" [46] (p. 7) in Guyana. These findings raise a warning flag. Consistent and documented failing to respect fundamental rights of a large number of people within the constructs of the LCDS is a serious concern.

5.4. Critique of the Model of Development-Distribution of Power

Asking why there have been such noted shortfalls in democratic engagement within the LCDS is important. We have put forward arguments which focus on the role of the Guyanese government, but Norway as the bi-lateral partner and financier is also accountable. An initial criticism, that the Norwegian aid department recognises, is that the selection of Guyana was done in haste as the opportunity to present the REDD+ model at COP 15 (2009) approached [77]. This was clearly a significant flaw as the Norwegian government entered a bi-lateral agreement not fully informed of the local political context and dynamics in Guyana. This lack of contextual understanding on the part of Norway is critical. Implementing REDD+ as a concept into Guyana is fundamentally challenging for some key reasons.

Firstly, the significance of mining within the Guyanese economy, particularly unreported and artisanal mining such as that prevalent in Chenapou, should not be underestimated [57,78]. Income from gold mining is integral to many in communities like Chenapou and so in effect, REDD+ must compete in order to substitute the income of gold. With little likelihood of this occurring, it seems unlikely that REDD+ could tackle even the recorded levels of deforestation associated with mining [39]. Additionally, of the four established partner countries that Norway has engaged with—Brazil, Indonesia, Tanzania and Guyana—Guyana is considered the least transparent, ranking lowest on the Corruption Perception Index [38].

Nevertheless, Norway signed an agreement worth US$ 250 million with almost no prior knowledge of the country and, importantly, with little to no on the ground presence. These are issues latterly acknowledged by the Norwegian development agency:

> "The NICFI [Norwegian climate and forest initiative] operations in two key partner countries (Guyana and Indonesia) were less well regarded, both in terms of staffing levels and operational experience with these country partners ... the number of staff is perceived as small, particularly the operational capacity in two countries with large bilateral programmes" [38] (p. xxviii)

The model adopted by Norway should be questioned. Norway has provided substantial financial promise with little oversight and no internal presence to a country recognised as having issues with political corruption and transparency. Guyanese commentators underline this, noting that the "likely principal result of Norwegian aid funds will be to consolidate further the political racialization of Guyana" [35] (p. 274).

The Norwegian aid department must carry some responsibility for providing such funding into a country where they had little understanding of the institutional capacity or history of engaging the Amerindian communities [11]. Norway has sought to take a removed role in funding these interim REDD+ projects, but, as has been presented, the costs of offering little effective operational staff and simply providing substantial funds are manifold. With no oversight, Norway has been unable to ensure that the LCDS would provide the participation and inclusivity of Amerindians as mandated in the Norway-Guyana agreement. Bulkan emphasises the potential consequences of such limited prior knowledge stating that: "REDD illustrates how dispensation of international aid, without robust checks and balances, can maintain and extend entrenched power" [35] (p. 249).

There is a failure on the part of planning and process awareness from Norway and NORAD, which is being felt most by those whose rights and opportunity to engage in the LCDS have been constrained. Marginalisation has been reinforced through the cultural bias [26] or status quo such that indigenous Amerindian communities are further disconnected from politics and power. It is deeply concerning to read admission of this lack of awareness within reviews of NORAD and yet observe the continuation of those same dynamics:

> "NICFI presence in some partner countries is perceived as being too limited. This is particularly so in Guyana where despite excellent technical progress, there is considerable dissent among wider stakeholders at the limited progress on enabling activities and a view that Norway has an incomplete view of how its funds are being spent. It is concluded that the staffing situation in Guyana requires deeper consideration of alternative options." [38] (p. xxxii)

The operation of power in an aid relationship (between donor and receiver) is significant and frequently glossed over and yet, as Eyben et al. note, "it is always the giver who has the power, stressing that there is no such thing as a free gift." [28] (p. 89). It is necessary to better understand the operations of power in local and global interactions that are involved in changing national policies and consequently local practices, such as the ability of Amerindian communities in Guyana to access and use forest resources. For instance, if NGOs and development agencies work without an understanding of contextual power structures, they may unknowingly perpetuate the inequity and injustice they are striving to change [28]. This holds for third party/state sponsored forest conservation or sustainable forest management, which should be aware of potentially continuing or even exacerbating unequal power relations and thereby contributing to disempowerment.

The "deeper consideration" that NORAD outlines does not provide improvement to the LCDS and indigenous communities unless it is supported by effective action. Implementing principles such as FPIC requires collaborative, early engagement with clearly defined criteria and a reliable system of accountability [79]. The Opt-In process in Guyana could be the key to providing a choice for indigenous communities, which is at the heart of FPIC. Participation is principally about information and the power to act, or not act, on that information [22]. For FPIC to be adequately addressed, it requires an understanding that engagement and information is needed on an ongoing cycle throughout the process. Lessons should be taken both from those situations where FPIC is inadequately observed and those who have shown greater success [10].

The importance of addressing this is outlined as Norway looks to play a pivotal role in REDD+ globally [38]. An understanding and consideration of the shortcomings of their approach in Guyana and its implications could, and should, provide reflection applicable to their broader development aid model.

6. Conclusions

At the inception of the pioneering LCDS, Bharrat Jagdeo and the government of Guyana set out a commitment to guarantee the participation of all stakeholders and to uphold the rights of indigenous peoples [49]. Guyana was to be a "global model for REDD+" [44] (p. 41), leading the way in achieving a low carbon development path for the rest of the world to follow.

Our findings call into question the extent to which those founding principles have been achieved. Through our case study analysis of Chenapou village, it is evident that the 'inclusive', 'broad-based' and 'participatory' [41] tenets of the LCDS have not been realised in this community. Participation in Chenapou can, at best, be considered an uneven one-way process of fairly insubstantial and tokenistic informing. At worst, the accounts reflected a situation which is aptly described as 'non-participation'. The dynamic observed appears to be a continuation and consolidation of the historic power asymmetry between the controlling coastlanders and the subordinated Amerindians in Guyana (Although, it should be noted, progress with community engagement in some practical monitoring aspects of REDD+ has been made in select locations within Guyana (e.g., [80]), this is of little value if not replicated across the country.).

Our findings should raise sincere concerns not only for the implementing bodies in Guyana, but also for the project funders in Norway. Denying Amerindian participation has consequences for the LCDS and REDD+ projects, for the governments of Guyana and Norway and for the indigenous communities. Frustration and distrust amongst those in Chenapou towards the process, and thus the government, were apparent. Whilst it may be suggested that the existing APNU + AFC administration look to be moving away from the LCDS concept [81], our findings of a democratic deficit apparent in the LCDS remain pertinent when considering any future resource governance steps by the Guyanese government.

More significantly, the failure of effective participation represents a stark omission and suppression of indigenous rights. Measures such as FPIC are written into extensive national and international charters of which both Guyana and Norway are signatories. By continuing, in light of prior warnings, to fail to respect these fundamental statutes, the LCDS, Guyana and Norway jeopardise their potential role as a model to the international community. This failure transgresses the democratic rights of Guyanese Amerindian's, treating them as a sub-set of citizen and exacerbating an existing socio-ethnic power divide.

To begin to address this we suggest that, amongst other steps, engaging in genuine two-way dialogue with Amerindian communities become a priority before further progression of LCDS activities. The dialogue must become more accessible, not relying on the internet or single community representatives to make the process cost effective as these have proved ineffective to date. It is necessary that the MSSC become significantly more inclusive, adopt a more transparent selection method, meet more often and be listened to, but it should not be the only platform for broad, participatory discussion and negotiation. Outreach programs are essential and these must alter from one-way channels of information to become opportunities for civil society to shape, mould and participate in the political process. The deficit of knowledge amongst Amerindians and growing frustrations are a clear indication of the need for change, and progress should be re-oriented to consider this a primary objective.

The Government of Norway and the operative institutions also have an important role and clear responsibilities. The lack of planning and contextual knowledge prior to establishing the Norway-Guyana agreement has had clear and damaging implications. Furthermore, the absence of in country presence has meant that oversight is minimal. Going forward, greater assurances need be made that indigenous communities are functionally involved within the LCDS process and that the transparency and multi-stakeholder engagement, which was applauded in the first articulation of the LCDS [43], be instated. The acknowledgement and conception of broad engagement in the LCDS existed on paper, but the practical application has not followed. Further research into safeguards and the democratic or participatory qualities of environmental policies, such as the LCDS and REDD+ in practice, is also required.

The failure to engage local actors and to respect indigenous rights in the case of Chenapou is significant. At best, it represents a negligence which is deleterious to both the Amerindian community and the facilitators of the LCDS mechanism – the Governments of Guyana and Norway. It produces frictions which need not exist and reifies power asymmetries which marginalise communities. At worst, it is a knowing transgression of fundamental rights which, if unchecked, could be corrosive to any future success of the LCDS and REDD+ models that follow it.

Acknowledgments: The authors are grateful to the community of Chenapou who are central to, and were central in, this research-*Tenki*. We are also grateful to all those who reviewed, supported and gave advice in the process of producing this paper, particularly the team at WWF-Guianas during the research phase and Lund University & Lund University Centre for Sustainability Studies for funding the open access publishing.

Author Contributions: Sam Airey designed and conducted the research in Guyana and analysed the data. This article was written in collaboration with Torsten Krause.

Conflicts of Interest: The authors declare no conflict of interest.

Appendix A

Table A1. Chenapou Village Guest book entries from 1998–2015 (no data available for 2010 & 2011).

Reason for Visit	1998	1999	2000	2001	2002	2003	2004	2005	2006	2007	2008	2009	2012	2013	2014	2015	Total
Tourists	17	52	20	7	17				2			20	11	2	2	2	148
Mining GGMC *	2	3		4		10			1			4	7				35
Researchers	11	1							3	1		5	4		8	3	36
Churches	1				1									5	4		11
Educational	4	2	12									3			4		25
Health	1	13		4	2			8		5	4	2		2	2		39
KNP (PAC) †	6	4				6	5		13	3		16	1	1		1	60
Central Government		6	2		4	6	2				6			2		6	34
WWF ‡													5	4	16	6	31
Activists		2						2									2
Police		1			1	2			1								7
Regional Govt.		1		1									1	9	7	8	27
Amalia Falls													1	1			2
Tourism workshop															3		3
Internet																1	1
Labour/business															4	2	6
Solar Installation							6										6
Totals	42	85	34	16	25	24	13	10	20	9	10	50	30	26	50	29	

* Guyana Geology and Mines Commission; † Kaieteur National Park—Protected Areas Commission; ‡ World Wide Fund for Nature.

References

1. Walker, W.; Baccini, A.; Schwartzman, S.; Ríos, S.; Oliveira-Miranda, M.A.; Augusto, C.; Ruiz, M.R.; Arrasco, C.S.; Ricardo, B.; Smith, R.; et al. Forest carbon in Amazonia: The unrecognized contribution of indigenous territories and protected natural areas. *Carbon Manag.* **2014**, *5*, 479–485. [CrossRef]

2. Bluffstone, R.; Robinson, E.; Guthiga, P. REDD+ and community-controlled forests in low-income countries: Any hope for a linkage? *Ecol. Econ.* **2013**, *87*, 43–52. [CrossRef]

3. Schroeder, H. Agency in international climate negotiations: The case of indigenous peoples and avoided deforestation. *Int. Environ. Agreem. Polit. Law Econ.* **2010**, *10*, 317–332. [CrossRef]

4. Sikor, T.; Stahl, J.; Enters, T.; Ribot, J.C.; Singh, N.M.; Sunderlin, W.D.; Wollenberg, E. REDD-plus, forest people's rights and nested climate governance. *Glob. Environ. Chang.* **2010**, *20*, 423–425. [CrossRef]

5. Ricketts, T.H.; Soares-Filho, B.; da Fonseca, G.A.B.; Nepstad, D.; Pfaff, A.; Petsonk, A.; Anderson, A.; Boucher, D.; Cattaneo, A.; Conte, M.; Creighton, K.; et al. Indigenous Lands, Protected Areas, and Slowing Climate Change. *PLoS Biol.* **2010**, *8*, e1000331. [CrossRef] [PubMed]

6. Van Dam, C. Indigenous territories and REDD in Latin America: Opportunity or threat? *Forests* **2011**, *2*, 394–414. [CrossRef]

7. United Nations Framework Convention on Climate Change. *Adoption of the Paris Agreement*; United Nations Framework Convention on Climate Change: New York, NY, USA, 2015; p. 32.

8. Thompson, M.C.; Baruah, M.; Carr, E.R. Seeing REDD+ as a project of environmental governance. *Environ. Sci. Policy* **2011**, *14*, 100–110. [CrossRef]

9. Larson, A.M.; Petkova, E. An introduction to forest governance, people and REDD+ in Latin America: Obstacles and opportunities. *Forests* **2011**, *2*, 86–111. [CrossRef]

10. Lawlor, K.; Madeira, E.M.; Blockhus, J.; Ganz, D.J. Community participation and benefits in REDD+: A review of initial outcomes and lessons. *Forests* **2013**, *4*, 296–318. [CrossRef]

11. Kronenberg, J.; Orligóra-Sankowska, E.; Czembrowski, P. REDD+ and Institutions. *Sustainability* **2015**, *7*, 10250–10263. [CrossRef]

12. McDermott, C.L.; Coad, L.; Helfgott, A.; Schroeder, H. Operationalizing social safeguards in REDD+: Actors, interests and ideas. *Environ. Sci. Policy* **2012**, *21*, 63–72. [CrossRef]

13. Visseren-Hamakers, I.J.; McDermott, C.; Vijge, M.J.; Cashore, B. Trade-offs, co-benefits and safeguards: Current debates on the breadth of REDD+. *Curr. Opin. Environ. Sustain.* **2012**, *4*, 646–653. [CrossRef]

14. Krause, T.; Collen, W.; Nicholas, K.A. Evaluating Safeguards in a Conservation Incentive Program: Participation, Consent, and Benefit Sharing in Indigenous Communities of the Ecuadorian Amazon. *Ecol. Soc.* **2013**, *18*, 1. [CrossRef]

15. Krause, T.; Nielsen, T.D. The legitimacy of incentive-based conservation and a critical account of social safeguards. *Environ. Sci. Policy* **2014**, *41*, 44–51. [CrossRef]

16. Caplow, S.; Jagger, P.; Lawlor, K.; Sills, E. Evaluating land use and livelihood impacts of early forest carbon projects: Lessons for learning about REDD+. *Environ. Sci. Policy* **2011**, *14*, 152–167. [CrossRef]

17. Jagger, P.; Lawlor, K.; Brockhaus, M.; Gebara, M.F.; Sonwa, D.J.; Resosudarmo, I.A.P. *REDD+ Safeguards in National Policy Discourse and Pilot Projects*; Center for International Forestry Research (CIFOR): Bogor, Indonesia, 2012; pp. 301–316.

18. Cancun Agreements. Available online: http://unfccc.int/meetings/cancun_nov_2010/items/6005.php (accessed on 23 October 2016).

19. Cooke, B.; Kothari, U. *Participation: The New Tyranny?* Zed Books: London, UK, 2001.

20. Sunderland, T.; Powell, B.; Ickowitz, A.; Foli, S.; Pinedo-Vasquez, M.; Nasi, R.; Padoch, C. *Food Security and Nutrition: The Role of Forests*; Center for International Forestry Research (CIFOR): Bogor, Indonesia, 2013.

21. Zenteno, M.; Zuidema, P.A.; de Jong, W.; Boot, R.G. Livelihood strategies and forest dependence: New insights from Bolivian forest communities. *For. Policy Econ.* **2013**, *26*, 12–21. [CrossRef]

22. Arnstein, S.R. A Ladder of Citizen Participation. *J. Am. Inst. Plan.* **1969**, *35*, 216–224. [CrossRef]

23. White, S.C. Depoliticising development: The uses and abuses of participation. *Dev. Pract.* **1996**, *6*, 6–15. [CrossRef]

24. Turnhout, E.; van Bommel, S.; Aarts, N. How participation creates citizens: Participatory governance as performative practice. *Ecol. Soc.* **2010**, *15*, 26. [CrossRef]

25. Fischer, F. Participatory Governance as Deliberative Empowerment: The Cultural Politics of Discursive Space. *Am. Rev. Public Adm.* **2006**, *36*, 19–40. [CrossRef]
26. Lukes, S. *Power: A Radical View*, 2nd ed.; Palgrave Macmillan: Houndmills, UK; New York, NY, USA, 2004.
27. Haugaard, M. Power: A 'family resemblance'concept. *Eur. J. Cult. Stud.* **2010**, *13*, 419–438. [CrossRef]
28. Eyben, R.; Harris, C.; Pettit, J. Introduction: Exploring power for change. *IDS Bull.* **2006**, *37*, 1–10. [CrossRef]
29. Dooley, K.; Griffiths, T. *Indigenous Peoples' Rights, Forests and Climate Policies in Guyana: A Special Report*; Amerindian Peoples Association: Bourda, Guyana; Forest Peoples Programme: Moreton-in-Marsh, UK, 2014.
30. Bulkan, J. *Hegemony in Guyana: Redd-Plus and State Control over Indigenous Peoples and Resources*; SSRN Scholarly Paper ID 2883671; Social Science Research Network: Rochester, NY, USA, 2016.
31. Office of the President, Republic of Guyana. *Creating Incentives to Avoid Deforestation*; Office of the President: Bourda, Guyana, 2008.
32. Gutman, P.; Aguilar-Amuchastegui, N. *Reference Levels and Payments for REDD+: Lessons from the Recent Guyana-Norway Agreement*; World Wild Fund: Gland, Switzerland, 2012; p. 16.
33. EU Parliament. *Synopsis of Feature Address made by his Excellency Bharrat Jagdeo, President of the Republic of Guyana at the Opening Ceremony of the Third Regional Meeting of the ACP-EU Joint Parliamentary Assembly (Caribbean), on February 25, 2009*; The Guyana International Conference Centre: Liliendaal, Guyana, 2009.
34. Office of the President. *Transforming Guyana's Economy While Combating Climate Change*; Office of the President: Georgetown, Guyana, 2010.
35. Bulkan, J. REDD letter days: Entrenching political racialization and State patronage through the Norway-Guyana REDD-plus agreement. *Soc. Econ. Stud.* **2014**, *63*, 4.
36. UN-REDD Programme. UN-REDD Programme. Available online: http://www.un-redd.org/ (accessed on 23 October 2016).
37. Angelsen, A.; Brockhaus, M.; Sunderlin, W.D.; Verchot, L.V. *Analysing. REDD+: Challenges and Choices*; Center for International Forestry Research: Bogor, Indonesia, 2012.
38. LTS International, Ecometrica, Indufor Oy, and Chr. Michelsen Institute. *Real-Time Evaluation of Norway's International Climate and Forest Initiative*; Synthesising Report 2007–2013; Norad: Oslo, Norway, 2014.
39. Guyana Forestry Commission. *Guyana REDD+ Monitoring Reporting and Verification System—Year 5 Summary Report*; Guyana Forestry Commission: Georgetown, Guyana, 2015; p. 27.
40. Bulkan, J. Forest Grabbing Through Forest Concession Practices: The Case of Guyana. *J. Sustain. For.* **2014**, *33*, 407–434. [CrossRef]
41. Government of Guyana. *Memorandum of Understanding between the Government of the Cooperative Republic of Guyana and the Government of the Kingdom of Norway Regarding Cooperation on Issues Related to the Fight against Climate Change, the Protection of Biodiversity and the Enhancement of Sustainable Development*; Government of Guyana: Georgetown, Guyana, 2009.
42. Bureau of Statistics. *2012 Census—Compendium 2: Population Composition*; Bureau of Statistics: Georgetown, Guyana, 2016.
43. Dow, J.; Radzik, V.; Macqueen, D. *Review of Guyana LCDS Consultation Process*; International Institute for Environment & Development: London, UK, 2009.
44. Office of the President. *Low Carbon Development Strategy: Transforming Guyana's Economy While Combating Climate Change*; Office of the President: Georgetown, Guyana, 2013.
45. Donovan, R.; Clarke, G.; Sloth, C. *Verification of Progress Related to Enabling Activities for the Guyana-Norway REDD+ Agreement*; Rainforest Alliance: Richmond, VT, USA, 2010; p. 40.
46. Donovan, R.Z.; Moore, K.; Stern, M. *Verification of Progress Related to Indicators for the Guyana-Norway REDD+ Agreement*; 2nd Verification Audit Covering the Period 1 October 2010–30 June 2012; Rainforest Alliance: Richmond, VT, USA, 2012.
47. Office of the President. *Amerindian Act*; Office of the President: Georgetown, Guyana, 2006.
48. Amerindian Land Titling. UNDP in Guyana. Available online: http://www.gy.undp.org/content/guyana/en/home/operations/projects/environment_and_energy/amerindian-land-titling.html (accessed on 23 October 2016).
49. Office of the President. *Joint Concept Note*; Office of the President: Georgetown, Guyana, 2012.
50. Office of the President, Republic of Guyana. Meeting 76 Multi-Stakeholder Steering Committee (MSSC). Available online: http://www.lcds.gov.gy/index.php/documents/minutes-of-mssc-and-briefing-sessions/158-meeting-76-multi-stakeholder-steering-committee-mssc (accessed on 22 December 2016).

51. United Nations Development Programme. *United Nations Development Programme Country: Guyana Project Document*; United Nations Development Programme: New York, NY, USA, 2013.

52. Project Management Office. *Update on Guyana REDD+ Investment Fund (GRIF) Projects*; Project Management Office: Georgetown, Guyana, 2016.

53. Office of Climate Change. *Low-Carbon Development Strategy Draft for Discussion Opt-In Mechanism Strategy*; Office of Climate Change: Georgetown, Guyana, 2014.

54. Foti, J.; de Silva, L.; World Resources Institute. *Voice and Choice: Opening the Door to Environmental Democracy*; World Resources Institute: Washington, DC, USA, 2008.

55. Stone, S.; León, M.C. *Climate Change & the Role of Forests—A Community Manual*; REDD+: Georgetown, Guyana, 2010.

56. MacKay, F. *Workshop on 'Indigenous Peoples, Forests and the World Bank: Policies and Practices' Held in Washington, DC, 9–10 May 2000*; Forest Peoples Programme: Moreton-in-Marsh, UK, 2000.

57. Colchester, M.; la Rose, J.; James, K. *Mining and Amerindians in Guyana*; Amerindian People's Association: Bourda, Guyana, 2002.

58. Davis, O.; Ragnauth, P.; Watkins, W.; Welch, V.; Drakes, O. *Kaieteur National Park Management Planning Process 2nd Community Consultation with Chenapau Village*; National Parks Commission (NPC): Georgetown, Guyana, 2009.

59. Parliament of the Republic of Guyana. The Kaieteur National Park (Amendment) Act 2000 | Parliament of Guyana. Available online: http://parliament.gov.gy/publications/acts-of-parliament/the-kaieteur-national-park-amendment-act-2000/ (accessed on 21 December 2016).

60. O'Brien, K. Responding to environmental change: A new age for human geography? *Prog. Hum. Geogr.* **2010**, *35*, 542–549. [CrossRef]

61. Flick, U. *An Introduction to Qualitative Research*; SAGE Publications Inc.: Thousands Oaks, CA, USA, 2009.

62. Creswell, J.W. *Qualitative Inquiry& Research Design*; SAGE Publications Inc.: Thousands Oaks, CA, USA, 2007.

63. Gubrium, J.F.; Holstein, J.A. *Handbook of Interview Research: Context and Method*; SAGE Publications, Inc.: Thousands Oaks, CA, USA, 2002.

64. Fabinyi, M.; Evans, L.; Foale, S.J. Social-ecological systems, social diversity, and power: Insights from anthropology and political ecology. *Ecol. Soc.* **2014**, *19*, 28. [CrossRef]

65. Utting, P. Social and political dimensions of environmental protection in Central America. *Dev. Chang.* **1994**, *25*, 231–259. [CrossRef]

66. Boodram, R. *Another Young Woman Jumps off Kaieteur Falls*; Kaieteur News: Georgetown, Guyana, 2015.

67. Bulkan, J. The Limitations of International Auditing: The Case of the Norway-Guyana REDD+ Agreement. In *The Carbon Fix: Forest Carbon, Social Justice and Environmental Governance*; Routledge: London, UK, 2016; p. 18.

68. Atkinson, S.; Wilson, D.; da Silva, A.; Benjamin, P.; Peters, C.; Williams, I.; Alfred, R.; Thomas, D. *Our Land, Our Life: A Participatory Assessment of the Land Tenure Situation of Indigenous Peoples in GUYANA*; Amerindian Peoples Association/Forest Peoples Programme: Demerara-Mahaica, Guyana, 2016.

69. Writer, S. *Gov't Seeking Final Opt-in Mechanism under Norway Forests Deal*; Stabroek News: Georgetown, Guyana, 2016.

70. Bulkan, J. The Struggle for Recognition of the Indigenous Voice: Amerindians in Guyanese Politics. *Round Table* **2013**, *102*, 367–380. [CrossRef]

71. Sanders, A. *The Powerless People: An Analysis of the Amerindians of the Corentyne River*; Macmillan Caribbean: London, UK, 1987.

72. Fujisaki, T.; Hyakumura, K.; Scheyvens, H.; Cadman, T. Does REDD+ Ensure Sectoral Coordination and Stakeholder Participation? A Comparative Analysis of REDD+ National Governance Structures in Countries of Asia-Pacific Region. *Forests* **2016**, *7*, 195. [CrossRef]

73. Reed, P. REDD+ and the indigenous question: A case study from Ecuador. *Forests* **2011**, *2*, 525–549. [CrossRef]

74. Twyman, C. Participatory conservation? Community-based natural resource management in Botswana. *Geogr. J.* **2000**, *166*, 323–335. [CrossRef]

75. Bäckstrand, K. Democratizing global environmental governance? Stakeholder democracy after the world summit on sustainable development. *Eur. J. Int. Relat.* **2006**, *12*, 467–498. [CrossRef]

76. Amerindian Community Slams LCDS Consultation. *Kaieteur News*; Georgetown, Guyana, 10 March 2010. Available online: http://www.kaieteurnewsonline.com/2010/03/10/amerindian-community-slams-lcds-consultation/ (accessed on 21 December 2016).

77. Bade, H. Aid in a Rush. A case study of the Norway-Guyana REDD+ partnership. *Foreign Policy Anal.* **2007**, *4*, 59.

78. Hammond, D.S.; Gond, V.; de Thoisy, B.; Forget, P.-M.; de Dijn, B.P.E. Causes and Consequences of a Tropical Forest Gold Rush in the Guiana Shield, South America. *AMBIO J. Hum. Environ.* **2007**, *36*, 661–670. [CrossRef]

79. Mahanty, S.; McDermott, C.L. How does 'Free, Prior and Informed Consent' (FPIC) impact social equity? Lessons from mining and forestry and their implications for REDD+. *Land Use Policy* **2013**, *35*, 406–416. [CrossRef]

80. Bellfield, H.; Sabogal, D.; Goodman, L.; Leggett, M. Case study report: Community-based monitoring systems for REDD+ in Guyana. *Forests* **2015**, *6*, 133–156. [CrossRef]

81. Eleazar, G. Jagdeo's 'limited' Low Carbon Strategy being expanded. *Demerara. Waves*, 20 July 2016. Available online: http://demerarawaves.com/2016/07/20/jagdeos-limited-low-carbon-strategy-being-expanded/ (accessed on 21 December 2016).

forests

MDPI

Article

Revitalizing REDD+ Policy Processes in Vietnam: The Roles of State and Non-State Actors

Thu Ba Huynh * and Rodney J. Keenan

School of Ecosystem and Forest Sciences, University of Melbourne, Parkville 3010 VIC, Australia;
rkeenan@unimelb.edu.au
* Correspondence: huynht@unimelb.edu.au; Tel.: +61-3-9035-3529

Academic Editors: Esteve Corbera and Heike Schroeder
Received: 31 October 2016; Accepted: 14 February 2017; Published: 24 February 2017

Abstract: Vietnam was one of the first countries to introduce the National REDD+ (Reduced Emissions from Deforestation and Forest Degradation) Action Program in 2012. The country has recently revised the Program to aim for a more inclusive 2016–2020 strategy and a vision to 2030. This study explores how Vietnam policy actors view REDD+ policy development and their influence in these processes. The results can contribute to the discussion on how policy actors can effectively influence policy processes in the evolving context of REDD+ and in the types of political arrangements represented in Vietnam. We examined the influence of state and non-state actors on the 2012 National REDD+ Action Program (NRAP) processes, and explored factors that may have shaped this influence, using a combination of document analysis and semi-structured interviews with 81 policy actors. It was found that non-state actors in REDD+ are still on the periphery of decision making, occupying "safe" positions, and have not taken either full advantage of their capacities, or of recent significant changes in the contemporary policy environment, to exert stronger influence on policy. We suggest that REDD+ policy processes in Vietnam need to be revitalized with key actors engaging collectively to promote the possibilities of REDD+ within a broader view of social change that reaches beyond the forestry sector.

Keywords: Vietnam; REDD+ policy; REDD+ governance; non-state actors in REDD+

1. Introduction

Reduced emissions from deforestation and forest degradation (REDD) aims to mitigate climate change through the application of conditional incentives for protection and enhancement of the carbon sequestration functions of forests. Adding a plus to become "REDD+", the concept goes beyond deforestation and forest degradation, and includes the role of conservation, sustainable management of forests, and enhancement of forest carbon stocks. Since its inception, REDD+ has been evolving and progressing while serving as a broad platform for a wide range of actors to pursue their own ideas and goals [1]. Currently, there are more than 300 REDD+ initiatives taking place in 47 countries [2]. However, REDD+ implementation at the national level has been slower than expected and political economic factors (i.e., institutions; interests; ideas and information [3]) are perhaps the biggest barriers to implementation [4]. A recent review of multi-national REDD+ studies suggested that REDD+ should further promote and support transformational change [5]. The Paris Agreement in 2015 was considered a major step forward in providing specific reference to the need to invest in efforts to reduce deforestation, sustainably manage forests, and enhance forest carbon stocks. The Agreement is expected to send strong signals to different actors across multiple landscapes and positively stimulate new policies and the provision of finance for REDD+ initiatives and sustainable forest management around the globe [6].

According to Ostrom [7], centralised governance is unlikely to effectively tackle the challenges of climate change, with polycentric forms of governance instead more suited to facilitating the experimental efforts required at multiple levels to successfully address many climate change issues. In the context of REDD+, it was found that while polycentric governance may offer benefits for learning, it has not proven valuable for enabling REDD+ implementation on the ground, in the absence of a binding international agreement [8]. Since REDD+ governance involves a range of actors at different levels in specific and sometimes unique political structures, multilevel governance in REDD+ has been identified as a key challenge [9]. Earlier research suggested that REDD+ progress can be realized if REDD+ policies are consistent with good forest governance [10]. Specifically, high levels of policy inclusiveness and ownership are key elements in ensuring effective and equitable REDD+ policy design and success [11]. It was noted that national circumstances are key to progress in REDD+ readiness (i.e., developing effective national policies, capacity building, and implementing subnational projects). Despite this, there has been little focus on assessment of REDD+ performance in most countries in order to suggest improvements in policies or practice [12].

Since 2009, Vietnam has been at the forefront of REDD+ readiness, having received support from both FCPF (Forest Carbon Partnership Facility of the World Bank) and the joint UN-REDD program (UNDP, United Nations Development Programme; UNEP, United Nations Environment Programme; and FAO, Food and Agriculture Organization of the United Nations), with a total financial commitment of over 84 million USD [13]. Vietnam was one of the first countries to introduce a National REDD+ Action Program (NRAP) in 2012. By 2016, ten Provincial REDD+ Action Plans (PRAP) and 35 Site-based REDD+ Implementation Plans (SiRAP) had been developed and approved. These processes put an emphasis on multi-stakeholder engagement, including with men and women from forest dependent communities [14]. In July 2016, the Participatory Self-Assessment of the REDD+ Readiness Package in Vietnam found that *"The approved NRAP fails to meet all expectations and requirements set out"* [15]. A number of challenges were identified in this report, including (i) weak inter-sectoral coordination and coordination with the private sector and civil society organizations, (ii) ineffective consultation and communications with ethnic minorities and vulnerable groups, and (iii) an incomplete REDD+ safeguard system. In terms of the sub-national REDD+ planning, PRAPs and SiRAPs have been formulated in the absence of a detailed NRAP, representing a challenge for implementation [14,16]. In November 2016, the country submitted the revised NRAP for the final approval from the Prime Minister. The revised version aims for a more inclusive approach and to provide more guidance on strategy and implementation during 2016–2020 period and a vision to 2030.

Vietnam is considered to have an authoritarian governance and political structure [17]. Due to its limited accessibility, London [17] claimed that *"Vietnam's politics are not widely understood"* and that they are complex and rapidly changing. However, this does not mean that it is not inclusive, and it has been argued that these "internally-inclusive" forms of participation in policy processes such as REDD+ are required for the resolution of the underlying conflicts and tensions amongst stakeholders that are evident in Vietnam [1]. If not managed carefully, the demands of donors and international non-governmental organizations (NGOs) (focused on a discourse on participation, benefit sharing, and tenure security) could upset this delicate internal balance and potentially undermine the development of Vietnam's national strategies for REDD+ and their implementation. Given the authoritarian structure, the Vietnamese government plays a dominating role in REDD+ discourse and, while this also may show the country's commitment to REDD+ implementation, there is a real need to engage with other political and institutional challenges associated with inclusion [18]. In an assessment of REDD+ Readiness in four countries (including Vietnam), different interpretations of participation between civil society and government were evident [12], and a more recent comparative study [19] (in this issue) indicated that non-state actors in Vietnam have had opportunities to participate actively in shaping REDD+ policy. However, these opportunities have tended to be available to those who can effect change rather than those affected by REDD+ policies.

This paper aims to explore how Vietnam's policy actors view REDD+ policy development and their influence in these processes. We were interested in understanding how the history of decision making in a "command and control" form of governance in Vietnam may affect the dynamics and approaches of different actors. We focused on the NRAP formulation process (2010–2012) at the national level policy domain and on Lam Dong Province as the only pilot province. Within this paper, REDD+ actors will refer to those who are involved in REDD+ policy formulation. It excludes indigenous people and local communities since most REDD+ policy development activities took place at the national level during this period (except for the pilot Free Prior Informed Consent exercise in Lam Dong Province). The study endeavours to develop new insights that could guide REDD+ actors in shaping climate change governance. The paper addresses three key research questions:

1. Who are the actors involved in REDD+ policy and what is their level of influence in the NRAP processes?
2. What are the factors shaping non-state actors' influence in REDD+?
3. What mechanisms and strategies might lead to better outcomes for REDD+ policy?

The paper will begin by providing framing of governance and policy-making. This is followed by a brief introduction to the history and development of civil society organizations in Vietnam and a description of the methods. Results are discussed in the context of concepts of collective action and social learning. The conclusion suggests a revitalization of REDD+ policy processes through changes to the interactions, tactics, and rhetoric of state and non-state actors in Vietnam.

1.1. Governance and Policy-Making

The term governance emerged during the 70s–80s, describing a broader concept that goes beyond actions of governments and states alone [20]. Notably, Foucault made a clear case that "good government" concerns more than just government by the state [21]. As reviewed by Knieling [22], governance therefore has an ambiguous nature, demonstrating the murky boundaries between state and society and its "complex reality". This paper uses Agrawal's definition of environmental governance as a starting point for analysis, in which political actors influence environmental outcomes within a set of regulatory processes, mechanisms, and an architecture of institutions and organizations [23]. Policy outcomes are framed as influence and changes in ideas, understandings, approaches, behaviors, strategies, policies, and legislations [24]. During the 1990s, in the context of globalization, there was considerable discussion about non-state governance, including the role of corporations and the market [25]. It is now widely recognized that governance systems involve the co-existence of, and interactions between, state and non-state actors, and that non-state governance is not independent from the state. Conventional approaches to understanding political outcomes have been challenged, with a new focus on the strong involvement and alternate approaches taken by non-governmental actors [20]. These forms of diversity of perspectives and strategies to handle controversies in policy-making are central in theories on collaborative governance [26] and networked governance [27]. Folke et al. [28] discussed adaptive governance systems in social-ecological systems and highlighted the importance of the processes of participation, collective action, and social learning. Reed et al. [29] defined social learning as a process that must: *"(1) demonstrate that a change in understanding has taken place in the individuals involved; (2) demonstrate that this change goes beyond the individual and becomes situated within wider social units or communities of practice; and (3) occur through social interactions and processes between actors within a social network"*. These key elements of social learning enable development of policies, which could be built upon various knowledge systems and experiences of teams and actor groups.

Policy-making can be seen as a complex and difficult process in which participation represents a powerful theme [30]. Arnstein's ladder of participation distinguishes different levels of participation in relation with power in the form of an eight rung framework. Here, relationships with publics range from consultation, manipulation, to control [31]. Though Arnstein highlights the multitudinous forms of participation in policy-making, this framework tends to over-simplify the nature of involvement [32] and does not capture the diversity of forms of participation [33], or how participation might be progressed as a collective process among all stakeholders [34].

Since a policy arena is often populated with actors driven by diverse agendas, Mayer and Bass [35] assert that effective policy-making needs to engage with various actors and not just authorities and elites. It has also been observed that the focus of policy-making around particular policy themes may be altered with irregular policymaking patterns when participation of relevant supporting actors declines [36]. There has been great progress in participation in policy processes, with increasing links with actors that were "once, outsiders to policy-making" [30]. The involvement of different interests and actors is a central determinant for achieving consensus or resolving conflict within policy subsystems [37]. However, it could also cause opposition and delay and thus make centralized governance more attractive [38]. Literature on policy has been dominated by focus on "state-centric" analysis [39] but NGOs are now considered a vital part of policy landscape [40] and there is a need for more research on the practical role that NGOs can play in policy-making [41].

Roberts [24] defined policy as a *"set of principles and intentions used to guide decision making"*. However, decision making is also a part of policy-making and there is a dynamic relationship between the two. Decision making often takes place within complex systems and structures. For example, in democracies elected politicians are generally responsible for policy decisions, while policy proposals are typically prepared by civil servants.

Policy-making around REDD+ has encountered no exception to the strong involvement of NGOs. Driven by western liberal democratic principles of inclusiveness and representation in decision making, participation in REDD+ processes is receiving a high level of attention, and programs such as UN-REDD and FCPF have been tailored to enhance participation. Participation is considered necessary for sound REDD+ policies and good governance [42] and the argument has been that the more inclusive REDD+ policies processes are, the more room there should be for considering equity and reducing the risk of conflict among policy actors and stakeholders [11]. Researchers have emphasized the need to engage with REDD+ actors both vertically and horizontally [3,9]. Creating mechanisms and processes for participation of different stakeholders can be complex and resource intensive. On-the-ground experience suggests there is still a high level of variability in the extent and depth of participation in REDD+ processes and highlights the importance of taking national contexts and priorities into account in approaching REDD+ governance [12,43].

In this context, Vietnam is an interesting case. The country has a comprehensive set of laws and policies relevant to REDD+ implementation, including the Forest Protection and Development Law, the National Forest Development Strategy for 2006–2020; Decree 99 on Payment for Forest Environmental Services; the National Strategy on Climate Change; the National Green Growth Strategy; the National Strategy on Biodiversity Conservation toward 2020 and vision to 2030; and Decision 83 (Vietnam Ministry of Agriculture and Rural Development (MARD)) on Sustainable Forest Management and Forest Certification Scheme. In June 2016, Vietnam's Prime Minister announced some significant new policies regarding forest management, targeting timber harvesting in natural forests and forest conversion.

However, Vietnam's political decision-making process is often described as either consensus-based or simply confusing and inexplicable. Officials in Vietnam do not necessarily know how decisions are made and who made them [44]. Participation in political, social, and economic arenas is a constitutional right, enshrined in the Grassroots Democracy Ordinance in 2007, but the political regime remains "solidly authoritarian" [17], with one-party and a highly internalised and autocratic decision making model. In 2013, the country revised its Constitution, which did not take into account calls for reform emerging from both within and outside the Party [17]. Within the REDD+ arena, government representatives dominated the processes, which raised concerns around inclusiveness and representation [45]. A policy analysis in six REDD+ countries found that the dominant REDD+ policy actors in Vietnam do not challenge business-as-usual discourses and the minority policy groups tend to focus on environmental justice issues and ignore politico-economic drivers of deforestation and degradation [18]. In addition, the structure of REDD+ payments at the national level may create risks for centralization of forest governance and thus limited involvement of different stakeholders, particularly non-state actors and others at the local level [46]. In Vietnam, the REDD+ NRAP formulation process was initiated in 2010 with a consultancy commissioned by UN-REDD. The process was heavily criticised by the government (GOV), and its final product was ultimately considered a background document. The NRAP underwent further consultations and was finally approved by the GOV in 2012. Formulation of the National REDD+ Program was led by state actors, but offered some entry points to non-state actors to engage in determining the policy formulation process, its content and measures.

1.2. Civil Society Organisations (CSOs) in Vietnam

Globalization and the rise of democratic governments are two factors that have triggered the dramatic expansion in size, scope, and capacity of civil society over the past few decades. With the emergence of new civil society actors, the boundaries between government and non-government sectors have become blurred and there has been considerable experimentation with the nature and structures of these organizations. This expansion of activity and the variety of forms of civil society organisations (CSOs) has been accompanied by their growing influence in shaping global public policy [47]. A study by the Norwegian Agency for Development Cooperation, evaluating their REDD+ support to CSOs in four countries, found that international NGOs have been directly involved in developing national REDD+ strategies. For example, in the Democratic Republic of Congo, national NGO platforms were instrumental in national REDD+ strategy development [48].

It has been suggested that NGOs in Vietnam are "virtually non-existent" [49]. Others argue that the notion of civil society in Vietnam is based on Marxist-Leninist ideology [50,51] with western liberal concepts of society being of limited value in this context [52]. The country's complex history and regulatory environment mean that it is difficult to clearly define CSOs in Vietnam. A broader understanding of civil society is not yet fully part of mainstream political thinking. While the term "civil society organization" is not found in legal documents, the Vietnamese translation of NGO (*phi chinh phu*) indicates a risk of decreased government control and could potentially provoke suspicion from local authorities. Currently, it could be considered that there are four broad types of CSOs operating in Vietnam: mass organizations, professional associations, Vietnamese NGOs, and community-based organizations (CBOs) [53].

Since the first National Congress of the Communist Party in 1935, much consideration was given to the development of mass organizations [53]. However, CSOs that may have been active during the revolution against the French in 1945 were integrated into the state in 1954 [50]. Consequently, civil society in the context of the one party state system prior to the "Doi Moi" reforms in 1986 was weak and limited. In addition, public administration reform has led to an accountability system within state institutions with a limited role for, or engagement of, non-state actors [54]. Since the 1990s, with normalization of relations with the USA and the presence of international development agencies, CSOs have flourished. By the early 2000s, scholars observed extended discussions on CSOs in the media and less state repression. There has been an increase in the number and diversity of these organizations, with 364 associations registered at the national level, despite incomplete legal structures [53]. Reviewing contemporary literature over the past two decades on Vietnam civil society, particularly the critical areas of environmental governance and anti-China demonstrations, Bui [55] observed a certain level of endorsement and tolerance (of CSOs) by the party state *"to fill a gap in the governance network"*. Well-Dang [56] claimed that there lies a *"vibrant reality of civil society"*, which exerts significant political influence. It is, however, unclear if influence is achieved either because of, or despite the existing political system [57].

Despite recent changes, there are still limitations on the capacity of CSOs to engage in key areas of policy, and CSOs are still *"deeply entangled with each other and the state"* [50]. CSOs engage in some forms of advocacy, but within bounds set by state authorities and, according to the 2005–2006 "Civil Society Index" study (CSI), little effort has been channelled to policy advocacy [58]. CSOs are faced with challenges in penetrating policy processes, given the ideological hegemony exercised by the state [55].

Given the distinctive nature and history of Vietnamese CSOs, their entangled relationship with state and constrained engagement in policy advocacy, it is important to study the role that local and international CSOs can play to enhance policy processes and outcomes in the contemporary context of Vietnam.

2. Research Methods

This study used a combination of document analysis, semi-structured interviews, and participant observation. The interviews and surveys were conducted within the framework of the Module 1 (M1) and Module 2 (M2) of the Global Comparative Study (GCS) on REDD+ by the Centre for International Forestry Research (CIFOR). M1 has a focus on REDD+ policies and processes analysis. M2 observes and documents the implementation of REDD+ project activities and their impacts.

Analysis of policy documents is an important part of policy study, although it cannot give a complete picture of policy development [35]. The documents used for analysis included government policies, strategic plans, and evaluation reports (from both GOV and non-GOV sources) in the field of sustainable development, forestry, and climate change. The study adopted steps for document analysis by Mayers and Bass [35], investigating the political context where national REDD+ strategies developed and obtaining background information, prior to conducting interviews. The researchers were aware of potential issues from using official documents (i.e., credibility) to construct reality. However, these documents could also be interesting, precisely because of the biases they could reveal [59]. In-depth interviews were conducted with 81 actors, classified as state (including media, research institutes, representatives from provincial, district, and communal governments) and non-state agencies (UN agencies, international and domestic NGOs, and private sector). Seventy percent of the total respondents were state actors and almost half of them were directly involved in policy-making at national and subnational levels. The interview was structured around four sections with 13 questions. Each interview lasted between 45 and 60 min and in person (with the exception of a few conducted via phone calls and follow-up emails). In the first set of questions, the respondents were asked to describe their organizational interests and activities in REDD+. The second section aimed to explore the respondent's perception on key challenges and opportunities for REDD+ implementation in Vietnam. The third section asked the respondents to comment on the NRAP consultation processes. Finally, they were asked to provide policy assessments of REDD+ outcomes (i.e., policy impact including emissions/removals; livelihoods; biodiversity; administrative and technical capacity [45]) in terms of effectiveness, cost-efficiency, equity, and other co-benefits. All interviews were recorded and transcribed for analysis using the NVivo 10 software package (2012) to facilitate the process of coding and identifying data patterns and themes (i.e., challenges around REDD+ consultation processes), and generating theory. Data was also collected via empirical participatory observations in policy events including workshops, meetings, and technical working groups' dialogues over the course of four years (2010–2014). In many cases, the researcher was an overt full member of the event/activity where the status was known. In other cases, the researcher assumed the role of a participating observer.

3. Results

3.1. Who Are the Actors Involved in REDD+ Policy Processes?

The Vietnam REDD+ policy landscape is structured according to three levels of policymaking, coordination, and implementation (Figure 1). The National REDD+ Steering Committee is the ultimate policy-making body chaired by the Ministry of Agriculture and Rural Development (MARD) Minister and constituted by members from various ministries. Immediately under the REDD+ Committee is the Vietnam REDD+ office overseeing all REDD+ activities. In 2010, a MARD ministerial decision to establish the REDD+ Network effectively "invited" NGOs to participate in REDD+. In the NRAP in 2012, the language changed: "non-governmental organizations are requested to participate in activities relating to the Program" [60]. The Vietnam REDD+ network started with members from 12 organizations (four government and eight international agencies and NGOs) in 2009, and has expanded to include 200 individuals from 56 organizations. The implementation level is comprised of six sub technical working groups (STWGs) in the areas of governance, private sector engagement, Measurement, Reporting and Verification (MRV), local implementation, and REDD+ financing benefit distribution system and safeguards, which are chaired by both state and non-state agencies. The composition of STWGs is diverse with participation of new stakeholders, beyond the list provided in the MARD decision for STWGs establishment. According to UN-REDD, REDD+ is the first government programme in Vietnam that has involved civil society organizations and NGOs to such an extent in decision-making processes [61]. The STWGs were active during 2010–2013 and contributed to a number of key technical documents for REDD+ development in Vietnam. In 2014, STWGs were largely inactive due to the unclear objectives and outcomes, and lack of capacity to leverage policy impacts [62]. Recently, concerted efforts to revitalize the STWGs have been observed. Some working groups (i.e., governance) are more active than others (Personal communications with UN-REDD professional staff).

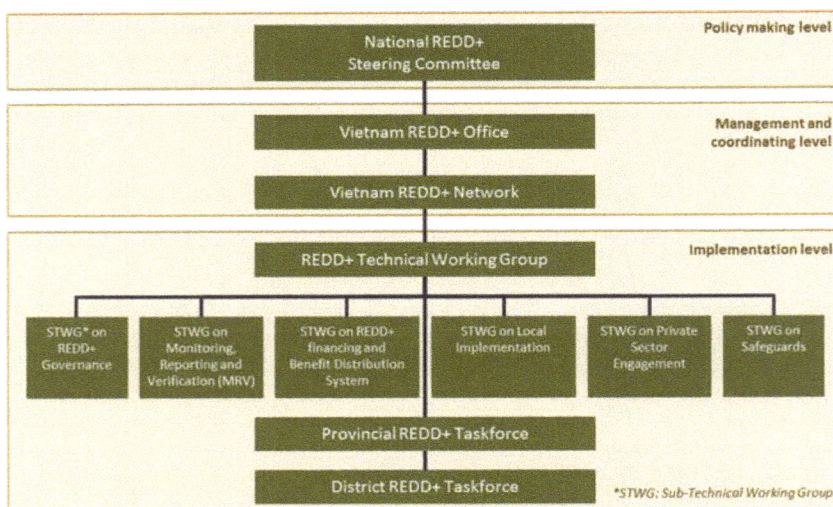

Figure 1. Vietnam REDD+ Institutional Arrangement (updated from the Vietnam REDD+ Website [63]). REDD+, Reduced Emissions from Deforestation and Forest Degradation; STWG, sub technical working group.

REDD+ actors were largely from forestry and development organizations. Almost 47% of non-state actors considered the main reason for their organizations' involvement in REDD+ was to align their organizational goals with REDD+ themes. Another 34% of the respondents considered their REDD+ involvement was initiated due to the possible linkages with their existing operational programs.

The respondents' organizational mandates and their REDD+ objectives were largely compatible and could be divided into three major categories (Figure 2): (i) policy development, (ii) capacity building, and (iii) project implementation (i.e., forestry and development). Approximately 50% of the interviewed non-state organizations were delivering projects/programs. Among the organizations with policy development mandate (i.e., 31% of the total), only a small number of actors mentioned work around improving legal framework and developing new policies. Overall, respondents did not clearly mention their roles in REDD+ policy development, except for one respondent from a Vietnamese NGO that used the term "*policy advocacy*". The phrase "*we support the government*" was the most frequent reference in the responses to this question. This support was being delivered in the forms of project design, demonstration of best practices, and thematic policy advice. Respondents from the international organizations used terms such as "*behind-the-scene assistance*" with a less assertive position in relation to influence "*we hope . . .* " and putting an emphasis on "*respecting what the GOV wants and supporting the GOV's position*".

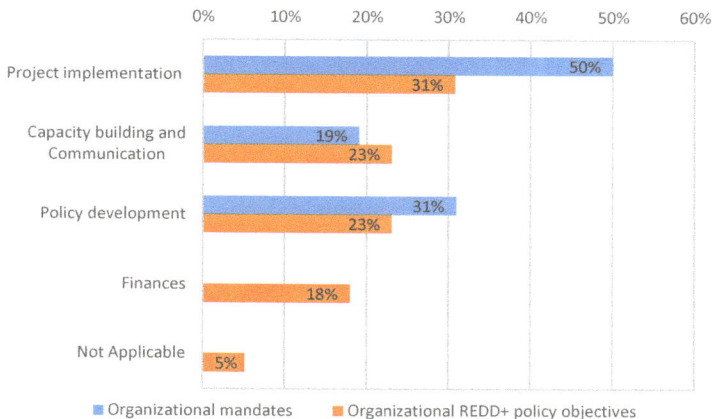

Figure 2. Organizational mandates and REDD+ objectives of non-state organizations.

In summary, REDD+ actors were mainly from the forestry sector or development agencies with a stronger focus on service delivery and implementation than policy development or advocacy. Their involvement in REDD+ activities was driven by the commonality between REDD+ and their existing mandates. Many were seeking opportunities to advance their organizational mandates through influencing the design and implementation of REDD+, and generally framed their organizational objectives for REDD+ around giving support to the government.

3.2. How Do Actors View Their Influence in the NRAP Processes?

REDD+ actors' views on the NRAP consultation process were grouped under two categories, negative and positive (Table 1). Views from state and non-state actors were fairly consistent. There were slightly more people in both groups who felt negatively about this process. Since data collection was carried out during the early phase of NRAP formulation, there were some respondents who were not aware of this process (i.e., 22% of state and 17% of non-state actors). A group of respondents did not clearly express a view, but provided general comments towards the process.

Respondents with positive views considered the REDD+ Network as an effective consultation forum and acknowledged the openness of the whole process. Within the negative category, the majority of references (i.e., 76%) remarked on the limited participation, top-down approach, and ineffective mechanisms, indicating the absence of clear goals and poor communication. Key stakeholders from the international REDD+ community were dissatisfied with the limited knowledge about how this formulation and approval processes evolved. Respondents from NGOs were frustrated with the ways UN-REDD "*dominated the consultation process*", and the state's tokenistic effort to seek inputs, claiming that "*asking for comments on already written drafts is not sufficient*". There is also disagreement and discontent among the NGOs over who is more connected and has influence. One respondent expressed: "*Those big NGOs that work closely with the GOV know what the GOV wants and together they determine the themes*".

Table 1. REDD+ actors' views on National REDD+ Action Program (NRAP) consultation processes.

Category	Expressed Themes	State Actors ($n = 56$) (%)	Non- State Actors ($n = 25$) (%)
Negative views	Limited participation and top down approach Ineffective mechanisms Lack of coordination and inadequate information sharing Absence of government (GOV) leadership	32	33
Positive views	Awareness raising and interesting debates via numerous workshops Open to various stakeholders REDD+ Network as an effective mechanism for consultation	30	30
Respondents do not know	Not aware of the process	22	17
Other comments	Neutrality is necessary A better mechanism is required Roles of REDD+ network, civil society organisations (CSOs,) and non-governmental organizations (NGOs) need to be strengthened and empowered Clear goals, specific agendas, and creative approach.	16	20

While the NRAP consultation process was subject to a range of criticisms, more than 50% of the respondents in both state and non-state sectors felt that the claims and positions of their organizations were considered seriously during the decision-making process (Figure 3). These respondents appreciate the GOV's *"open-mindedness"* and the fact that *"the GOV listens"*. Many among the 20% of the non-state actors and 16% state actors who did not think that their organizations had any influence on REDD+ process cited the absence of effective mechanisms and channels to approach/influence the GOV. Some respondents seriously doubted that their voices were heard, pointing to the GOV's limited capacity and resources to process their recommendations. There was a number of people (i.e., 24% of non-state and 31% of state actors), who did not articulate a view on the influence. It is interesting that there were more state than non-state actors within this category. These respondents may be (i) those who were involved in the process but unsure about the outcomes, (ii) those who were not directly involved in the process, and hence unable to give concrete answers, and (iii) those who did not want to comment.

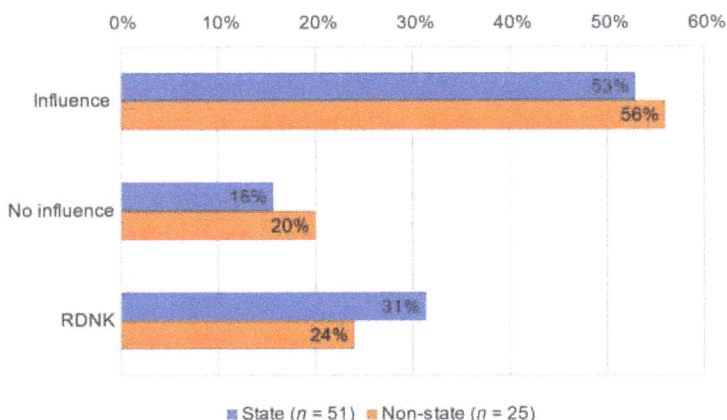

Figure 3. Organizational influence on REDD+ decision making process.

In short, the interviewees indicated that there is a lot of room for improvement around the NRAP formulation process. Both state and non-state actors pointed to the absence of a clear goal and effective mechanisms for consultations, which ultimately led to limited participation. However, a level of increasing openness in REDD+ policy-making was also observed.

4. Analysis and Discussion

4.1. What Are Factors Shaping Non-State Actors' Influence in REDD+ Policy Processes?

From the analysis in this study, there were two factors that hamper Vietnamese NGOs (VNGOs) effectively influencing REDD+ policy processes: (i) organizational inertia resulting from the long history of entanglement with the state and (ii) limited practical experience in policy work and constrained engagement with policy processes.

Driving and restraining forces that may have shaped REDD+ actors' influence in policy processes were analysed in the context of four factors—namely: (i) strong REDD+ momentum at the global level, which has resulted in the Prime Minister's decision to develop a REDD+ National Program in 2010; (ii) available resources (finances and time) allowed for NRAP formulation; (iii) a shift towards more open policy-making within REDD+; and (iv) the contemporary change of government's views towards CSOs that endorses them as participants in a new form of governance (Figure 4).

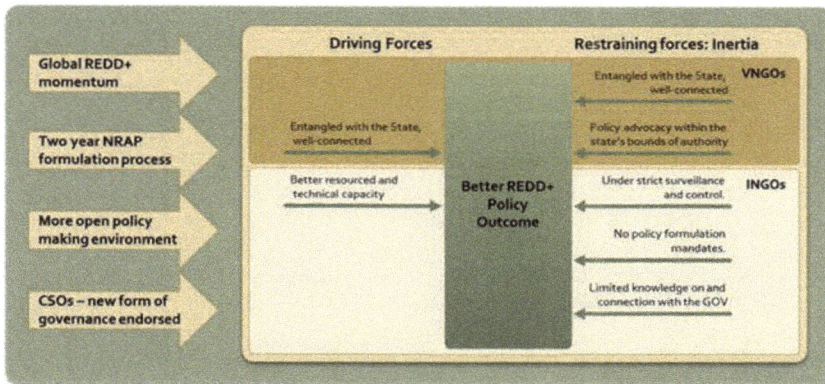

Figure 4. Force Field Analysis of Vietnamese NGOs (VNGOs) and international NGOs (INGOs) in Vietnam REDD+ policy arena.

The long history of "democratic centralism" under Marxist-Leninist philosophy has shaped Vietnamese decision-making culture. Members of the Communist Party are required to give full attention to effective implementation of decisions and there are potentially significant negative consequences if a person attempts to question and criticize the decisions made [64]. This focus on implementation was partly shaped by the pressure and expectations from REDD+ donors on Vietnam as one of the first UN-REDD and FCPF countries to deliver and set an example for REDD+ globally. Others have identified that consultations on development and implementation of REDD+ in Vietnam have often been inadequate due to pressures (i.e., time, donor's priorities, and cost) on the intermediaries carrying out the consultation [65].

Bui [55] considered that *"Vietnamese NGOs are not as well connected and organized as in many other countries"*. This is perhaps true when VNGO's operations are examined from the viewpoints of resource mobilization, partnership building, or connection with local constituencies. However, domestic NGOs in Vietnam are distinctively characterized by their entanglement with the state both on administrative and operational levels, so in effect, they are very well connected. This leads to a lack of autonomy that may be considered inconsistent with western theories of civil society, but which may actually give them more power and influence within Vietnamese decision making. A small number of newly-established local NGOs have used different channels (personal connections or relationships) to influence policy-making at a higher level, but their contribution is still ad hoc. Hannah [52] cited VNGOs' access to *"political knowledge elite"* with intimate familiarity with the rules, procedures, and relationships with GOV staff as enabling stronger participation in policy. This type of *"mutual colonization"* (between state and VNGOs) is a successful tactic that VNGOs are employing to achieve social and political goals [52], but this process has also bred a level of organizational inertia because of their strong entanglement. For example, VNGOs were well-represented in and sometimes chaired the STWGs. However, this representation did not give them a strong weight in REDD+ decision making. In general, VNGOs are still considered by the government as service providers with little real agency, and thus less influence. For example, a respondent from a VNGO noted: "[the] *GOV has a narrow band of interest in what VNGOs have to say"*.

Despite the authoritarian policy environment, consultation for policy formulation is not new in Vietnam's forestry sector. While policy makers may have different ideas on consultation, for example, one key REDD+ policy maker in MARD noted: *"Vietnam cannot conduct consultations in the same way as other countries"*, they are also willing to tailor the approach for REDD+. The NRAP formulation process was fully funded by the UN-REDD program and took place over the course of two years. While there was an increasing level of openness with participation of more non-state actors, criticism and dissatisfaction with NRAP processes was also evident. There was very little action to improve the NRAP formulation process, despite the more open policy environment and the resources available for REDD+ policy-making. It was evident that there was a stronger focus on implementation than formulation of policies (Figure 2).

The situation is different for international NGOs (INGOs) and organizations. These are often financially resourced by Western donors, who consider the involvement of civil society organisations as a key ingredient in promoting good governance [66]. These organizations are active in REDD+ and making serious efforts to work in collaboration with the state in the area of policy advocacy. However, this focus and approach to policy advocacy is influenced by their history and operations in Vietnam, which has occurred under cumbersome registration, approval procedures, and relatively strict surveillance. The perception of the political system as monolithic and authoritarian is common and has thus created another type of inertia. While these INGOs are staffed with and sometimes headed by Vietnamese nationals, there seems to be a gap between their notion of CSOs, their perception of political systems, and the dynamic political realities. In a forum for CSO in 2010, a senior GOV representative stated *"INGOs do not understand how GOV works"*. This observation was consistent with the views of REDD+ policy makers interviewed for this study.

Ultimately, good policy process is more likely to result in positive policy outcomes. It is important that policy actors understand politics and know how to handle policy processes strategically. In this context, political intelligence can be both a benefit and a disadvantage for both VNGOs and INGOs in different ways. VNGOs have strong state connections that may facilitate their local operations and manoeuvring through the system. INGOs can plead ignorance of internal politics while challenging the state on sensitive issues.

4.2. Mechanisms and Strategies to Enhance REDD+ Policy Processes

Policy revitalisation is *"recognising and unblocking counterproductive patterns in policy processes"*, noting that policy actors are caught in and often not aware of these *"stagnated patterns"* [38]. This study shows that REDD+ state and non-state actors are preoccupied with discussions on highly technical issues—finding solutions for their own operational issues and pursuing their own agendas. Similar patterns were found in another study, where Vietnam's REDD+ actors relied extensively on scientific and technical justifications to promote REDD+ discourse [67]. The symptom of policy stagnation here is manifested in the absence of open dialogues on how NRAP formulation process should take place to allow for different views and innovative ideas while seeking common goals and securing commitments. Neither state nor non-state actors questioned the absence of these open dialogues which, in a way, has undermined their chances to achieve positive outcomes. In conceptualizing the current dynamics in natural resource management, Carr [32] found that "top-down" and "bottom-up" focuses do not hold much value and called for a "middle ground" where vertical and horizontal links are sought. We argue that REDD+ policy processes in Vietnam need to be revitalized, starting with REDD+ actors' recognition of the causes and patterns of policy stagnation, and creation of this "middle ground" to nurture political intelligence, new ideas, linkages, and eventually enhance social learning. Wells-Dang [68] suggested that CSOs in Vietnam need to unite and employ a diversity of strategies and tactics to be effective. In the political context of Vietnam, collective action could offer ways to achieve these.

Vietnamese history has consistently shown that collective actions have been considered the key elements to success in response to natural disasters or foreign invasions [69]. The Sixth National Congress of the Communist Party in 1986 reaffirmed that *"the government is the tool for the country's socialist collective mastery"*. Ensuring high levels of consensus and striving for solidarity and collective action are the fundamental premises in official Communist Party documents and are strongly reflected in decision-making processes.

VNGOs and INGOs in Vietnam's REDD+ arena could overcome inertia associated with their history and close ties to government by forming strong alliances between the groups active within and beyond REDD+ networks. This requires a commitment to seek solutions, the right set of skills to facilitate stronger networking, developing "action maps" to explore dynamics, learning how mutual actions may reinforce each other, and unlocking learning opportunities [70].

As things stand, the prospect of a broad alliance forming via collective action is not promising. The STWGs generally focus on technical issues and the REDD+ focal ministry does not show a great deal of initiative and lacks experience in dealing with external stakeholders (for example, in the private sector). There is also a disconnect between the REDD+ Steering Committee and the STWGs. The feedback loops between these two, which are key to building knowledge and effective policy-making, are absent.

If Vietnam wants to take concrete steps towards implementing REDD+ at a wider scale, both state and non-state actors need to pause from implementation and switch attention to reflect and discuss how to revitalize current policy processes. Specifically, local NGOs and international organizations need to re-evaluate their strategies and consider the possible impact of collective power in order to aim for more fundamental changes. The room to "manoeuver" and improve policy designs may not be as limited as people think. The challenge remains to identify where this room is located and the entry points for policy influence. While MARD may be inclined to employ a limited range of traditional consultation processes, REDD+ can potentially provide new learning opportunities for both state and non-state organisations.

Beutz [71] suggested that a functional democracy for a one-party state may need two elements of (i) accountability and (ii) openness. In Vietnam, vertical accountability has declined and citizen participation at the local level remains limited compared to 2011 [72]. Currently, there is public outcry for a more transparent and accountable political system, manifested via the visible crisis around recent environmental disasters and the demonstrations against Chinese occupation of the South China Sea [73]. Taking this wider political context into account, REDD+ actors need to strive for accountability and legitimacy. This level of increasing openness experienced in the REDD+ policy process may serve as a platform for further pursuing accountability. In addition, the proposed draft NRAP institutional arrangements will position the National REDD+ Steering Committee under the chairmanship of a Deputy Prime Minister, with strengthened mechanisms for accountability. We suggest that both the "policy community" (REDD+ Steering Committee) and "epistemic community" (i.e., a knowledge-based network of experts, including Vietnamese STWGs and members of the REDD+ network that assists policy makers to identify interests, frame the issues, and develop specific policies [74]) needs to enhance their interactions and reinforce their sense of collective accountability to achieve better policy outcomes. If the state outsources certain functions to escape scrutiny [20], the non-state actors have the chance to work collectively, treating REDD+ policy processes as a social learning process, in which each sector is held accountable by the other.

5. Conclusions

Despite the rhetoric of more open policy-making for REDD+ in Vietnam, this study found that the process is still far from inclusive and does not provide a strong basis for addressing key policy objectives. While REDD+ policy is progressing from the 'readiness' phase to create a more inclusive and dynamic strategy, without revitalization of policy processes there are risks in the coming phase of the NRAP of little substantive policy development and limited incorporation of different interests, resulting in policy that does not clearly address the drivers of deforestation and forest degradation in Vietnam.

Due to their history of development and often close association with the state, non-state actors and organisations are still on the periphery of decision making regarding REDD+, or they occupy "safe" positions that align with current government policies. This results in a high level of policy inertia, with few voices arguing for significant change.

By acting collectively and moving beyond their "comfort zones" as service providers and intermediaries, both Vietnamese and international NGOs could overcome this inertia and push for fundamental changes. These actors can extend beyond a limited thematic focus and aim for incorporating policies to support programs like REDD+ in a broader view of social change. Using their collective political intelligence, building a broad alliance via collective actions, and creating opportunities for social learning, where actors can share knowledge and experience and hold each other accountable, they can operate to bring about greater benefits from REDD+ for wider society and more focused policy on the key drivers of forest loss or restoration.

In the Vietnamese political system, room to "manoeuver" can be opened up with the right tactics. Policy makers in Vietnam are increasingly seeking better information and ideas on best practices from various actors in order to reinforce the ruling party's legitimacy. It is important for policy actors to recognize policy stagnation, understand its patterns and root causes, and find the right tactics for addressing it. REDD+ state actors should seek to expand and build extra-sectoral partnerships across multiple organisational boundaries, including those beyond the forestry sector.

The World Economic Forum claimed in 2013, "*Civil society's time has come*" [47]. Professor Dang Huu, president of the Vietnam Institute of Development Studies said in 2006: "*As the reform process moves forward, unique opportunities are created for Vietnamese policy and lawmakers to promote an enabling environment for the establishment and growth of non-state organizations*". This enabling environment can be created and facilitated through the joint actions of Vietnamese policy makers and non-state policy actors in REDD+ policy processes.

Acknowledgments: Data collection for a component of this research was undertaken under the Module 1 and Module 2 of the Global Comparative Study (GCS) on REDD+, managed by CIFOR (the Center for International Forestry Research). Maria Brockhaus and William Sunderlin led the design of M1 and M2 methodologies. Pham Thu Thuy (CIFOR), Nguyen Tuan Viet, and Bui Minh Nguyet were involved in data collection and transcription for this paper. Our sincere thanks for their support and cooperation. Thanks also to Dr Adam Bumpus for his contribution to the design and review of the outputs of this study.

Author Contributions: Thu Ba Huynh and Rodney J Keenan jointly designed the research for this paper. Thu Ba Huynh undertook the data collection, analysis and coordinated the writing process. Rodney J Keenan contributed to the writing.

Conflicts of Interest: The authors declare no conflict of interest.

Abbreviations

Acronym	Definition
CBO	Community-Based Organization
CIFOR	Center for International Forestry Research
CSI	Civil Society Index
CSO	Civil Society Organization
FAO	Food and Agriculture Organization of the United Nations
FCPF	Forest Carbon Partnership Facility of the World Bank
GCS	Global Comparative Study
GOV	Government
INGO	International Non-Governmental Organization
MARD	Vietnam Ministry of Agriculture and Rural Development
MRV	Measurement, Reporting and Verification
NGO	Non-Governmental Organization
NRAP	National REDD+ Action Program
RDKN	Respondent Does Not Know
REDD	Reducing Emissions From Deforestation and Forest Degradation
STWG	Sub-Technical Working Group
UNDP	United Nations Development Programme
UNEP	United Nations Environment Programme
USA	United States of America
USD	US Dollar
VIC	Victoria
VNGO	Vietnam Non-Governmental Organization

References

1. Angelsen, A.; Brockhaus, M.; Sunderlin, W.D.; Verchot, L.V. *Analysing REDD+: Challenges and Choices*; Center for International Forestry Research (CIFOR): Bogor, Indonesia, 2012; p. 426.
2. Simonet, G.; Karsenty, A.; de Perthuis, C.; Newton, P.; Schaap, B.; Seyller, C. REDD+ Projects in **2014**: An Overview Based on a New Database and Typology. Available online: http://www.chaireeconomieduclimat.org/en/publications-en/information-debates/id-32-redd-projects-in-2014-an-overview-based-on-a-new-database-and-typology/ (accessed on 15 January 2017).
3. Brockhaus, M.; Angelsen, A. Seeing REDD+ through 4is: A political economy framework. In *Analysing REDD+: Challenges and Choices*; Center for International Forestry Research: Bogor, Indonesia, 2012; pp. 15–30.
4. Sills, E.O.; Atmadja, S.; de Sassi, C.; Duchelle, A.E.; Kweka, D.; Resosudarmo, I.A.P.; Sunderlin, W.D. *REDD+ on the Ground: A Case Book of Subnational Initiatives across the Globe*; Center for International Forestry Research (CIFOR): Bogor, Indonesia, 2014.
5. Fischer, R.; Hargita, Y.; Günter, S. Insights from the ground level? A content analysis review of multi-national REDD+ studies since 2010. *For. Policy Econ.* **2016**, *66*, 47–58. [CrossRef]
6. Keenan, R.J. The paris climate agreement and forests: Will the cop21 agreement encourage growth in investment in sustainably-managed forests? *Asia Pac. Policy Soc.* **2016**. Available online: www.policyforum.net/the-paris-climate-agreement-and-forests/ (accessed on 15 January 2017).
7. Ostrom, E. A polycentric approach for coping with climate change. In *World Bank Policy Research Working Paper 5095*; World Bank: Washington, DC, USA, 2009.
8. Sunderlin, W.; Sills, E.; Duchelle, A.; Ekaputri, A.; Kweka, D.; Toniolo, M.; Ball, S.; Doggart, N.; Pratama, C.; Padilla, J. REDD+ at a critical juncture: Assessing the limits of polycentric governance for achieving climate change mitigation. *Int. For. Rev.* **2015**, *17*, 400–413. [CrossRef]
9. Korhonen-Kurki, K.; Brockhaus, M.; Bushley, B.R.; Babon, A.; Gebara, M.F.; Kengoum Djiegni, F.; Pham, T.T.; Rantala, S.; Moeliono, M.; Dwisatrio, B.; et al. Coordination and Cross-Sectoral Integration in REDD+: Experiences from Seven Countries. *Clim. Dev.* **2015**, *8*, 458–471. [CrossRef]
10. Kanowski, P.J.; McDermott, C.L.; Cashore, B.W. Implementing REDD+: Lessons from analysis of forest governance. *Environ. Sci. Policy* **2011**, *14*, 111–117. [CrossRef]

11. Humphreys, D. *Logjam: Deforestation and the Crisis of Global Governance*; Earthscan: London, UK; Sterling, VA, USA, 2006; p. 302.

12. Minang, P.A.; Van Noordwijk, M.; Duguma, L.A.; Alemagi, D.; Do, T.H.; Bernard, F.; Agung, P.; Robiglio, V.; Catacutan, D.; Suyanto, S. REDD+ readiness progress across countries: Time for reconsideration. *Clim. Policy* **2014**, *14*, 685–708. [CrossRef]

13. Cường, L.V.; Quang, Đ.V.; Đơ, T.T. *Báo Cáo Dòng Tài Chính REDD+ Tại Việt Nam Giai Đoạn 2009–2014*; REDD+ Vietnam and Forest Trend: Hanoi, Vietnam, 2015.

14. UN-REDD. *Operationalizing REDD+ in Viet Nam through Provincial REDD+ Action Plans (PRAP) REDD+ and the Rationale for Sub-National Planning*; UN-REDD Vietnam: Hanoi, Vietnam, 2016.

15. MARD. *Participatory Self-Assessment of the REDD+ Readiness Package in Vietnam*; MARD: Hanoi, Vietnam, 2016.

16. UN-REDD. *Site-Based REDD+ Implementation Plan (Sirap) in Viet Nam*; UN-REDD Vietnam Phase II Program: Hanoi, Vietnam, 2016.

17. London, J.D. Politics in contemporary vietnam. In *Politics in Contemporary Vietnam*; Palgrave Macmillan: Hampshire, UK; New York, NY, USA, 2014; pp. 1–20.

18. Brockhaus, M.; Di Gregorio, M.; Mardiah, S. Governing the design of national Redd+: An analysis of the power of agency. *For. Policy Econ.* **2014**, *49*, 23–33. [CrossRef]

19. Fujisaki, T.; Hyakumura, K.; Scheyvens, H.; Cadman, T. Does REDD+ ensure sectoral coordination and stakeholder participation? A comparative analysis of Redd+ national governance structures in countries of asia-pacific region. *Forests* **2016**, *7*, 195. [CrossRef]

20. Michaels, R. The mirage of non-state governance. *Utah Law Rev.* **2010**, *2010*, 31–45.

21. Foucault, M. *Discipline and Punish: The Birth of the Prison/Michel Foucault*; Translated from the French by Alan Sheridan; Vintage Books: New York, NY, USA, 1979.

22. Knieling, J.; Leal Filho, W. *Climate Change Governance*; Springer Science & Business Media: Berlin, Germany, 2012.

23. Lemos, M.C.; Agrawal, A. Environmental governance. *Annu. Rev. Environ. Resour.* **2006**, *31*, 297. [CrossRef]

24. Roberts, J. *Environmental Policy*; Taylor and Francis: Hoboken, NJ, USA, 2004.

25. Cashore, B.; Vertinsky, I. Policy networks and firm behaviours: Governance systems and firm reponses to external demands for sustainable forest management. *Policy Sci.* **2000**, *33*, 1–30. [CrossRef]

26. Gray, B. *Collaborating: Finding Common Ground for Multiparty Problems*; Jossey-Bass: San Francisco, CA, USA, 1989.

27. Roberts, N. Wicked problems and network approaches to resolution. *Int. Public Manag. Rev.* **2000**, *1*, 1–19.

28. Folke, C.; Hahn, T.; Olsson, P.; Norberg, J. Adaptive governance of social-ecological systems. *Annu. Rev. Environ. Resour.* **2005**, *30*, 441–473. [CrossRef]

29. Reed, M.; Evely, A.C.; Cundill, G.; Fazey, I.R.A.; Glass, J.; Laing, A.; Newig, J.; Parrish, B.; Prell, C.; Raymond, C. What is social learning? *Ecol. Soc.* **2010**, *14*. [CrossRef]

30. Colebatch, H.K. *Beyond the Policy Cycle: The Policy Process in Australia*; Colebatch, H.K., Ed.; Allen & Unwin: Crows Nest, Australia, 2006.

31. Arnstein, S. A ladder of citizen participation. *J. Am. Plan. Assoc.* **1969**, *35*, 216–224. [CrossRef]

32. Carr, A. *Grass Roots and Green Tape: Principles and Practices of Environmental Stewardship*; Federation Press: Leichhardt, Australia, 2002.

33. Ross, H.; Buchy, M.; Proctor, W. Laying down the ladder: A typology of public participation in australian natural resource management. *Aust. J. Environ. Manag.* **2002**, *9*, 205–217. [CrossRef]

34. Collins, K.; Ison, R. Jumping off arnstein's ladder: Social learning as a new policy paradigm for climate change adaptation. *Environ. Policy Gov.* **2009**, *19*, 358–373. [CrossRef]

35. Mayers, J.; Bass, S. *Policy That Works for Forests and People/Authors, James Mayers and Stephen Bass*; International Institute for Environment and Development: London, UK, 1999.

36. Jochim, A.E.; May, P.J. Beyond subsystems: Policy regimes and governance. *Policy Stud. J.* **2010**, *38*, 303–327. [CrossRef]

37. Weible, C.M. Expert-based information and policy subsystems: A review and synthesis. *Policy Stud. J.* **2008**, *36*, 615–635. [CrossRef]

38. Termeer, C.; Dewulf, A.; Breeman, G. Governance of wicked climate adaptation problems. In *Climate Change Governance*; Springer: Berlin, Germany; New York, USA, 2013; pp. 27–39.

39. Newell, P. *Climate for Change: Non-State Actors and the Global Politics of the Greenhouse*; Cambridge University Press: Cambridge, UK, 2006.
40. Keen, S. Non-governmental organitions in policy. In *Beyond the Policy Cycle: The Policy Process in Australia*; Colebatch, H.K., Ed.; Allen & Unwin: Crows Nest, NSW Australia, 2006.
41. Dombrowski, K. Filling the gap? An analysis of non-governmental organizations responses to participation and representation deficits in global climate governance. *Int. Environ. Agreem. Politics Law Econ.* **2010**, *10*, 397–416. [CrossRef]
42. Angelsen, A. *Moving ahead with REDD: Issues, Options and Implications*; Center for International Forestry Research (CIFOR): Bogor, Indonesia, 2008; p. 156.
43. Cadman, T.; Maraseni, T.; Breakey, H.; López-Casero, F.; Ma, H.O. Governance values in the climate change regime: Stakeholder perceptions of Redd+ legitimacy at the national level. *Forests* **2016**, *7*, 212. [CrossRef]
44. Lucius, C. *Vietnam's Political Process: How Education Shapes Political Decision Making*; Routledge: New York, NY, USA, 2009.
45. Angelsen, A.; Brockhaus, M.; Kanninen, M.; Sills, E.; Sunderlin, W.D.; Wertz-Kanounnikoff, S. *Realising Redd+: National Strategy and Policy Options*; Center for International Forestry Research (CIFOR): Bogor, Indonesia, 2009; p. 361.
46. Vijge, M.J.; Brockhaus, M.; Di Gregorio, M.; Muharrom, E. Framing national Redd+ benefits, monitoring, governance and finance: A comparative analysis of seven countries. *Glob. Environ. Chang.* **2016**, *39*, 57–68. [CrossRef]
47. World Economic Forum. *The Future Role of Civil Society*; World Economic Forum: Cologny, Switzerland, 2013.
48. NORAD. *Real-Time Evaluation of Norway's International Climate and Forest Initiative*; Lessons Learned from Support to Civil Society Organisations; NORAD: Oslo, Norway, 2012.
49. Potter, D. *Ngos and Environmental Policies: Asia and Africa*; David, P., Ed.; FRANK Cass: Portland, OR, USA, 1996.
50. Norlund, I. Civil society in Vietnam. Social organisations and approaches to new concepts. *Asien* **2007**, *105*, 68–90.
51. Fforde, A. Civil society, the state, and the business sector–protagonists of democratization processes? Insights from vietnam. In *Towards Good Society: Civil Society Actors, the State and the Business Class in Southeast Asia–Facilitators of or Impediments to a Strong, Democratic, and Fair Society*; Agit-Druck: Berlin, Germany, 2005; pp. 173–192.
52. Hannah, J. *Local Non-Government Organizations in Vietnam: Development, Civil Society and State-Society Relations*; University of Washington: Seatle, WA, USA, 2007.
53. William, T.; Nguyễn, T.H.; Phạm, Q.T.; Tuyết, H.T.N. *Civil Society in Vietnam: A Comparative Study of Civil Society Organizations in Hanoi and Ho Chi Minh City*; The Asia Foundation: Hanoi, Vietnam, 2012.
54. Thaveeporn, V. Authoritarianism reconfigured: Evolving accountability relations within Vietnam's one-party rule. In *Politics in Contemporary Vietnam: Party, State, and Authority Relation*; London, J.D., Ed.; Palgrave Macmillan: Houndmills, UK; New York, NY, USA, 2014.
55. Bui, T.H. The development of civil society and dynamics of governance in Vietnam's one party rule. *Glob. Chang.Peace Secur.* **2013**, *25*, 77–93. [CrossRef]
56. Wells-Dang, A. The political influence of civil society in vietnam. In *Politics in Contemporary Vietnam: Party, State, and Authority Relations*; London, J.D., Ed.; Palgrave Macmillan: Houndmills, UK; New York, NY, USA, 2014.
57. Morris-Jung, J. Politics in contemporary Vietnam: Party, state and authority relations. *Contemp. Southeast Asia* **2014**, *36*, 473–476.
58. Norlund, I.; SNV; UNDP Viet Nam; Viet Nam, L.H.H.K.H. *Filling the Gap: The Emerging Civil Society in Viet Nam*; Viet Nam Union of Science and Technology Associations: Hanoi, Vietnam, 2007.
59. Bryman, A. *Social Research Methods*, 4th ed.; Oxford University Press: Oxford, UK, 2012; p. 766.
60. Government of Vietnam. *National REDD+ Action Program*; MARD: Hanoi, Vietnam, 2012.
61. UN-REDD. *Lessons Learned Viet Nam Un-REDD Programme, Phase 1*; UN-REDD: Hanoi, Vietnam, 2012.
62. Huynh, T.B.; Bumpus, A. *Stakeholder Analysis & Stakeholder Engagement for the Implementation of National REDD Action Plan in Viet Nam*; UN-REDD: Hanoi, Vietnam, 2014.

63. Vietnam, R. Institutional Arrangement for REDD+ in Viet Nam. Available online: http://www.vietnam-Redd.org/Web/Default.aspx?tab=introdetail&zoneid=106&itemid=428&lang=en-US (accessed on 1 September 2016).

64. Duong, M.N. *Grassroots Democracy in Vietnamese Communes*; The Centre for Democratic Institutions, Research School of Social Sciences, The Australian National University: Canberra, Australia, 2004.

65. Pham, T.T.; Moeliono, M.; Nguyen, T.H.; Nguyen, H.T.; Vu, T.H. *The Context of REDD+ in Vietnam: Drivers, Agents and Institutions*; Center for International Forestry Research (CIFOR): Bogor, Indonesia, 2012.

66. Howell, J.; Pearce, J. *Civil Society and Development: A Critical Exploration*; L. Rienner Publishers: Boulder, CO, USA, 2001.

67. Di Gregorio, M.; Brockhaus, M.; Cronin, T.; Muharrom, E.; Mardiah, S.; Santoso, L. Deadlock or transformational change? Exploring public discourse on REDD+ across seven countries. *Glob. Environ. Politics* **2015**, *15*, 63–84. [CrossRef]

68. Wells-Dang, A.; Wells-Dang, G. Civil society in asean: A healthy development? *Lancet* **2011**, *377*, 792–793. [CrossRef]

69. Kelly, P.M.; Hien, H.M.; Lien, T.V. Responding to el nino and la nina: Averting tropical cyclone impacts. In *Living with Environmental Change: Social Vulnerability, Adaptation and Resilience in Vietnam*; Adger, N., Kelly, P.M., Nguyen, H.N., Ebrary, I.N.C., Eds.; Routledge: New York, NY, USA, 2001; p. 314.

70. Putnam, R. Unlocking organizational routines that prevent learning. *Syst. Think.* **1993**, *4*, 2–4.

71. Beutz, M. Functional democracy: Responding to failures of accountability. *Harv. Int. Law J.* **2003**, *44*, 387.

72. CECODES; VFF-CRT; UNDP. *The Viet Nam Governance and Public Administration Performance Index (PAPI) 2015: Measuring Citizens' Experiences*; Centre for Community Support and Development Studies (CECODES), Centre for Research and Training of the Viet Nam Fatherland Front (VFF-CRT), and United Nations Development Programme (UNDP): Hanoi, Vietnam, 2016.

73. Morris-Jung, J. Vietnam's New Environmental Politics: A Fish out of Water? Available online: http://thediplomat.com/2016/05/vietnams-new-environmental-politics-a-fish-out-of-water/ (accessed on 3 October 2016).

74. Haas, P.M. Introduction: Epistemic communities and international policy coordination. *Int. Org.* **1992**, *46*, 1–35. [CrossRef]

forests

MDPI

Article

Resources and Rules of the Game: Participation of Civil Society in REDD+ and FLEGT-VPA Processes in Lao PDR

Irmeli Mustalahti [1,*], Mathias Cramm [1], Sabaheta Ramcilovic-Suominen [1] and Yitagesu T. Tegegne [2]

[1] Department of Geographical and Historical Studies, University of Eastern Finland, Yliopistokatu 7, 80100 Joensuu, Finland; m.cramm@gmail.com (M.C.); sabaheta.ramcilovik-suominen@uef.fi (S.R.-S.)
[2] European Forest Institute, Yliopistokatu 6, 80100 Joensuu, Finland; yitagesu.tekle@efi.int
* Correspondence: irmeli.mustalahti@uef.fi; Tel.: +358-505-632-071

Academic Editors: Esteve Corbera and Heike Schroeder
Received: 31 October 2016; Accepted: 14 February 2017; Published: 21 February 2017

Abstract: Reducing Emissions from Deforestation and Forest Degradation (REDD+) aims to achieve its purpose by working across multiple sectors and involving multilevel actors in reducing deforestation and forest degradation in tropical countries. By contrast, the European Union (EU) Action Plan on Forest Law Enforcement, Governance and Trade (FLEGT) and its Voluntary Partnership Agreements (VPAs) focus on forestry and functions at a bilateral state level. The FLEGT Action Plan specifically aims to tackle illegal logging and improve forest governance in countries exporting tropical timber to the EU. Since illegal logging is just one driver of forest degradation, and legalisation of logging does not necessarily reduce deforestation and forest degradation, the two instruments differ in scope. However, by addressing the causes of forest degradation and their underlying governance issues, the FLEGT VPAs and REDD+ share many functional linkages at higher levels of forest policy and forest governance. The contribution and participation of civil society organisations (CSOs) and other actors are imperative to both processes. Our study is based on a survey of key actors (national and international) in REDD+ and FLEGT VPA processes in the Lao People's Democratic Republic (Lao PDR). Our analysis was guided by the theoretical perspectives of the policy arrangement approach and examination of two specific dimensions of this approach, namely resources and rules of the game. This paper argues that participation of CSOs in both processes is crucial because it facilitates and nurtures much needed cooperation between other national and international actors. The paper concludes that participation of CSOs could bring valuable information and knowledge into REDD+ and FLEGT VPA processes, thus contributing to increased legitimacy, justice and transparency.

Keywords: REDD+; FLEGT VPA; civil society organisations; participation; resources; rules of the game; policy arrangement approach; Lao PDR

1. Introduction

Climate change induced by human activities is well substantiated, for instance as demonstrated by the Intergovernmental Panel on Climate Change in the Fifth Assessment Report [1]. Deforestation and forest degradation are a major concern globally because they are one of the drivers of global warming. As much as 10% of the annual anthropogenic greenhouse gas emissions is estimated to be due to tropical deforestation and forest degradation [2]. Indeed, tropical forests showed a significant trend of increased annual forest loss between 2000 and 2012 [3]. Therefore, sustainable forest management (SFM) and conservation can play a key role in reducing global anthropogenic greenhouse gas emissions [4,5].

Scholars argue that countries in the tropics often have institutional, economic and political disadvantages in responding effectively to multifaceted and complex challenges, such as those related to society-nature interactions [6–9]. The international community has therefore taken a substantial interest in facilitating the national efforts of tropical countries to reduce deforestation and forest degradation because such interventions are of interest and benefit at a global scale. Different approaches have arisen in the wake of this interest. Two such approaches are (i) Reducing Emissions from Deforestation and Forest Degradation in Developing Countries, and the role of conservation, sustainable management of forests and enhancement of forest carbon stocks in developing countries (REDD+) and (ii) the European Union's Forest Law Enforcement, Governance and Trade (FLEGT) Action Plan and Voluntary Partnership Agreements (VPAs). REDD+ is being negotiated under the United Nations Framework Convention on Climate Change (UNFCCC), and developing countries are being supported by several donor countries, especially Norway, as well as international organisations such as the UN and the World Bank. The scope of REDD+ is to address the direct and underlying drivers of deforestation and forest degradation, and under this approach, countries are paid based on their performance in forest carbon sequestration. The latest climate agreement negotiated in Paris features an explicit stand-alone article [10] (Article 5) devoted to REDD+ that sends a clear message that REDD+ and forests will play a role in the fight against climate change. Meanwhile, the EU FLEGT Action Plan was developed in 2003 in response to imports of illegally sourced timber and vast illegal logging operations in supplying timber products [11]. The overall aim of the FLEGT VPA is to limit the import of illegal timber into the EU and reduce illegal logging, while strengthening land tenure, access and rights, and increasing the participation of different actors [11]. This should help lead the way to SFM (sustainable forest management), a principle that is adopted by most countries through the Rio Conventions [12]. FLEGT VPAs are negotiated at a bilateral level between the European Commission (EC) representing the EU member states and an individual partner country [13].

REDD+ works across multiple sectors and thus involves many different actors. Incorporating the views of all of these actors in REDD+ planning and implementation processes is important to ensure that REDD+ processes and finance will be accessible and fair for the different actors [14–16]. The FLEGT VPA process is negotiated and implemented as an agreement between the EC and each specific country individually. The VPA agreements focus solely on forestry and illegal logging, and also affect numerous actors at the national scale that are either directly or indirectly in touch with forests and involved in forestry related matters [17,18]. By addressing the causes of forest degradation and their underlying governance issues, negotiation and implementation of the FLEGT VPAs share many functional linkages with REDD+ processes [19,20]. It is argued that REDD+ should include safeguards that recognise and protect the continuity of multipurpose functions of the forest for local people and avoid dependence on external payments [15,21,22]. In a similar fashion, VPAs in principle have to include provisions for justice and sustainability matters [13]. In this context, the participation of CSOs is essential in terms of developing such safeguards for REDD+ and FLEGT.

The aim of this paper is to understand how and to what extent CSOs are engaged in REDD+ and in the emerging FLEGT VPA process in the Lao People's Democratic Republic (PDR). In addition, the paper aims to clarify the possible role which CSO participation plays in facilitating the implementation of the two processes. REDD+ in Lao PDR is at the stage of developing the national REDD+ strategy, while the VPA process is at the pre-negotiation phase. In this paper, we aim to explore these two processes through three research questions: (i) how are CSOs currently participating in REDD+ and VPA processes? (existing participation); (ii) how can the role of CSOs be strengthened? (potential participation); (iii) how can CSOs foster cooperation between the REDD+ and VPA processes?

In addressing these questions, the paper contributes to the literature on the role of non-state actors in environmental and forest governance processes and interventions in general terms as well as to the emerging literature on the same issue in the context of FLEGT and REDD+. While the literature on the non-state actors' role and participation is rather well developed in both general terms and with regard to FLEGT and REDD+ processes, such analysis in the case of Lao PDR is a novelty and is much needed. Considering the country's single party system, the involvement and role of non-state actors, such as CSOs, is an issue that has lagged behind, both in practice and in academic work. A limitation of this study is that we had difficulty in accessing information and objective opinion on the issues in focus, which in particular reflected the recent emergence of domestic CSOs and that their operations and involvement is highly regulated in Lao PDR. This situation is also clearly seen in the study sample which was dominated by international NGOs due to the difficulty of identifying domestic CSOs working on FLEGT and REDD+ in Lao PDR.

We use the policy arrangement approach (PAA) proposed by Bas Arts and colleagues to analyse the involvement of CSOs in FLEGT and REDD+ processes and their role in facilitating implementation of and interactions between the processes in Lao PDR. One of the key reasons for adopting this framework for our study is the concept of "political modernisation", which is the "building block" on which the PAA rests. The concept of political modernisation, as further explained in the section describing the theoretical framework, is about the change in forest governance represented by the emergence of new actors. Therefore, we felt that the PAA was particularly suited even though there are various other frameworks for facilitating analyses of environmental governance. We also opted for the PAA framework due to its strong emphasis on power and power relations—a dimension which is largely overlooked in the analytical frameworks emerging from the schools of institutionalism and institutional economics. Our conceptual understanding of PAA and its relevance are further explained in Section 3 which introduces the theoretical framework of the study. The last part of the article is organised in the following sections: Section 4 presents research methods; Section 5 presents and discusses the results; and Section 6 draws some conclusions and offers recommendations. However, first, in Section 2 we discuss the case study (REDD+ and FLEGT-VPA processes) in Lao PDR.

2. Case Study in Lao PDR: Forestry Governance, CSOs, REDD+ and FLEGT VPA

2.1. Forest Governance and Participatory Forestry in Lao PDR

Lao PDR is a tropical country in Southeast Asia that shares most of its border with Vietnam and Thailand. Forest cover in the country is high and is estimated at around 40%, but deforestation levels have been alarming. Lao PDR has vast natural resources that attract foreign direct investment with the aim of developing commercial agriculture and forestry production [23]. At the same time, the government of Lao PDR and donor countries have agreed to pursue forest conservation through global mechanisms such as REDD+ and FLEGT. Consequently, the various types of investments and conservation practices for managing natural resources and land are creating more pressure on rural communities but also opportunities to which these communities must adapt in order to secure their livelihoods [24,25].

Although Lao PDR is well-endowed with natural resources that can make a major contribution to the country's long-term economic development, the country has experienced an alarming deforestation rate in the last two decades. From 1982, the forest area has been decreasing dramatically, at an estimated annual average net deforestation rate of 0.7%, or forest loss of about 76,000 ha per annum [26]. The immediate and underlying factors causing deforestation and forest degradation in Lao PDR are well covered in [26,27]. Lao PDR state sources name the following activities as the principal causes of deforestation: shifting cultivation, commercial agriculture, forest fires, and mining operations in forested areas. Extraction of wood (legal and illegal) and unsustainable non-timber forest product harvesting are named as the major causes of forest degradation. Factors underlying these causes include population growth and infrastructure expansion, urbanisation, construction of roads, hydropower plants and increasing price and demand for raw materials (e.g., minerals, timber) in the regional and global markets. However, scholars increasingly question these factors and sources, and point to state-supported large-scale land concessions for fast-growing plantations and development projects as the main reasons for declining forest cover in Lao PDR, rather than shifting cultivation practiced by upland minorities [28–30].

In Lao PDR, villagers have been involved in so-called participatory forestry since the 1990's [31,32]. The first participatory forest management (PFM) intervention was the Joint Forest Management approach introduced by the Lao Swedish Forestry Programme [33]. In the 1990's, the long-term rights for use of natural forest could be allocated to individuals and organisations [34] (art.5, pp. 48–54). The law made it possible to develop another participatory forestry model with a deeper involvement of villagers, the Forest Management and Conservation Programme (FOMACOP). FOMACOP, supported by World Bank and Finnish development cooperation, operated as a pilot project of participatory sustainable production forestry. This project paid special attention to the building of villager organisations and entrepreneurial development as well as to the technical aspects associated with forest management carried out by villagers [35].

The concept of practising participatory forestry in production forests through village organisations lost political support in Lao PDR and FOMACOP ended in 2001 [36]. The Forest Sector Strategy [37] instead adopted a timber production approach under sustainable forest management (SFM) in cooperation with local villages. The stated goal was decentralisation of land and forest resources. However, from a critical point of view, rather than decentralising natural resources and providing greater rights and control to local villages, the initiative at the time also led to re-centralisation of power [38]. In more recent years, there has been renewed interest in and political acceptance of village forestry, which is, however, highly regulated. In this vein, the former FOMACOP was replaced by the Sustainable Forestry for Rural Development Scaling Up (SUFORD-SU) Project (another project supported by the Finnish development aid) that follows more closely the political decisions and regulations of Lao PDR. Production forest areas under SUFORD-SU are no longer allocated to a particular village locality. Instead, management of production forest areas must be implemented jointly by several villages, their village forest organisations and district forest management units [39–41].

Currently, the involvement and roles of villagers in SFM are embedded in the existing laws and regulations and in the National Growth and Poverty Eradication Strategy of Government of Lao PDR [42]. The most relevant laws dealing with CSO participation include: (i) the Forest Law [34]; (ii) Prime Minister's (PM) Decree No. 59 [43]; (iii) Regulation No. 0204 from the Ministry of Agriculture and Forestry (2003) and (iv) the Forestry Strategy [37]. These laws have apparently strengthened the involvement of CSOs in community-based forestry interventions in Lao PDR, and provide more freedom to develop village forestry models based on decentralised natural resources rights and control by local villages. In summary, the last two decades have seen a shift towards decentralisation, a shift back to recentralisation, and now recently again a shift towards decentralisation of natural resources in the Lao PDR.

2.2. Civil Society Organisations (CSOs) in Lao PDR

In 2009, the Government of Lao PDR [44] passed decree No. 115/PM which introduced national CSOs in the country, or non-profit associations (NPAs) as they are called. According to the decree, NPAs should contribute to socio-economic development and poverty eradication. In this paper, "CSOs" refers to Lao NGOs/NPAs, community-based organisations and mass organisations as well as international NGOs, which may have a role to play in the REDD+/VPA processes in the country. However, it should be pointed out that the Lao-based organisations are highly regulated by the state and their freedom of action and expression is still tightly monitored. In fact, the so-called "mass organisations" are one of the four key institutions where leading figures are party members [45]. During this study, we found that the national CSOs were not yet actively involved in the REDD+/FLEGT processes. Therefore, the data and results presented in this study are dominated by information from international NGOs.

In the context of Lao PDR's single-party and authoritarian ruling system, it is interpreted as quasi normal that the NPAs would be under governmental control and in line with government policies and goals [46,47]. For instance, the decree on NPAs stipulates that "undermining the national, collective and individual interest is forbidden" for NPAs [44]. Another decree [48] further constrains the activities of NPAs by imposing stricter rules for obtaining finance and by limiting the fields in which they may work. Besides this "new civil society" initiative, Lao PDR has four so-called mass organisations (including the Farmers' Union and Women's Union) that have close ties to the Party. These mass organisations have extensive organisational networks stretching from the top of the party hierarchy down to the village level in order to disseminate information and scale up activities [47,49]. In the case of new development activities, international NGOs, on the other hand, are comparatively more active in Lao PDR, and have recently been recognised by the government as "important contributors to national socio-economic advancements" [50]. However, their functional freedom also hangs in the balance, because the new draft decree places them under more stringent supervision by the Ministry of Foreign Affairs [51]. With the constant change in political position from centralised to decentralised, market to state regulated and an ad-hoc manner of institutional and regulation building, it remains to be seen what role CSOs will play in the country's development and in the processes of FLEGT and REDD+. However, there are hopes that the situation might change for the better with the appointment of the new prime minister in January, 2016, who apparently has adopted a more "open approach" to development and natural resource governance.

2.3. National REDD+ and FLEGT VPA Processes and Organisational Structures

The key challenges in forest and land governance in Lao PDR have been well addressed by several scholars (see for example references [26–32]) who all discuss two key challenges: the ambiguous and uncertain laws and regulations in land-use planning and allocations that further create opportunities for unsound resource use practices, and poor law enforcement performance. To address the challenges facing forest resources in the country, Lao PDR is participating in REDD+ and FLEGT VPA processes. Lao PDR was one of the first 14 countries to become a REDD+ country participant under the Forest Carbon Partnership Facility (FCPF) of the World Bank in 2008. The country's Readiness Preparation Proposal (R-PP), which is the most central national-level document that defines how a country actually wants to implement REDD+, stems from 2010. Official communications from Lao PDR requesting entry into VPA negotiations date from April 2012. By entering into a VPA with the European Commission (EC), Lao PDR, at least in theory, aims to address the problem of illegal logging activities, improve governance in the forest sector, guarantee access of timber products to the EU markets, build capacities, and increase revenue from timber exports.

2.3.1. REDD+ Process in Lao PDR

Since 2007, Lao PDR has been involved in international negotiations through a REDD+ mechanism, and the government has voiced support for an internationally binding agreement. REDD+ preparations are being supported by FCPF, the Forest Investment Programme, and UN-REDD. The Lao PDR R-PP was submitted in December 2010 and accepted in 2011 [26]. Currently, the country is gaining experience from different REDD+ pilot projects and other readiness activities and setting up REDD+ institutions. In 2011–2012, the country started an institutional reform which reorganised the responsibilities of government bodies in relation to REDD+. A new Ministry of Natural Resources and Environment (MoNRE) was established which also led to changes in the structure of the existing Ministry of Agriculture and Forestry (MAF). Prior to the establishment of MoNRE, all forestry and therefore REDD+ and FLEGT related matters fell under the MAF or, more precisely, under its Department of Forestry (DoF). Once MoNRE was established in 2012, the government merged the former Division of Forest Conservation and the Division of Forest Protection and Restoration at MAF's Department of Forestry (DoF) into the Department of Forest Resource Management (DFRM) at MoNRE [52]; the Division of Forest Production remained under the DoF at MAF.

The REDD+ Division was placed under the DFRM at MoNRE, while the REDD+ Office was under DoF at MAF. These unclear mandates between the ministries caused a lot of ambiguity and overlap in REDD+ institutional settings, and the projects related to REDD+ and production and those related to conservation and protection are managed by two different ministries (MAF and MoNRE, respectively). These overlapping roles have caused conflicts, but more importantly they have provided opportunities for the private sector to establish land investments and timber extraction under a grey area of legality and with little or no participation of local communities [49,53–55]. Furthermore, REDD+ activities were transferred back to MoNRE [56]. However, in late April 2016, the new government of Lao PDR decided to restructure some ministries and to reconsolidate the forest sector under one ministry—the Ministry of Agriculture and Forestry. The Minister of MoNRE officially transferred the DFRM to the Minister of MAF on 19 August 2016. According to the official "Handing over Note", the DFRM under the MoNRE was returned to MAF. The two national REDD+ offices will be merged into one office under DoF.

The R-PP defines an institutional setup for national and subnational levels for governing REDD+ in Lao PDR [57]. In accordance with this setup, a National REDD+ Task Force was established in 2008 under the leadership of the Director-General of Forestry, and composed at that time of 12 members from various organisations and an array of legal mandates [56]. Following the establishment of MoNRE and under the ongoing institutional reforms during 2012 and 2013, the Task Force was moved from MAF to MoNRE, and was expanded to include first 24 members, from 18 ministries and various cross-sectoral organisations [58,59], and then later 30 members. Initially, the Director-General of DFRM chaired the Task Force, then the Vice-Minister of MONRE took over the chairmanship. Now that forest management is again solely the responsibility of MAF, the leadership of the National REDD+ Task Force has returned again to MAF. Now MAF is considering who will lead the Task Force, and how to convene and make decisions. The National REDD+ Office acts as a secretariat to the National REDD+ Task Force. Similarly, provincial REDD+ Task Forces (PRTFs) are supported by Provincial REDD+ Offices (PROs). To date, PRTFs and PROs have been established in three provinces—Houaphan, Luang Prabang, and Champasack. Four provinces—Xayaboury, Luang Namtha, Bokeo, and Oudomxay—are also to establish PRTFs and PROs before the end of the 2016.

2.3.2. FLEGT-VPA Process in Lao PDR

Voluntary Partnership Agreements (VPAs) are a key instrument of the EU FLEGT Action Plan. The VPA is a voluntary bilateral trade agreement between the EU and the government of a timber producing and trading partner country. The VPA becomes legally binding upon signature and ratification by both parties [11]. In Lao PDR, MAF—specifically its Department of Forest Inspection (DoFI)—is the ministry with lead responsibility for negotiating the VPA. The FLEGT VPA informal negotiation started in April 2012, although it was not until June 2015 that the Government Office of Laos approved the start of negotiations. A FLEGT Standing Office was opened by the MAF in October 2013 [60]. In July 2014, a workshop on actor engagement involving participants from the government, private sector, CSOs and academic institutions was held in Lao PDR. At the conclusion of the workshop, participants identified the following as the most essential issues to be resolved in order to advance the VPA process: (1) establishing a national steering committee for the VPA process; (2) building the capacity of actors; (3) strengthening communication on FLEGT among actor groups [61]. During 2014, a number of workshops and events organised by the EU and German agency for international cooperation (GIZ) were held to promote EU FLEGT in Lao PDR and in Southeast Asia in general. The options for inclusive participation, building and involvement of civil society in the process and legal enforcement were reviewed and discussed. The conclusion was that more work and a willingness to foster an open and transparent approach was needed for the FLEGT VPA to be successfully implemented in Lao PDR. These options were revisited almost three years later in December 2016 when a draft Work Plan for Forest Law Enforcement and Governance (FLEG) in ASEAN (Association of South East Asian Nations) was developed.

At the present time, Lao PDR is finalising its legality definition and—as the head of the Lao FLEGT Standing Office stated at the ASEAN meeting in December 2016—the aim is to conclude the VPA negotiations with the EU in 2018. The FLEGT VPA meetings are still ongoing and according to EUFLEGT news (accessed on 25 January 2017), the recent meeting on FLEGT VPA in Lao PDR acknowledged that engagement of private sector and civil society is the main challenge.

Concerning activities on the ground, significant support from the EU is currently organised under the auspices of the GIZ and its ProFLEGT project which aims to conclude a VPA between the EU and the government of Lao PDR. The project activities have so far focussed on support and capacity building for governmental agencies, as well as policy coordination. However, activities on the ground related to testing and implementation of the system in the three pilot provinces—including Sayaboury, Khammouane and Attapeu—have not taken place as originally planned. In addition to the Pro-FLEGT project, the EU-FAO FLEGT Programme was established in 2012. This Programme aims to improve awareness and strengthen the participation of CSOs and forest communities in the FLEGT VPA countries, and has since been active in the Lao FLEGT policy process.

For the government of Lao PDR, a VPA with the EU could provide an opportunity to resolve some of these issues and privileged access to the European market. Nevertheless, a lot will depend on whether FLEGT VPA will catalyse fundamental governance and institutional reforms, and on whether such reforms will address illegal logging and deforestation outside of the forestry sector, such as that which originates from agricultural and other development sectors. However, such a major shift away from the current governance norms is not likely to happen in the near future. This argument is based for example [62]. The main VPA impacts in the country are therefore expected to be in terms of strengthening forest governance generally.

3. Theoretical and Analytical Framework: Political Modernisation and the Policy Arrangement Approach (PAA)

Our analysis for exploring the participation of CSOs in REDD+ and FLEGT VPA processes in Lao PDR is guided by the *policy arrangement approach* (PAA). The PAA is defined as "the temporary stabilisation of the content and organisation of a policy domain" [63] (p. 96). It is helpful to examine some of the theoretical and conceptual underpinnings behind the PAA framework in order to better understand the REDD+ and FLEGT-VPA processes and the roles of CSOs. The concept of *political modernisation*, which is defined as "*a comprehensive process of changes in the political domain of society*" [63] (p. 101), may serve as a useful starting point. According to this concept, political modernisation refers to the process of shifting and redefining the relationships between the state, the market and civil society. These processes are characterised by the relative increase of political power of non-state actors, such as CSOs and private-sector actors [63–65].

Political modernisation consists of two distinct phases: *early modernity* and *late modernity*. Early modernity relates to the politics revolving around a strong nation-state model and the belief in a highly manageable society and nature [65] (p. 343). Early modernity is characterised by a regulatory state that is dominant over the market and civil society and optimism over such a state's capability to solve societal problems by rational policy making and comprehensive planning [65] (p. 343), [66] (p. 29). Late modernity is characterised by a discourse of governance, interdependence and the inevitability of the need for cooperation between government, market and society where there can be "no monopoly of knowledge, problem-solving, or steering capacity", as explained in [63] (pp. 101–102).

One way in which late modernity manifests itself is through what is understood as multilevel governance, or multiactor governance. Multilevel implies that governance takes place on multiple levels from local to global, and that these levels are not 'separated' from each other; rather, their boundaries are blurred and they are in constant interaction with each other. Multiactor refers to the number of different actors besides the state that can be found at the different levels and which constitute what is understood as governance [67,68]. Some examples of these actors in forest governance include the private sector, civil society, international actors, local communities and forest users.

The PAA framework comprises four interrelated analytical dimensions: (i) actors and their coalitions; (ii) resources and power; (iii) discourses; and (iv) rules of the game [64]. Bas Arts and colleagues state in their publications on the PAA that all four dimensions should be considered together in the analysis and that change in one of them means a change in one or more of the other three dimensions. While all these dimensions link to our key research question related to participation, we decided, however, to focus on two dimensions of PAA because those were the most crucial issues based on our pre-study analysis: *rules of the game* and *resources*, which refer to actors' capacities—economic, political, cultural, knowledge and beyond—and the extent to which actors are able to exert influence over other actors (i.e., actors' power relationships). The *rules of the game* can be understood by the following essential questions [63,65]: Who decides the agenda? Who makes the decisions? Due to the interrelatedness of the dimensions of a policy arrangement, these issues of "rules of the game" are closely related to actors' resources and thus also to power [69]. By contrast, resources are defined as assets that policy actors have or can mobilise to achieve their policy goals. The division of power and influence between these actors determines the degree to which actors can influence the policy outcomes. While different types of power capacities are certainly interwoven with each other and they can all be relevant, two of these power types may be of special interest in the context of this study: the political and the knowledge-based power capacities that are a central part of the rules of the game in any governance process.

For participation and further cooperation to be realised, actors can be presupposed to have and hold power, because power is required for actions such as formulating new rules, influencing decisions or reframing discourses that contribute to making participation/cooperation happen. Further, the achieved outcomes depend on the types of resources actors possess and how they use them, and on how actors play the "policy game" and are affected by the rules. Rules of the game decide the (non) existence of participation—i.e., who may participate and who decides this; the actors with the "strongest" resources (e.g., political and knowledge-based power) are the ones who make the decisions and achieve outcomes. The results of this study (Section 5) are presented in two sections based on our interview findings on the CSOs' resources and role in the 'rules of the game': (1) Resources of CSOs and challenges to CSO participation in REDD+ and FLEGT VPA processes; and (2) Rules of the game from the perspectives of CSOs facing challenges in fostering cooperation in REDD+ and FLEGT-VPA processes. Prior to the results and discussion, the research materials and methods are presented.

4. Materials and Methods

The first step to collect the data for this study in 2014 was an extensive review of existing scientific and grey literature, including policy documents, briefing notes and policy updates. The literature review provided important contextual information that was then used for creating an online questionnaire-based survey. This enabled the collection of more specific and relevant data for the study.

The second step in 2014 was identification of key national and international policy actors (representatives of international and national organisations, bilateral and multilateral donor agencies and national level ministries and relevant departments) leading and representing their respective organisations in the REDD+ and FLEGT related processes. In total, 51 contacts were identified and contacted (named in Table 1 as pre-survey interviews) with the help of international and national CSOs and bilateral project staff active in Laos. However, only 39 could be considered to have relevant information related to more recent development in REDD+ and FLEGT processes.

In the third step in August 2014, the identified 39 key respondents were invited to take part in an online survey. These organisations and actors were first contacted by email and then were either interviewed personally by phone or in face-to-face interviews in Lao PDR. Following the interviews, eventually 17 out of 39 respondents (response rate of 44%) took a part in the on-line survey.

In the fourth and final step, key-informant interviews were carried out with selected actors to triangulate the data and generate understanding of the historical context of forest governance in Lao PDR. Three of these informants participated in the survey, but the seven additional informants did not. The face-to-face interviews took place in Vientiane in January and April 2015. In 2016, five of informants were re-contacted by email in order to up-date the information. The distribution of organisational actors in different data collection phases is presented in Table 1.

Table 1. Distribution of organisational actors in the survey and interviews.

Organisational Type	*n* (pre-Survey Interviews 2014)	*n* (Survey 2014)	*n* (Key Informant Interviews 2015)
Government	3	2	
National NGO	1	2	1
Community-based organisation		2	
International NGO	4	5	2
Intergovernmental organisation		1	2
Foreign government agency	5	3	4
Private sector		1	
Academia		1	1
Total	13	17	10

In the analysis of the survey data, distinct approaches were employed for answers to fixed-choice questions and open-ended questions. Responses to fixed-choice questions were graphically displayed in an easily understandable format as quantitative data using Microsoft Excel, and these graphs were then transferred to Microsoft Word. For open-ended questions, analysis was conducted in Microsoft Word where responses by individual respondents were examined to identify main themes or topics in their responses. The data from the interviews conducted in Lao PDR were used to support and complement the survey data so as to provide a deeper understanding and a more refined overall context of the topic.

5. Results and Discussion

5.1. Resources of CSOs and Challenges to CSO Participation in REDD+ and FLEGT VPA Processes

This section addresses the dimension of resources of PAA, interpreted and used hereafter as the capacities of CSOs (as one policy actor) to exert influence over other actors (i.e., actors' power relationships). Figure 1 summarises the types of roles that the CSOs have in the REDD+ process in Lao PDR according to respondents. It is interesting to note that none of the respondents thought that the CSOs—although participating—were actually affecting any decisions surrounding the REDD+ process. Rather, CSOs were seen to have other kinds of roles. Most of the respondents (13 out of 17) acknowledged that the CSOs' role has been to raise awareness and engage in building capacities. For instance, a respondent from a national CSO commented that CSOs had been raising awareness on both REDD+ and FLEGT VPA processes among CSO members. Figure 1 also indicates that the CSOs had a role in terms of implementing projects on the ground. This correlates with the many different REDD+ readiness activities and pilot projects in areas of climate change, deforestation and SFM. Only one respondent (of a national CSO) mentioned that CSOs had been conducting FPIC (Free Prior Informed Consent) activities. FPIC is the principle that a community has the right to give or withhold its consent to proposed REDD+ projects that may affect the lands they customarily own, occupy or otherwise use.

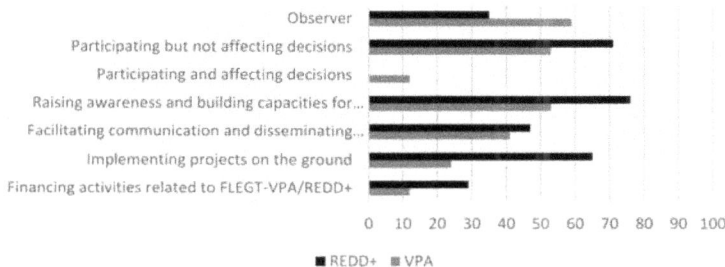

Figure 1. Role of CSOs in REDD+ and FLEGT VPA processes in Lao PDR (number of respondents = 17).

Figure 1 summarises the types of roles played by CSOs in the FLEGT VPA process in Lao PDR. The category "participating and affecting decisions" had a very low score, with only two respondents indicating this option. Even so, the result was not as drastic as in the case of REDD+ where none of the respondents indicated this option.

Nine respondents thought that the role of CSOs is to raise awareness and help build capacities for others. For instance, a respondent from a national CSO stated that they were providing information to their CSO network, which is an informal network between collaborating actors. Furthermore, as shown in Figure 1, six respondents indicated that "facilitating communication and disseminating information between different levels" was a role of CSOs in the FLEGT VPA process. Later, during face-to-face interviews, a Lao CSO respondent commented that they had been facilitating communication and disseminating information concerning FLEGT-VPA issues for local authorities and local communities and sharing lessons learnt at regional and national level. Lastly, only four of the respondents, in comparison to 11 in the case of REDD+, mentioned that CSOs were "implementing projects on the ground". The difference could partly be explained by the fact that the FLEGT VPA process was still at a very early stage, and CSOs are currently not officially involved in forest law enforcement activities. In reality, however, via their networks, social media and newspapers, they have an important role in monitoring and reporting for example illegal logging activities in Lao PDR, as commented on by an international CSO representative during pre-study interviews.

Much of the wood in Lao PDR originates from conversions of forests to other land uses, such as teak and rubber plantations, hydropower plants and mining and other infrastructure projects. The government of Laos (GoL) has been experimenting with bans on logging and/or log exports from as early as 1991 [26,49,62] when the nationwide moratorium on all logging concessions was adopted (Decree No. 67, from August 1991). However, logging—legal and illegal—has continued ever since; in the 1980s and 1990s logging was undertaken exclusively by the Lao Army and state owned enterprises, and later by international investments under the pretext of nation-wide development projects.

Several scholars have shown that private investors have so far successfully circumvented the logging ban by using the willingness of the government to grant land concessions for development projects, such as hydropower installations, plantations and mining (see for example [26,49,54]). Regrettably, as scholars and practitioners have by now concluded, many of these development projects never take place because many private investors abuse the development concessions by selling their development concessions to a third party after having logged the land and profited from timber sales. Therefore, large-scale logging has continued to a greater or lesser extent in Lao PDR despite the formal logging moratorium.

Separate questions were posed about challenges related to the participation of CSOs in REDD+ and in FLEGT VPA processes, but the data show noticeable similarities between both cases. Therefore, these results are presented jointly. Based on the responses, four main categories of challenges were identified: (i) governmental control and lack of an enabling political environment for CSOs to operate; (ii) limited understanding about REDD+ and FLEGT-VPA processes and their complexity; (iii) limited human and financial resources within the CSOs; and (iv) few national efforts in enabling REDD+ and VPA implementation in the country.

The first and main challenge raised during the survey, but also during face-to-face interviews among the respondents, is that the political system does not support or facilitate the operation of an independent civil society. Respondents also expressed a perception that the government exercises strict central control over REDD+ and related issues, as well as over the FLEGT-VPA process. One of the national key informants argued during pre-study face-to-face interviews in 2014 that it is generally difficult for non-state actors, including CSOs, to influence or affect government policy processes. These views are in line with the country's political reality, which limits the functioning of a free and independent civil society. During the survey, another national CSO respondent claimed that there is no free speech and that it will never be possible in the current political context. Moreover, the government is intending to pass new decrees that would restrict the operation of international CSOs as well as the establishment and registration of national CSOs in the country [48]. Consequently, media with a focus on the region, such as South China Morning Post (17 September 2014) and Radio Free Asia (2 October 2014), have been worried that this would curb the space for open dialogue in Lao PDR society. The new decrees would further strengthen government control over CSOs in Lao PDR, and possibly impede the funding of CSOs with foreign money.

More specifically for REDD+ and FLEGT VPA processes, during the survey, a government representative explained that there is a lack of an enabling regulatory and institutional framework for CSO participation. Progress in national REDD+ development has remained fairly limited, and this is reflected in the survey responses. REDD+ as well as FLEGT require an institutional setup that promotes and facilitates multiactor participation in order to have all relevant actors involved in the process. Additionally, a survey respondent of a foreign government agency argued that a CSO representative should be a member of the national REDD+ Task Force and participate in the decision making process on REDD+. The same respondent also noted that such a development is not likely to happen in the near future. Moreover, for the FLEGT VPA process, a survey respondent of an international NGO explained that not many CSOs work on this because the topic of FLEGT VPA is a sensitive issue in Laos. The respondent further explained that forest law enforcement and monitoring of illegal logging will be difficult to implement at a local level because such actions would run counter to national and subnational income from the unlicensed logging activities. Indeed, based on several documents and articles, illegal logging and related trade are controversial issues in the country because it is argued that influential actors from the private and the public sector are involved in these trades and powerful vested interests exist [26,49,70].

The second and equally important fundamental challenge concerns the technical complexity of REDD+ and lack of knowledge and understanding about FLEGT VP. This challenge holds true for most of the actors involved in both processes. While it is important for participating actors to understand the processes and how the processes relate to each other and other factors (such as forest monitoring, forest management planning, logging licencing), there has been little information dissemination and capacity building to address the issue. Ambiguity about REDD+ will drive actors' interests down and it will be difficult for the process to gain momentum. One respondent of an international NGO stated that it is difficult to see how REDD+ relates to the communities, i.e., to their participation and benefits. In addition, on several occasions, the lack of REDD+ structures, pilot project structures and progress were pointed out as obstacles and barriers for REDD+ implementation. This implies that four years after the approval of the country's R-PP's, there is still a need for extensive capacity building among actors "outside of the country's capital". As one respondent of an international organisation put it, there is need to simplify REDD+ and make it understandable for a wider range of actors. Arguably, this is challenging when overall progress with REDD+ in Lao PDR has been slow, and institutional and regulatory developments within the process have been lagging behind. The same logic applies to the VPA process; without sufficient understanding, actors cannot properly engage in the VPA process. Moreover, awareness of the VPA process among Lao CSOs and communities is still low, as pointed out by some respondents from private sector and community-based organisations. A key informant also stated that awareness of VPA is even lower than that of REDD+.

The third significant challenge is a lack of capacities. The Lao NPAs were specifically pointed out as lacking human, financial and project management capacities in relation to REDD+ and VPA processes. According to a key informant's interviews, the most pressing issues are the lack of financial and properly trained human resources in the NPAs. A respondent of a foreign government agency also stated that NPAs do not have adequate knowledge about international policy processes, and the "big picture" surrounding those processes. Lack of these capacities and resources undoubtedly has to do with the very recent emergence of Lao civil society organisations and the limited political space for their activities and growth. This issue will need to be addressed for Lao NPAs to be able to effectively engage in REDD+ and VPA processes. However, in the current setting, this is only possible with facilitation by the national government. This is because the government, by passing laws and regulations, ultimately determines the degree to which NPAs may or may not be allowed to develop themselves.

The fourth challenge reflects the government's perceived lack of national efforts in enabling REDD+ and VPA implementation in the country. One respondent from the national level said that there is a contradiction in the national vision of development. The issues of conservation and of contradictory government policies were again mentioned by another respondent and an informant in 2016, as also evidenced by [71]. Further, a respondent of an international NGO stated that there is a lack of initiative with regard to the REDD+ process within the government. Instead of national authorities or national CBOs, the key players have been the bilateral and multilateral donor projects (such as CliPAD, and JICA's F-REDD project). The CSOs and the INGOs do not have a clear role in the REDD+ process in Lao PDR. SNV (an INGO founded in the Netherlands in 1965, with local presence in many countries in Asia, Africa and Latin America) has been active through their REDD+ pilot project, but this support was available only until October 2016.

A key informant explained that the green light from the government to involve CSOs in the FLEGT-VPA process has been pending. Understandably, it can become difficult to involve CSOs in the REDD+ or VPA process if the process lacks clear structure, direction and momentum [72], and the government is first and foremost responsible for all of these. For instance, in 2012, in the case of REDD+, the division of DoF between two ministries as a result of institutional reform was seen to cause weaknesses in its capacities; the lack of clarity in institutional mandates thus caused delays in the processing of FCPF grant financing by the World Bank [73]. However, in 2016, structural changes and integration of international experts under DoF were again taking place. This is the result of transfer of the DFRM from the Ministry of Natural Resources and Environment back to the Ministry of Agriculture and Forestry, and the REDD+ office is now being consolidated under DoF.

5.2. The "Rules of the Game" and Challenges of Realising Cooperation between REDD+ and FLEGT VPA Processes

Most respondents, in answering the question of future potential cooperation between REDD+ and FLEGT-VPA process, identified various opportunities such as matters concerning governance of forest, the law reform and enforcement. These three aspects relate to the interlinkages between the rules of the game and resources (the two analytical dimensions included in the PAA; see for example [63,65]). The "rules of the game" are closely related to actors' resources and therefore to power during the policy game: Who participates in the policy development related to REDD+ and FLEGT-VPA? Are CSOs excluded or will they participate in the future? If CSOs participate in the policy game, they need resources, which refers to actors' capacities—economic, political, cultural, knowledge and beyond – and also to the extent to which they are able to exert influence over other actors (i.e., actors' power relationships; see for example [63,65]. While different types of power capacities are certainly interwoven with access to resources, two capacities—the political and the knowledge-based power capacities—may be of special interest in the context of potential participation of CSOs in governance of forests, law reform and enforcement. These two capacities—the political and the knowledge-based power capacities—are central to the rules of the game in REDD+ and FLEGT-VPA processes.

For example, six respondents highlighted the overlap in the institutional structures of the two initiatives and in the potential participation of CSOs to foster cooperation. Foreign government agency respondents pointed out that under the process of REDD+ policy development, key actors such as CSOs should be involved so that duplications and contradictions within VPA process can be avoided. Further, the same respondents said that the work of the technical working groups of both processes should be synchronised. Misunderstanding and miscommunication were said to be the wider issues currently and the organisational setup of both MAF and MONRE were described as very complex. Due to the fact that causes of deforestation and forest degradation largely originate from outside the forest sector [74] and that REDD+ (and also VPA) requires effective cross-sectoral coordination and

CSOs participation, advanced communication mechanisms within and among ministries would play a key role.

Three respondents also acknowledged that there is potential for cooperation between the two processes in terms of addressing causes of deforestation and forest degradation. The two processes were regarded as providing opportunities for strengthened roles and rights of local communities. A foreign government agency respondent argued, for example, that REDD+ and FLEGT-VPA processes should adopt common approaches for empowering local communities to have rights and means to engage in law enforcement around forests, especially in combating poaching and illegal logging, and to receive real incentives for doing so. This view was shared by a key informant, who also noted CSOs' possible role in training local communities for these actions. A government respondent echoed this view. Beyond deforestation and forest degradation, six different respondents pointed to forest governance in general as an important topic for both processes; some areas for cooperation include access to information and greater transparency, and land law reform and enforcement. As one international expert noted: "Both (processes) are strongly linked to governance issues, so there could be scope for using both processes to improve forest governance." The respondent continued, however, that "(this discussion) would have to involve the policy sector in a broad sense—not just the forestry sector, which operates independently of for example land development sectors."

Figure 2 presents potential challenges to creating cooperation between the two processes. The two options clearly chosen most often were "insufficient capacities" (14 respondents) and 'lack of political will/interest' (13 respondents). The lack of capacities appears to be omnipresent and may undermine many efforts. These findings are linked to the concept of *resources* and refer to actors' capacities—economic, social and political—as well as to the concept of power.

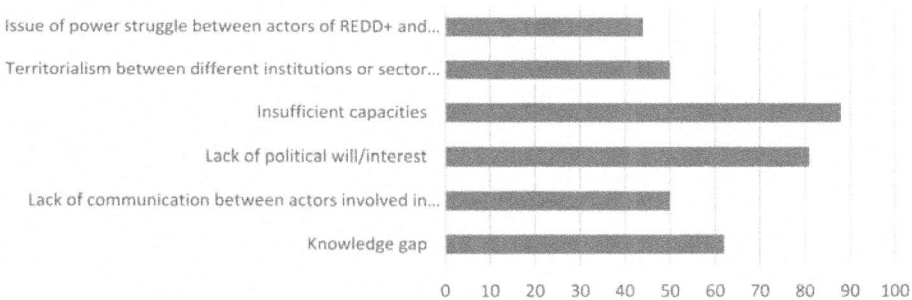

Figure 2. Challenges to cooperation between REDD+ and VPA processes (number of respondents = 16).

For example, the interviewed foreign government agency representative highlighted both processes as requiring significant investments from the Lao government, and that coordinated cooperation between REDD+ and VPA processes could lead to increased efficiency. There could consequently be a reduced need for financial and human resources, therefore such cooperation could be something to truly strive for in the country context of Lao PDR. A foreign government agency respondent explained the lack of political will or interest as related to the fact that (in the case of the VPA process) very little timber from Lao PDR goes to Europe and that benefits from REDD+ are still unclear: "So, what is the incentive for the Lao government to take them seriously?" This lack of interest could partly be explained by the third most chosen option (10 respondents), namely the "knowledge gap" about the processes and possible cooperation between them. Understanding opportunities for cooperation and its potential benefits first requires in-depth knowledge of each process, which is not yet the case for many actors in Lao PDR. An international NGO respondent argued that REDD+ and VPA concepts are new to Lao PDR and capacities of relevant government and private actors in relation to these concepts were still very low. A Lao CSO respondent explained that many people were saying that these issues were new for them and very difficult to understand, so they did not become interested in them in the first place. This indicates the need for awareness raising and capacity building.

Respondents also acknowledged that the "issue of power struggle between actors of REDD+ and VPA processes" (7 respondents), "territorialism between different institutions or sector organisations" (8 respondents) and "lack of communication between actors involved in REDD+ and FLEGT-VPA" (8 respondents) also pose challenges to bringing about cooperation between the processes. Each of these difficulties also relates to the same cross-sectoral aspect of different ministries and organisations that are involved in the same governance matters. These issues are not unique to Lao PDR (see e.g., [20] Tegegne et al., 2014 for similar findings in Cameroon and Congo Basin countries); both REDD+ and FLEGT-VPA require improved governance involving multiple actors in order to be successful. Therefore, strong leadership and coordination across sectors and institutions is instrumental to progress.

One of the respondents (a representative of a foreign government agency) explained the current lack of cooperation as being due to several factors. According to another respondent discussions and coordination between the processes cannot be constructive when the institutional setup for REDD+ is still ambiguous and the collaboration among actors generally weak. The respondent further added that the agency in charge had not shown strong leadership in coordination. Moreover, institutional and human capacity is quite limited and actors cannot understand well the requirements of both initiatives, and therefore it is quite difficult to discuss the duplication, demarcation and cooperation of the initiatives among actors.

Finally, respondents were also asked to evaluate at what organisational/administration/interest level(s) they believed that challenges to cooperation would occur (Figure 3). Here, a trend of increased "likelihood" of challenges is observed to occur from lower towards higher levels of governance. The implication is that decision-making power is concentrated at higher levels of governance rather than at lower levels, as in a 'top-down' system. This is because challenges would occur at the level where actual decisions are being made and authority is located. On the other hand, a foreign government agency respondent argued that 'once you have a successful case at village level, district government will become your ambassador to the higher levels of government'. Challenges that prevent or obstruct cooperation would still need to be tackled at all levels, with sufficient attention being paid to central and provincial levels.

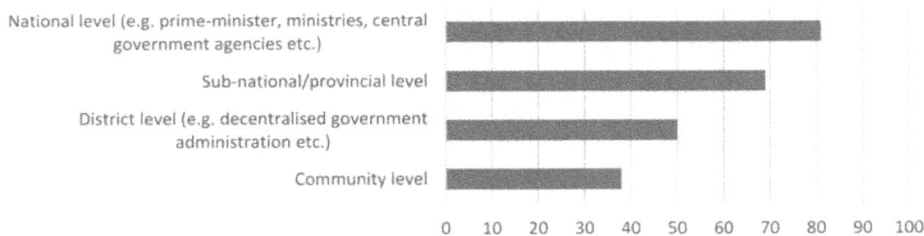

Figure 3. Perceived occurrence of challenges related to cooperation at different levels (number of respondents = 16).

The literature describes how for one process or regime to be able to "learn" from another process and apply the lessons learnt, their institutions and/or membership must overlap [19,20,75]. In the REDD+ and FLEGT-VPA processes in Lao PDR, as described in detail in Section 2.3, several albeit shifting and therefore relatively complicated institutional connections and overlaps do exist between actors and institutions involved. Based on interviews as well as national documents (such as [57]), it is possible that more and new overlaps will come into effect as both processes take shape and evolve. Currently, existing overlaps imply that the REDD+ and FLEGT-VPA processes already have interlinks in memberships and thus are able to "learn" from each other and apply lessons learnt. However, both of these processes are centrally controlled, and although CSOs could have *resources* to support the implementation of both processes, currently CSOs have limited access to set the *rules of the game*, as we further elaborate in Section 6.

6. Conclusions and Recommendations

The national CSOs in Lao PDR are composed of the newly emerging Non-Profit Associations (NPAs). However, the government does not view these organisations as independent players nor as a balancing force or counterweight to government policies, as is the case in many other countries [46]. In other words, the government does not share power with these actors (nor any other) but retains central control and power for itself [46,76]. Instead, the government recognises that CSOs should play a role in advancing development in the country, but this role should not contradict government policies and goals [48]. The political environment in Lao PDR provides the frame for the conceptual understanding of *resources*, which refers to the extent to which CSOs are able to exert influence over central control and power (actors' power relationships, as referred to in Bas Arts' conceptual frame [69]). Based on this interpretation, the *rules of the game* in REDD+ and FLEGT-VPA processes can be understood through the following crucial questions [63,65]: Who decides on the policy agenda and processes? Who may participate? Who decides the implementation?

The results of this study suggest that participation of international and local CSOs in REDD+ and FLEGT-VPA processes in Lao PDR is limited and significantly constrained. Where the CSOs are able to participate, the nature of that participation seems not to actually affect decisions that steer the processes, but serves merely to raise awareness and to a lesser degree to bring about implementation of local pilot project activities. Moreover, for the most part, the same kinds of challenges exist and hinder meaningful participation of local CSOs in both REDD+ and FLEGT-VPA; i.e., awareness and knowledge of the processes are low, and local CSOs lack capacities to a significant extent. The political system and governance structures of Lao PDR are also unsupportive of open and free participation by local as well as international CSOs. The respondents felt that local CSOs need better knowledge of REDD+ as well as of FLEGT-VPA and that experiences from other countries could be utilised to link these processes together and provide meaningful ways for CSOs to participate in them. While evidence of existing forms of cooperation between REDD+ as well as of FLEGT-VPA was limited, respondents identified a potential for bringing about cooperation in the future. For this to happen, the respondents again stressed the importance of building the capacities of all the actors involved. Three respondents also acknowledged that there is potential for cooperation between the two processes/international regimes in terms of addressing causes of deforestation and forest degradation (c.f., similar experiences related to the two international regimes in Cameroon and the Republic of Congo [20]).

Interaction of international regimes is premised on the fact that these regimes influence the development and performance of each other in several ways [77]. The interactions between policies depend on the overall political and institutional setting in which policies or regimes exist and interact [78]. This paper contributes to the literature on these interactions by investigating *rules of the game* and how CSOs could strengthen the interactions between FLEGT and REDD+ in Lao PDF country context.

Country context is important when it comes to effectiveness and pace of REDD+ readiness [79] or other natural resources governance-related processes. Indeed, the seemingly slow progress of Lao PDR in REDD+ and FLEGT-VPA can be partly explained by overall weak governance, the current political environment, and lack of *resources*. Thus, we see the multifaceted nature of power in relation to availability of resources and the capacity to achieve outcomes [69]. Participation and cooperation can be presupposed to require that specific actors exercise power if they are to perform these functions. But weak capacity or willingness of the government institutions and/or other "duty bearer" agencies will hinder the formulation of the rules etc. needed to make participation/cooperation happen. For example, in Lao PDR, the issue of weak forest governance is partly seen as a result of lack of a regulatory framework and legislation. However, while some inconsistencies and loopholes do exist, the enforcement and implementation in particular of laws and policies in Lao PDR is also acknowledged to be poor [32,49,57].

The latest in over two decades of GoL logging and/or log export bans is the new Prime Minister's Order (PM's Order 15) ban on timber exports, issued in May 2016. The moratorium does not exempt project developers or infrastructure-concession operators, and requires them not to use timber to pay for infrastructure development projects. In addition, the government is in charge of logging and selling wood directly to project developers and concession operators. In December 2016, the Laotian Times reported that the PM's Order 15 had successfully curbed harmful practices in the forestry sector and timber industry. The scientific community has therefore been led to believe the implementation of the Order has so far been taken seriously. There have as yet been no formal studies to support the merits of this Order and immediate and urgent research into these claims is required.

Thompson, Baruah and Carr [16] point out that it is problematic that states are typically treated as the main parties responsible for implementing REDD+, when they do not have sufficient capacities to successfully enforce rules and regulations. The kind of outcomes that are actually achieved depend, however, on the types of resources actors possess and how they use them, and on how actors play the whole "policy game", i.e., are influenced by which rules. Rules of the game decide the (non)existence of cooperation, who may participate and who decides this. Those who make the decisions and achieve outcomes are the actors with the greatest resources (e.g., political and knowledge-based power). A recent study [79] across REDD+ countries also found that not much attention is paid to subnational-level processes. As the REDD+ and FLEGT-VPA processes in Lao PDR evolve, state actors will need to consider how CSOs can be better involved in the processes, and in what roles, so as to utilise their capacities and for the CSOs themselves to benefit from the processes.

Currently, in Lao PDR, international and national CSOs are to a great extent concerned about and affected by their opportunities to participate in natural resources governance. In this situation, where local communities are adjusting to REDD+ and FLEGT-VPA to be able to benefit from global flows of values, CSOs are being placed in an increasingly important role, serving as the voice of the community. Traditionally, CSOs have been the weakest actors in natural resources governance in Lao PDR, but CSOs now perceive their role to be growing stronger. In the case of REDD+ and FLEGT-VPA, CSOs (both national and international) are acting as the mediators between global contexts and local conditions.

In Lao PDR, international and national CSOs should and do have a role in facilitating both REDD+ and FLEGT-VPA processes. However, it is important to remember this basic rule of the game in life in general: the actors with the greatest resources can easily overrule the decision making of those with fewer resources. There is also therefore a risk that externally funded and hence greater resourced CSOs "speak on behalf" of local communities and that local governance structures are bypassed in the process. There is further research needed on how the district and provincial level governance and village organisations, such as mass organisations (e.g., youth and women's unions) and forest management units at village level, will be integrated in the REDD+ and FLEGT-VPA processes, because they are key actors in the reduction of deforestation and forest degradation and enforcement of forest law.

This study has important implications for the countries in the Southeast Asian region involved in the FLEGT VPA and REDD+ processes, and in particular those such as Vietnam and Cambodia with similar socio-political environments. The finding that CSOs have a key role to play in strengthening cooperation and interactions between REDD+ and FELGT VPA, as well as in awareness raising and pilot project implementation, make it imperative to address further the participation challenges faced by CSOs in the region.

Limited understanding of REDD+ and FELGT VPA processes and limited human resources were identified as key challenges to CSO participation. On the other hand, according to the study, CSO understanding and resources are likely to improve in step with stronger CSOs which have a key role in raising awareness about FLEGT and REDD+. In other words, creating and promoting enabling conditions for a better involvement of CSOs is an important strategy to address barriers to CSO participation. In Lao PDR, the authoritarian regime would surely benefit from a more open and inclusive approach. Such an approach would in turn increase the legitimacy of government operations as well as acceptance of external interventions, such as FLEGT VPA and REDD+ processes. Social resistance—either hidden or more strategic and organised—continues to obstruct both acceptance of state authority and implementation of international interventions in Southeast Asia [80]. We therefore argue in favour of participation of non-state actors as a strategy for smoother implementation of environmental governance interventions in general and of FLEGT and REDD+ in particular.

Acknowledgments: We would like to acknowledge all respondents in Lao PDR who took part in the study survey. This work was part of a long-term study funded by the Academy of Finland (grant number 265,159) called "REDD+: The new regime to enhance or reduce equity in global environmental governance? A comparative study in Tanzania, Mexico and Laos". The authors closely cooperated within this comparative research project. We would especially like to acknowledge Salla Rantala and Paula Williams for all their valuable comments, and a warm thanks is extended to Yidnekachew Biza for his research assistance. The European Forest Institute is also acknowledged for institutional support. We also acknowledge the proofreading support of Nick Quist Nathaniels.

Author Contributions: Irmeli Mustalahti, main author; Mathias Cramm, collecting and analyzing data; Sabaheta Ramciloviv-Suominen, REDD+ issues and literature analysis; Yitagesu T. Tegegne, FLEGT VPA issues.

Conflicts of Interest: The authors declare no conflict of interest.

References

1. Intergovernmental Panel on Climate Change (IPCC). Summary for Policymakers. In *Climate Change 2013: The Physical Science Basis*; Contribution of Working Group I to the Fifth Assessment Report of the Intergovernmental Panel on Climate Change; Stocker, T.F., Qin, D., Plattner, G.-K., Tignor, M., Allen, S.K., Boschung, J., Nails, A., Xia, Y., Bex, V., Midgley, P.M., Eds.; Cambridge University Press: Cambridge, UK; New York, NY, USA, 2013.
2. Achard, F.; Beuchle, R.; Mayaux, P.; Stibig, H.J.; Bodart, C.; Brink, A.; Carboni, S.; Desclee, B.; Donnay, F.; Eva, H.D.; et al. Proximate causes and underlng driving forces of tropical deforestation. *Glob. Chang. Biol.* **2014**, *20*, 2540–2554. [PubMed]
3. Hansen, M.C.; Potapov, P.V.; Moore, R.; Hancher, M.; Turubanova, S.A.; Tyukavina, A.; Thau, D.; Stehman, S.V.; Goetz, S.J.; Loveland, T.R.; et al. High-resolution global maps of 21st-century forest cover change. *Science* **2013**, *342*, 850–853. [CrossRef] [PubMed]
4. Lima, M.G.B.; Braña-Varela, J.; Kleymann, H.; Carter, S. *Contribution of Forests and Land Use to Closing the Gigatonne Emissions Gap by 2020*; WWF-WUR policy brief; Wageningen University and Research: Wageningen, The Netherland, 2015.
5. United Nations Framework Convention on Climate Change (UNFCCC). Key decisions relevant for reducing emissions from deforestation and forest degradation in developing countries (REDD+). Available online: http://unfccc.int/6917.php (accessed on 15 December 2016).
6. Conca, K.; Dabelko, G.D. *Green Planet Blues: Critical Perspectives on Global Environmental Politics*; Conca, K., Dabelko, G.D., Eds.; Westview Press: Boulder, CO, USA, 2014.
7. Ebeling, J.; Yasue, M. Generating carbon finance through avoided deforestation and its potential to create climatic, conservation and human development benefits. *Philos. Trans. R. Soc. B Biol. Sci.* **2008**, *363*, 1917–1924. [CrossRef]
8. Grieg-Gran, M.; Porras, I.; Wunder, S. How can market mechanisms for forest environmental services help the poor? Preliminary lessons from Latin America. *World Dev.* **2005**, *33*, 1511–1527. [CrossRef]
9. Gupta, A. Transparency in Global Environmental Governance: A Coming of Age? *Glob. Environ. Politics* **2010**, *10*, 1–9. [CrossRef]
10. The United Nations Framework Convention on Climate Change (UNFCCC). Adoption of the Paris Agreement. Available online: https://unfccc.int/resource/docs/2015/cop21/eng/l09r01.pdf (accessed on 15 December 2016).

11. European Commission. Communication from the Commission to the Council and European Parliament: Forest Law Enforcement, Governance and Trade (FLEGT) Proposal for an EU Action Plan. Available online: http://eur-lex.europa.eu/legal-content/EN/TXT/?uri=CELEX%3A52003DC0251 (accessed on 15 December 2016).

12. United Nations. Non-Legally Binding Authoritative Statement of Principles for a Global Consensus on the Management, Conservation and Sustainable Development of All Types of Forests. A/CONF.151/26 (Vol. III). Annex III. Available online: http://www.un.org/documents/ga/conf151/aconf15126-3annex3.htm (accessed on 15 December 2016).

13. Ramcilovic-Suominen, S.; Gritten, D.; Saastamoinen, D. Concept of livelihood in the FLEGT voluntary partnership agreement and the expected impacts on the livelihood of forest communities in Ghana. *Int. For. Rev.* **2010**, *12*, 361–369. [CrossRef]

14. Rica, C.; Corbera, E.; Estrada, M.; May, P.; Navarro, G.; Pacheco, P. Rights to Land, Forests and Carbon in REDD+: Insights from Mexico, Brazil and Costa Rica. *Forests* **2011**, *2*, 301–342.

15. Mustalahti, I.; Bolin, A.; Paavola, J.; Boyd, E. REDD+ reconcile local priorities and needs with global mitigation benefits? Lessons from Angai Forest, Tanzania. *Ecol. Soc.* **2012**, *17*, 16. [CrossRef]

16. Thompson, M.C.; Baruah, M.; Carr, E.R. Seeing REDD+ as a project of environmental governance. *Environ. Sci. Policy* **2011**, *14*, 100–110. [CrossRef]

17. Beeko, C.; Arts, B. The EU-Ghana VPA: A comprehensive policy analysis of its design. *Int. For. Rev.* **2010**, *12*, 221–230. [CrossRef]

18. Dooley, K.; Ozinga, S. Building on Forest Governance Reforms through FLEGT: The Best Way of Controlling Forests' Contribution to Climate Change? *Rev. Eur. Community Int. Environ. Law R* **2011**, *20*, 163–170. [CrossRef]

19. Ochieng, R.M.; Visseren-hamakers, I.J.; Nketiah, K.S. Interaction between the FLEGT-VPA and REDD+ in Ghana: Recommendations for interaction management. *For. Policy Econ.* **2013**, *32*, 32–39. [CrossRef]

20. Tegegne, Y.; Ochieng, R.; Visseren-Hamakers, I.; Lindner, M.; Fobissie, K. Comparative analysis of the interactions between the FLEGT and REDD+ regimes in Cameroon and the Republic of Congo. *Int. For. Rev.* **2014**, *6*, 602–614. [CrossRef]

21. Mustalahti, I.; Taku, T. Forest management in REDD+: New opportunity or more risks? *Scand. J. For. Res.* **2012**, *27*, 200–209. [CrossRef]

22. Mustalahti, I.; Rakotonarivo, S. REDD+ and Empowered Deliberative Democracy: Learning from Tanzania. *World Dev.* **2014**, *59*, 199–211. [CrossRef]

23. Ministry of Planning and Investment; World Bank. *Lao People's Democratic Republic—Investment and Access to Land and Natural Resources: Challenges in Promoting Sustainable Development, a Think Piece (A Basis for Dialogue)*; World Bank: Washington, DC, USA, 2011.

24. Baird, I.G. Turning Land into Capital, Turning People into Labor: Primitive Accumulation and the Arrival of Large-Scale Economic Land Concessions in the Lao People's Democratic Republic. *New Propos. J. Marx. Interdiscip. Inq.* **2011**, *5*, 10–26.

25. Wong, G.Y.; Darachanthara, S.; Soukkhamthat, T. Economic Valuation of Land Uses in Oudomxay Province, Lao PDR: Can REDD+ be Effective in Maintaining Forests? *Land* **2014**, *3*, 1059–1074. [CrossRef]

26. Lestrelin, G.; Trockenbrodt, M.; Phanvilay, K.; Thongmanivong, S.; Vongvisoul, T.; Pham, T.T.; Castella, J.C. *The Context of REDD+ in the Lao People's Democratic Republic Drivers, Agents and Institutions*; Center for International Forestry Research (CIFOR): Bogor, Indonesia, 2013.

27. Mission Aviation Fellowship. *Annual Review of REDD+ Activities in Lao PDR 2012–2013*; Mission Aviation Fellowship: Vientiane, Laos, 2013.

28. Barney, K. China and the Production of Forestlands in Lao PDR: A political ecology of transnational enclosure. In *Taking Southeast Asia to Market: Commodities, Nature, and People in the Neoliberal Age*; Cornell University Press: Ithaca, NY, USA, 2008; pp. 91–107.

29. Lestrelin, G. Land degradation in the Lao PDR: Discourses and policy. *Land Use Policy* **2010**, *27*, 424–439. [CrossRef]

30. Vongvisouk, T.; Mertz, O.; Thongmanivong, S.; Heinimann, A.; Phanvilay, K. Shifting cultivation stability and change: Contrasting pathways of land use and livelihood change in Laos. *Appl. Geogr.* **2014**, *46*, 1–10. [CrossRef]

31. Daoroung, P. Community forests in Lao PDR: The new era of participation? *Watershed* **1997**, *3*, 1–8.

32. Fujita, Y.; Phengsopha, K. The Gap between Policy and Practice in Lao PDR. In *Lessons from Forest Decentralization: Money, Justice and the Quest for Good Governance in Asia-Pacific*; Colfer, C.J.P., Dahal, G.R., Capistrano, D., Eds.; Earthscan: London, UK, 2008; pp. 117–131.

33. Program, L.F.; Makarabhirom, P.; Raintree, J.; Lao, T.; Forestry, S. *Comparison of Village Forestry Planning Models Used in Laos*; Pearmsak Makarabhirom and John Raintree Regional Community Forestry Training Center (RECOFTC): Bangkok, Thailand, 1999.

34. Government of Lao PDR. *The Forestry Law*; Government of Lao PDR: Vientiane, Laos, 1996.

35. Samountry, X.; Bounphasaisol, T.; Leuangkhamma, T.; Phiathep, O.; Wayakone, S.; Williams, P. *Evaluation of Forest Management and Conservation Programme. FOMACOP Village Forestry Pilot Model (1995–2000). Technical Report*; FOMACOP: Vientiane, Laos, 2001.

36. Mustalahti, I. Participatory forestry in the crossroads in Laos and Vietnam. Two participatory forestry case studies. In *Contextualising Natural Resource Management in the South*; Vihemäki, H., Ed.; Institute of Development Studies, University of Helsinki: Helsinki, Finland, 2007; pp. 193–236.

37. Government of Lao PDR. *Decree on Endorsement and Declaration of the Forestry Strategy to the Year 2020 of the Lao PDR*; Government of Lao PDR: Vientiane, Laos, 2005.

38. Mustalahti, I.; Lund, J.F. Where and How Can Participatory Forest Management Succeed? Learning From Tanzania, Mozambique, and Laos. *Soc. Nat. Resour. Int. J.* **2009**, *23*, 31–44. [CrossRef]

39. Phanthanousy, B.; Sayakoummane, S. *The Lao PDR—Community Forestry in Production Forest 2005*; Community Forestry Policy Forum, RECOFTC: Bangkok, Thailand, 2005.

40. Ministry of Agriculture and Forestry. SUFORD Scaling Up. Available online: http://www.suford.org/wp-content/uploads/2015/01/sufor_scaling_up_brochure.pdf (accessed on 15 December 2016).

41. World Bank. *Project Appraisal Document. Scaling–up Participatory Sustainable Forest Management Project*; Report No: 75632-LA; World Bank: Washington, DC, USA, 2013.

42. Government of Lao PDR. *National Growth and Poverty Eradication Strategy*; Government of Lao PDR: Vientiane, Lao People's Democratic Republic, 2004.

43. Government of Lao PDR. *Decree on Sustainable Management of Production Forest Areas*; Government of Lao PDR: Vientiane, Lao People's Democratic Republic, 2002.

44. Government of Lao PDR. *Decree on Associations, No. 115/PM.*; Government of Lao PDR: Vientiane, Lao People's Democratic Republic, 2009.

45. Stuart-Fox, M. The Political Culture of Corruption in Lao PDR. *Asian Stud. Rev.* **2006**, *30*, 59–75. [CrossRef]

46. Belloni, R. Development in Practice Building civil society in Lao PDR: The Decree on Associations. *Dev. Pract.* **2014**, *24*, 353–365. [CrossRef]

47. KEPA. Reflections on Lao Civil Society. Available online: www.kepa.fi/tiedostot/lao_cs_2013.pdf (accessed on 10 November 2014).

48. Government of Lao PDR. *Draft Decree on Associations and Foundations*; Government of Lao PDR: Vientiane, Lao People's Democratic Republic, 2014.

49. Barney, K.; Canby, K. *Baseline Study 2, Lao PDR: Overview of Forest Governance, Markets and Trade*; EU FLEGT Facility: Kuala Lumpur, Malaysia, 2011.

50. LIWG. INGOs Important Contributors to Development. Land issues working group: Vientiane, Laos PDR. Available online: http://www.laolandissues.org/2014/10/29/ingos-important-contributors-to-development/ (accessed on 15 December 2016).

51. Morning Post. Laos NGO Restrictions Threaten Development, Say Non-Profit Groups. Available online: http://www.scmp.com/news/asia/article/1594490/laos-ngo-restrictions-threaten-development-say-non-profit-groups (accessed 15 December 2016).

52. Ministry of Natural Resources and Environment. *MoNRE Regarding the Organization and Activities of the Department of Forest Resource Management*; Decision No. 3121; Ministry of Natural Resources and Environment: Vientiane, Lao People's Democratic Republic, 2012.

53. Barney, B.K. Power, Progress and Impoverishment: Plantations, Hydropower, Ecological Change and Community Transformation in Hinboun District, Lao PDR. Available online: http://www.cifor.org/publications/pdf_files/books/bbarney0701.pdf (accessed on 15 December 2016).

54. Baird, I. *Quotas, Powers, Patronage and Illegal Rentseeking: The Political Economy of Logging and the Timber Trade in Southern Laos*; Forest Trends: Vientiane, Lao People's Democratic Republic, 2009.

55. Kenney-Lazar, M. Land Concession, Land Tenure, and Livelihood Change: Plantation Development in Attapeu Provinc, Southern Laos. Available online: http://www.laolandissues.org/wp-content/uploads/2012/01/Kenney-Lazar-Land-Concessions-Attapeu.pdf (accessed on 15 December 2016).
56. Vongvisouk, T.; Lestrelin, G.; Castella, J.; Mertz, O.; Broegaard, R.B.; Thongmanivong, S. REDD+ on hold: Lessons from an emerging institutional setup in Laos. *Asia Pac. Viewp.* **2016**, *57*, 393–405. [CrossRef]
57. Readiness Preparation Proposal; Lao PDR. Readiness Preparation Proposal (R-PP). Available online: https://www.forestcarbonpartnership.org/sites/forestcarbonpartnership.org/files/Documents/PDF/Sep2011/LaoR-PPFinaldraftrevised21DEC2010-CLEAN.pdf (accessed on 15 October 2014).
58. Forest Carbon Asia. New National REDD+ Task Force Appointed in Laos. Available online: http://www.forestcarbonasia.org/in-the-media/new-national-redd-task-force-appointed-laos (accessed on 15 October 2014).
59. Ministry of Natural Resources and Environment. *MoNRE on REDD+ Task Force for Implementation of Reducing Emission from Deforestation and Forest Degradation*; MoNRE Agreement No. 7176/2013; Ministry of Natural Resources and Environment: Vientiane, Lao People's Democratic Republic, 2013.
60. FERN. Forest Watch FLEGT Update November 2013. Available online: http://www.fern.org/sites/fern.org/files/FLEGT%20update_0.pdf (accessed on 2 November 2014).
61. EU FLEGT Facility. Stakeholders in Lao PDR Plan Ways to Engage in VPA Processes. Available online: http://www.euflegt.efi.int/laos-news/-/asset_publisher/FWJBfN3Zu1f6/content/stakeholders-in-lao-pdr-plan-ways-to-engage-in-vpa-processes (accessed on 1 May 2016).
62. Saunders, J. *Illegal Logging and Related Trade: The Response in Lao PDR*; Chatham House: London, UK, 2014.
63. Arts, B.; Leroy, P.; van Tatenhove, J. Political modernisation and policy arrangements: A framework for understanding environmental policy change. *Public Organ. Rev.* **2006**, *6*, 93–106. [CrossRef]
64. Arts, B.; van Tatenhove, J. Political modernisation. In *Institutional Dynamics in Environmental Governance*; Leroy, P., Arts, B., Eds.; Springer: Dordrecht, The Netherlands, 2006; pp. 21–43.
65. Arts, B.; Tatenhove, J. Van Policy and power: A conceptual framework between the "old" and "new" policy idioms. *Policy Sci.* **2004**, *37*, 339–356. [CrossRef]
66. Arts, B. "Green alliances" of business and NGOs. New styles of self-regulation or "dead-end roads"? *Corp. Soc. Responsib. Environ. Manag.* **2002**, *9*, 26–36. [CrossRef]
67. Arts, B. Assessing forest governance from a "Triple G" perspective: Government, governance, governmentality. *For. Policy Econ.* **2014**, *49*, 17–22. [CrossRef]
68. Forsyth, T. Multilevel, Multiactor Governance in REDD+: Participation, integration and coordination. In *Realising REDD+: National Strategy and Policy Options*; Angelsen, A., Ed.; CIFOR: Bogor, Indonesia, 2009; pp. 113–122.
69. Arts, B. Non-State Actors in Global Governance. Three Faces of Power. Available online: http://hdl.handle.net/11858/00-001M-0000-0028-6C87-B (accessed on 15 December 2016).
70. Vientiane Times. Logging, Tax Avoidance Top 2013 Corrupt Activities in Laos. Available online: https://www.travel-impact-newswire.com/2014/03/logging-tax-avoidance-top-2013-corrupt-activities-in-laos/ (accessed on 21 October 2014).
71. Hanssen, C. Lao land concessions, development for the people? Available online: http://www.condesan.org/mtnforum/sites/default/files/publication/files/4948.pdf (accessed on 15 December 2016).
72. Thongmanivong, S.; Phanvilay, K.; Vongvisouk, T. How Laos Is Moving Forward With REDD+ Schemes. Available online: http://epress-dev.lib.uts.edu.au/journals/index.php/ijrlp/article/view/3355 (accessed on 15 December 2016).
73. FCPF. REDD Readiness Progress Fact Sheet, June 2012. Available online: https://www.forestcarbonpartnership.org/sites/forestcarbonpartnership.org/files/Documents/PDF/June2012/Lao PDR REDD Readiness Progress Sheet_June 2012.pdf (accessed on 20 October 2014).
74. Geist, H.J.; Lambin, E.F. Proximate causes and undelying driving forces of tropical deforestation. *Bioscience* **2002**, *52*, 143–150. [CrossRef]
75. Gehring, T.; Oberthür, S. The Causal Mechanisms of Interaction between International Institutions 1. *Eur. J. Int. Relat.* **2009**, *15*, 125–156. [CrossRef]
76. Jersild, A.; Shroff, R. Feasibility of Various Responses and Interventions to Build Capacity of Local Civil Society Organizations (CSOs) in the Lao PDR. Available online: https://www.adb.org/sites/default/files/publication/28968/csb-lao.pdf (accessed on 15 December 2016).

77. Van Asselt, H.; Zelli, F. Connect the dots: Managing the fragmentation of global climate governance. *Environ. Econ. Policy Stud.* **2014**, *16*, 137–155. [CrossRef]

78. Gupta, A.; Pistorius, T.; Vijge, M.J. Managing fragmentation in global environmental governance: The REDD+ Partnership as bridge organization. *Int. Environ. Agreem. Polit. Law Econ.* **2016**, *16*, 355–374.

79. Minang, P.A.; van Noordwijk, M.; Duguma, L.A.; Alemagi, D.; Do, T.H.; Bernard, F.; Agung, P.; Robiglio, V.; Catacutan, D.; Suyanto, S. REDD+ Readiness progress across countries: Time for reconsideration. *Clim. Policy* **2014**, *14*, 685–708. [CrossRef]

80. Singh, S. *Natural Potency and Political Power: Forests and State Authority in Contemporary Laos*; Series of Southeast Asia: Politics, Meaning, and Memory; University of Hawai'i Press: Honolulu, HI, USA, 2012; p. 224.

Article

REDD+ in West Africa: Politics of Design and Implementation in Ghana and Nigeria

Adeniyi P. Asiyanbi [1], Albert A. Arhin [2,*] and Usman Isyaku [3,4]

1   Department of Development Studies, The School of Oriental and African Studies (SOAS), University of London, London WC1H 0XG, UK; aa158@soas.ac.uk

2   Department of Geography, University of Cambridge, Cambridge CB2 3EN, UK

3   Department of Geography, University of Leicester, Leicester LE1 7RH, UK; ui9@le.ac.uk

4   Department of Geography, Ahmadu Bello University, Zaria P.M.B 1045 Kaduna State, Nigeria

*   Correspondence: aaa72@cam.ac.uk; Tel.: +44-758-624-445

Academic Editors: Esteve Corbera and Heike Schroeder
Received: 2 November 2016; Accepted: 7 March 2017; Published: 11 March 2017

Abstract: This paper analyses the design and implementation of Reducing Emissions from Deforestation and Degradation, conserving and enhancing forest carbon stocks, and sustainably managing forests (REDD+) in the West African region, an important global biodiversity area. Drawing on in-depth interviews, analysis of policy documents and observation of everyday activities, we sought to understand how REDD+ has been designed and implemented in Nigeria and Ghana. We draw on political ecology to examine how, and why REDD+ takes the form it does in these countries. We structure our discussion around three key dimensions that emerged as strong areas of common emphasis in our case studies—capacity building, carbon visibility, and property rights. First, we show that while REDD+ design generally foregrounds an ostensible inclusionary politics, its implementation is driven through various forms of exclusion. This contradictory inclusion–exclusion politics, which is partly emblematic of the neoliberal provenance of the REDD+ policy, is also a contingent reality and a strategy for navigating complexities and pursuing certain interests. Second, we show that though the emergent foci of REDD+ implementation in our case studies align with global REDD+ expectations, they still manifest as historically and geographically contingent processes that reflect negotiated and contested relations among actors that constitute the specific national circumstance of each country. We conclude by reflecting on the importance of our findings for understanding REDD+ projects in other tropical countries.

Keywords: climate change; forests; Ghana; Nigeria; political ecology; REDD+; West Africa

1. Introduction

Reducing Emissions from Deforestation and forest Degradation plus sustainable forest management, conservation and carbon stock enhancement (REDD+) has continued to inspire climate policy optimism over the last decade. Yet, growing evidence from the implementation of this scheme across tropical countries reveals inherent complexities that warrant close scrutiny. This is crucial not only to understand the extent to which REDD+ does or does not deliver on its promises but also to provide insights and lessons from the very processes of designing and implementing such an ambitious scheme. Such insights and the promises of REDD+ are even more important in the wake of the newly agreed Sustainable Development Goals (SDGs) which aim to, among other things, urgently combat climate change, while sustainably managing forests and halting land degradation and biodiversity loss [1].

In this paper, we analyse the politics of design and implementation of REDD+ in the West African context, focusing on Nigeria and Ghana. Compared to other regions, the REDD+ literature on West

Africa is still relatively nascent [2]. Yet Arhin and Atela [3] have argued that unlike previous global climate change dispatches (e.g., the Clean Development Mechanisms) where African countries lagged other regions as project hosts, the advent of REDD+ has seen significant participation from the region, and from West Africa in particular. This represents a shift from the early REDD+ "bias against Africa and toward Latin America" [4]. For instance, all but one (Mauritania) of the continental coastal West African countries stretching from Mauritania to Nigeria are involved in REDD+. Host to a major global biodiversity hotspot [5,6], this region has a strong and diverse socio-cultural heritage, well-developed traditional ecological knowledge, as well as a significant rural population who rely directly on the forest for their livelihoods [2,7,8]. Besides, the rainforest (and to some extent, the transition zone) vegetation belt extending from the Congo Basin to Senegal (breaking at the Dahomey Gap) has a well-studied history of colonial and post-colonial forest development that reflect both regional continuities and a variety of inter-country specificities in terms of colonial legacies, political-administrative structures and geographies [9–15]. Since these socio-cultural, political, ecological, and historical dimensions significantly shape the prospect, nature and impact of REDD+, detailed studies are required to further our understanding of country specificities and regional patterns, thereby generating the much-needed debates on REDD+ in this region.

In contributing to such studies, we draw on insights from political ecology to analyse the politics of design and implementation of REDD+ in our case study areas. Political ecology emphasises questions of interests and power as actors engage in unequal relations over the environment [10,16–18]. We find this perspective useful to foreground the politics in REDD+ design and implementation by scrutinising the convergence of actors, the interplay of multiple interests, and the interactions of power, histories and geographies that underpin the framing and the implementation of REDD+ [19]. We have allowed our empirical findings to guide the overall structure of our discussion, choosing to only foreground political ecology's critical ethos and its attentiveness to fields of interest-laden relations. Through our case studies, we demonstrate how REDD+ design reflects and is underpinned by ostensibly inclusive visions that are malleable, optimistic and all-encompassing, promising a win-win scenario for all parties [20–23]. Conversely, implementation of REDD+ has proceeded precisely through various forms of trade-off and exclusion of certain actors, interests, knowledges, practices, forest uses, and claims to resources [24–30]. We note that both the ostensibly inclusionary vision of REDD+ design and the failure of this vision to translate into reality must be understood partly in terms of the neoliberal provenance of this scheme [31–33]. Scholars of neoliberal environmental governance have analysed the participatory and perpetually optimistic framings of neoliberal conservation projects, and their repeated failure to realise such visions [34–38].

Yet, exclusion in REDD+ implementation is not merely an unintended failure or ineffectiveness of the participatory vision, as Špirić and co-authors [39] seem to suggest in the case of Mexico. Rather, we argue that it is also a deliberate strategy, a political tool for pragmatically rendering socio-ecological complexities governable and for furthering certain interests. For instance, the technicality and complexity of REDD+, which foreclose autonomous local and national actions have been linked precisely to the "approach taken by government officials, consultants, forestry, and development experts to operationalise the idea of REDD+" [24,27,37,40,41]. Both discursive inclusion at the level of policy design and exclusion at the level of implementation are also partly inherent to the REDD+ policy itself. On the one hand, REDD+ platforms such as the Forest Carbon Partnership Facility and the United Nations-Reducing Emissions from Deforestation and Degradation (UN-REDD) dispatch guidelines and socio-environmental safeguards (such as the Free Prior and Informed Consent) to, among other things, foster a participatory approach, even as they carefully review REDD+ proposals to ensure adherence to a participatory ethos [42–44]. On the other hand, the various processes entailed in rendering forests visible as carbon, applying certain kinds of expertise, securing REDD+ forests, and even selecting pilot case studies always entail various forms of exclusion. For instance, the REDD+ requirement to guarantee property rights and ensure the permanence of carbon forests has seen the use of promised incentives and/or force to exclude other forest uses and resource claims—notwithstanding co-benefit claims [28–30,45]. Clearly, since these processes of inclusion and exclusion do not occur in vacuums, they are also necessarily shaped by contextual histories, geographies and socio-politics [21,25]. To be clear, while some studies have emphasised low participation and inclusion in REDD+ design and discourses, see [39,41], we are more interested in juxtaposing tendencies of inclusion/exclusion between design and implementation processes.

Through this combination of inclusion and exclusion, proponents of REDD+ in Nigeria and Ghana emphasise three foci of action: building institutional capacity, rendering carbon visible and clarifying property rights. While these programmatic goals of building institutional capacity, rendering carbon visible and clarifying property rights align with expected REDD+ activities common to most REDD+ projects globally, the actual implementation of these goals is shaped by a variety of factors. We thus suggest that to fully understand how and why REDD+ design and implementation proceed the way they do in the West African context, one must pay attention to (1) the politics of inclusion and exclusion at play in the design and implementation of REDD+; (2) the historically and geographically contingent nature of REDD+ design and implementation; and (3) the contested relations of interest that constitute the specific context within which global guidelines are being adapted. In so doing, we contribute to the burgeoning body of critical work on REDD+ and carbon forestry in Africa [3,25,28–30,32,46,47].

We have structured the paper as follows: we begin by describing the methods adopted for the research. This is followed by an analysis of the politics of REDD+ design in the two countries. We then describe efforts to implement these programme designs under three major headings: capacity building, visualising carbon, and defining property rights. In the last section, we draw some conclusions and highlight the general implications of our findings for understanding the design and implementation of REDD+ broadly.

2. Materials and Methods

The data for this article was obtained over 9 months of fieldwork in both Nigeria and Ghana between 2013 and 2014. We adopted a qualitative research method. We conducted several semi-structured interviews with a diverse set of actors who are directly or indirectly involved or affected in the design and implementation of REDD+ at both national and sub-national levels. In Ghana, the second author led the data collection. Here, 27 national-level stakeholders from government institutions, national and international non-governmental organisations (NGOs), development partners, research organisations and the private sector were interviewed. These respondents were selected purposively based on their involvement and interests in the REDD+ process. Purposive sampling is one of the non-probability sampling techniques involving deliberate target of specific units of interest from usually a small population to form a representative sample [48,49]. Similarly, 58 REDD+ stakeholders were purposively selected and interviewed in Nigeria. The first and the third authors carried out the interviews. The respondents include state officials, local NGOs, international REDD+ consultants, and community leaders. The difference in sample size between the two case study countries is a reflection of their distinctive political-administrative structures and institutional arrangements for REDD+ with that of Nigeria being more expansive. Some stakeholders such as the international NGOs operate in both countries as technical partners under the UN-REDD program. All interviews (except one on Ghana conducted via Skype) were conducted face-to-face using a conversational approach to allow for deeper probing of issues. The interviews generally lasted for an average of 40 min or more depending on the participants' willingness to articulate their views. All the interviews were tape-recorded and stored according to the dates and locations they were conducted. The Skype interview was carried out in a similar fashion. Interview questions consist of relevant thematic issues that relate to politics, power relations, multi-level governance arrangement, tenure and other information linked to design and implementation of REDD+ in both countries.

The qualitative interviews were complemented with a detailed review of relevant reports on REDD+ in both countries. These include the REDD+ Project Idea Note (R-PIN) and the REDD+ Readiness Programme Proposal (R-PP) in the case of Ghana; and the National Programme Documents

(NPD), the (REDD+ Readiness Programme Proposal) R-PP and the (Preliminary Assessment Report) PAR in the case of Nigeria. Other documents reviewed include REDD+ progress reports for both countries, Terms of Reference for consultants, consultancy reports, REDD+ project documents and the National REDD+ Strategy document of Ghana. Additionally, we drew on observation of selected REDD+ meetings and events, and extensive documentation of REDD+ events. These policy events include environmental summits, workshops, round-table discussions and project meetings. We also obtained data at REDD+ sites through participant observation of everyday activities of communities. In Nigeria, further data was obtained from Ekuri, Iko-Esai, Katabang, and Ikang communities which were purposively sampled from each of the three REDD+ pilot clusters. These communities were selected based on the relative significance attached to them by the project proponents. The collected data sets were analysed using qualitative methods of content analysis, grounded theory, and discourse analysis [50,51]. First, the interviews were transcribed verbatim and coded. Second, drawing on the suggestions of Strauss and Corbin, and Saldana [52,53], coding was done manually using pen and highlighters to assign descriptive and attribute information to sections of the transcripts and secondary documents. This is followed by the axial coding process of re-organising and categorising similar or related codes into themes that constitute the structure of our findings. This process was repeated until a saturation point was reached where no new themes emerge from the data. Discourse analysis involves the use of a critical narrative approach for the interpretation of meanings emerging from the data in relation to our written memos of observations during the data gathering processes.

3. Results

In what follows, we discuss the results of our study under two broad headings: Politics of REDD+ Design and Politics of Implementation. The latter is further divided into three subsections.

3.1. Politics of Design

In Nigeria, REDD+ emerged partly in response to the recent economic drawback of Cross River State (one of Nigeria's 37 federating units) due to fast-declining oil revenues and growing public debt (see Figure 1 for a map of Cross River). It was also partly a culmination of efforts especially among civil society actors and state bureaucrats to address what they considered a "catastrophic" level of deforestation in Cross River State which is an important global biodiversity hotspot and a historically significant socio-ecological area [5,6,54]. These were considerable efforts to supposedly preserve what is now widely regarded as "Nigeria's last remaining rainforest" in Cross River State, building on decades of conservation and development interventions that peaked with the decentralised conservation of the late 1980s [55–57]. Similar motivations underlie the REDD+ process in Ghana as well (see Figure 2 for a map of REDD+ pilots in Ghana). Here, REDD+ policy proceeded on two main premises. The first is that the country's forests were fast disappearing (estimated at approximately 2% annually) and needed a transformational change from how forests have been managed in the past. REDD+ presented this opportunity to "get things right" (Interview with Senior Forestry Official). The second premise was the government's expectation to mobilise significant financial resources through the performance-based payments promise of REDD+ as well as the proliferating multilateral REDD+ finance arrangements. As a senior official at the Ghana Forestry Commission described: "because [lack of] funding has affected implementation of some of our beautiful policies, we strategically positioned ourselves to tap into the opportunities of funding that REDD+ promised to offer ... [including] the various funding mechanisms promoted by the global community to support country-level efforts". By 2014, Ghana had developed an Emission Reduction Project Idea Note (ER-PIN) under the Carbon Fund of the World Bank, which was expected to lead to full-scale implementation of REDD+ (see Figure 3 for progress and key timelines on REDD+ in both countries). Under the carbon fund, the World Bank guarantees to purchase the full magnitude of emissions reductions produced by participating countries. As such, early REDD+ ideas began to emerge within this alignment of the aspirations of the financially-stressed government (and especially forestry) departments in both countries and the emergent conservation interests to protect the forest through creative carbon finance which is based not on cutting the forest (e.g., timber extraction) but on a seemingly compelling idea of "doing nothing" [58,59].

For Nigeria, this early alignment of interests found full expression in an environment summit convened in June 2008 which brought together stakeholders, including state bureaucrats, local and international conservation NGOs, environmental entrepreneurs and expert, forest communities, and businesses including bankers, even if the interest of two major coalitions—one for carbon forestry and the other for biodiversity conservation—dominated the summit. The communiqué of this crucial summit recommended that the state government "halt revenue target based on timber exploitation and focus on forest conservation and regeneration for possible carbon finance", "declare a two-year moratorium on logging" and "initiate action to take advantage of the carbon credit market" [60]. These measures found favour with a government that was keen to attract international finance. The Cross River State government, thus, declared a total logging ban, halted revenue generation from timber and initiated early REDD+ consultations. Following a series of reconnaissance surveys, preliminary assessments, and interactions with international REDD+ partners at regional and international events such as Conference of the Parties to the Kyoto Protocol in Copenhagen, the country's first REDD+ proposal, the National Programme Document was approved by the UN-REDD. The UN-REDD is the United Nations platform for REDD+ implementation support, jointly led by the United Nations Development Programme (UNDP), United Nations Environment Programme (UNEP) and the Food and Agricultural Organisation (FAO). An essential pre-condition for such approval is a demonstration by Nigeria's programme proponents of broad-based stakeholder consultation and consensus over the largely expert-written proposals. As such, the circumstances within which REDD+ emerged in Nigeria are those that mobilised a variety of interests in ways that are partly contingent and partly strategic in meeting international requirements for approval.

Figure 1. Map of Cross River State, Nigeria.

Figure 2. Map of REDD+ pilot areas in Ghana (2014). Note: The project in Ankasaho Amuni is an extension of the pilot in Bedum operated by the Portal Limited. REDD+: Reducing Emissions from Deforestation and forest Degradation plus sustainable forest management, conservation and carbon stock enhancement.

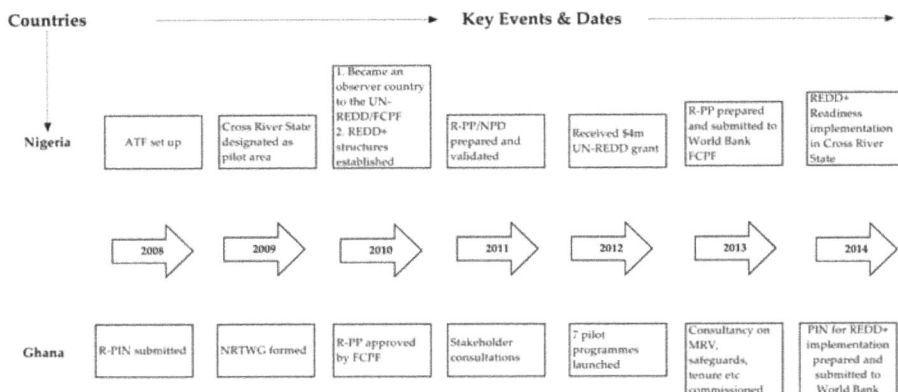

Figure 3. Key timelines for REDD+ in Nigeria and Ghana. ATF: Anti-Deforestation Task Force; UN-REDD: United Nations-Reducing Emission from Deforestation and Degradation; FCPF: Forest Carbon Partnership Facility; R-PP: REDD+ Readiness Programme Proposal; NPD: National Programme Document; R-PIN: REDD+ Project Idea Note; NRTWG: National REDD+ Technical Working Group; MRV: Measurement, Reporting and Verification.

Similarly, in Ghana, the origin of REDD+ is traced to the country's response to the World Bank's call for proposal from interested tropical countries to participate in the then newly-launched Forest Carbon Partnership Facilities (FCPF). Under the FCPF, countries were required to prepare a REDD Project Idea Note (R-PIN), after which they received a grant for the preparation of a detailed Readiness Preparation Proposal (R-PP). Although the development of the R-PIN was almost exclusively carried out by the Ghana Forest Commission, the development of the R-PP drew on the experiences and participation of a wide range of actors from government ministries, the private sector, NGOs, community structures and international development partners. Using existing stakeholder participation structures, the Forestry Commission invited representatives from the National House of Chiefs, Ghana Timber and Milling Organisations, Civil Society Coalition of Forest Watch. These representatives formed the core National REDD+ Technical Working Group, which drafted the R-PP.

The inclusionary politics underlying the design of REDD+ is evidenced in at least three areas. One is in the adoption of a nested approach to REDD+ in Nigeria and Ghana. The nested approach with its growing popularity entails simultaneous national and sub-national (or pilot) level implementation [5,61]. In Nigeria, the nested approach was partly a programmatic imperative, since REDD+ required a national level carbon accounting, whereas, the project had been championed at the sub-national (Cross River) level. This approach was also partly necessitated by an important contextual factor: the federal government in Nigeria has no direct claim to land and forests. Nigeria's regionalisation in 1954 under the British colonial administration brought about a complete transfer of all forest-related powers from the central government to the regions, and later, to the states into which the regions split from the 1960s onwards. The post-independent Land Use Decree (later Land Use Act, LUA under the civilian rule) promulgated in 1978, only entrenched this colonial legacy, by vesting authority over all land in the governor of each state, who holds it in trust for the people [62]. Even the National Forest Policy (NFP) passed in 2006 only deferred to the LUA in specifying forest ownership: "the 1978 Land Use Act gives the lead on questions of land ownership and tenure. All land is owned, including trees growing on it either by the government or private owners." [5].

In Ghana, the nested approach proceeded slightly differently. Early inclusionary approaches were concretised in the devolution of piloting activities to various local constituencies and non-state actors. While the Ghana Forestry Commission took responsibility for policy actions and overall coordination, early REDD+ proposals sought actual piloting and demonstration activities among NGOs, communities and private forestry enterprises. This early form of nesting was cross-sectional, allowing for the participation of different non-state actor groups. As such, early design of REDD+ laid out plans to support and reward these largely independent demonstration activities—seven in all (See Figure 2). However, in late 2015, proponents, made a change to this model in the recently finalised national strategy. A decision was reached to ditch individual-based pilots in favour of a landscape approach. This was justified on the basis that, first, all the seven devolved pilot projects were located in the High-Forest (HFZ) and transitional ecological zones, which make up the southernmost third of the country. The recent change was to allow REDD+ programmes to be implemented in the Northern Savannah zone as well. Secondly, each of the three ecological zones in Ghana has historically experienced distinct drivers of deforestation (i.e., agricultural activities in the HFZ and charcoal production in the others). The landscape approach was thought to be more inclusive in terms of the scale of operations and regarding its focus on landscape-specific drivers of deforestation rather than at project levels. Thus, substantive nesting in Ghana is ecologically based, unlike Nigeria's political-administrative nesting. As such, the nested approach is clearly specified and encoded in programme documents of both countries as an "innovative" approach which is fitting for the contextual complexities of both countries [63,64]. Nesting thus represents a critical moment in the inclusive politics that underpins REDD+ design in Nigeria and Ghana.

The second aspect of REDD+ design where an inclusionary politics is apparent is in the aims and visions of REDD+ as expounded in project documents and in proponents' discourses. The programmatic aims are multiple and cross-sectoral, foregrounding co-benefits in both countries. In Nigeria, the REDD+ National Programme Document specified the overall goal of REDD+ as follows: "to contribute to climate change mitigation through improved forest conservation and enhancing sustainable community livelihoods." [63] (p. 11). The objectives of the project are unpacked into 14 outputs, which are then refined into "core and indicative activities, all structured into a coherent and detailed results framework" [63] (p. 11). Similarly, the declared vision of the REDD+ processes in Ghana is: "to significantly reduce emissions from deforestation and forest degradation over the next twenty years, while at the same time addressing threats that undermine ecosystem services and environmental integrity so as to maximise the co-benefits of the forests. By so doing, REDD+ will become a pillar of action for the national climate change agenda and a leading pathway towards sustainable, low emissions development" [64] (p. 17).

Furthermore, visions held by experts and other key proponents are often far more grandiose and ambitious than documented aims. In Nigeria, proponents not only seek an intersectoral basis for REDD+, they also maintain that REDD+ has "something in it for everyone" (Int. Cross River State REDD+ Coordinator). Moreover, so great are the potentials in REDD+ that one of the proponents declares: "currently, there is no alternative to REDD+" (Int. 1, UK-based REDD+ Consultant; also Int. 42, International NGO Executive). In Ghana, the recent National REDD+ Strategy captures those ambitious and cross-sectoral visions and aims of proponents to (i) reduce emissions from deforestation and forest degradation over the next twenty years; (ii) preserve Ghana's forests in order to sustain their ecosystem services, conserve biological diversity, and maintain a cultural heritage for generations to come; (iii) transform Ghana's major agricultural commodities and Non-Timber Forest Products (NTFPs) into climate-smart production systems and landscapes; (iv) expand platforms for cross-sector and public–private collaboration and sustainable economic development; and (v) generate substantial and sustainable economic and non-economic incentives and benefits to improve livelihoods across all regions. These visions are deliberately ambitious, broad, and all-inclusive, seeking to summon different actor groups: different state jurisdictions and departments, different forest users, NGOs, investors, academics, the military, among others. Thus, these declared visions of REDD+ also instantiate the inclusive politics that underpins REDD+ in Nigeria and Ghana. While these optimistic, and inclusionary tendencies are partly neoliberal in origin [34,36], they also reflect a rhetorical commitment to delivering REDD+ "co-benefits" and a universal inclusivity which Nuesiri [65] shows is a key attribute of UN-REDD policy in general and in Nigeria specifically.

The third and perhaps the most vivid representation of the inclusionary politics underpinning REDD+ is evidenced by the proposed institutional framework for REDD+ in both countries as presented in Figures 4 and 5 below. These institutional frameworks portray the interactions among the various stakeholders (including government departments, civil society, international organisations, forest communities, and the academia, among others). World Bank assessors considered Nigeria's framework "far too complex", despite their affinity for complex bureaucratic systems [66]. The frameworks were so inclusionary that, in the case of Nigeria, proponents listed multinational oil companies operating in Nigeria (such as Shell and Chevron) as partners and potential buyers of Nigeria's REDD+ carbon offset [5]. Driven by this inclusionary strategy, Nigeria's REDD+ has proceeded so that by the end of 2013, Nigeria's National Programme Document (NPD) had been accepted by the UNREDD with a US$4.2 million readiness fund. The World Bank's FCPF supported the country's Readiness Preparation Proposal (R-PP) with US$3.6 million, and the California-led Governors' Climate and Forests Task Force (GCF) provided additional support. In Ghana, following the approval of the R-PP, about $3.4 million were provided by the World Bank. As of 2016, about $100 million has been mobilised in commitment from multilateral, bilateral and other philanthropic sources although just about $25 million has been disbursed from the respective donors to the Ghanaian authorities [67].

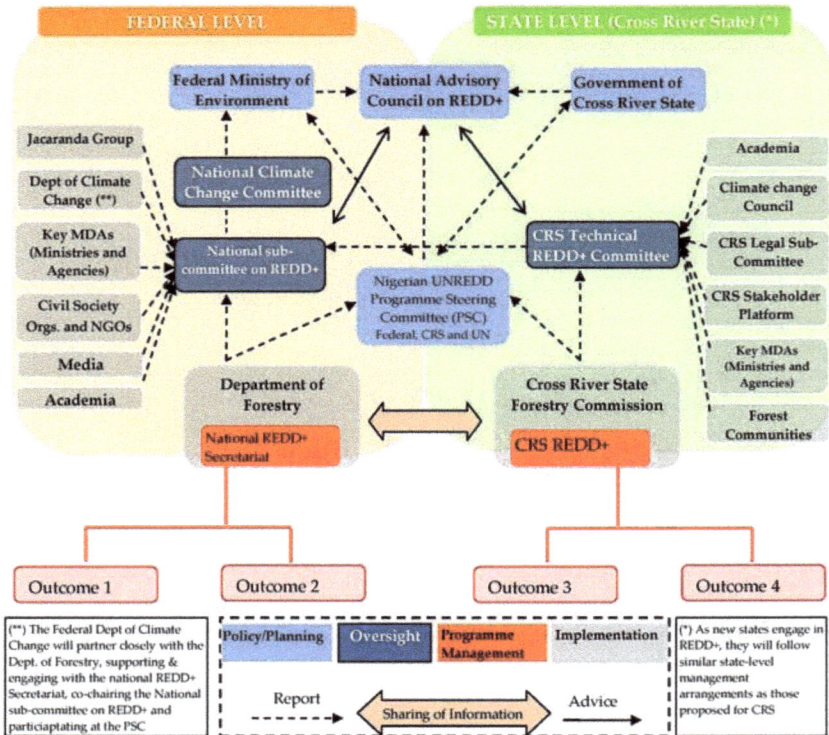

Figure 4. REDD+ Institutional Framework for Nigeria. MDA: Ministries, Departments and Agencies; NGOs: Non-governmental Organisations; CRS: Cross River State; UN: the United States. Source: Adapted from Nigeria R-PP, 2013, p. 11 [66].

This section has, so far, analysed the design of REDD+ in Nigeria and Ghana, and how a certain inclusionary drive underpinned design processes in the two countries—to varying extent. Details of some key design elements, however, differ between the two countries as partly evident in the discussion so far, and as shown in the summary of comparative characteristics of REDD+ in Ghana and Nigeria in Table 1. Nevertheless, in both cases, it is an interplay of nationally-adapted global REDD+ guidelines, interactions with international REDD+ partners (such as the World Bank and the UN-REDD), specific histories, and the negotiation of actors' interests that underpin the design of REDD+. These elements interact to forge a tentative inclusionary approach both as a contingent and a strategic dimension of the design process. An inclusionary approach, among other things, serves to lend the projects some legitimacy, minimise opposition in its early stages, and guarantee the technical and financial support of international partners. Yet, this is not to suggest that REDD+ design was totally and always effectively inclusive; it is to emphasise the ways in which inclusion was partly a deliberate strategy which contrasts with the exclusionary politics through which implementation was pursued. The following section examines the implementation of these proposals, focusing on three major themes that emerged as common areas of emphasis in both countries, i.e., capacity building, visualising carbon, and defining property.

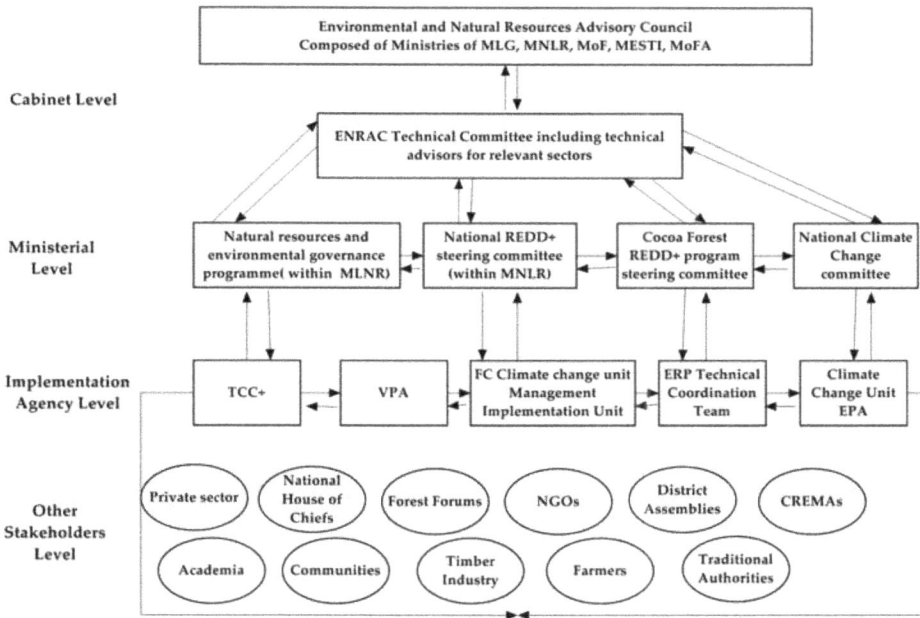

Figure 5. REDD+ Institutional Framework for Ghana. MLG: Ministry of Local Government; MLNR: Ministry of Lands and Natural Resource Management; MoF: Ministry of Finance; MESTI: Ministry of Environment, Science, Technology and Innovation; MoFA: Ministry of Food and Agriculture; TCC+: Technical Coordination Committee; VPA, Voluntary Partnership Agreements; FC: Forestry Commission; ERP: Emission Reduction Programme; EPA: Environmental Protection Agency; NGOs: Non-Governmental Organisations; CREMAs: Community Resource Management Areas. Source: Adapted from Forestry Commission, 2015 [64].

Table 1. Characteristics of REDD+ in Ghana and Nigeria.

Design/Implementation Elements	Ghana	Nigeria
Scale of REDD+ planning	National accounting with an ecological landscape approach to implementation	National and sub-national (state) implementation
Basis for nesting	Ecologically driven, based on zoning of the landscape and unique drivers of deforestation	Politically driven, based on pre-existing federal structure and the power differentials between federal and state control of land and forests
Total area	8 million hectares spread across five administrative regions.	14.4 million hectares in Cross River State
Institutional arrangement	Design based on a multi-stakeholder principle that situates actors across national, operational and local levels	Design based on the "all-affected"/multi-stakeholder principle
Land, forest and carbon tenure	Forest is owned by communities, but the National Forestry Commission holds the rights to manage forests. Dtermination of carbon tenure ongoing	The state governor in each state holds land in trust for the people. Therefore, there is a formal state trusteeship over land and an overlapping customary ownership. Carbon ownership claims are still indeterminate, but there are indications of state control over how carbon rights might be distributed
Implementing agency	National—The Ghana Forestry Commission	National—National REDD+ Secretariat Sub-national—Cross River State REDD+ Unit in the State Forestry Commission
Landscapes/ecosystems under focus	Rainforest, Agricultural land (Cocoa farm)	Rainforest; Mangrove
Securing forest property for REDD+	Through negotiations with cocoa farmers (in agricultural lands) and through protection of existing state Protected Areas	Through state-wide forest protection (a logging moratorium)
Funding Allocation	Approximately $100 million committed between 2009 and 2014 although actual disbursement stands at about $30 million	A sum of US$4.2 million from UN-REDD; $3.6 million from FCPF; and $132,000 from Green Climate Fund; $466,000 from Small Grants Programme
Expansion	To be expanded into transition and northern savannah zones	To be expanded to other Nigerian states
Stated drivers of Deforestation	Agricultural expansion, illegal logging, fuelwood harvesting, mining, infrastructure development	Agriculture, state timber policies, mining, and infrastructural development

REDD+: Reducing Emissions from Deforestation and forest Degradation plus sustainable forest management, conservation and carbon stock enhancement; UN-REDD: United Nations-Reducing Emission from Deforestation and Degradation; FCPF: Forest Carbon Partnership Facility.

3.2. Politics of Implementation (UNREDD, United Nations-REDD Project; FCPF, Forest Carbon Partnership Facility)

Having discussed the politics of REDD+ design, this section highlights how elements of project design codified in key project documents are pursued, contested and even transformed as they enter the realm of everyday politics of implementation. We show that not only did the earlier inclusionary approach change in implementation, but various forms of exclusion and narrowing were also strategically wielded to render social and ecological complexities governable. This section proceeds along three major areas of common implementation focus in Nigeria and Ghana. These are capacity building, carbon visibility, and property rights.

3.2.1. Capacity Building

Capacity building has come to be identified as a central aspect of REDD+ readiness in many countries [37]. Weak institutional and technical capacities to implement REDD+ and specifically to monitor forest carbon have been a raging concern among global proponents of REDD+. For instance, in one extensive review of the capacity of 99 countries to engage in REDD+, [68] found that "very large capacity gaps were observed in forty-nine countries, mostly in Africa". This and many other studies have called for institutional and technical capacity for countries as a key focus of implementation of REDD+ [69,70]. In line with this global emphasis, both Nigeria and Ghana have pursued the goal of improving institutional and technical capacity at the national and state levels. Yet, the implementation of this goal was more than a mere technical intervention. Rather it entailed evaluating, discounting, and re-working existing institutional arrangements as well as authorising and legitimising new institutional arrangements, actors, knowledge, and practices which are claimed to be REDD+ enabling [71,72]. In both countries, the processes of institutional restructuring involved blending of old and new institutional units; overhaul of forestry law and forestry policy; and everyday efforts to impart technical know-how in expertise areas (such as remote sensing, participatory governance) that are considered critical for REDD+.

In Nigeria, capacity building was pursued through complex cross–scalar efforts to rework institutions (specifically forest laws and institutional structures). One aspect through which this was to be achieved involved translating the proposed complex institutional structure (depicted in Figures 4 and 5 above) into reality through the creation of a whole REDD+ institutional network, composed of old and new state and non-state institutions. At the national level, new departments and units were created, for instance, the national REDD+ Secretariat, the National Advisory Council on REDD+, and the National sub-committee on REDD+. This is an addition to efforts to restructure existing departments such as the Federal Forestry Department and the Federal Climate Change Department (Formerly Special Climate Change Unit). Restructuring efforts were most intense at the state level in Cross River since it is at this level that the practical REDD+ demonstration would take place—in line with the nested approach—and it is at this level that forestry laws have significant bearings on landscapes and people, as earlier noted. Similarly, the REDD+ processes in Ghana focused on creating a new institutional network, comprising of old and new public institutions. Here, the Forestry Commission set up the Climate Change Unit to be the secretariat of the REDD+ processes and created a four-tier system of institutional arrangement. At the cabinet level, the Natural Resources Advisory Council (ENRAC), headed by the Vice President, was set up to provide high-level backing to the REDD+ processes. At the ministerial level, an existing inter-sectoral Technical Coordinating Committee-Plus (TCC+) was enrolled into the institutional architecture of REDD+ while a completely new multi-stakeholder National REDD+ Working Group (NRWG) was created. The Climate Change Unit, which serves as a secretariat to REDD+ Secretariat was also created at the operational/implementation levels. However, unlike Nigeria where much implementation continues at the state level, Ghana's implementation focus was more on the national level.

A starting point for proponents pursuing restructuring in both countries was to review the old forestry law and existing institutional framework for forest policy. In Nigeria, the 1956 Eastern Nigerian forest law (which governed forests in Cross River State) was reviewed and approved by the legislature and the executive in 2010. This was followed by the reworking of the Cross River State Forestry Commission to grant more powers to non-state actors (notably NGOs), while orienting the goals of the Commission, its missions, reporting systems and administrative frameworks towards a new vision of carbon forestry [73,74]. In Ghana, the existing 1994 Forest and Wildlife Policy was also revised and replaced with a 2011 Forest and Wildlife Policy [75]. In both countries, this wave of restructuring was rationalised partly through the problematisation of "the long-standing system of viewing the forest as a source of revenue for government [which] is an outdated, colonial and pre-oil mentality . . . and civil servants charged with forest management responsibilities who are involved in illegal logging activities" [76] (pp. 3–4). While REDD+ proponents, both in Ghana and in the Cross River State of Nigeria, claimed to replace timber forestry with carbon forestry, on the one hand, they also grappled with the reality that a carbon regime could not be built ex nihilo. Consequently, restructuring in both countries progressed through the selective exclusion of certain timber forestry practices, knowledge, actors, and interests.

In Nigeria for instance, such restructuring entailed the disciplining of timber forestry actors, partly by steering everyday forestry practice and knowledge away from those required for timber forestry to those most useful for carbon forestry. A major aim was to equip foresters and other REDD+ proponents with the tools and capacity to render forests visible in new ways. Recognising the lack of such knowledge and the need for it in the emergent dispensation, the Cross River State Forestry Board intervened. The Chair of the Board recounts: "to monitor the forest you must have the capacity. When we came in here, most people had not seen a GPS, not to talk of knowing the relevance of Geographic information system (GIS). So we had to purchase GPS units … and trained them" (Chairman Forestry Board and State REDD+ Coordinator). Forestry staff members in the various outposts were henceforth required to report monthly GPS readings of their activities to the Commission headquarters. This intervention failed to achieve the desired results as no forestry outpost reported any GIS reading. Nevertheless, MRV laboratories (one at the national REDD+ Secretariat in Abuja and one at the Forestry Commission in Calabar, Cross River) have been established. Experts and international consultants have been deployed to foster these new forms of knowledge and practices. Specifically, a FAO MRV expert was seconded to the Forestry Commission in Cross River State, while several other consultants from the FAO, UNDP and UNEP continued ad-hoc consultancies. Failing efforts to integrate remote sensing into the total forestry structure, REDD+ proponents began narrowing their focus on select units (e.g., the cartographic unit), and individuals who were strategic to the carbon forestry regime became the targets of REDD+ capacity building. In Ghana, a technology transfer initiative was instituted where more preference was given to foreign consultants partnering local firms and staff of the Commission to work on deliverables such as MRV and Reference Emission Level (REL). The Commission also initiated "various capacity building activities … to sharpen the skills of these bodies to make them more effective at their roles" [61]. These include various training programmes on designing REDD+ projects, carbon stocks assessments, MRV and REL establishment, etc. Thus, a path dependency manifests in Ghana's REDD+ design where the continuous capacity building is emphasised in the implementation of REDD+.

Important in this wave of institutional re-organisation aimed at building capacity are the assumptions underpinning it and how these serve to legitimise the exclusion of traditional forestry actors. In Nigeria for instance, existing forestry institutions have consistently been described as lacking technical capabilities, old-fashioned, corrupt and aiding illegal logging. For instance, the bulky preliminary assessment report which formed the basis for the project proposals had noted: "In most states, management capacity of the state forestry departments and local organisations is low, with poor funding, low staff morale, limited technical training and often high levels of government corruption" [5]. However, these evaluations often ignore the longer history of declining state support for forestry, increasing state forestry revenue target, and recent spread of industrial timber and agricultural concessions—factors which serving and retired foresters blamed for widespread corruption and decline of Cross River forests. Nevertheless, these assumptions of deficiency have served to justify the selective exclusion of forestry bureaucrats in the REDD+ processes. As a consequence, NGO actors and international consultants dominate the emergent institutional structures for REDD+ [29,33]. This dominance should also be understood within the historical context of the post-1989 decentralisation efforts which began with the constitution of the Cross River National Park in 1991 and fostered the rise of an NGO sector and the growth of socio-environmental entrepreneurs some of whom are now prominent in REDD+ design and implementation [11,43]. Though this allowed the implementation of REDD+ to progress since these NGO actors claim know-how in carbon forestry, it has also stoked a tension between traditional state foresters who increasingly feel professionally marginalised and excluded and the members of the NGO-led REDD+ coalition who lead REDD+ processes, having been appointed to state positions.

Also remarkable is the nature of the emergent institutional arrangement under REDD+, presented in Figures 4 and 5 above. If the representation of the arrangement is complex, translating the representation into reality was even more complex, thereby limiting the extent to which implementers could achieve smooth national/state nesting. First, the ambitiously inclusive institutional architecture proposed in design documents in both countries quickly gave in to highly selective coalitions, with actors in the state forestry institutions and in the NGO sector decrying the lack of participation and transparency in the implementation process (Interview Forestry Director; Interview local NGO; Interview International NGO). Second, as implementation progressed, it became clear that many of the old and new state agencies overlap and even compete. As such, on the ground, institutional arrangements are complex and far less inclusive than design documents suggest. Indeed, the constant clash and negotiation of interests among actors in the state and beyond the state reflect a condition marked by "micro-politics, in which actors pursue various overt and covert negotiating strategies to achieve personal ends" [29]. In Nigeria for instance, conservation NGOs leading state anti-deforestation efforts continue to appropriate the logging moratorium for the protection of primates and other wildlife. Also, some members of the Anti-Deforestation Task Force often strike illegal deals with loggers and timber merchants. Some foresters also withdrew support for REDD+ to defend their professional interests and public rights to forest products. In Ghana, although the Forestry Commission began the REDD+ processes through the FCPF, organised the consultations and led the development of the REDD+ Strategy, the Ministry of Lands (which is its parent institution) also led almost parallel processes of consultations, piloting and other processes under the Forest Investment Project (FIP) of the World Bank. A key aim of capacity building is to enable a socio-technical institutional formation that can render forest carbon visible and amenable to accounting. This visualisation of carbon is the focus of the next section.

3.2.2. Visualising Carbon

In line with increasing emphasis at the global level for tropical countries to make carbon visible and tractable through credible measurement, reporting and verification (MRV) processes, a second major focus of implementation in both countries has been to render the forest visible as carbon [27] . This also reflects ongoing effort to grapple with the requirements and capacity for MRV in both countries.

The foundation for this aspect of REDD+ implementation was laid by REDD+ experts who determine the overall biomass carbon potential in both Nigeria and Ghana. For example, a network of international experts from the United Nations Environment Programme's (UNEP) World Conservation Monitoring Centre (WCMC) conducted a carbon survey in 2010 using remote sensing and global soil charts to estimate Nigeria's biomass carbon as 7.5 gigatons and demonstrating the national spread of carbon The survey also suggested focus areas for optimum REDD+ co-benefits by overlaying the carbon map with biodiversity areas of interest. Similarly, through a collaboration between the Ghana Forestry Commission and Forest Trends, Nature Conservation Research Centre (NCRC) and some researchers from the University of Oxford and the National Aeronautics and Space Administration (NASA), a biomass map was produced for Ghana. The map estimated total above ground national carbon stocks to be 1.75 gigaton of carbon (Gt.). Rendering carbon visible in this way requires excluding a range of other things from view. Excluded from view are areas of local importance to communities—fertile farmlands, areas rich in non-timber forest products, sacred forests, and community settlements. Such carbon visibility exercises which were championed by local and international consultants in Nigeria and Ghana, as elsewhere [37,77] have not emphasised local knowledge and capacity for measuring carbon. This is in spite of a growing literature showing the importance and effectiveness of locally trained forest communities in accurately monitoring carbon in ways that safeguard local rights [78–81]. Though implementers in both countries have organised series of workshops on remote sensing, including at least one on-the-ground carbon estimation exercise at the community level in Nigeria, these exercises were rather symbolic as they were not significant enough to help communities crack open and sustainably engage with the black box of technical carbon estimations procedures.

Closely linked to these are efforts to territorialise these emergent carbon visibilities. Given the lack of capacity in both countries, as in most tropical countries, to monitor the forest nationally and near-real time [66], implementer's narrowed territorialisation of national carbon visibilities to sub-national pilots. Difficulty in monitoring also relates to challenges in establishing a plausible and widely agreeable reference baseline for emissions reduction, which REDD+ implementers and the World Bank review committees jointly agreed was fundamental to the project [5]. Generally, not only is baseline determination a technically challenging endeavour, it is also a political one [55,82]. This means that although carbon is rendered visible at the national level, actual monitoring and capturing of value would proceed at the sub-national level, which does not merely align with the nested approach, but also represents a narrowing of the scope of what ought to be national carbon monitoring.

At the sub-national level in Cross River State and in the project areas in Ghana, pilot areas for REDD+ and the extents of forests are neatly demarcated on maps. However, on the ground, the situation is different: there seem to be no credible information on the definite extent of the forest (loss) and the definite areas marked out for REDD+. A retired senior forester in Nigeria observed: "Obviously since, say Independence, there hasn't been any detailed inventory survey. So, whatever we are even claiming about the boundaries and the sizes of the Cross River forests I think is guess-work" (Interview Retired Forestry Director). An official of the Anti-deforestation Task Force charged with forest protection also observed that "there are no clear boundaries till today. How do you do REDD+ when you do not even have a boundary you can claim to be your own" (Interview Task Force Official). If uncertainties with forest extent and boundaries pose technical challenges for REDD+, this challenge takes on a political form with respect to communities whose forests are also being constituted into REDD+ pilots. For instance, the Project Idea Notes (PINs) for Nigeria's REDD+ pilots had noted: "the project is viable and attractive to carbon finance only if the project area includes the multiple community forests and forest reserves. A project considering only one of these areas would not be viable on its own" [5]. Thus, to make Nigeria's carbon forests marketable and finance-able, they needed to be rendered visible as "clusters" which are based mainly on forest contiguity. Once carbon forests are rendered as clusters, socio-political jurisdiction must be re-arranged as such: supra-community governance levels that correspond to clustered forests. What is excluded from such a process are the various ways through which local forest governance had been pursued prior to REDD+ and the various patterns of inter-community resource relations that pre-date current interventions. Yet, such imposed institutional arrangements exemplified by clustering, scholars of institutional bricolage warn [83,84], often fail to grasp the ways in which community resource governance institutions are rather more organic, multipurpose and representing layered imbrications of new and pre-existing institutions.

Further work on carbon visualisation specific to the Ghana case involved mass mobilisation and education. Dubbed the "REDDeye road show", the format for the procession over the years has involved mobilisation of a cross-section of actors including celebrities, school children, private firms, government agencies and communities into street marches in different locations across multiple regions in Ghana. Per the government, the central objective has been to raise awareness and "open their (i.e., youth, school children) eyes on the importance of the REDD+ as a mechanism to reduce the devastating effects of climate change" (Interview with official of Forestry Commission, April 2014). Accordingly, these street processions interspersed with dancing, drumming and drama promote messages centred on behavioural changes such as changes such as "before you cut a tree, think twice", "save the earth against the removal of trees" and "let's protect the environment together". However, far from the mere raising of awareness, the mass education and processions form part of the larger efforts to visualise carbon as a fictitious product that can be controlled, calculated and managed [58]. At the same time, the exclusionary politics in the REDD+ Roadshow is seen on its key targets (i.e., general public, school children, youth). Here, timber merchants (both legal and illegal) understood to have driven deforestation are given insufficient attention as targets. These carbon visualisation processes are also linked to processes of defining carbon rights and tenure, which is the focus of the next section.

3.2.3. Defining Property Rights

In addition to building capacity and visualising carbon, defining property rights is the third major focus of implementation in our case studies. Questions of tenure are so central to REDD+, as literature in this area has demonstrated, since this relates to the permanence of emission reduction, benefit-sharing, access to forest resources under REDD+, and overall project effectiveness [41,85]. It is also around these aspects of REDD+ in West Africa that the literature on justice in ecosystem governance is recently gaining traction, see [86]. In both Ghana and Nigeria, proponents of REDD+ have sought to define property rights through mutually reinforcing practices of invoking existing legal-institutional framework for forest and land, and through material practices of formulating new rules and controlling access to the carbon forest [5,64,87]. However, determination of property right is even more complex, involving discursive and practical negotiation and contestations between REDD+ proponents, forest communities and other stakeholders. As in other post-colonial tropical countries where formal and customary land claims overlap, there are complexities over land rights in both countries, and project proponents show an awareness of this. For instance, a National Validation Workshop of REDD+ stakeholders in Nigeria called for "due clarification and definition of carbon rights and land tenure matters as they affect REDD+" [64]. In Ghana, the National REDD+ Strategy recognised that tenure rights are ambiguous, contested and "poses major challenges to Ghana's REDD+ process" [75].

Implementers in both countries are trying to respond to this imperative in various ways. In Nigeria, REDD+ proposals have linked carbon rights to forest and land rights by invoking the National Forest Policy, which itself referred to the Land Use Act (LUA). The National Forest Policy (NFP), passed in 2006, only deferred to the LUA in specifying forest ownership: "the 1978 Land Use Act gives the lead on questions of land ownership and tenure. All land is owned, including trees growing on it either by government or private owner." [64]. In Ghana, the government has initiated consultancies about tenure reviews which is expected to lead to legislations that will clarify and secure land tenure, tree tenure, carbon rights and benefit-sharing frameworks for REDD+. At the same time, there are divided opinions among the implementers in both countries. In Nigeria, some maintain that nobody owns carbon and that though there are legislations on the ownership of timber and land, there is currently no document specifying any pattern for carbon ownership (Interview, State REDD+ Coordinator; Interview, REDD+ Consultant). These warn that if current tenure arrangement which puts all land under the state control is strictly translated into carbon rights "that would be a disaster", since it will marginalise communities and other non-state claimants of carbon benefits (Interview, UK-based Nigerian REDD+ Consultant). Furthermore, currently in Ghana, communities own forest but they do not have management rights over the forest. Thus, different tenure and benefit sharing frameworks govern trees and lands on which the trees are located respectively. Here, opinions are divided on whether carbon should be treated as a natural resource such as timber or should be treated as a non-timber forest product, whose extraction is not tied to trees but to the lands, for more on this complexity see [88]. Either way, the existing tenure and benefit-sharing framework sharing framework are recognised by stakeholders as unfair, inequitable and community-marginalising as individual farmers are excluded from benefits from trees. Meanwhile, some respondents in both countries suggest that carbon rights be focused more on sharing rights to benefit from carbon among relevant actors including communities. This aligns with the view of carbon right as a bundle of rights with actors having different rights within the bundle of rights [88]. However, even this still entails a clarification of the "owner" in whom the cumulative rights are vested, including the right to exclude others and protect the resource. Indeed, it is through the exclusion of others that the "owner" can secure property right to the carbon forests and guarantee permanence.

While discussions continue around ownership of forest carbon, the expression of the State's power to exclude other land and forest claimants in Nigeria (i.e., through the total logging ban) indicates who currently wields the important rights. The logging ban is sustained by a government-constituted militarised Anti-deforestation Task Force, led by a conservationist. The moratorium is now considered an important means of demonstrating "political will" to international REDD+ partners. This is, in turn, ensuring continued technical and financial support from international REDD+ partners. However, local communities have continued to contest these efforts to secure the forest for REDD+ through a total logging ban that threaten long-standing forest management practices, local farming, local livelihood, local resource governance and resource rights. For instance, one petition written by a forest community on the 28 November 2011 was addressed to the State Governor. The community made a desperate plea for rescue from "the untold hardship [which] the Chair of the Anti-deforestation Task Force is impacting on us outside of the law" [89] (p. 1). They petitioned that "after having obtained a valid document from the Forestry Commission to officially evacuate processed chewing stick from Agbokim village, he impounded our vehicle" [89] (p. 1). Failing such petitions, communities have resorted to everyday acts of resistance including defying the repressive ban to access NTFPs—acts that the ATF would regard as "pilfering" and other "forest offences" [90,91]. In other cases, communities are reclaiming land areas long donated to forestry for reforestation purposes. Meanwhile, by contrast, Ghana is integrating tree conservation on farms. Yet, by existing laws, Ghanaian communities could own the trees but would not necessarily have the carbon management rights.

Thus, efforts to implement REDD+ in both countries proceed under conditions of exclusion, tension and contestations, although in different forms. Ultimately, efforts to clarify and define property rights in both Nigeria and Ghana are reinforcing state control over land, forest and carbon at the expense of community rights and public access to timber and non-timber forest products [41]. Under such circumstances, in Nigeria, safeguard tools such as the Free Prior and Informed Consent (FPIC) were largely reduced to another technical subject for NGO training and discussion. The application of FPIC bears no significance in the unilaterally-imposed ban on exploitation of timber and non-timber forest produce (e.g., chewing stick, shepherding staff) in state reserves, community forests and private forests. Meanwhile, these tensions and exclusion are also driving increased deforestation. As data from the Global Forest Watch show in Nigeria, deforestation has increased steadily since 2012, reaching a 14-year peak in 2014—a period when the Anti-deforestation Task Force was most active [92]. Therefore, exclusionary policies (like the moratorium) which undermine local property rights also tend to exacerbate deforestation and degradation Therefore, exclusionary policies (like the moratorium) which undermine local property rights also tend to exacerbate deforestation and degradation [93,94]. While some studies [95] claim that devolution of forest control may not necessarily lead to improved local and regional forest conditions, we argue that devolution also requires that we reframe such questions as: what constitutes forest improvement and who gets to define it?

4. Discussion and Conclusions

So, what shapes the design and implementation of REDD+ in Nigeria and Ghana? While a combination of in-country processes and international negotiations kick-started the project in Nigeria, Ghana initiated its REDD+ processes through early engagement with international REDD+ partners. On one hand, REDD+ design was clearly path-dependent and historically contingent, reflecting, in Nigeria, in the leading role of Cross River State and the subsequent adoption of the nested approach. This is linked to the decentralisation of colonial administration, and the Land Use Act that invests state governors with power over land, the series of conservation and development interventions that began with the constitution of the Cross River National Park and the more recently financial and ecological challenges in Cross River State. This path dependency manifests differently in Ghana's REDD+ design where a strong and relatively well-resourced Forestry Commission took the lead on REDD+ at the national level. Important here are historical factors such as historically distinct drivers of deforestation in the different ecological zones, which favoured an ecologically aligned nesting unlike Nigeria's political-administrative nesting of REDD+. Ultimately, while Nigeria's REDD+ is organised around strict protection of forest, in Ghana, REDD+ is organised around a variety of strategies (including cocoa intensification and on-farm tree planting) suited to the different socio-ecological areas. Design processes in both countries converge around a particular inclusionary ethos which was partly contingent and partly pragmatic and strategic. Overall, it is the interplay of nationally-adapted global REDD+ guidelines, specific histories, situated geographies and actors' (including state and non-state actors including NGOs and local communities) interests that underpin the design of REDD+ in such an inclusionary manner.

A similarly complex bricolage of heterogeneous factors shaped the implementation of REDD+ in the two contexts. First, the overarching focus on the three domains of capacity building, visualising carbon, and clarifying property rights in both Nigeria and Ghana apparently derives from the international REDD+ dispatches (e.g., United Nations Framework Convention on Climate Change (UNFCCC) and international proponents such as World Bank). The manifestation of these guidelines takes different forms in the two countries, as these guidelines interact with specific histories, institutional formation, interests of different stakeholders, existing property rights, and the various goals being pursued by proponents of these projects. Notably, efforts to build institutional capacity in both cases entailed processes of institutional restructuring involving the blending of old and new institutional units; overhaul of forestry law and forestry policy; and everyday efforts to impart technical know-how in disciplines and expertise areas (such as remote sensing, participatory governance) that are considered critical for REDD+. Reviews of REDD+ cases across several countries have shown capacity building to be a central objective which issues from international guidelines on REDD+ [4]. However, literature also shows that the great deal of financial, technical and administrative resources being devoted to capacity building for REDD+ is not translating to commensurate improvement in local and national capacity, thereby pointing to the motley of factors that shape capacity building processes [33,37,96]. Part of the problem, [37] argue in their analysis of Tanzania, is the inherent and insidiously alienating technicality and complexity of REDD+ which "did not fall from the sky" but has been produced through the self-interested and self-reproducing ways in which actors in the state, civil society and international organisations project REDD+ [24,40,97]. Indeed, the hegemonic carbon measurement approach, notably through remote sensing, does not only exclude other forms of mensuration and valuation, it also represents a regime of power which disciplines bearers of other knowledges [27].

An important goal of institutional capacity development is focused on equipping implementers to engage in processes of mensuration, calculation, representation and transaction of carbon as a resource. This process of rendering carbon visible as a resource and a commodity constitutes a substantive domain of action on its own, given its centrality to demonstrating forest-based emission savings and generating tradable offsets. A fundamental part of this in both Nigeria and Ghana is expert work at rendering carbon visible through maps, figures, charts. Often, these activities render carbon visible by rendering other things invisible, drawing on similar historic, simplifying forestry logic that produced bio-diverse landscapes as timber [98,99]. While these representations are informing mass education and mobilisation in Ghana, in Nigeria carbon representations are being uneasily territorialised through superimposition of significantly simplified images of carbon forests on actual, dynamic forest landscapes. As we have shown, these processes of territorialisation often stoke significant tension, and they are often limited, fractured and transformed through local agency [58,100].

Efforts to clarify property right is another central implementation goal in both Ghana and Nigeria, one which has also been identified in REDD+ literature as crucial for effective and equitable REDD+ [101,102]. In both cases, proponents of REDD+ realise the inadequacy of current legal-institutional frameworks for carbon rights determination. They are also aware of the failure of current land and forest tenure arrangements to guarantee community ownership and access rights, and the need to address these vulnerabilities to an equitable basis for carbon property right. What makes the difference are the ways in which proponents pursue these goals. In Ghana, proponents are pursuing legal and policy reviews to address these weaknesses, even if the history of similar processes to provide secure property rights for communities gives no basis for optimism. In Nigeria, there are a variety of opinions among proponents as to how to link emergent carbon rights to existing land and forest rights, though these mixed opinions still somewhat evade the need to guarantee rights to communities.

Meanwhile, through the moratorium in the Cross River, REDD+ is reinforcing existing land and forest rights regimes that privilege state control [30,71]. Tenure complexities and lack of political will to implement significant tenure reforms has been one of the most widely reported challenge to REDD+ in Africa [94,103–105] and elsewhere [106,107]. In the light of the political cost of tenure reform and the failure of REDD+ to incentivise real reform, there is a growing, if problematic, accommodation of intensified law enforcement, moratorium, and forest militarisation as "alternative policy options" for pursuing REDD+ [108]. These measures, especially when deployed in a totalizing manner, often further complicate resource relations, leading to further deforestation and marginalisation of forest communities and local populations [29,30,109].

As such, our findings make two major contributions to the literature on REDD+ design and implementation in tropical countries. First, we suggest that REDD+ projects progress through creative combination and iteration of inclusionary and exclusionary tendencies, and that project design and project implementation each respectively reflect these contrasting tendencies. REDD+ cannot be inclusive all-through, as REDD+ proponents and some analysts suggest [110], neither can it be all about lack of or inadequate inclusivity [39,41]. Only a combination of inclusion and exclusion better explains how certain claims, knowledges, aims, actors, practices, are centred while others are ignored [21,37]. Indeed, only such would explain the tentative persistence of REDD+ projects even under the most problematic conditions. At the same time, the tendency to include and exclude along the path from design to implementation cannot be reduced to a strategic practice of proponents alone. These tendencies are also path-dependent, they are contingent upon the geographies and histories of particular places, even as they reflect the problematic character of the neoliberal mode of environmental governance. Nevertheless, an analytical approach that is sensitive to power relations and strategic interests proves productive to scrutinise these imbrications of inclusion and exclusion.

Our second contribution, which flows from the first, is that analysis of REDD+ design and implementation will have to grapple, at once, with the globalist, homogenising nature of REDD+ guidelines and the specificities of place and history that condition REDD+ as a contingent process. In their review, Turnhout et al. [111] reflect the sense in which REDD+ manifests differently and generates different impacts in different contexts, thereby suggesting that REDD+ is a "patchwork of projects and practices with different foci". Writing on Indonesia, McGregor and colleagues show REDD+ to be comprised of "a heterogeneous regime of disjointed practices that reflect the existing political ecologies and interest of differently located actors" [21]. Despite the malleability of REDD+ and the differences observed across contexts, the homogenising tendency in a global dispatch such as REDD+ should not be downplayed. This relates partly to its neoliberal provenance on the one hand, and on the other hand, the mode of practice of international development institutions with their will to render REDD+ governable partly through the dispatch of guidelines, blueprints and principles across different context [112–114]. As we have shown above through the three major areas of implementation and similarities in the contradictory inclusion–exclusion politics from design to implementation, homogenisation persists as a direct effect of the application of similar project criteria, centrally dispersed standards, and similar neoliberal ethos. This homogenising tendency is itself political, insofar as these international guidelines either ignore local specificities or they appear amenable to appropriation for various purposes including those that reinforce existing power imbalance. For instance, the default endorsement of national carbon accounting and the use of state power to secure REDD+ forests are reinforcing state control of forests, and to some extent, carbon rights—a situation that poses serious challenges for REDD+ [28,115,116].

Nevertheless, the difference in REDD+ projects between the two countries has been explained in terms of several factors. Our study found that imbrications of different histories, actors' interests, relations of power, and local socio-ecologies explain much of the differences in the unfolding of REDD+ between the two countries. This understanding foregrounds the nature of REDD+ as a global policy but also as a situated project, anchored in locales. Not only is this understanding critical to the scholarly understanding of REDD+ projects across countries, it is also vital for international project proponents who must increasingly reflect on the potentials and limits of standard guidelines and engage with context-specific complexities.

Acknowledgments: This work is part of PhD studies of the authors. Adeniyi Asiyanbi's studies were financed by the Kings International Scholarship. Albert Arhin's work has been supported by the Gates Cambridge Scholarship while Usman Isyaku's work has been supported by the Petroleum Technology Development Fund (PTDF) Nigeria, and Ahmadu Bello University Zaria, Nigeria. We are grateful for these funding bodies and our supervisors.

Author Contributions: This paper, which is part of a larger study for a PhD, was designed by all the three authors. Adeniyi Asiyanbi and Usman Isyaku wrote the sections on Nigeria while Albert Arhin wrote that of Ghana. All authors contributed equally in editing and producing the final output.

Conflicts of Interest: The authors declare no conflict of interest.

References

1. United Nations. *Transforming Our World: The 2030 Agenda for Sustainable Development*; Resolution Adopted by the General Assembly on 25 September 2015; United Nations: New York, NY, USA, 2005.
2. Gunilla, E.; Olsson, A.; Ouattara, S. Opportunities and challenges to capturing the multiple potential benefits of REDD+ in a traditional transnational savanna-woodland region in west Africa. *Ambio* **2013**, *42*, 309–319.
3. Arhin, A.; Atela, J. Forest carbon projects and policies in Africa. In *Carbon Conflicts and Forest Landscapes in Africa*; Routledge: London, UK, 2015; p. 43.
4. Cerbu, G.A.; Swallow, B.M.; Thompson, D.Y. Locating REDD: A global survey and analysis of REDD readiness and demonstration activities. *Environ. Sci. Policy* **2011**, *14*, 168–180. [CrossRef]
5. National Forest Policy. National Forest Policy, The Federal Ministry of Environment Abuja. 2006. Available online: http://www.fao.org/forestry/15148--0c4acebeb8e7e45af360ec63fcc4c1678.pdf (accessed on 6 May 2014).
6. Myers, N.; Mittermeier, R.A.; Mittermeier, C.G.; Da Fonseca, G.A.; Kent, J. Biodiversity hotspots for conservation priorities. *Nature* **2000**, *403*, 853–858. [CrossRef] [PubMed]
7. Fairhead, J.; Leach, M. *Misreading the African Landscape: Society and Ecology in a Forest-Savanna Mosaic*; Cambridge University Press: Cambridge, UK, 1996; Volume 90.
8. Okali, D.; Eyog-Matig, O. *Rain Forest Management for Wood Production in West and Central Africa*; A Report Prepared for the Project Lessons Learnt on Sustainable Forest Management in Africa for The African Forest Research Network (AFORNET), Nairobi, Kenya; The Royal Swedish Academy of Agriculture and Forestry (KSLA): Stockholm, Sweden; The Food and Agriculture Organisation of United Nations (FAO): Rome, Italy, 2004.
9. Leach, M.; Mearns, R. *The Lie of the Land: Challenging Received Wisdom on the African Environment*; International African Institute in Association with James Currey Ltd.: London, UK, 1996.
10. Cline Cole, R.; Madge, C. *Contesting Forestry in West Africa*; Ashgate: Farnham, UK, 2000.
11. Hochschild, A. *King Leopold's Ghost: A Story of Greed, Terror, and Heroism in Colonial Africa*; Houghton Mifflin Harcourt: Boston, MA, USA, 1999.
12. Grove, R.; Falola, T. Chiefs, boundaries, and sacred woodlands: Early nationalism and the defeat of colonial conservationism in the gold coast and Nigeria, 1870–1916. *Afr. Econ. Hist.* **1996**, *24*, 1–23. [CrossRef]
13. Schroeder, R.A. Community, forestry and conditionality in the Gambia. *Africa* **1999**, *69*, 1–22. [CrossRef]
14. Amanor, K.S. The new frontier: Farmers' response to land degradation—A west African study. In *Revisiting Sustainable Development*; United Nations Research Institute for Social Development (UNRISD): Geneva, Switzerland, 1994; Volume 159.
15. Leach, M.; Mearns, R.; Scoones, I. Environmental entitlements: Dynamics and institutions in community-based natural resource management. *World Dev.* **1999**, *27*, 225–247. [CrossRef]
16. Robbins, P. *Political Ecology: A Critical Introduction*; John Wiley & Sons: Sussex, UK, 2011; Volume 16.
17. Death, C. *Critical Environmental Politics*; Routledge: Oxon, UK; New York, NY, USA, 2014.
18. Bryant, R.L. *The International Handbook of Political Ecology*; Edward Elgar Publishing: Cheltenham, UK, 2015.
19. Fairhead, J.; Leach, M. *Reframing Deforestation: Global Analyses and Local Realities with Studies in West Africa*; Psychology Press: London, UK, 1998.
20. Corbera, E. Problematizing REDD+ as an experiment in payments for ecosystem services. *Curr. Opin. Environ. Sustain.* **2012**, *4*, 612–619. [CrossRef]

21. McGregor, A.; Challies, E.; Howson, P.; Astuti, R.; Dixon, R.; Haalboom, B.; Gavin, M.; Tacconi, L.; Afiff, S. Beyond carbon, more than forest? REDD+ governmentality in Indonesia. *Environ. Plan. A* **2015**, *47*, 138–155. [CrossRef]
22. Visseren-Hamakers, I.J.; McDermott, C.; Vijge, M.J.; Cashore, B. Trade-offs, co-benefits and safeguards: Current debates on the breadth of REDD+. *Curr. Opin. Environ. Sustain.* **2012**, *4*, 646–653. [CrossRef]
23. Phelps, J.; Friess, D.; Webb, E. Win–win REDD+ approaches belie carbon–biodiversity trade-offs. *Biol. Conserv.* **2012**, *154*, 53–60. [CrossRef]
24. Leach, M.; Scoones, I. Carbon forestry in west Africa: The politics of models, measures and verification processes. *Glob. Environ. Chang.* **2013**, *23*, 957–967. [CrossRef]
25. Leach, M.; Scoones, I. *Carbon Conflicts and Forest Landscapes in Africa*; Routledge: Oxon, UK; New York, NY, USA, 2015.
26. Nel, A.; Hill, D. Constructing walls of carbon–the complexities of community, carbon sequestration and protected areas in Uganda. *J. Contemp. Afr. Stud.* **2013**, *31*, 421–440. [CrossRef]
27. Gupta, A.; Lövbrand, E.; Turnhout, E.; Vijge, M.J. In pursuit of carbon accountability: The politics of REDD+ measuring, reporting and verification systems. *Curr. Opin. Environ. Sustain.* **2012**, *4*, 726–731. [CrossRef]
28. Cavanagh, C.J.; Vedeld, P.O.; Trædal, L.T. Securitizing REDD+? Problematizing the emerging illegal timber trade and forest carbon interface in east Africa. *Geoforum* **2015**, *60*, 72–82. [CrossRef]
29. Asiyanbi, A.P. A political ecology of REDD+: Property rights, militarised protectionism, and carbonised exclusion in cross river. *Geoforum* **2016**, *77*, 146–156. [CrossRef]
30. Beymer-Farris, B.A.; Bassett, T.J. The REDD menace: Resurgent protectionism in Tanzania's mangrove forests. *Glob. Environ. Chang.* **2012**, *22*, 332–341. [CrossRef]
31. McAfee, K. The contradictory logic of global ecosystem services markets. *Dev. Chang.* **2012**, *43*, 105–131. [CrossRef]
32. Cavanagh, C.; Benjaminsen, T.A. Virtual nature, violent accumulation: The 'spectacular failure'of carbon offsetting at a ugandan national park. *Geoforum* **2014**, *56*, 55–65. [CrossRef]
33. Asiyanbi, A. Mind the gap: Global truths, local complexities in emergent green initiatives. In *The International Handbook of Political Ecology*; Edward Elgar Publishing: Cheltenham, UK, 2015; Volume 274.
34. Büscher, B. *Transforming the Frontier: Peace Parks and the Politics of Neoliberal Conservation in Southern Africa*; Duke University Press: Durham, UK, 2013.
35. Büscher, B.; Dressler, W.; Fletcher, R. *Nature Inc.: Environmental Conservation in the Neoliberal Age*; University of Arizona Press: Tucson, AZ, USA, 2014.
36. Fletcher, R. How I learned to stop worrying and love the market: Virtualism, disavowal, and public secrecy in neoliberal environmental conservation. *Environ. Plan. D Soc. Space* **2013**, *31*, 796–812. [CrossRef]
37. Lund, J.F.; Sungusia, E.; Mabele, M.B.; Scheba, A. Promising change, delivering continuity: REDD+ as conservation fad. *World Dev.* **2017**, *89*, 124–139. [CrossRef]
38. Fletcher, R.; Dressler, W.; Büscher, B.; Anderson, Z.R. Questioning REDD+ and the future of market-based conservation. *Conserv. Biol.* **2016**, *30*, 673–675. [CrossRef] [PubMed]
39. Špirić, J.; Corbera, E.; Reyes-García, V.; Porter-Bolland, L. *Uncovering REDD+ readiness in Mexico*; Universitat Autònoma de Barcelona: Bellaterra, Spain, 2016.
40. Lohmann, L. Carbon trading, climate justice and the production of ignorance: Ten examples. *Development* **2008**, *51*, 359–365. [CrossRef]
41. Bastakoti, R.R.; Davidsen, C. Nepal's REDD+ readiness preparation and multi-stakeholder consultation challenges. *J. For. Livelihood* **2016**, *13*, 30–43. [CrossRef]
42. Arhin, A.A. Safeguards and dangerguards: A framework for unpacking the black box of safeguards for REDD+. *For. Policy Econ.* **2014**, *45*, 24–31. [CrossRef]
43. McDermott, C.L.; Coad, L.; Helfgott, A.; Schroeder, H. Operationalizing social safeguards in REDD+: Actors, interests and ideas. *Environ. Sci. Policy* **2012**, *21*, 63–72. [CrossRef]
44. Nuesiri, E. Local government authority and representation in REDD+: A case study from Nigeria. *Int. For. Rev.* **2016**, *18*, 306–318. [CrossRef]
45. Thompson, M.C.; Baruah, M.; Carr, E.R. Seeing REDD+ as a project of environmental governance. *Environ. Sci. Policy* **2011**, *14*, 100–110. [CrossRef]
46. Leach, M.; Fairhead, J.; Fraser, J. Green grabs and biochar: Revaluing African soils and farming in the new carbon economy. *J. Peasant Stud.* **2012**, *39*, 285–307. [CrossRef]

47. Lyons, K.; Westoby, P. Carbon colonialism and the new land grab: Plantation forestry in Uganda and its livelihood impacts. *J. Rural Stud.* **2014**, *36*, 13–21. [CrossRef]
48. Kothari, C.R. *Research Methodology: Methods and Techniques*; New Age International: New Delhi, India, 2004.
49. Creswell, J.W. *Research Design: Qualitative, Quantitative, and Mixed Methods Approaches*; Sage Publications: Thousand Oaks, CA, USA, 2013.
50. Hajer, M.; Versteeg, W. A decade of discourse analysis of environmental politics: Achievements, challenges, perspectives. *J. Environ. Policy. Plan.* **2005**, *7*, 175–184. [CrossRef]
51. Li, T.M. *The Will to Improve: Governmentality, Development, and the Practice of Politics*; Duke University Press: Durham, UK, 2007.
52. Strauss, A.; Corbin, J. *Basics of Qualitative Research*; Sage: Newbury Park, CA, USA, 1990; Volume 15.
53. Saldaña, J. *The Coding Manual for Qualitative Researchers*; Sage: Newbury Park, CA, USA, 2015.
54. Oates, J.F. *Myth and Reality in the Rain Forest: How Conservation Strategies Are Failing in West Africa*; Univ of California Press: Oakland, CA, USA, 1999.
55. Abua, S.; Spencer, R.; Spencer, D. Design and outcomes of community forest conservation initiatives in cross river state of Nigeria: A foundation for REDD+? In *Conservation Biology: Voices from the Tropics*; John Wiley & Sons, Ltd.: Sussex, UK, 2013; pp. 51–58.
56. Ite, U.; Adams, W. Expectations, impacts and attitudes: Conservation and development in cross river national park, Nigeria. *J. Int. Dev.* **2000**, *12*, 325. [CrossRef]
57. Alashi, S.A. National parks & biodiversity conservation: Problems with participatory forestry management. *Rev. Afr. Political Econ.* **1999**, *26*, 140–144.
58. Stephan, B. How to trade 'not cutting down trees'. In *Interpretive Approaches to Global Climate Governance: (De)constructing the Greenhouse*; Routledge: Oxon, UK; New York, NY, USA, 2013; Volume 57.
59. Paterson, M.; Stripple, J. My space: Governing individuals' carbon emissions. *Environ. Plan. D Soc. Space* **2010**, *28*, 341–362. [CrossRef]
60. Summit Communiqué. *Communiqué If the Stakeholders' Summit on the Environment 25th–28th June, 2008*; Ministry of Environment: Calabar, Nigeria, 2008.
61. Ghazoul, J.; Butler, R.A.; Mateo-Vega, J.; Koh, L.P. REDD: A reckoning of environment and development implications. *Trends Ecol. Evol.* **2010**, *25*, 396–402. [CrossRef]
62. Land Use Act. Land Use Act, Chapter 202, Laws of the Federation of Nigeria. 1990. Available online: http://www.nigeria-law.org/Land%20Use%20Act.htm (accessed on 20 May 2016).
63. National Programme Document: Nigeria (2011). Available online: http://www.unredd.org/AboutUNREDDProgramme/NationalProgrammes/Nigeria/tabid/992/Default.aspx (accessed on 26 May 2014).
64. Ghana Forestry Commission. The National REDD+ Strategy (Final Draft). 2015. Available online: https://www.forestcarbonpartnership.org/sites/fcp/files/2015/April/Ghana%20National%20REDD%2B%20Strategy%20Final.pdf (accessed on 9 March 2016).
65. Nuesiri, E.O. Representation in REDD: NGOs and Chiefs Privileged over Elected Local Government in Cross River State, Nigeria. RFGI Working Paper No. 11. Available online: https://sdep.earth.illinois.edu/files/RFGI_Working_Papers/11_Emmanuel%20Nuesiri.pdf (accessed on 9 March 2017).
66. Kojwang, H.O. Review synthesis of Nigeria R-PP. 2013. Available online: https://www.forestcarbonpartnership.org/sites/fcp/files/2013/Nigeria%20PC16%20R-PP%20Synthesis%20Review%2015-Nov-2013.pdf (accessed on 9 March 2017).
67. Adjei, K.; Asare, R.A. Ghana: Mapping REDD+ Finance Flows 2009–2014. A Forest Trends REDDX Report. Available online: http://forest-trends.org/publication_details.php?publicationID=50302016 (accessed on 22 September 2016).
68. Romijn, E.; Herold, M.; Kooistra, L.; Murdiyarso, D.; Verchot, L. Assessing capacities of non-annex i countries for national forest monitoring in the context of REDD+. *Environ. Sci. Policy* **2012**, *19*, 33–48. [CrossRef]
69. Angelsen, A. *Realising REDD+: National Strategy and Policy Options*; Center for International Forestry Research (CIFOR): Bogor, Indonesia, 2009.
70. Angelsen, A.; Brockhaus, M.; Sunderlin, W.D.; Verchot, L.V. *Analysing REDD+: Challenges and Choices*; CIFOR: Bogor, Indonesia, 2012.
71. Biddulph, R. Geographies of Evasion. The Development Industry and Property Rights Interventions in Early 21st Century Cambodia. Ph.D. thesis, University of Gothenburg Series, Gothenburg, Sweden, May 2010.

72. Sunderlin, W.D.; Larson, A.M.; Duchelle, A.E.; Resosudarmo, I.A.P.; Huynh, T.B.; Awono, A.; Dokken, T. How are REDD+ proponents addressing tenure problems? Evidence from Brazil, Cameroon, Tanzania, Indonesia, and Vietnam. *World Dev.* **2014**, *55*, 37–52. [CrossRef]

73. CRSFC. *Annual Report of the Cross River State Forestry Commission*; Cross River State Forestry Commission Library: Calabar, Cross River, Nigeria, 2011.

74. CRSFC. *Annual Report of the Cross River State Forestry Commission*; Cross River State Forestry Commission Library: Calabar, Cross River, Nigeria, 2012.

75. Ministry of Lands and Natural Resources (MLNR). *Ghana Forest And Wildlife Policy*; Ministry of Lands and Natural Resources: Accra, Ghana, 2012.

76. Cross River State Government. *Pre-Summit Note: Proposed Summary of Cross River State Environmental Agenda and Action Plan: Thematic Areas for Discussion at the Pre-Summit*; Cross River State Ministry of Environment: Calabar, Nigeria, 2008 (in press).

77. Karsenty, A. The world bank's endeavours to reform the forest concessions' regime in central Africa: Lessons from 25 years of efforts. *Int. For. Rev.* **2016**, *18*, 16.

78. Fry, B.P. Community forest monitoring in REDD+: The 'm'in MRV? *Environ. Sci. Policy* **2011**, *14*, 181–187.

79. Danielsen, F.; Skutsch, M.; Burgess, N.D.; Jensen, P.M.; Andrianandrasana, H.; Karky, B.; Lewis, R.; Lovett, J.C.; Massao, J.; Ngaga, Y. At the heart of REDD+: A role for local people in monitoring forests? *Conserv. Lett.* **2011**, *4*, 158–167. [CrossRef]

80. Skutsch, M. *Community Forest Monitoring for the Carbon Market: Opportunities under Redd*; Earthscan/Routledge: Oxon, UK; New York, NY, USA, 2012.

81. Larrazábal, A.; McCall, M.K.; Mwampamba, T.H.; Skutsch, M. The role of community carbon monitoring for REDD+: A review of experiences. *Curr. Opin. Environ. Sustain.* **2012**, *4*, 707–716. [CrossRef]

82. Ite, U.E.; Adams, W.M. Forest conversion, conservation and forestry in cross river state, Nigeria. *Appl. Geogr.* **1998**, *18*, 301–314. [CrossRef]

83. Cleaver, F. *Development through Bricolage: Rethinking Institutions for Natural Resource Management*; Routledge: Oxon, UK, 2012.

84. De Koning, J.; Cleaver, F. Institutional bricolage in community forestry: An agenda for future research. In *Forest-People Interfaces*; Wageningen Academic Publishers: Wageningen, The Netherlands, 2012; pp. 277–290.

85. Cotula, L.; Mayers, J. *Tenure in Redd: Start-Point or Afterthought?* International Institute for Environment and Development (IIED): London, UK, 2009.

86. Isyaku, U.; Arhin, A.A.; Asiyanbi, A.P. Framing justice in REDD+ governance: Centring transparency, equity and legitimacy in readiness implementation in West Africa. *Environ. Conserv.* **2017**. [CrossRef]

87. Marfo, E.; Acheampong, E.; Opuni-Frimpong, E. Fractured tenure, unaccountable authority, and benefit capture: Constraints to improving community benefits under climate change mitigation schemes in Ghana. *Conserv. Soc.* **2012**, *10*, 161. [CrossRef]

88. Karsenty, A.; Vogel, A.; Castell, F. "Carbon rights", REDD+ and payments for environmental services. *Environ. Sci. Policy* **2014**, *35*, 20–29. [CrossRef]

89. Cross River State Forestry Commission. A Petition to the Cross River State Governor by Agbokim Waterfall Community. Unpublished work, 2011.

90. Bryant, R.L. *The Political Ecology of Forestry in Burma: 1824–1994*; University of Hawaii Press: Honolulu, HI, USA, 1997.

91. Scott, J.C. *Weapons of the Weak: Everyday Forms of Resistance*; Yale University Press: New Haven, CT, USA, 1985.

92. Global Forest Watch. Tree Cover Loss, Cross River Nigeria. 2016. Available online: http://climate. globalforestwatch.org/ (accessed on 23 March 2016).

93. Poudel, M.; Thwaites, R.; Race, D.; Dahal, G.R. Social equity and livelihood implications of REDD+ in rural communities–a case study from Nepal. *Int. J. Commons* **2015**, *9*, 177–208. [CrossRef]

94. Karsenty, A.; Ongolo, S. Can "fragile states" decide to reduce their deforestation? The inappropriate use of the theory of incentives with respect to the redd mechanism. *For. Policy Econ.* **2012**, *18*, 38–45. [CrossRef]

95. Yin, R.; Zulu, L.; Qi, J.; Freudenberger, M.; Sommerville, M. Empirical linkages between devolved tenure systems and forest conditions: Primary evidence. *For. Policy Econ.* **2016**, *73*, 277–285. [CrossRef]

96. Burgess, N.D.; Bahane, B.; Clairs, T.; Danielsen, F.; Dalsgaard, S.; Funder, M.; Hagelberg, N.; Harrison, P.; Haule, C.; Kabalimu, K. Getting ready for REDD+ in Tanzania: A case study of progress and challenges. *Oryx* **2010**, *44*, 339–351. [CrossRef]

97. Tienhaara, K. The potential perils of forest carbon contracts for developing countries: Cases from Africa. *J. Peasant Stud.* **2012**, *39*, 551–572. [CrossRef]

98. Putz, F.E.; Redford, K.H. The importance of defining 'forest': Tropical forest degradation, deforestation, long-term phase shifts, and further transitions. *Biotropica* **2010**, *42*, 10–20. [CrossRef]

99. Scott, C.T. Sampling methods for estimating change in forest resources. *Ecol. Appl.* **1998**, *8*, 228–233. [CrossRef]

100. McAfee, K.; Shapiro, E.N. Payments for ecosystem services in Mexico: Nature, neoliberalism, social movements, and the state. *Ann. Assoc. Am. Geogr.* **2010**, *100*, 579–599. [CrossRef]

101. Resosudarmo, I.A.P.; Atmadja, S.; Ekaputri, A.D.; Intarini, D.Y.; Indriatmoko, Y.; Astri, P. Does tenure security lead to REDD+ project effectiveness? Reflections from five emerging sites in Indonesia. *World Dev.* **2014**, *55*, 68–83. [CrossRef]

102. Sunderlin, W.D.; Ekaputri, A.D.; Sills, E.O.; Duchelle, A.E.; Kweka, D.; Diprose, R.; Doggart, N.; Ball, S.; Lima, R.; Enright, A. *The Challenge of Establishing REDD+ on the Ground: Insights from 23 Subnational Initiatives in Six Countries*; Center for International Forestry Research (CIFOR): Bogor, Indonesia, 2014; Volume 104.

103. Hoare, A.L. *Community-Based Forest Management In the Democratic Republic of Congo a Fairytale or a Viable Redd Strategy*; Forest Monitor: Cambridge, UK, 2010.

104. Dulal, H.B.; Shah, K.U.; Sapkota, C. Reducing emissions from deforestation and forest degradation (REDD) projects: Lessons for future policy design and implementation. *Int. J. Sustain. Dev. World Ecol.* **2012**, *19*, 116–129. [CrossRef]

105. Fobissie, K.; Alemagi, D.; Minang, P.A. REDD+ policy approaches in the Congo basin: A comparative analysis of cameroon and the democratic republic of congo (DRC). *Forests* **2014**, *5*, 2400–2424. [CrossRef]

106. Larson, A.M.; Brockhaus, M.; Sunderlin, W.D.; Duchelle, A.; Babon, A.; Dokken, T.; Pham, T.T.; Resosudarmo, I.A.P.; Selaya, G.; Awono, A.; et al. Land tenure and REDD+: The good, the bad and the ugly. *Glob. Environ. Chang.* **2013**, *23*, 678–689. [CrossRef]

107. Larson, A.M.; Brockhaus, M.; Sunderlin, W.D.; Duchelle, A.; Babon, A.; Dokken, T.; Pham, T.T.; Resosudarmo, I.; Selaya, G.; Awono, A. Land tenure and REDD+: The good, the bad and the ugly. *Glob. Environ. Chang.* **2013**, *23*, 678–689. [CrossRef]

108. Bolin, A.; Lawrence, L.; Leggett, M. *Land Tenure and Fast-Tracking REDD+: Time to Reframe the Debate?* Analytical Paper; Global Canopy Programme: Oxford, UK, 2013.

109. Tollefson, J. International media spotlight on the amazon roams, but rarely enlightens. *Elem. Sci. Anthr.* **2015**, *3*, 58. [CrossRef]

110. Baroudy Ellysar. *Foreword for Forest Carbon Partnership Facility Annual Report*; World Bank: Washington, DC, USA, 2015.

111. Turnhout, E.; Gupta, A.; Weatherley-Singh, J.; Vijge, M.J.; De Koning, J.; Visseren-Hamakers, I.J.; Herold, M.; Lederer, M. Envisioning REDD+ in a post-Paris era: between evolving expectations and current practice. *Wiley Interdiscip. Rev. Clim. Chang.* **2016**. [CrossRef]

112. Schroeder, H.; McDermott, C. Beyond carbon: Enabling justice and equity in REDD+ across levels of governance. *Ecol. Soc.* **2014**, *19*, 31. [CrossRef]

113. Goldman, M. *Imperial Nature: The World Bank and Struggles for Social Justice in the Age of Globalization*; Yale University Press: New Haven, CT, USA, 2006.

114. Ferguson, J. *Global Shadows: Africa in the Neoliberal World Order*; Duke University Press: Durham, NC, USA, 2006.

115. Phelps, J.; Webb, E.L.; Agrawal, A. Does REDD+ threaten to recentralize forest governance. *Science* **2010**, *328*, 312–313. [CrossRef] [PubMed]

116. Sandbrook, C.; Nelson, F.; Adams, W.M.; Agrawal, A. Carbon, forests and the redd paradox. *Oryx* **2010**, *44*, 330–334. [CrossRef]

forests

Article

Costs of Lost opportunities: Applying Non-Market Valuation Techniques to Potential REDD+ Participants in Cameroon

Dara Y. Thompson, Brent M. Swallow * and Martin K. Luckert

Department of Resource Economics and Environmental Sociology, University of Alberta, Edmonton, Alberta, T6G-2H1, Canada; dara@ualberta.ca (D.Y.T.); luckert@ualberta.ca (M.K.L.)
* Correspondence: brent.swallow@ualberta.ca; Tel.: +1-780-492-6656

Academic Editors: Esteve Corbera and Heike Schroeder
Received: 30 October 2016; Accepted: 28 February 2017; Published: 3 March 2017

Abstract: Reduced Emissions from Deforestation and Forest Degradation (REDD+) has been systematically advanced within the UN Framework Convention on Climate Change (UNFCCC). However, implementing REDD+ in a populated landscape requires information on local costs and acceptability of changed practices. To supply such information, many studies have adopted approaches that explore the opportunity cost of maintaining land as forest rather than converting it to agricultural uses. These approaches typically assume that the costs to the smallholder are borne exclusively through the loss or gain of the production values associated with specific categories of land use. However, evaluating the value of land to smallholders in incomplete and messy institutional and economic contexts entails other considerations, such as varying portfolios of land holdings, tenure arrangements, restricted access to capital, and unreliable food markets. We suggest that contingent valuation (CV) methods may provide a more complete reflection of the viability of REDD+ in multiple-use landscapes than do opportunity cost approaches. The CV approach eliminates the need to assume a homogenous smallholder, and instead assumes heterogeneity around social, economic and institutional contexts. We apply this approach in a southern rural Cameroonian context, through the lens of a hypothetical REDD+ contract. Our findings suggest local costs of REDD+ contracts to be higher and much more variable than opportunity cost estimates.

Keywords: reduced emissions from deforestation and forest degradation; carbon retention; avoided deforestation; willingness to accept; payments for ecosystem services; contingent valuation.

---

1. Introduction

Reduced Emissions from Deforestation and Forest Degradation (REDD+) has been systematically advanced within the UN Framework Convention on Climate Change (UNFCCC), and is now embraced by the Paris Climate Agreement [1]. The early conception of REDD+ was that it would be primarily implemented through actions of national governments, which would enact instruments and policies most appropriate to their particular national circumstances. This conception prompted concerns that REDD+ would reverse general trends toward devolution of forest governance to local users. Local user groups have been shown to be effective managers of local forests, while individual farmers have been shown to be involved in planting trees and clearing forests for agriculture. REDD+ was thus seen as a paradox; on one hand recognizing that improvements in forest governance require devolution, while on the other hand encouraging a centralization of the finance and accountability for REDD+ toward national governments [2]. Over time, however, it has been recognized that REDD+ will need to be adapted to the governance context of particular countries where it is implemented [3]. To the extent that forest governance involves a diversity of government levels and user groups, REDD+ will need to engage with that same diversity of interests [4], particularly the indigenous peoples and local communities who may be most affected [5,6]. This local engagement is particularly important in 'fragile states' where central governments exert minimal control over local land use [7,8]. Local users of land and forests need to be involved in order for REDD+ to be viable, and to fulfill the principle of Free, Prior and Informed Consent (FPIC) that is promoted as the international norm for developments that potentially impinge upon the land or resource rights of local and indigenous peoples [9].

However, many of the studies that have investigated how local economic conditions could affect the support of REDD+ have ignored various local concerns. Economic studies of REDD+ have focused on the opportunity costs of leaving land in high carbon uses, compared to clearing trees and converting the land into lower carbon land uses. Evidence on the farm-level returns and carbon fluxes associated with alternative land uses, including intact primary forests, secondary forests, tree crop systems, and annual food crops, showed that large areas of tropical forests were being converted into alternative uses that generate very low returns per tonne of carbon emitted [10,11]. Influential reports such as Stern (2008) and Eliasch and Office of Climate Change (2009) drew upon that evidence in their support of REDD+ [12,13]. McKinsey and Company (2009) also drew upon that evidence to generate their well-known Global Greenhouse Gas Abatement Cost Curve, which indicates reduced slash and burn agriculture conversion to be one of the lowest cost approaches to abating greenhouse gas emissions [14].

Previous economic studies of the opportunity costs of avoided deforestation have implicitly assumed that financial compensation for foregone income would be sufficient to garner local support for land use restrictions that maintain high carbon land uses. For example, Swallow et al. (2007) consider the costs in humid forest areas of Cameroon, Peru and Indonesia [15]. Within Akok village in Cameroon, the authors estimate that emissions released have resulted in returns of around 8 USD/t CO_2e, using the social net present value approach [16]. For smallholders, conversions from mixed food crops fallow systems to shade-cocoa agroforestry practices were found to be a 'win-win' situation; both carbon sequestration and incomes to farmers would increase [12]. The methods used in the Swallow et al. (2007) study have been endorsed by the Carbon Finance Unit of the World Bank and are detailed in a training manual published by the World Bank Institute (2011) [13].

Such opportunity cost approaches may suffer from a number of shortcomings. First, returns are based on production values of specified categories of land. They do not consider the many nuanced combinations of land holdings that households possess that could influence their willingness to participate in REDD+ programs. Second, these approaches lack considerations of fluid, complex property rights. Gregerson et al (2010) note that property rights to forest lands and products are often insecure and contested, and thus not easily disaggregated or made subject to sale or lease [17]. Cerbu et al (2013) and Sunderlin et al (2014) both found tenure insecurity and the customary practice of 'the right of the axe' (securing land rights through land clearing) to be major constraints to implementation of REDD+ in the forest areas of Cameroon [18,19]. Besides tenure insecurity, Cameroon suffers from other characteristics of a fragile state, and is rated by the Fund for Peace as being in the "alert" category according to its fragile states index [20]. Third, investments may be impeded by a lack of capital. For example, despite the win-win situation identified In the Swallow et al. (2007) study [15], conversions to shade-cocoa agroforestry are rare, perhaps because there is a minimum amount of capital investment required to develop a cocoa agro-forest. Many smallholders are unable to come up with this investment, and are reliant on the mixed-food crops for subsistence [21]. Fourth, dynamic considerations of uncertain futures are frequently ignored. For example, Swallow et al. (2007) [15] is an ex post study where opportunity costs are based on conditions in the recent past, and sensitivity analysis used to assess the impacts of various social discount rates.

The purpose of this study is to explore contingent valuation (CV) methods as a means to account for local contexts in influencing opportunity costs of avoided deforestation. The general idea is that when a household is considering how much they are willing to accept for conserving trees, they consider the myriad of considerations, referred to above, so that their choices reflect local contexts.

The application of our approach eliminates the notion of a homogenous smallholder, and instead assumes household-level heterogeneity around social, economic and institutional contexts of land holdings. We argue that this method is suitable for informing the design of contracts that reduce property rights to land when focused on situations that include complex and inconsistent institutional parameters, as well as imperfect, unreliable and missing economic contexts. Despite the potential advantages of this approach, we are only aware of one study in Cambodia that has used CV in a REDD+ setting [22].

We explore the value of lost opportunities through a study of farmers' willingness to accept compensation for REDD+ in a multiple land use context in southern Cameroon. We present three types of results. First, we use contingent valuation to assess the willingness of individual land users to accept compensation for the lost opportunities that REDD+ would create. Second, we investigate factors affecting that willingness. Third, by aggregating willingness to accept compensation (WTA) amounts across the sample, we generate a potential abatement cost curve for avoided emissions.

2. Materials and Methods

2.1. Approaches to Measuring Costs of Avoided Deforestation

Studies that estimate opportunity costs from farm-level data have been described as "bottom up." Those studies have considered three main factors: (1) the returns that forest users could obtain from clearing and converting forest land into different types of agricultural uses compared to the alternative of leaving the forest intact; (2) the time-averaged carbon stock associated with alternative uses; and (3) the efficiency of targeting payments to those who actually would convert any particular area from forest to agriculture in the absence of a compensation payment [16]. Bottom-up studies such as Swallow et al. (2007) [15] implicitly assume perfect targeting: if payments could be targeted only to those land users who converted land within a particular period, what amount of money would actually compensate them for the lost earnings associated with that conversion? Other bottom-up opportunity cost studies have made the efficiency of targeting more explicit [23] and considered other types of opportunity costs such as the costs of lost value-added income that may accrue at the regional level [24]. On the other hand, "top-down" studies of opportunity costs use aggregate models of forest product supply and demand and treat carbon storage as an additional demand for forest products. The top-down studies are not able to consider how that demand would be expressed at the local level [25].

An alternative approach is to directly elicit the values that land users place on foregoing the opportunity to convert their land into low carbon uses with stated preferences. Contingent valuation (CV) approaches allow for the integrated nature of household decision making in the context of both missing markets and integrated production–consumption decision-making [26]. Further, CV approaches can capture heterogeneity in production, consumption, and perceptions of the losses associated with irreversibility [26,27].

CV approaches frequently ask respondents the amounts that they would be willing to pay (WTP) to avoid a loss. An alternative approach is to elicit the amounts that they would be willing to accept (WTA) to incur a loss. Although basic economic theory suggests that the amounts should be very similar, empirical studies show that WTA tends to be considerably higher than WTP for the same loss. A substantial literature [26–29] has investigated possible reasons for these differences. In our application, we elicit WTA measures because of the nature of REDD+. Implementation of REDD+ with smallholders potentially involves offering payments to individual land users who commit to avoid deforestation. Therefore, we are interested in knowing how much these contracts would cost in terms of payments to local smallholders.

Despite their potential strengths, contingent valuation approaches for estimation of WTA are controversial for several reasons [26]. First, CV estimates can suffer from hypothetical bias because participants typically have no experience placing monetary values on the goods in question.

This situation is certainly a possibility in the case of payments for avoided deforestation. Second, respondents may feel the need to say yes to adopting a hypothetical contract (i.e., yea-saying) because of the natural human inclination to please others, thus creating an enumerator bias. Residents of the study site were involved in several agronomic and land use studies over the previous 10–15 years and may perceive the possibility of future benefits or threats from participating in research. Third, respondents may tend to anchor their responses to the first value that the enumerator mentions, especially if they have little experience in valuing the proposed change. This problem could occur in our case because farmers do not seem to have experiences in receiving payments for a future commitment. Rather, their experiences are with payments for specific delivered actions or products. Fourth, the nature and amount of information presented as background to the hypothetical scenario may bias the respondents' responses and undermine the validity of the study [30,31]. Fifth, there may be an embedding or scope effect depending upon whether respondents view the good on its own or as part of a larger inclusive package. Below we describe the steps taken to minimize the effects of these potential biases.

2.2. Study Site and Context

The field study was conducted in the village of Akok, a collaborative of eight sub-villages, located in the South Province of Cameroon. This area was selected as it had been previously studied through the Alternative to Slash and Burn (ASB) Partnership for the Tropical Forest Margins, with research conducted by scientists associated with the International Institute for Tropical Agriculture, the World Agroforestry Centre and other research partners [32]. The ASB studies provide descriptions of the main land uses [33], spatial analysis of changes in land use [34], institutional capacity for REDD+ [18], as well as estimates of the carbon stock [35] and private economic returns associated with those land uses [15]. Most residents are of the Bulu ethnic group, one of the Beti-Pahuin ethnicities occupying that part of the Congo basin.

The main agricultural production system in this area, like other parts of the Congo basin, is a shifting cultivation system [33]. Generally, shifting cultivation progresses from land being in a high-carbon state, primary forest, to a low-carbon state, such as groundnut or cassava production. Because the REDD+ contract is based on avoiding deforestation, these land-use transitions are important for WTA estimates. Figure 1 provides a summary of these transitions. A forest area that has not been used for agriculture in recent years is the starting point. Underbrush and larger trees are selectively cleared and burned to produce nutrient-rich ash. Farmers then plant crops that do well in that type of soil, particularly forest melon (*esëp*). Once *esëp* has matured for approximately one year, the produce is harvested and the field burned. The result is a forest fallow field (*fulu*), which is left to dry. The next step is to plant a mixture of food crops (*afub owondo*) which is the most important land type for subsistence. For about three years, crops such as cassava, cocoyam, plantain and peanut are cultivated on a semi-annual basis. Another group of food crops, known as *asan*, is planted after vegetation is cleared from lowland peat fields [33].

After the food crop phase, farmers allow the land to lay fallow in order for underbrush and soil nutrients to re-establish. There are four classifications of fallow within the *Bulu* tenure system: young (1–4 years, *nyengue*), old (4–10 years, *ekotok*), very old (10–20 years, *nfos ekotok*) and degraded secondary forest (20–40 years, *nfos afan*). Any fallow may be used to create an *afub owondo* field or other crop. However, the longer the land lays fallow, the more fertile the soil. Cocoa is typically introduced into the traditional shifting cultivation system after longer-term fallow, or even along with an *esëp* field, which can be considered a 'cocoa with fruit' field. As cocoa fields can be created from any stage of forest or very old fallow, these are not included in the illustration of the shifting cultivation transition shown in Figure 1.

Shifting cultivation in the Akok area primarily occurs in the context of customary clan-lineage tenure agreements. The traditional *Bulu* system employs a provisional, cyclic method of household land allocation [36]. An individual or household does not hold title to the land in the *Bulu* system.

However, while an individual or household has cleared and is cultivating an area, that particular area is recognized as that household's possession. There are three broad categories of possession within the *Bulu* system: (1) individual and/or household; (2) lineage; and (3) community or clan. A household holds individual/domestic possession during cultivation and through to the old fallow stage (*ekotok*), which is approximately 10 years after the last crop is harvested. Once this land transforms into a very old fallow or secondary forest (approximately *nfos ekotok* and *nfos afan*), the land reverts to the possession of the entire lineage and anyone within the lineage can utilize it. Finally, if the land is left uncultivated long enough to grow back into secondary forest, it is available to all members of the community or clan [36].

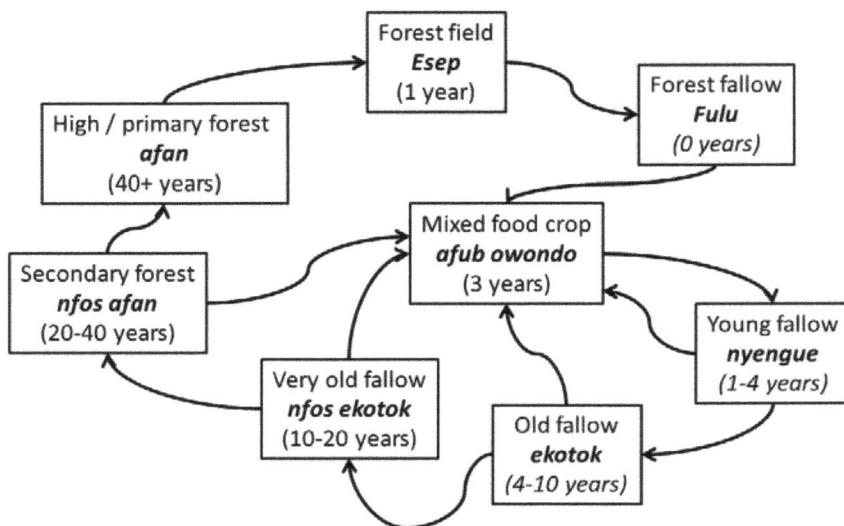

Figure 1. The typical *Bulu* shifting cultivation system in Southern Cameroon (adapted from Brown, 2006 [33]).

The Forestry Law of 1994 has complicated tenure arrangements in the area. There are three important differences between *Bulu* forest tenure and the 1994 Forest Law. First, the Forest Law makes a clear distinction between agricultural land and forest land through the evidence of human presence through cultivation. Second, the *Bulu* system simply equates labour applied to land to the creation of rights, while the 1994 Forest Law is much more complex in the granting of legal title. Third, the fundamental concept of ownership differs; *Bulu* tenure does not identify exclusive property title to land, while the 1994 Forest Law allows for common, private and state property rights. This challenging, messy and inconsistent institutional context can potentially influence each smallholders' behaviour in different ways. Indeed, as we designed our survey, we had to investigate whether the local context allowed individuals to express WTA for constraints on land use. Results of key informant interviews, focus groups and pretests indicated that individuals were willing to make such land use decisions (with an exception regarding *afan* that we discuss below). However, as a result of complications to property rights, we expect that individuals may differ in their perceptions of the security of their tenure to fallow land, and thus the opportunities that they would be foregoing if they accepted the proposed contract. These differences in perceptions could be a source of heterogeneity among the respondents to this survey, and would also be a source of heterogeneity among actual contract holders [37].

Rural smallholders in this region of Cameroon, and many other areas globally, are dependent on land and forest resources for meeting both their income generation and subsistence needs. By entering into a contractual obligation to preserve forests, a smallholder would be making a conscious choice to reduce the amount of land that they could convert to other uses in the near future. Smallholders in Cameroon, as elsewhere, make daily decisions to ensure that there is enough food on the table, that their families are taken care of financially, and that there is hope for the future. Land available for cultivation can be crucial to all of these goals. In situations with missing markets or limited access to substitutes, there are few opportunities for smallholder farmers to find alternatives to traditional food production systems [38]. Moreover, in subsistence-based economies, it may be rare for farmers to have cash to pay for food and food markets may be thinly traded. Yet, in entering into a contract to avoid deforestation, essentially, the smallholder is choosing between the potential to use land versus future cash payments from the contract. The importance of land to livelihoods, and the rarity of cash and prices in such economies could cause smallholders to be hesitant to accept payments for the loss of use of land.

2.3. Data Collection

The field study in Akok was conducted in February and March of 2011 and began with a half-day focus group discussion. Participants provided information about local farming and the *Bulu* tenure system, which helped to clarify the plausibility of specific forest conservation contracts in that local context. Participants expressed concerns that a forest conservation contract would lead to the Government of Cameroon exercising full control over their shifting cultivation lands (as per the Forest Law of 1994). If they could no longer rely on clearing long-duration fallow to maintain soil fertility on their fields, farmers were concerned that they would not be able to meet the subsistence food needs of their families.

A questionnaire was pre-tested with 29 Akok households. From previous studies in the area, we anticipated finding approximately 200 households in the eight sub-villages and thus chose to implement a census sampling approach. Excluding the 29 households involved in the pre-test, the survey was administered to all of the 169 other households. Only two of the 169 households were unable to complete the survey.

The survey was designed to be conversational, as inconsistent literacy levels across the village required the survey to be administered as face-to-face interviews. Each survey took approximately one hour to complete, and households were compensated for their time with a bar of soap and a pen. Responses were kept anonymous, and respondents were reminded that they were free to stop the interview at any time. Informed consent was obtained through an information/consent form as stipulated by the University of Alberta Ethics Board.

The survey included basic information about the household including demographic information and land holdings. We also asked a number of questions about perceptions related to tenure security and the role of cash in their livelihoods. The WTA questions were framed with the context of a hypothetical forest conservation contract. The contract was described as a ten-year obligation to maintain any forest and fallow lands that were more than ten years old. In other words, the household would forego use of *nfos ekotok* and *nfos afan* for ten years, but *ekotok* (old fallow between 4 and 10 years) would be available for clearing. The household would be directly paid as compensation for forgone use. To illustrate the effects of this undertaking, the enumerators showed respondents a version of Figure 1 without the boxes for *esëp* and *fulu*. We noted that participants would receive assistance in the use of new agronomic techniques to sustain food production without the use of long-duration fallows

One challenge in implementing the study was the explanation of REDD+ and payments for ecosystem services to the community. For example, Whittington (1998) note how the concept of payments for ecosystem services is fundamentally difficult to communicate [39]. Our approach was to explain the context at a general level; that an organization would like to develop a program that encourages farmers to keep trees on the land, and that the household would be compensated to make sure that the trees were conserved.

To reduce hypothetical bias, the preamble to the WTA section of the survey contained a 'cheap talk' section that reminded respondents of the need to avoid hypothetical bias (Lusk, 2003) [40]. Also, the lead researchers did not accompany the local *Bulu*-speaking research assistants in order to minimize 'yea saying' and to maintain consistency between respondents' questionnaires.

The WTA question asked was "Would you be willing to accept (some amount) CFA (Central African franc) per hectare per year for 10 years to participate in the contract?" As the values had to be read aloud, all respondents were asked to respond with a "yes" or "no" to each value, beginning with the payment amount set to zero. We used a single-bounded continuous bid sequence presented in a payment card format. A payment card with increasing bids (in CFA francs) was utilized rather than a single-bounded approach in order to reduce starting point bias (Chien et al., 2005), and the potential misunderstandings that could arise in a tight-knit community by administering different double-bounded bids among households [41]. By starting every individual at zero and increasing by the same increments, each smallholder was able to identify their unique bid, rather than being anchored to the first value presented to them.

2.4. Measures and Models

Parametric and non-parametric approaches were used to analyze the WTA data. An advantage of non-parametric methods is that they do not rely on a specific functional form [35]. Furthermore, non-parametric approaches may be simple to calculate [42], which was a substantial advantage during pre-testing for establishing an upper bound bid of 1,000,000 CFA/ha. WTA results were calculated using the non-parametric Turnbull upper bound method. The Turnbull method may be used to capture a monotonically increasing bid continuum [43]. Assuming a normal distribution for the cumulative distribution function (F^*_j), the expected WTA is:

$$E(WTA) = \sum_{j=0}^{M*} t_j \, f^*_{j+1} \tag{1}$$

where t_j is the bid level ($ amount) posed to each respondent and f^*_{j+1} is the probability distribution function for the next highest bid level [44].

For the parametric approach, we estimate a binary logit model that describes the effects of determinants on the probability of accepting a bid:

$$
\begin{aligned}
P\,(YES) = {}& \alpha + \beta_1 BID + \beta_2 CASH\&FOOD + \beta_3\ PRICE + \beta_4\ INSECURE + \beta_5\ CASH\&LAND \\
& + \beta_6\ VERYOLDFALLOW + \beta_7\ YOUNG\&OLDFALLOW + \beta_8\ AREACOCOA \\
& + \beta_9 AREACULTIVATED + \beta_{10} DEPENDENTS + \beta_{11}\ GENDERHOUSEHOLDHEAD \\
& + \beta_{12}\ AGEHOUSEHOLDHEAD + \beta_{13}\ MARRIAGE + \varepsilon
\end{aligned}
\tag{2}
$$

where βs are coefficients, ε is the error term, and the variables are defined in Table 1. We have expected signs for some variables. We expect that the higher the *BID* value, the more likely a respondent is to accept a bid. For *CASH&FOOD*, we expect that stronger agreement with being able to substitute market food with home-grown food will increase the probability of accepting the contract which provides cash for purchasing food. For *PRICE*, we expect that stronger agreement with clearing more land in response to higher prices will decrease the probability of accepting a contract. The other perception variables were also thought to be potentially important, but they could be important for varying reasons causing their expected signs to be ambiguous. For example, *INSECURE* could have a negative influence on accepting a contract if respondents believed that the land had to be cultivated to maintain rights, so the contract would create insecurity. However, this variable could have the opposite effect if respondents believed that the contract would alleviate the need for them to cultivate the land in order to maintain secure rights. The characteristics of land holdings and households are included as controls for which we have no expectation of signs.

Table 1. Variables in the binary logit model describing participation in the contract.

Variables	Description of Variables
Dependent Variable	
P(YES)	The probability of a smallholder agreeing to enter into a contract at a given bid level
	Independent Variables
BID	Payment level that would be paid to the smallholder on an annual basis per hectare (i.e., CFA/ha/year)
Perceptions (higher rating corresponds with agree more)	
CASH&FOOD	**Subsistence vs. market food** *"If I cannot produce enough food to feed everyone in my family, I can buy it at the market"*
PRICE	**Market sensitivity** *"If the price of one of the crops that I cultivate goes up, I will clear more of my fallow that is 10 years or older"*
INSECURE	**Risk perceptions of tenure security** *"If I do not want someone else to be able to cultivate my fallows that are 10 years old and older, I will have to cultivate them"*
CASH&LAND	**Cash and land investments** *"If I receive additional revenue, I will invest in agroforestry"*
Characteristics of Land Holdings	
VERYOLDFALLOW	Total area of currently fallowed land that is older than 10 years (ha)—*nfos ekoyok*
YOUNG&OLDFALLOW	Total area of currently fallowed land that is 1–10 years (ha)—*nyengue* and *ekotok*
AREACOCOA	Total area of current cocoa agro-forests (ha)
AREACULTIVATED	Total area of dedicated food crops that the household is currently cultivating—*afub owondo* and *esëp* (ha)
Household Characteristics	
DEPENDENTS	Number of dependents in the household (0–14 years old)
GENDERHOUSEHOLDHEAD	Gender of head of household (1 = male, 0 = female)
AGEHOUSEHOLDHEAD	Age of head of household (years)
MARRIAGE	Marital status of head of household (1 = married, 0 = single or widowed)

To estimate the costs of avoided carbon dioxide emissions associated with the hypothetical forest conservation contracts, the above-ground time-averaged carbon stocks of land use systems within the ASB benchmark site were applied to the smallholders' estimated hectares of each land type, and those carbon measures translated into carbon dioxide equivalents. As described in [11], the original ASB team working in Cameroon estimated time-average carbon stocks using the carbon density measurements of [35] and the conversion factor of 3.67 to convert tonnes of carbon into tonnes of carbon dioxide equivalent: tonnes C × 3.67 = tonnes CO_2e. Total CO_2e stock for the household was calculated for the situation at the time of the survey, and for an avoided future situation in which the areas of *nfos ekotok* and *nfos afan* would be replaced with *afub owondo*. The difference was assumed to the amount of avoided CO_2e due to the contract. Each individual household's lowest value of 'yes' for the WTA statement was applied to this estimate of avoided CO_2e.

A potential supply curve of land offered into forest conservation contracts for the Akok households was generated from the individual WTA estimates, combined with the number of hectares of fallow managed by each household. Results for the 167 households were ordered from lowest (0 CFA/ha/annum) to highest WTA (>1,000,000 CFA/ha/annum) to create the potential supply curve.

To be comparable to other studies of the costs of REDD+, we translated the WTA estimates from CFA per hectare per year into US Dollar net present value per tonne of CO_2e over the 10 years of the hypothetical contract. For each survey respondent who held some area of fallow land at the time of the survey, and accepted one of the bids offered (1,000,000 CFA/hectare or less), we estimated the tonnes of CO_2e that they currently hold in fallow and the net present value in USD/tonne of CO_2e conserved for the 10 years of the hypothetical contract. To calculate the present value of future annual payments, we followed Swallow et al (2007) [15] in applying discount rates of 0.1 percent (following Stern, 2008 [12]) and 15 percent (following earlier ASB studies).

3. Results

We begin our presentation of results with an overview of some descriptive statistics that provide a characterization of the households' landholdings. We then discuss the WTA results in terms of dollar amounts and numbers of households that accepted and rejected contracts. Determinants of WTA are then investigated with the binomial logit model. Our results conclude with a discussion of the derived carbon abatement curve.

Respondents were asked to identify the origin of each cultivated field to compare to the shifting cultivation model presented by Brown [33]. These results are summarized in Table 2. Most smallholders convert high forest into transitionary forest fields (60%), however many also use secondary forest (21%) and very old fallow (13%) to create forest fields (Table 2). Mixed food crops were created from several different types of fallow, but most frequently from old fallow (63%). Another type of field (*asan*) was included, which are "off-season fields that are cultivated in lower lying areas where there is sufficient moisture to carry a crop through the long dry season" (Brown, 2006, p.76) [33]. Although the majority (66%) of these wetlands are cultivated from swamps (*merecage*), smallholders identified that they are also created from other types of forest and fallow. Cocoa was planted on every original land type, with 47% originating from lands that were fallow for at least 10 years prior to being converted to cocoa (high forest, secondary forest, very old fallow and old fallow).

Table 2. Shifting cultivation transitions between land types.

		Current land types			
		Forest Field (*esep*)	Mixed Food Crops (*afub owondo*)	Wetland (*asan*)	Cocoa
		% of small holders transitioning between land types			
Original land type	High Primary Forest (*afan*)	60	1	7	18
	Secondary Forest (*nfos afan*)	21	1	5	11
	Very Old Fallow (*nfos ekotok*)	13	14	7	3
	Old Fallow (*ekotok*)	6	63	9	15
	Young Fallow (*nyegue*)	1	9	7	3
	Forest Fallow (*fulu*)	0	7	0	0
	Forest Field (*esep*)	-	6	0	17
	Swamp (*merecage*)	0	0	66	33
Number of households with this field type		134	142	40	133

The mean total land holdings for households in Akok village was 29.0 ha, of which 24.4 ha was fallow and available for future cultivation. The 24.4 ha of fallow was comprised of 11.1 ha of secondary forest, 5.5 ha of very old fallow, 4.4 ha of old fallow, and 3.5 ha of young fallow. For the average household, therefore, the forest conservation contract would mean that only the 3.5 ha of young fallow and 4.4 ha of old fallow would be available for clearing for shifting cultivation within the next ten years. The average household reported to be actively cultivating 4.9 ha of land, including 2.3 ha of cocoa agroforestry, 1.4 ha of forest field, 0.8 ha of mixed food crops, and 0.1 ha of wetland.

Survey results indicate that 94 percent of all respondents would accept the contract within the range of bids offered. However, 10 percent of these respondents accepted a bid of 0. These respondents were then asked whether they were willing to pay to participate in the contract, and all responded "yes." In contrast, 6% of the sample was not willing to accept the contract for any of the amounts offered.

Figure 2 shows the proportion of respondents that indicated they would be willing to accept a bid amount for each bid level. Results indicate an acceptance rate of greater than 50 percent if compensation value was approximately greater than or equal to 200,000 CFA/ha/annum. The mean WTA was 226,047 CFA/ha/annum, while the median value was 222,416 CFA/ha/annum. At higher bid values, the curve becomes flatter indicating that a relatively small percentage of the respondents (i.e., approximately 10%) required bids of between 500,000 and 1,000,000 CFA/ha/annum to accept the contract.

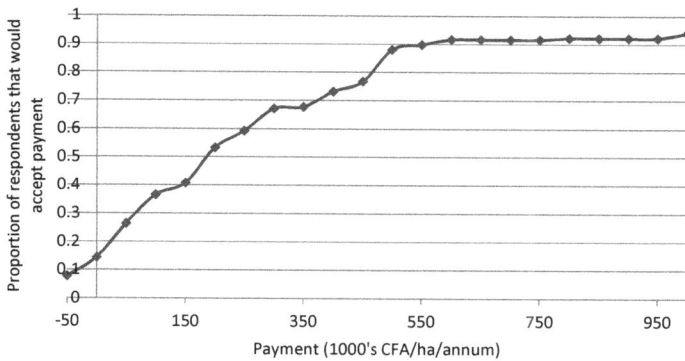

Figure 2. Turnbull upper bound non-parametric WTA (willingness to accept) results for smallholders accepting a deforestation restriction contract in Akok, Cameroon.

Table 3 contains the results of the binary logit model on the probability of accepting the contract. Six of the variables are found to be statistically significant at the 5% level of better. *BID* is shown to be highly significant and displays the expected sign, where higher bids increase the probability of accepting the contract. A number of perception variables are also significant. *CASHANDFOOD* is negative and significant, indicating that people who are more in agreement with their ability to buy food at the market will be less inclined to accept the contract. This result is contrary to our initial expectations, but could be caused by income effects for which we are not able to control. People who state that they are able to feed their family with market purchased food are likely to have higher incomes, which could decrease their willingness to accept cash for a contract. PRICE is also negative and significant indicating, as expected, that stronger agreement with clearing more land in response to higher prices will decrease the probability of accepting a contract. INSECURE is also significant with a positive sign, perhaps indicating that people tend to believe that the contract will alleviate the need for them to cultivate the land in order to maintain rights.

Table 3. Results of a binary logit model of smallholders' willingness to accept a bid for a hypothetical deforestation restriction contract.

Explanatory Variables	Estimated Coefficient	Standard Error
BID	0.0000050797 ***	0.0000005464
CASH&FOOD	−0.00325 ***	0.00073
PRICE	−0.00152 ***	0.00034
INSECURE	0.00106 ***	0.00024
CASH&LAND	0.09343	0.13574
VERYOLDFALLOW	−0.00025	0.00046
YOUNG&OLDFALLOW	−0.00100**	0.00052
AREACOCOA	−0.00034	0.00064
AREACULTIVATED	0.00028	0.00074
DEPENDENTS	−0.01368	0.04407
GENDERHOUSEHOLDHEAD	−0.36618	0.34160
AGEHOUSEHOLDHEAD	0.00148 ***	0.00038
MARRIAGE	−0.00020	0.00031
Constant	−1.1186 ***	0.43354
Log likelihood	−1479.06	
χ2 (9 d.f.)	1129.07	
McFadden Pseudo R^2	0.2762	
Observations (21 × 167)		
Median WTA	222416 CFA/ha/annum	
	(467.27 USD/ha/annum)	

***, **, * Significant, respectively, at the 1%, 5% and 10% levels.

With respect to types of land holdings, people who had larger areas of *YOUNG&OLDFALLOW* (fallow less than 10 years) were less likely to accept the contract. Our first hypothesis was that larger areas of young and old fallow would provide farmers with more options for renewing their crop cultivation land if they did enter the contract. This contrary result suggests a different rationale that is consistent with the results from the focus group and survey. As shown in Table 2, forest fields are rarely created from young or old fallow, but instead are created by slash and burn of primary or secondary forests. Respondents may have been worried that young and old fallow areas could be drawn into the contract, thus removing those areas as future options for slash and burn transition into forest fields.

Finally, *AGEHOUSEHOLDHEAD* was positive and significant, perhaps because older people were more inclined to prefer cash, rather than expending labor to grow their crops.

Figure 3 shows marginal costs of carbon abatement for various amounts of carbon derived from conserving areas of land. Due to missing data for some households, this supply curve is based on 135 the possible 169 households. We estimate that the amount of carbon held in fallows by those households equates to about one million tonnes of CO_2e. Across the 135 households, the mean net present value of ten annual payments of the WTA is USD 29.94/t CO_2e at a 15% discount rate and USD 55.67/t CO_2e at a 0.01% discount rate. However, accounting for the sizes of the fallow areas on each of the 135 farms, the weighted mean net present value of the ten annual payments was USD 11.32/t CO_2e at a 15% discount rate and USD 29.94 / t CO_2e at a 0.01% discount rate, implying that households with larger areas had lower WTAs. However, the shape of the curve (Figure 3), assuming the 15% discount rate indicates that farmers would be willing to conserve 61% of the standing carbon stock for payments equating to a net present value of less than 10 USD/t CO_2e. Lower payments would still attract considerable interest from area farmers. For example, about 27% of the carbon could be conserved for less than payments equating to a net present value of less than 5 USD / t CO_2e. At the other extreme end of the figure, the 18 highest observations of costs per tonne, which range from USD 55 to 675/t CO_2e, in the right hand panel of Figure 3, add up to less than 2% of the carbon stock.

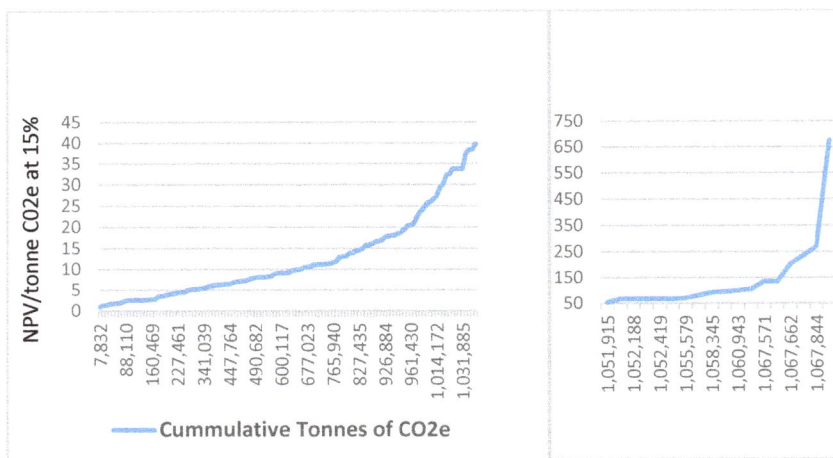

Figure 3. Marginal carbon abatement cost curve of land offered into REDD contracts based on smallholder land holdings and the proposed REDD-like contract in Akok village, Cameroon (*n* = 135). The right hand segment of the graph represents the highest 18 observations.

4. Discussion

With its high levels of forest carbon stocks, and potential threats of deforestation, Cameroon could make substantial contributions toward realizing the promise of REDD+. Like other African countries with relatively weak governance indicators, however, Cameroon has not attracted large investments in REDD+ [45]. To be more attractive destinations for REDD+ investments, countries such as Cameroon must conserve forested areas in a manner consistent with local institutions and acceptable to those with a stake in the way land is used. Previous studies [18,46] have focused on the institutional arrangements for land and forest governance in Cameroon.

Here, our primary focus is on the farm-level economics of REDD+. In identifying the potential for conserving lands as part of REDD+ programs, costs of lost opportunities are frequently measured by forgone production values for given types of land. However, our results using a WTA approach suggest that there are a number of features of local contexts that could cause opportunity cost approaches to be misleading.

Our estimates, derived from WTA's, tend to be somewhat higher than previous estimates based on opportunity cost approaches. For example, using a bottom up approach and comparing opportunity costs across locations in Indonesia, Peru and Cameroon, Swallow et al. (2007) [15] found a range of costs of 2 to 10 USD/t CO_2e for 20 years of avoided deforestation using an interest rate of 15%. Their estimate for Akok village is around 8 USD/t CO_2e. Similarly, Boucher (2008) [47] found a range of costs of 1.84–5.18 USD/t CO_2e for 20 years of sequestration and reported results from top-down studies ranging from 7.77 to 18.86 USD/t CO_2e for 20 years of sequestration. In contrast, our estimated range of values for a 10-year contract is between 15.79 and 5533.15 USD/t CO_2e at a discount rate of 0.1%, and between 7.97 and 2792.25 USD/t CO_2e at a discount rate of 15%. Our lower bound estimate of WTA is roughly equal to the estimated opportunity cost for households in Cameroon calculated by Swallow et al. (2017) [15]. We have not adjusted for inflation because the CFA to USD exchange rate changed by less than 10% over this period.

The possibility of low costs per tonne CO_2e was one of the factors that originally drew attention to the use of REDD+ as a mitigation option. The results presented in this paper suggest that previous estimates of opportunity costs may have seriously under-estimated the amounts that land users would need to be paid in exchange for farmers giving up discretion over future land use. The relatively high value of opportunity cost derived through the CV stated preference method is also consistent with the general hypothesis posited in this study, that smallholders consider more than the forgone production when considering REDD contract acceptance.

The drivers behind probabilities of accepting a contract provide some insights into local contexts that could be driving up WTA's. Though higher *BID* amounts do increase the probability of acceptance, these amounts are having to overcome a number of other drivers that reduce acceptance, likely because households wanted to maintain options to work their land. Local challenging, incomplete and messy institutional and economic contexts are apparent in influencing decisions. Specifically, people who believe more strongly that they have sufficient cash to consider markets as means of contributing to food security (i.e., *CASH&FOOD*) were less likely to accept a contract. Likewise, people who agreed more that they would want to clear more land if prices improved (i.e., *PRICE*) were also less likely to lock up their lands in a contract. Moreover, people who agreed less with the need work their land in order to maintain their rights were also less likely to accept a contract. There were also some characteristics of households, and their land holdings, which were found to be impediments. Younger household heads (AGEHOUSEHOLDHEAD) and households with more areas of young and old fallow (YOUNG&OLDFALLOW) were also less likely to accept a contract.

One striking finding is the degree of heterogeneity among values of land to households. The CV approach leads to a diverse range of acceptance for forest carbon conservation. The sample is split roughly into three groups: individuals willing to participate for no compensation, the majority of the sample willing to participate given some compensation, and individuals not willing to participate in the program, or only willing at extremely high levels of compensation relative to the rest of the sample. In the first group, 10% of the sample required no payment, perhaps because they saw the contract as a means of securing tenure to their land. However, there was also 5% of the sample that would not accept any offered amount to accept the contract. Nonetheless, the carbon abatement costs Figure 3) suggests a clear population subset that is willing to enter the proposed deforestation mitigation contract for a relatively low payment. Policy makers would be able to prevent nearly half of the available carbon stock (i.e., approximately 5,000,000 t CO_2e) from being emitted for a 10-year period for approximately 8 USD/t CO_2e CO_2, considering an interest rate of 15%.

No matter what amount would need to be paid to smallholders to avoid deforestation, an underlying principal, as discussed above, is that of Free, Prior and Informed Consent (FPIC) [9]. Consistent with this principal are contingent valuation approaches that solicit consensual (i.e., willing) values. The heterogeneity among households that arises in identifying consent presents both problems and promise for structuring REDD+ contracts. Promise arises from identifying large numbers of households who would willingly pay cost-effective amounts (relative to other options) to sequester carbon. Problems arise in developing payment schemes that could reflect the heterogeneity in WTA, but that would be deemed equitable among participants.

Acknowledgments: The authors acknowledge generous financial assistance from the University of Alberta and the World Agroforestry Centre. Caitlin Schmidt from the University of Alberta, Doug Brown from World Vision, Zac Tchoundjeu and Serge Ngendakumana from the World Agroforestry Centre, and Valentina Robiglio and Guillaume Lescuyer from the International Institute for Tropical Agriculture (IITA) provided valuable input and guidance for the field studies in Cameroon. Arlette, Chistiane, Yannick and Dominique all supported implementation of the survey in Akok. Two anonymous reviewers provided very helpful comments and suggestions. Of course, the people of Akok are thanked for their participation in the focus group and household survey.

Author Contributions: B.S. conceived this study and secured support from the funders and host organizations. DT., B.S. and M.L. designed the survey with inputs from C.S. D.T. implemented the field studies with those listed in the acknowledgements. D.T. did most of the data analysis with support from B.S. and M.L. D.T. wrote the first draft of the manuscript. B.S. and M.L. revised and refined the manuscript. B.S. is the corresponding author.

Conflicts of Interest: The founding sponsors had no role in the design of the study; in the collection, analyses, or interpretation of data; in the writing of the manuscript, and in the decision to publish the results.

References

1. Harris, N.; Stolle, F. Forests Are in the Paris Agreement! Now What? World Resources Institute Blog. 5 January 2016. Available online: http://www.wri.org/blog/2016/01/forests-are-paris-agreement-now-what (accessed on 18 May 2016).
2. Sandbrook, C.; Nelson, F.; Adams, W.M.; Agrawal, A. Carbon, forests and the REDD paradox. *Oryx* **2010**, *44*, 330–334. [CrossRef]
3. Corbera, E.; Schroeder, H. Governing and implementing REDD+. *Environ. Sci. Policy* **2011**, *14*, 89–99. [CrossRef]
4. Desbureaux, S.; Brimont, L. Between economic loss and social identity: The multi-dimensional cost of avoiding deforestation in eastern Madagascar. *Ecol. Econ.* **2015**, *118*, 10–20. [CrossRef]
5. Griffiths, T. *Seeing 'REDD'? Forests, Climate Change Mitigation and the Rights of Indigenous Peoples and Local Communities*; Forest Peoples Programme: Moreton-in-Marsh, UK, 2008.
6. Peskett, L.; Huberman, D.; Bowen-Jones, E.; Edwards, G.; Brown, J. *Making REDD Work for the Poor. Prepared on Behalf of the Poverty Environment Partnership*; Overseas Development Institute: London, UK, 2008.
7. Karsenty, A.; Ongolo, S. Can 'fragile states' decide to reduce their deforestation? The inappropriate use of the theory of incentives with respect to the REDD mechanism. *For. Policy Econ.* **2012**, *18*, 38–45. [CrossRef]
8. Ongolo, S. On the banality of forest governance fragmentation: Exploring "gecko politics" as a bureaucratic behaviour in limited statehood. *For. Policy Econ.* **2015**, *53*, 12–20. [CrossRef]
9. Mahanty, S.; McDermott, C.L. How does 'Free, Prior and Informed Consent' (FPIC) impact social equity? Lessons from mining and forestry and their implications for REDD+. *Land Use Policy* **2013**, *35*, 406–416. [CrossRef]
10. Palm, C.A.; Vosti, S.A.; Sanchez, P.A.; Ericksen, P.J. *Slash-and-Burn Agriculture: The Search for Alternatives*; Columbia University Press: New York, NY, USA, 2005.
11. Grieg-Gran, M. *The Cost of Avoiding Deforestation: Update of the Report Prepared for the Stern Review of the Economics of Climate Change*; ACM Digital Library: New York, NY, USA, 2008.

12. Stern, N. *The Economics of Climate Change: The Stern Review*; Cambridge University Press: New York, NY, USA, 2006.
13. Eliasch & Office of Climate Change. *The Eliasch Review: Climate Change: Financing Global Forests*; H.M. Government: London, UK, 2008. Available online: https://www.gov.uk/government/uploads/system/uploads/attachment_data/file/228833/9780108507632.pdf (accessed on 1 March 2017).
14. McKinsey & Company. Pathways to a Low-Carbon Economy: Version 2 of Global Greenhouse Gas Abatement Cost Curve. 2009, p. 7. Available online: http://www.mckinsey.com/business-functions/sustainability-and-resource-productivity/our-insights/pathways-to-a-low-carbon-economy (accessed on 3 February 2016).
15. Swallow, B.; van Noordwijk, M.; Dewi, S.; Murdiyarso, D.; White, D.; Gockowski, J.; Hyman, G.; Budidarsono, S.; Robiglio, V.; Meadu, V.; et al. *Opportunities for Avoided Deforestation with Sustainable Benefits: An Interim Report of the ASB Partnership for the Tropical Forest Margins*; ASB Partnership for the Tropical Forest Margins: Nairobi, Kenya, 2007.
16. World Bank Institute. *Estimating the Opportunity Costs of REDD+: A Training Manual*; World Bank Institute, Version 1.3; World Bank: Washington, DC, USA, 2011.
17. Gregersen, H.; El Lakany, H.; Karsenty, A.; White, A. *Does the Opportunity Cost Approach Indicate the Real Cost of REDD+?* Rights and Resources Initiative: Washington, DC, USA, 2010.
18. Cerbu, G.A.; Sonwa, D.J.; Pokorny, B. Opportunities for and capacity barriers to the implementation of REDD+ projects with smallholder farmers: Case study of Awae and Akok, Centre and South Regions, Cameroon. *For. Policy Econ.* **2013**, *36*, 60–70. [CrossRef]
19. Sunderlin, W.D.; Larson, A.M.; Duchelle, A.E.; Resosudarmo, I.A.P.; Huynh, T.B.; Awono, A.; Dokken, T. How are REDD+ proponents addressing tenure problems? Evidence from Brazil, Cameroon, Tanzania, Indonesia, and Vietnam. *World Dev.* **2014**, *55*, 37–52. [CrossRef]
20. Fund for Peace. Fragile States Index 2016. Available online: http://fsi.fundforpeace.org/ (accessed on 1 March 2017).
21. Jagoret, P.; Michel-Dounias, I.; Malézieux, E. Long-term dynamics of cocoa agroforests: A case study in central Cameroon. *Agrofor. Syst.* **2011**, *81*, 267–278. [CrossRef]
22. Caravani, A.; Graham, K. Market and Non-Market Costs of REDD+ Perceived by Local Communities: A Case Study in East Cambodia Cambodia. REDDnet Case Study. 2011. Available online: http://theredddesk.org/sites/default/files/resources/pdf/2011/case_study_5_-_cambodia_alice_final.pdf (accessed on 4 January 2017).
23. Wünscher, T.; Engel, S.; Wunder, S. Spatial targeting of payments for environmental services: A tool for boosting conservation benefits. *Ecol. Econ.* **2008**, *65*, 822–833. [CrossRef]
24. Ghazoul, J.; Butler, R.; Mateo-Vega, J.; Pin Koh, L. REDD: A reckoning of environment and development implications. *Trends Ecol. Evolut.* **2010**, *25*, 396–402. [CrossRef] [PubMed]
25. Kindermann, G.; Obersteiner, M.; Sohngen, B.; Sathaye, J.; Andrasko, K.; Rametsteiner, E.; Beach, R. Global cost estimates of reducing carbon emissions through avoided deforestation. *Proc. Natl. Acad. Sci.* **2008**, *105*, 10302–10307. [CrossRef] [PubMed]
26. Whittington, D.; Pagiola, S. Using contingent valuation in the design of payments for environmental services mechanisms: A review and assessment. *World Bank Res. Obs.* **2012**, *27*, 261–287. [CrossRef]
27. Jack, B.K.; Beria, L.; Ferraro, P.J. A revealed preference approach to estimating supply curves for ecosystem services: Use of auctions to set payments for soil erosion control in Indonesia. *Conserv. Biol.* **2009**, *23*, 359–367. [CrossRef] [PubMed]
28. Kling, C.L.; List, J.A.; Zhao, J. A dynamic explanation of the willingness to pay and willingness to accept disparity. *Econ. Inq.* **2013**, *51*, 909–921. [CrossRef]
29. Kahneman, D.; Tversky, A. Prospect theory: An analysis of decision under risk. *Econometrica* **1979**, *47*, 263–291. [CrossRef]
30. Kahneman, D.; Knetch, J.L.; Thaler, R.H. Experimental tests of the endowment effect and the Coase theorem. *J. Political Econ.* **1990**, *98*, 1325–1348. [CrossRef]
31. Venkatachalam, L. The contingent valuation method: A review. *Environ. Impact Assess. Rev.* **2004**, *24*, 89–124. [CrossRef]

32. Palm, C.A.; Tomich, T.; van Noordwijk, M.; Vosti, S.; Gockowski, J.; Alegre, J.; Verchot, L. Mitigating GHG emissions in the humid tropics: Case studies from the Alternatives to Slash-and-Burn Program (ASB). *Environ. Dev. Sustain.* **2004**, *6*, 145–162. [CrossRef]

33. Brown, D.R. Personal preferences and intensification of land use: Their impact on southern Cameroonian slash-and-burn agroforestry systems. *Agrofor. Syst.* **2006**, *68*, 53–67. [CrossRef]

34. Robiglio, V.; Mala, W.A.; Diaw, M.C. Mapping landscapes: Integrating GIS and social science methods to model human-nature relationships in southern Cameroon. *Small-Scale For. Econ. Manag. Policy* **2003**, *2*, 171–184.

35. Sonwa, D.J. *Biomass Management and Diversification within Cocoa Agroforests in the Humid Forest Zone of Southern Cameroon*; Cuvillier Verlag: Gottingen, Germany, 2004.

36. Gerber, J.F.; Veuthey, S. Possession Versus Property in a Tree Plantation Socioenvironmental Conflict in Southern Cameroon. *Soc. Nat. Resour.* **2011**, *24*, 831–848. [CrossRef]

37. Van den Berg, J. Sustainable Exploitation and Management of Forest Resources: Diverging Perceptions on the Forest. In Forest Management Related Studies of the Tropenbos-Cameroon Programme; Papers Presented at a Joint WAU-Tropenbos Workshop. Available online: www.tropenbos.org/file.php/316/workshop-berg.pdf (accessed on 4 January 2017).

38. Ndoye, O.; Kaimowitz, D. Macro-economics, markets and the humid forests of Cameroon, 1967–1997. *J. Mod. Afr. Stud.* **2000**, *38*, 225–253. [CrossRef]

39. Whittington, D. Administering contingent valuation surveys in developing countries. *World Dev.* **1998**, *26*, 21–30. [CrossRef]

40. Lust, J.L. Effects of cheap talk on consumer willingness-to-pay for golden rice. *Am. J. Agric. Econ.* **2003**, *85*, 840–856.

41. Chien, Y.L.; Huang, C.J.; Shaw, D. A general model of starting point bias in double-bounded dichotomous contingent valuation surveys. *J. Environ. Econ. Manag.* **2005**, *50*, 362–377. [CrossRef]

42. Tambour, M.; Zethraeus, N. Nonparametric willingness-to-pay measures and confidence statements. *Med. Decis. Mak.* **1998**, *18*, 330–336. [CrossRef] [PubMed]

43. Kriström, B. A non-parametric approach to the estimation of welfare measures in discrete response valuation studies. *Land Econ.* **1990**, *66*, 135–139. [CrossRef]

44. Haab, T.C.; McConnell, K.E. *Valuing Environmental and Natural Resources: The Econometrics of Non-Market Valuation*; Edward Elgar: Cheltenham, UK, 2002.

45. Cerbu, G.A.; Swallow, B.M.; Thompson, D.Y. Locating REDD: A global survey and analysis of REDD readiness and demonstration activities. *Environ. Sci. Policy* **2011**, *14*, 168–180. [CrossRef]

46. Boucher, D.H. *Estimating the Cost and Potential of Reducing Emissions from Deforestation. Tropical Forests and Climate, Briefing #1*; Union of Concerned Scientists: Washington, DC, USA, 2008.

47. Awono, A.; Somorin, O.A.; Eba'a Atyi, R.; Levang, P. Tenure and Participation in Local REDD+ Projects: Insights from Southern Cameroon. *Environ. Sci. Policy* **2014**, *35*, 76–86. [CrossRef]

forests

MDPI

Article

Transforming Justice in REDD+ through a Politics of Difference Approach

Kimberly R. Marion Suiseeya

Department of Political Science, Northwestern University, Evanston, IL 60208, USA;
kimberly.suiseeya@northwestern.edu; Tel.: +1-847-491-8985

Academic Editors: Esteve Corbera and Heike Schroeder
Received: 30 September 2016; Accepted: 22 November 2016; Published: 30 November 2016

Abstract: Since Reduced Emissions from Deforestation and Degradation "Plus" (REDD+) starting gaining traction in the UN climate negotiations in 2007, its architects and scholars have grappled with its community-level justice implications. On the one hand, supporters argue that REDD+ will help the environment and forest-dependent communities by generating payments for forest carbon services from industrialized countries seeking lower cost emissions reductions. Critics, by contrast, increasingly argue that REDD+ is a new form of colonization through capitalism, producing injustice by stripping forest communities of their rights, denying them capabilities for wellbeing, and rendering forest peoples voiceless in forest governance. This paper argues that current REDD+ debates are too focused on relatively simple visions of either distributive or procedural justice, and pay too little attention to the core recognitional justice concerns of REDD+ critics, namely questions of what values, worldviews, rights, and identities are privileged or displaced in the emergence, design, and implementation of REDD+ and with what effects. This paper examines the tensions that emerge when designing institutions to promote multi-scalar, multivalent justice in REDD+ to ask: what are the justice demands that REDD+ architects face when designing REDD+ institutions? Complexifying the concepts of justice as deployed in the debates on REDD+ can illuminate the possibilities for a diversity of alternative perspectives to generate new institutional design ideas for REDD+.

Keywords: REDD+; justice; institutions; forest peoples

---

1. Introduction

Climate change scientists and policy makers have discussed the significance of forest loss and degradation as a source of carbon emissions for decades. Recent studies estimate that forest loss and degradation contribute approximately 10% of global carbon emissions—a significant contributor to climate change [1]. In 2007, at the 13th Conference of Parties to the UN Framework Convention on Climate Change in Bali, Indonesia, policy makers formally introduced an approach to address forests as an emissions source. The mechanism, REDD—Reduced Emissions from Deforestation and Degradation—sought to harness the power of markets to keep carbon in forests and trees by paying forest owners (mostly in developing countries) to reduce forest loss and degradation. Shortly after its introduction, REDD evolved into REDD+ to recognize the importance of enhancing carbon stocks [2]. Debates regarding REDD+'s potential impacts on forest communities were immediately contentious. On the one hand, REDD+ proponents argue it is a win-win solution that can help the environment, forest-dependent communities, and developed countries seeking lower cost emissions reductions [2,3]. Critics, by contrast, argue that REDD+ is a new form of colonization that strengthens state control over forests and produces injustice by stripping forest communities of their rights, denying them capabilities for wellbeing, and rendering forest peoples voiceless in forest governance [4,5]. While debates over approaches to forest governance are common and often productive, this contentious debate between REDD+ proponents and critics has persisted with little progress towards resolution, despite efforts on both sides. The growing urgency to conserve forests for people, nature, as well as climate change mitigation, requires that these conflicts over REDD+ are transformed into more productive engagement before the collective opportunities to conserve forests are lost.

In this article I offer an alternative approach to understanding the justice concerns of forest peoples (forest peoples are "peoples who live in and have customary rights to their forests, and have developed ways of life and traditional knowledge that are attuned to their forest environments. Forest peoples depend primarily and directly on the forest both for subsistence and trade in the form of fishing, hunting, shifting agriculture, the gathering of wild forest products and other activities" [6]). I argue that the current REDD+ debate as articulated by REDD+ proponents is too focused on relatively simple visions of either distributive or procedural justice, and pays too little attention to the related but distinct idea of recognitional justice. When addressing social impacts, REDD+ architects—those actors designing REDD+ frameworks, mechanisms, and interventions—emphasize questions of distributive and procedural justice such as: how will the costs and benefits of REDD+ be distributed, who decides who are the winners and losers, and to what extent can REDD+ interventions minimize or mitigate potential negative social impacts [7,8]? Although these are important considerations for REDD+ design, they neglect the core recognitional justice concerns of REDD+ critics: what values, worldviews, rights, and identities are privileged or displaced in the emergence, design, and implementation of REDD+ and with what effects (e.g., [9])? By embracing a politics of difference approach to examine current REDD+ debates, this paper complexifies the concepts of justice deployed in REDD+ and illuminates the possibilities for a diversity of alternative perspectives that can help generate new institutional design ideas for REDD+.

To reframe the existing justice debates in REDD+, I draw from two distinct bodies of literature. Section 2 begins with a justice primer that draws from the growing body of scholarship on environmental justice to outline this paper's approach to justice. Through this review, I develop a politics of difference analytical lens that facilitates exposure of dominant norms of justice that constrain how policy makers and practitioners think about justice [10]. In Section 3, I review the scholarly literature on REDD+ to identify four main themes or contentious points in ongoing REDD+ debates. I then apply the politics of difference lens to these common justice concerns to highlight the nested and correlational nature of those concerns and demonstrate how REDD+ architects narrowly scope the debate to focus solutions on a predefined understanding of the problem of forest loss and degradation and advance technocratic, state-centered solutions. I conclude with a discussion that encourages policy architects to attend to the politics of difference when designing REDD+.

I begin the paper with a final introductory note. Policy makers, practitioners, and scholars searching for specific solutions to REDD+ justice problems in this paper will likely be disappointed. Central to this paper's approach is the implicit understanding that justice in REDD+ demands a rejection of prescriptive approaches to justice. There is no singular institutional design that will lead to more just forest carbon interventions. Instead, the approach offered in this paper provides an analytical lens through which to bring to light new insights in the REDD+ justice debates. My aim is to highlight how approaching REDD+ through an explicit politics of difference lens can both illuminate critical recognitional concerns that are embedded in REDD+ institutions while simultaneously uncovering their relationships with distributive and procedural injustices. By exploring and understanding justice claims though a lens of difference rather than through a lens of distribution [11], I aim to offer REDD+ constituencies a more productive avenue for engaging in REDD+ debates. Through this recognitional lens, policy makers and scholars can envision new, and perhaps more effective, possibilities for alternative institutional designs to address some of those criticisms and move the debates on REDD+ past their current polarized form.

2. The Multiple Dimensions and Meanings of Justice

Environmental justice scholars and practitioners have difficulty pinpointing a universal definition of justice. Instead, the meaning of environmental justice is primarily known through injustices that emerge [12]. In the 1980s, environmental justice cases in the United States began to garner public attention [13]. Researchers focused on the disproportionate negative environmental health impacts that communities of color experienced related to toxic dumping [14] and siting of noxious facilities such as landfills and factories [15]. Similarly, early political ecologists emphasized the unfair cost burdens that resource-dependent communities shouldered related to natural resource conservation projects, such as forest preservation [16] and wildlife conservation [17], resulting in reduced access to traditional livelihood and cultural resources. These early examples of environmental injustice emphasize the distributive dimensions of justice, where distributive justice refers to an equitable distribution of costs and benefits, harms, and goods related to environmental governance [18].

Beginning in in the late 1980s and early 1990s, scholars expanded research on environmental justice to highlight the procedural aspects of environmental injustices, much as early environmental justice activists had stressed from the outset of the movement [19]. This line of research documented how communities were not only denied access to resources or disproportionately exposed to environmental harms, but also excluded from decision-making processes that impact how they interact with their environment (e.g., [20–22]). Procedural justice refers to the ability of all individuals impacted by a decision to meaningfully participate in the decision-making process and therewith shape the potential outcomes of the process [23–25]. Procedural justice has long been a focus of environmental policy especially in the United States and Europe where public input in decision-making processes is formally facilitated by various government agencies through open comment periods and grievance mechanisms (e.g., the National Environmental Policy Act (NEPA) and the Aarhus Convention are two examples of legal instruments that facilitate procedural justice in environmental policy-making in the US and Europe, respectively).

In more recent years, scholars aiming to effectively analyze environmental justice have emphasized the need to move beyond primarily distributive and procedural understandings of justice. Building upon global justice theorists such as Nancy Fraser, Iris Young, and others, and through an empirical examination of environmental justice cases, David Schlosberg [18] explicitly draws attention to the multidimensional nature of environmental justice, first through three main dimensions—distributive, procedural, and recognitional. These dimensions capture the multiple avenues through which justice can be produced. The third dimension, recognitional, refers to the ability to participate in and benefit from environmental governance without being required to assimilate to dominant cultural norms [26]. In other words, recognitional justice, or justice as recognition, requires that difference in cultures, lifeways, and ways of knowing are recognized, respected, and appropriately incorporated in environmental policy processes from conception to design, implementation, monitoring, and evaluation. Where distributive and procedural justice and injustice are often readily and somewhat objectively observed, for example, through an examination of the actual distribution of benefits and costs related to a project or documentation of a consultation process, respectively, recognitional justice requires scholars and policy makers to identify and confront the ways in and pathways through which cultural norms infuse and shape environmental institutions to perpetuate the dominance of these norms.

Schlosberg [25] later added a fourth dimension, capabilities, reflecting influences from Amartya Sen and Martha Nussbaum, to identify the importance of possessing the freedoms to realize one's aspirations [27]. Thus, even in cases where distributive, procedural, and recognitional justice concerns are attended to, if the subjects of justice—the rights-holders—do not have the capabilities to benefit from or participate in environmental governance initiatives, then injustices may persist. A related body of literature that examines equity similarly advances a multi-dimensional framework that includes distributive and procedural equity as well as contextual equity to captures both the recognitional and capabilities dimensions articulated by Schlosberg [28]. This broad consensus on the multidimensional nature of justice leads to the first consideration that scholars examining the justice effects of REDD+ should make explicit: to fully assess and understand environmental justice and more accurately identify the mechanisms of injustice, scholars have to consider the multiple dimensions of justice [12].

Three additional considerations should inform how scholars analyze environmental justice. First, scholars interested in analyzing environmental justice should attend to the nested nature of the multiple dimensions of justice. For example, studies on environmental justice in water governance in Australia demonstrate that efforts to promote distributive justice are more likely to succeed when procedural justice is also facilitated [29]. Other studies demonstrate that any positive benefits from more just distributive outcomes could be diminished if procedural injustice exists [30,31]. Policy designs that include Free Prior Informed Consent (FPIC), for example, seek to ensure that communities are consulted in the design and implementation of projects in order to promote procedural justice and ensure that communities can more effectively understand the costs and benefits of these projects [32]. The idea is that projects require community buy-in and input in order to most effectively design the processes and structures for just distribution of costs and benefits related to the project.

Second, scholars should attend to the multivalent nature of justice [33]. This means that environmental justice scholarship needs to consider the multiple ways that individuals and communities experience, understand, and conceptualize justice and injustice [9,34]. In other words, what is just for one individual or community may not be just for another. Although there are multiple justice principles and theories, often international environmental policy approaches tend to draw on more liberal notions of justice. Programs and projects that assume universal acceptance of particular justice principles threaten to undermine the pursuit of justice. Schroeder and Pogge [35], for example, point out that the Convention on Biological Diversity advances a concept of *justice-in-exchange* for managing genetic resources, which may not reflect the conceptualization of justice held by the custodians of genetic resources (e.g., see [24]). Alternative conceptions of justice may invoke diverse principles of justice, for example, *justice-as-needs*, *egalitarian*, or *equality* or *equity* [36], all conceptions that draw on different allocation rules (see also [37]). Attending to the multivalent nature of justice is difficult both for scholars and policy makers: it suggests that there may be no clear path towards justice. However, to ignore the multiple meanings of justice is to ensure the perpetuation of injustice for some.

Third, analyses of environmental justice should consider how jurisdictional, geographic, and temporal scales interact with justice concerns [33,38]. While many REDD+ proponents engage in debates oriented towards both *interstate* and *intergenerational* justice concerns (i.e., questions related to rights to pollute and rights to develop and questions on impacts on future generations), many REDD+ critics are concerned with *intrastate* justice concerns that emerge from a *transnational* context (and include both *inter-* and *intragenerational* concerns). This means that while these justice concerns center on the relationships between communities and the state within which a community is located *(intrastate)*, they emerge from a transnational context in which states opt into interstate arrangements to govern intrastate affairs. Examining these justice tensions within and across scales is important for identifying the scales at which and by whom justice concerns emerge and need to be addressed. While it may seem simple to suggest that concerns between states and their communities should be dealt with domestically, as sometimes asserted by REDD+ proponents, such an approach ultimately ignores the role that an initiative like REDD+ plays in generating or exacerbating these justice concerns.

To summarize, uncovering the mechanisms that lead to injustice requires studies of environmental justice to consider: (1) its multiple dimensions (distributive, procedural, and recognitional); (2) the nested nature of these multiple dimensions; (3) the multivalent nature of justice; and (4) the scalar interactions of justice concerns. In this paper, I direct attention to justice concerns that manifest at the community or individual level *(intrastate)*, rather than at the state level, but are transnational in nature. I also assume that justice requires all three dimensions of justice (distributive, procedural, recognitional) with the explicit acknowledgement that efforts towards procedural and distributive justice can be undermined if recognitional dimensions are not first addressed [10]. Similarly, efforts to improve recognitional justice can also have positive procedural or distributive justice outcomes. As I argue in the next sections of this paper, more explicit attention to recognitional justice in particular can facilitate and streamline such examinations and help identify opportunities for facilitating justice [39]. This is particularly important in the context of REDD+, which has been a focal point for justice-based critiques of environmental governance from global to local scales. In the next sections of this paper I demonstrate how explicit attention to the recognitional justice dimensions of the most common social justice critiques of REDD+ can provide pathways for effectively analyzing the multidimensional, nested, multivalent, and scalar nature of justice in REDD+, thereby providing a useful tool for analyzing potential policy options and institutional designs for a more just REDD+.

3. Results: Examining REDD+ Through a Politics of Difference Lens

While there is an emerging body of scholarship calling for more explicit consideration of the recognitional dimensions of justice in environmental governance [27,40], much of the current literature on justice and REDD+ focuses primarily on distributive and procedural justice [41–51]. Additionally, there is limited (although growing) engagement among scholars and practitioners on how to practically engage recognitional justice in institutional designs [52,53]. In what follows, I introduce four common justice concerns that emerge in popular and scholarly debates on REDD+ (notably, although this paper focuses on REDD+, the primary themes that emerged from the literature largely reflect the justice concerns in broader global forest governance efforts [34], with some exceptions noted). These are drawn from an in-depth review of the literature from 2007 to 2015. In the presentation of these issues, I aim to elucidate the core justice concerns related to REDD+ beyond the dominant soundbites on burden-sharing and rights to develop and pollute. I discuss these issues through a politics of difference lens that parses out concerns related to three primary dimensions of justice: distributive, procedural, and recognitional. Table 1 details such an approach to REDD+ justice critiques. This lens facilitates a deeper understanding of the complex nature of justice by emphasizing that justice is not simply a question of the distribution of costs and benefits (distributive justice) or access to decision-making processes (procedural justice), but also about the extent to which *difference* is recognized, respected, and included in institutions such that cultural assimilation is not a precondition for participation in and receiving benefits from REDD+ initiatives (e.g., an operational definition of recognitional justice). This recognitional dimension of REDD+ has been largely absent from the justice debates on REDD+ in the scholarly literature. Through this discussion I aim to demonstrate how these concerns are often discussed through distributive and/or procedural lenses, as introduced above, but are fundamentally recognitional in nature. In other words, I argue here that attention to recognitional justice in the design of REDD+ could alleviate or mitigate many of these concerns. At a minimum, the recognitional lens offers new insights into the following four justice critiques of REDD+.

3.1. As Currently Implemented, REDD+ Places Authority for Forest Governance in the Hands of the State and, to a Lesser Extent, Markets

The main objective of REDD+ is to reduce *global* carbon emissions, and while the climate benefits of reduced emissions are accrued globally, the costs of doing so are highly localized: the price of reducing one ton of carbon in Indonesia may significantly diverge from the costs of the same ton of carbon in Brazil, but the climate benefits are equal. A market-oriented approach could promote efficiency by incentivizing emissions reductions in the places where it is the most cost efficient to do so. Although its underlying principles seem straightforward, policymakers and REDD+ architects identified five core technical challenges that could undermine its effectiveness: measurement, reporting, verification, permanence, and leakage [54]. Effective REDD+ initiatives need to credibly ensure that the emissions reductions are accurately measured and verified in order to be sold as carbon credits, that these reductions are permanent, and do not simply displace emissions to another forest. To meet these demands, programmatic requirements are highly technical and complex, and generally not amenable to purely local or community administration. Instead, some form of coordination with governments is required. To support the development of the institutional infrastructures and capacity required for REDD+, multilateral and bilateral donors have invested millions of dollars in "REDD+ Readiness" activities for target participating countries. Thus, as currently implemented, REDD+ operates more like traditional overseas development assistance than the market-based approach initially envisioned and centers states as the core actors for enacting REDD+.

Table 1. Applying a politics of difference lens to analyze Reduced Emissions from Deforestation and Degradation "Plus" (REDD+) justice critiques.

	Distributive	Procedural	Recognitional
State-centered	Who has authority to determine allocations of costs/efforts and benefits from REDD+ activities? *Concern: states will capture the benefits and communities will bear the costs.*	What is the process by which allocation decisions are made and who has the authority to make decisions? *Concern: states will make decisions about how to allocate costs and benefits and fail to incorporate community concerns and ideas.*	Who do states represent and how inclusive and legitimate are states' claims of representation? *Concern: representation by states requires assimilation and secession of some self-determination rights.*
Property Rights	Who should receive payments for reduced emissions? *Concern: weak tenure arrangements will prevent some rights-holders from receiving benefits.*	What is the process by which resource tenure rights are allocated and enforced? *Concern: in contexts where formal resource tenure rights are absent or incomplete, some rights-holders will be excluded.*	Who has access and rights to claim allocations? What are other ways of thinking about ownership? Who has authority/delegates authority/recognizes authority? *Concern: recognition of customary and/or traditional resource rights that are not codified or recognized by state legal authorities.*
Values	What ecological and environmental values should be conserved and compensated for? *Concern: compensation and/or benefits may not be adequate in both aggregate and distributive terms.*	What is the process by which decisions regarding prioritization of values and determination of worth are made? *Concern: compensation and/or benefits may not reach the appropriate beneficiaries or may be viewed as insufficient in return for effort.*	To what extent and how are diverse values of forests understood, articulated, and integrated into forest governance approaches? By whom? *Concern: narrow understandings of forests can lead to reductionist valuations of forests, exclude diverse values, and render forest peoples invisible.*
Decision-making	How many seats need to be at the decision-making table? *Concern: compensation and/or benefits may not reach the appropriate beneficiaries or may be viewed as insufficient in return for effort.*	Do stakeholders have a seat at the table? *Concern: compensation and/or benefits may not reach the appropriate beneficiaries.*	What does the decision-making table look like, whose values and cultural norms are reflected in the design of the process, and to what extent is the process inclusive in its structure and implementation? *Concern: pre-determined agendas and structures for engagement and decision-making may render some forest peoples invisible or exclude them from meaningful engagement.*

The REDD+ literature most frequently discusses the justice concerns that result from a (re)concentration of state power and authority over forest governance in terms of rent-seeking and democratic deficits (distributive and procedural dimensions of justice). In other words, risks of injustice emerge through two pathways: (1) if states capture the bulk of the payments from REDD+ activities [55]; and (2) if states do not meaningfully facilitate and integrate community concerns and ideas [45].

When viewed through a politics of difference lens, however, the recognitional dimensions of forest peoples' justice concerns emerge. By placing states at the center of REDD+ decision-making, REDD+ activities may serve to legitimize state claims of authority over forests and REDD+ activities and recentralize forest governance, therewith diminishing the power and agency of local communities to determine their fates and lifeways [5]. The concerns herein are threefold: first, while governments in many developing countries legally (by statute) own the vast majority of forests (for example, in Indonesia, the government owns 100% of forests, in Brazil 77%: "extrapolated to the global forest estate of 3.9 billion hectares, these data suggest that approximately 77 percent of the world's forest is—according to national laws—owned and administered by governments, at least 4 percent is reserved for communities, at least 7 percent is owned by local communities, and approximately 12 percent is owned by individuals" [56], (p. 6)), nearly 1.6 billion people worldwide are dependent on these forests [6]. In many cases, these forest communities are the de facto managers of these resources, albeit sometimes with limited to no authority to develop and/or enforce regulations. Indigenous Peoples, whose population is estimated at up to 550 million globally, govern 65% of the world's land, yet only 10% of this authority over land is recognized by states [57]. Thus, the critical question for understanding justice concerns around state centralization is about who states do and do not represent and how inclusive and legitimate states' claims of representation of diverse peoples are.

Second, access to REDD+ benefits mediated by states may exclude populations whose citizenship and/or territories are not formally recognized by states, therewith promoting continued subordination of marginalized groups to state dominance and authority. Under such circumstances, while the impacts may include distributive injustices, they emerge from a recognitional injustice, namely a lack of recognition of their claims to forestlands by the state.

Third, and related to the second concern, in some cases non- or under-recognized groups may not seek or desire state recognition because they are ultimately pursuing self-determination and sovereignty and resist state authority over their communities and territories. Formal state recognition could require some forms of assimilation as well as require these groups to submit to state authority, in direct contradiction to their self-determination objectives.

3.2. Related to the First Concern, as Currently Implemented, REDD+ Architects Embrace a Narrow and Formal Understanding of State-Sanctioned and Largely Unrestricted Resource and Land Tenure Rights

Numerous scholars and policy makers have called for stronger property rights (e.g., transferrable and legally enforceable) as a prerequisite for REDD+ success [58]. Conventionally, scholars and policy makers who discuss concerns related to strong, neoliberal private property rights and market transactions frame them in terms of distributive justice. By creating powerful market incentives, REDD+ could incentivize unfettered market transactions that economically displace forest peoples and other resource-dependent communities from their traditional livelihood sources while more politically and economically powerful capture the financial rewards of REDD+. In the most basic terms, REDD+ proponents frame the underlying justice concern as distributive in nature: those without secure property rights would lose their forests and benefits to market forces, bearing the costs of REDD+ without receiving an equitable share of the benefits [59].

To address these concerns, REDD+ architects have promoted the establishment of secure, transferable, and legally enforceable property rights to forest resources. In practice, because tenure arrangements in many REDD+ countries are often informal and property rights systems non-existent or weak, this has largely translated into supporting the development of individual property rights systems [60]. Although stronger property rights can be an important precondition for poverty alleviation [56], and most forest peoples welcome, if not demand, strong and enforceable recognition of their traditional land and resource rights, the current emphasis on individual, neoliberal property rights raises important recognitional justice concerns.

First, because most forests in most REDD+ countries are "owned" by governments [56], states can further legitimize claims for authority over REDD+ forestlands. This failure to recognize traditional, community, and/or non-state governance results in displacement of these "informal" institutions and ultimately confers primary beneficiary status to states. When benefits are distributed based on rights and not responsibilities, communities that have traditionally governed certain forest areas may experience distributive injustices.

Second, as noted above, forest peoples and communities support strong tenure arrangements, although, as related to concerns in Section 3.1, this does not necessarily require state recognition of rights. Tenure systems take a variety of forms that may be individual, collective, or hybrid. For example, among the Makgong people, an ethnic minority in Laos who practice shifting cultivation as part of their forest-based livelihoods, families rotate among plots based upon length of the required fallow period for the plot and what crops need to be planted that season [34]. By requiring communities with traditionally collective or flexible forms of ownership and governance to transition to individual property, policy makers are redefining peoples' relationships with land and resources, as well as how communities organize themselves to meet their needs. Moreover, individual property rights privilege transferability functions of rights over access functions, the latter being critical for forest peoples' identities and lifeways.

3.3. As currently Implemented, REDD+ Artificially Narrows the Definition, and Therewith Value, of Forests by Prioritizing One Service (Carbon Sequestration) over other Values, Uses, and Ways of Understanding Forests

In recent years, international attention to climate change mitigation has transformed discussions on global forest governance to focus on the carbon sequestration services of forests, a value previously absent from global forest debates. Critics of REDD+ argue that the overemphasis on carbon services instrumentalizes forests and reduces them to carbon sinks—resources that capture and store carbon—therewith potentially displacing other values [61–63]. In other words, although managing forests for carbon does not necessarily require displacing other values, the functional value of forests embraced in REDD+ becomes reliant upon the capacity and rate with which trees uptake carbon. Plantation forests become more valuable than old growth forests. Although REDD+ proponents argue that assigning an economic value for the carbon in forests provides a funding lifeline for forest conservation characterized by precipitous declines over the last 15 years, and that safeguards can prevent loss of other important ecosystem services from protected forests, REDD+ critics fear that the reductionist pressures of REDD+' underlying market logics displace, devalue, and exclude forests—and their peoples—beyond carbon [64] (the importance of moving beyond carbon in REDD+ was the central theme of a 2012 conference organized by scholars at Oxford [65]).

The justice concerns that emerge from the ongoing struggles to define forests are nested and three-fold: first, the process of narrowing the definition of forests to forests-as-carbon has taken place in venues and by actors distant from these forests. The embedded exclusion in such processes is not simply a form of procedural injustice, whereby those communities (e.g., forest peoples) most impacted by policy decisions are absent in decision-making processes, but is both rooted in and exacerbates recognitional injustices. The scope of engagement is predetermined and narrow, thereby silencing alternative visions and understandings of forests.

Second, because REDD+ prioritizes the carbon values of forests over other values, forest peoples who do engage in REDD+ programs and projects must assent to the vision of forests embedded in those processes. In other words, these processes presume that the carbon services of forests hold universal value. To engage in these processes requires that participants accept these hegemonic presumptions in order to engage in and benefit from REDD+ interventions—the very definition of recognitional injustice.

Third, once crystallized, forests-as-carbon forms the foundation of REDD+ benefits-sharing systems. Forest values are transformed into payments for ecosystem services (here, carbon) and distributed to recognized beneficiaries. While the resulting justice concerns are difficult to separate, the potential distributive injustices include insufficient replacement value of forests (i.e., the costs outweigh the benefits), unfair and/or inequitable distribution of benefits among beneficiaries, and insufficient protection of other ecosystems services and values provided by forests for local communities.

3.4. Current REDD+ Modes of Engagement Are not Aligned with Diverse Models of Decision-Making and May Exclude a Variety of Voices and Interests

Stakeholder engagement, participation, and community-centered approaches have been staples in environmental governance initiatives for nearly thirty years. Community engagement has been mainstreamed to the point where virtually all forest governance initiatives in REDD+ countries require some degree of participation from local communities. REDD+ has similarly emphasized community engagement: initiatives funded by the United Nations REDD+ (UN-REDD) program (one of the largest REDD+ programs), for example, require that impacted communities provide their free, prior, and informed consent (FPIC). FPIC is an approach increasingly used by development agencies, donors, and governments to ensure that communities only engage in projects they consent to freely and before projects have launched. FPIC is intended to protect communities from bearing the costs of projects they have little or no interest in or that they view as harmful to their communities [66,67]. Under current REDD+ initiatives, project proponents implement FPIC processes in areas they have identified as feasible for REDD+. This usually entails a series of meetings and information sessions informing a community of what REDD+ is followed by a request for consent by individuals and/or households within the community.

While in principle FPIC could promote both procedural and recognitional justice, current practices raise concerns. Although UN-REDD specifies that each country should design contextually relevant ways of adhering to their safeguards, in practice many REDD+ proponents have adopted a blueprint approach to FPIC [68]. The process is often highly formalized and dominated by western modes of engagement that may be unfamiliar, inaccessible, or uncomfortable for different communities [69]. In these cases, the process of community engagement may not facilitate effective voice of diverse stakeholders. Alternatively, in some contexts free consent is practically infeasible. For example, reports across Cambodia and Laos suggest that consent may be coerced: attendance at village meetings is often mandatory, yet attendance conveys consent [70]. Under such circumstances FPIC is impossible. In this sense, REDD+ may require that communities engage in a prescribed way that may not be comfortable or appropriate and thus there is actually some harm done or, at a minimum, no real change in voice in these communities. Whereas procedural justice lets us see the importance of including impacted communities in decision-making processes, recognitional justice lets us see that the modes of engagement are critical for communities to have a voice and perhaps even influence these decision-making processes.

3.5. Summary

To summarize, there are four important types of justice challenges in the design of REDD+ that emerge from discussions of recognitional justice principles. First is an overly *narrow conception of property rights* as the fundamental institutions in the design and implementation of REDD+ policies. This tendency to emphasize strong, freely transferable, and relatively unqualified ownership rights over stored carbon risks promoting idealized (and unrealistic) visions of unfettered carbon markets that may favor wealthy, high emitting nations as well as national governments and elites in developing countries (e.g., [53,71]). This raises the second major concern: that REDD+ gives *too much authority to the nation state*, thereby failing to recognize the views and interests of local groups or individuals who may not be well-represented by their national governments [31,72–74]. More attention to recognitional justice ideas should suggest alternative arrangements giving more autonomy to local communities in

REDD+ programs to create enabling environments by helping them to work directly with civil society groups or international actors. Third, attention to recognitional justice highlights how REDD+ risks overlooking the *diversity of values provided by forests* (and indeed the diversity of definitions of forests) beyond carbon storage. In this sense, recognizing the legitimacy of other uses and values of forests by using a recognitional justice lens should help limit the risk of privileging carbon storage over all other forest values and services. Finally, recognitional justice ideas illustrate the inadequacy of *current modes of engagement with local communities in REDD+ programs*, relying on relatively rigid and formalized meetings and information sessions with local groups that fail to recognize and solicit the full range of voices and perspectives in a local community.

4. Discussion and Conclusions

The analysis herein highlights the nested and correlational nature of justice concerns and demonstrates how REDD+ architects narrowly scope the debate to focus solutions on a predefined understanding of the problem and advance technocratic, state-centered solutions. These solutions fail to account for the complex and nuanced nature of justice and injustice as experienced, envisioned, or anticipated by forest peoples across the globe. Similarly, global forest loss and degradation has serious, compound justice implications that are not fully captured in REDD+ debates. There is a risk that, unless the current REDD+ conflicts are transformed into more productive debates that capture the multi-scalar, multivalent nature of justice, it will be difficult to channel resources and energy towards addressing the broader issues in forest loss and degradation [27].

In this paper I have sought to challenge scholars and policymakers engaged in REDD+ debates to move beyond the dominant sound bites on burden-sharing and rights to develop and pollute. Instead I argue that the justice implications of REDD+ as currently implemented are much more nuanced and complex than the current neoliberal framing of the debates suggests. This analysis demonstrates that, when viewed through a recognitional lens, the justice concerns related to REDD+ are not simply its distributive and procedural impacts—problems often readily tackled through technocratic solutions. Instead, the challenges REDD+ faces are centered on the extent to which REDD+ initiatives can account for, accommodate, and embrace difference. In other words, the central justice questions that REDD+ architects should consider are recognitional in nature: what values, identities, lifeways, and voices are displaced and what are the most appropriate ways of engaging and preserving these to advance a global common goal of reduced forest loss and degradation? By shifting the focus to recognitional justice, this paper opens up new spaces for innovating policy solutions. In particular, I suggest that by attending to one of more of the criteria outlined in Section 2 of this paper (multidimensional justice, nested justice, multivalent justice, and scalar dynamics of justice) that policy makers and practitioners can begin to see why their attempts to address tenure security and participation through dominant neoliberal frameworks are met with resistance. Despite the best intentions of some REDD+ proponents, when property rights and participation are approached through a narrowly scoped set of values and rights, many forest peoples may be further marginalized or rendered invisible. Once these complexities are illuminated, forest governance actors can begin to work towards solutions that attend to the recognitional justice concerns that must be addressed in order to move REDD+, or any other forest governance initiative, forward.

Many of the recognitional justice concerns discussed above emerge, in part, because of the weak institutional contexts of REDD+ participating countries. While REDD+ proponents view its objectives as primarily a mitigation scheme, REDD+ critics see REDD+ as a state-building enterprise and the strengthening of the state is a threat to the lifeways and wellbeing of many forest and Indigenous Peoples. Thus, the challenge for policy makers concerned about the climate implications of forest loss and degradation is to first address the recognitional concerns of forest peoples before addressing both procedural and distributive justice concerns. In doing so, special attention must be directed to the challenges of the relationship between REDD+ and the possibilities for rights and representation in the political economic context of target communities.

In closing, for those interested in advancing justice in REDD+, future research should consider the question of voice and choice in REDD+. Greater attention to the interlinked visions of justice, including recognitional justice, is a useful way to widen the range of possible policy options and principles to incorporate exciting new developments in other areas of climate change policy (including emissions trading policy designs) where new limits on private property rights, national government's discretion, and protections for local communities and all resource users are being debated and developed. For example, policy reforms grounded in principles found in the Public Trust Doctrine, a centuries-old common law limitation on the power of governments and private individuals to limit public uses of or benefits from many natural resources [75]. Emerging initiatives to advance deliberative democracy in environmental governance also offer some potential new avenues for recognitional justice in REDD+. One core feature of deliberative democracy is that decisions are reason-based, which is to say that demands and views need to be responded to and not simply tabled or ignored [76,77]. Although there are limited examples of deliberative democracy in practice in global environmental governance, new initiatives in biodiversity conservation are emerging that provide opportunities for rethinking institutions. Bio-cultural Community Protocols, for example, are an experimental approach used to govern access and benefits sharing for genetic resources that seek to strengthen the capacity and agency of local communities to implement international and national laws [78,79]. They allow communities to "provide clear terms and conditions to regulate access to their knowledge and resources" [80]. Such innovations provide alternative ways of understanding how REDD+ decisions unfold and the extent to which institutional designs can be more inclusive and facilitate effective voice and representation of the diversity within and between forest communities.

Acknowledgments: The author wishes to thank Leigh Raymond for his guidance and considerable input on this project and Savannah Schulze for research assistance. I am also grateful to two anonymous reviewers for their helpful comments.

Conflicts of Interest: The author declares no conflict of interest.

References

1. Intergovernmental Panel on Climate Change (IPCC). *Climate Change 2014: Synthesis Report. Contribution of Working Groups I, II and III to the Fifth Assessment Report of the Intergovernmental Panel on Climate Change*; IPCC: Geneva, Switzerland, 2014; Volume 2014, p. 151.
2. Agrawal, A.; Nepstad, D.; Chhatre, A. Reducing emissions from deforestation and forest degradation. *Annu. Rev. Environ. Resour.* **2011**, *36*, 373–396. [CrossRef]
3. Gupta, J. Glocal forest and REDD+ governance: Win–win or lose–lose? *Curr. Opin. Environ. Sustain.* **2012**, *4*, 620–627. [CrossRef]
4. Birrell, K.; Godden, L.; Tehan, M. Climate change and REDD+: Property as a prism for conceiving Indigenous Peoples' engagement. *J. Hum. Rights Environ.* **2012**, 196–216. [CrossRef]
5. Phelps, J.; Webb, E.L.; Agrawal, A. Does REDD+ threaten to recentralize forest governance? *Science* **2010**, *328*, 312–313. [CrossRef] [PubMed]
6. Chao, S. *Forest Peoples: Numbers Across the World*; Forest Peoples Programme: Moreton-in-Marsh, UK, 2012.
7. Kanowski, P.J.; McDermott, C.; Cashore, B.W. Implementing REDD+: Lessons from analysis of forest governance. *Environ. Sci. Policy* **2011**, *14*, 111–117. [CrossRef]
8. McDermott, M.; Mahanty, S.; Schreckenberg, K. Examining equity: A multidimensional framework for assessing equity in payments for ecosystem services. *Environ. Sci. Policy* **2013**, *33*, 416–427. [CrossRef]
9. Sikor, T.; Cầm, H. REDD+ on the rocks? Conflict over forest and politics of justice in vietnam. *Hum. Ecol.* **2016**, *44*, 217–227. [CrossRef] [PubMed]
10. Marion Suiseeya, K.R. Displacing difference and the barriers to environmental justice. *Politics Groups Identities* **2015**, *3*, 697–702. [CrossRef]
11. Young, I.M. *Justice and the Politics of Difference*; Princeton University Press: Princeton, NJ, USA, 2011.
12. Walker, G. *Environmental Justice: Concepts, Evidence and Politics*; Routledge: London, UK, 2012.
13. Mohai, P.; Pellow, D.; Roberts, J.T. Environmental justice. *Annu. Rev. Environ. Resour.* **2009**, *34*, 405–430. [CrossRef]

14. Bullard, R.D. *Dumping in Dixie: Race, Class, and Environmental Quality*; Westview Press: Boulder, CO, USA, 1990; p. 262.

15. Taylor, D.E. *Toxic Communities: Environmental Racism, Industrial Pollution, and Residential Mobility*; NYU Press: New York, NY, USA, 2014; p. 356.

16. Blaikie, P.; Brookfield, H. *Land Degradation and Society*; Methuen & Co. Ltd.: London, UK, 1987.

17. Peluso, N.L. Coercing conservation: The politics of state resource control. *Glob. Environ. Chang.* **1993**, *3*, 199–217. [CrossRef]

18. Schlosberg, D. Reconceiving environmental justice: Global movements and political theories. *Environ. Politics* **2004**, *13*, 517–540. [CrossRef]

19. United Church of Christ Commission for Racial Justice. In *'Principles of Environmental Justice'*, Proceedings of the the First National People of Color Environmental Leadership Summit, Washington, DC, USA, 27 October 1991; United Church of Christ Commission for Racial Justice: New York, NY, USA, 1991.

20. Smith, P.D.; McDonough, M.H. Beyond public participation: Fairness in natural resource decision making. *Soc. Nat. Resour.* **2001**, *14*, 239–249. [CrossRef]

21. Shrader-Frechette, K. *Environmental Justice: Creating Equality, Reclaiming Democracy*; Oxford University Press: New York, NY, USA, 2002.

22. Heiman, M.K. Race, waste, and class: New perspectives on environmental justice. *Antipode* **1996**, *28*, 111–121. [CrossRef]

23. Clayton, S. Preference for macrojustice versus microjustice in environmental decisions. *Environ. Behav.* **1998**, *30*, 162–183. [CrossRef]

24. Marion Suiseeya, K.R. Negotiating the Nagoya Protocol: Indigenous demands for justice. *Glob. Environ. Politics* **2014**, *14*, 102–124. [CrossRef]

25. Schlosberg, D. *Defining Environmental Justice: Theories, Movements and Nature*; Oxford University Press: Oxford, UK, 2007.

26. Martin, A.; Akol, A.; Phillips, J. Just conservation? On the fairness of sharing benefits. In *The Justices and Injustices of Ecosystem Services*; Sikor, T., Ed.; Routledge: New York, NY, USA, 2013.

27. Martin, A.; Coolsaet, B.; Corbera, E.; Dawson, N.; Fraser, J.A.; Lehman, I.; Rodriguez, I. Justice and conservation: The need to incorporate recognition. *Biol. Conserv.* **2016**, *197*, 254–261. [CrossRef]

28. McDermott, C.; Schreckenberg, K. Examining Equity: A Multidimensional Framework for Assessing Equity in the Context of REDD+. In Proceedings of the Oxford Conference 2012 Beyond Carbon: Ensuring Justice and Equity in REDD+ Across Levels of Governance, Oxford, UK, 23–24 March 2012.

29. Patrick, M.J. The cycles and spirals of justice in water-allocation decision making. *Water Int.* **2014**, *39*, 63–80. [CrossRef]

30. Clayton, S. New ways of thinking about environmentalism: Models of justice in the environmental debate. *J. Soc. Issues* **2000**, *56*, 459–474. [CrossRef]

31. Bolin, A.; Tassa, D.T. Exploring climate justice for forest communities engaging in REDD+: Experiences from tanzania. *Forum Dev. Stud.* **2012**, *39*, 5–29. [CrossRef]

32. Mahanty, S.; McDermott, C.L. How does 'free, prior and informed consent' (FPIC) impact social equity? Lessons from mining and forestry and their implications for REDD+. *Land Use Policy* **2013**, *35*, 406–416.

33. Walker, G. Beyond distribution and proximity: Exploring the multiple spatialities of environmental justice. *Antipode* **2009**, *41*, 614–636. [CrossRef]

34. Marion Suiseeya, K.R. The justice gap in global forest governance. Duke University: Durham, NC, USA, 2014.

35. Schroeder, D.; Pogge, T. Justice and the convention on biological diversity. *Ethics Int. Aff.* **2009**, *23*, 267–280. [CrossRef]

36. Okereke, C. Equity norms in global environmental governance. *Glob. Environ. Politics* **2008**, *8*, 25–50. [CrossRef]

37. Elster, J. Local justice. *Eur. J. Sociol.* **1990**, *31*, 117–140. [CrossRef]

38. Walker, G.; Bulkeley, H. Geographies of environmental justice. *Geoforum* **2006**, *37*, 655–659. [CrossRef]

39. Martin, A.; Gross-Camp, N.; Kebede, B.; McGuire, S.; Munyarukaza, J. Whose environmental justice? Exploring local and global perspectives in a payments for ecosystem services scheme in rwanda. *Geoforum* **2014**, *54*, 167–177. [CrossRef]

40. Schlosberg, D.; Carruthers, D. Indigenous struggles, environmental justice, and community capabilities. *Glob. Environ. Politics* **2010**, *10*, 12–35. [CrossRef]

41. Emuge, A.L. Transcending Challenges of Attaining Distributive Justice in Pro-Poor Activities of REDD-Plus (REDD+): Justice in Brazil, Vietnam and Tanzania. Ph.D. Thesis, Norwegian University of Life Sciences, Ås, Norway, 2013.

42. Putra, J.D. *Distributing REDD-Plus Benefits to Indigenous Peoples: Important Role of Distributive Justice in REDD-Plus Policy*. Available online: https://ssrn.com/abstract=2602039 (accessed on 22 April 2015).

43. Skutsch, M. Slicing the REDD+ pie: Controversies around the distribution of benefits. *CAB Rev.* **2013**, *8*, 1–10. [CrossRef]

44. Mathur, V.N.; Afionis, S.; Paavola, J.; Dougill, A.J.; Stringer, L.C. Experiences of host communities with carbon market projects: Towards multi-level climate justice. *Clim. Policy* **2014**, *14*, 42–62. [CrossRef]

45. Marion Suiseeya, K.R.; Caplow, S. In pursuit of procedural justice: Lessons from an analysis of 56 forest carbon project designs. *Glob. Environ. Chang.* **2013**, *23*, 968–979. [CrossRef]

46. Mohammed, A.J.; Inoue, M. Understanding REDD+ with actor-centered power approach: A review. *J. Biodivers. Manag. For.* **2016**, *5*, 1. [CrossRef]

47. Roe, S.; Streck, C.; Pritchard, L.; Costenbader, J. *Safeguards in REDD+ and Forest Carbon Standards: A Review of Social, Environmental and Procedural Concepts and Application*; Climate Focus: Washington, DC, USA, 2013.

48. Dunlop, T.; Corbera, E. Incentivizing REDD+: How developing countries are laying the groundwork for benefit-sharing. *Environ. Sci. Policy* **2016**, *63*, 44–54. [CrossRef]

49. Luttrell, C.; Loft, L.; Gebara, M.F.; Kweka, D.; Brockhaus, M.; Angelsen, A.; Sunderlin, W.D. Who Should Benefit from REDD+? Rationales and Realities. *Ecol. Soc.* **2013**, *18*, 52. [CrossRef]

50. Jagger, P.; Lawlor, K.; Brockhaus, M.; Fernanda Gebara, M.; Sonwa, D.J.; Resosudarmo, I.A.P. REDD+ safeguards in national policy discourse and pilot projects. In *Analysing REDD+: Challenges and Choices*; Angelsen, A., Brockhaus, M., Sunderlin, W.D., Verchot, L., Eds.; CIFOR: Bogor, Indonesia, 2012.

51. Brown, D.; Seymour, F.; Peskett, L. How do we achieve REDD co-benefits and avoid doing harm? In *Moving ahead with redd: Issues, options and implications*; Angelsen, A., Ed.; CIFOR: Bogor, Indonesia, 2008; p. 156.

52. Pokorny, B.; Scholz, I.; de Jong, W. REDD+ for the poor or the poor for REDD+? About the limitations of environmental policies in the amazon and the potential of achieving environmental goals through pro-poor policies. *Ecol. Soc.* **2013**, *18*, 3. [CrossRef]

53. Sikor, T.; Martin, A.; Fisher, J.; He, J. Toward an empirical analysis of justice in ecosystem governance. *Conserv. Lett.* **2014**, *7*, 524–532. [CrossRef]

54. Angelsen, A.; Brockhaus, M.; Sunderlin, W.D.; Verchot, L.V. *Analysing REDD+: Challenges and Choices*; Center for International Forestry Research (CIFOR): Bogor, Indonesia, 2012; p. 456.

55. Karsenty, A.; Vogel, A.; Castell, F. "Carbon rights", REDD+ and payments for environmental services. *Environ. Sci. Policy* **2014**, *35*, 20–29. [CrossRef]

56. White, A.; Martin, A. *Who Owns the World's Forests*; Forest Trends: Washington, DC, USA, 2002.

57. Alden Wily, L. *The Tragedy of Public Lands: The Fate of the Commons under Global Commercial Pressure*; International Land Coalition: Rome, Italy, 2011.

58. Gumbo, D.; Mfune, O. The forest governance challenge in REDD+: Core governance issues that must be addressed for REDD+ success in zambia. *Nat. Faune* **2013**, *27*, 49–53.

59. Bachram, H. Climate fraud and carbon colonialism: The new trade in greenhouse gases. *Capital. Nat. Social.* **2004**, *15*, 5–20. [CrossRef]

60. Doherty, E.; Schroeder, H. Forest tenure and multi-level governance in avoiding deforestation under REDD+. *Glob. Environ. Politics* **2011**, *11*, 66–88. [CrossRef]

61. Visseren-Hamakers, I.J.; McDermott, C.; Vijge, M.J.; Cashore, B. Trade-offs, co-benefits and safeguards: Current debates on the breadth of REDD+. *Curr. Opin. Environ. Sustain.* **2012**, *4*, 646–653. [CrossRef]

62. Phelps, J.; Friess, D.; Webb, E. Win–win REDD+ approaches belie carbon–biodiversity trade-offs. *Biol. Conserv.* **2012**, *154*, 53–60. [CrossRef]

63. Corbera, E.; Estrada, M.; May, P.; Navarro, G.; Pacheco, P. Rights to land, forests and carbon in REDD+: Insights from mexico, brazil and costa rica. *Forests* **2011**, *2*, 301–342. [CrossRef]

64. Corbera, E.; Schroeder, H. Governing and implementing REDD+. *Environ. Sci. Policy* **2011**, *14*, 89–99. [CrossRef]

65. Tyndall Centre for Climate Change Research. Proceedings of "Oxford Conference 2012 - 'Beyond Carbon: Ensuring Justice and Equity in REDD+ Across Levels of Governance'." 23–24 March 2012, Oxford, England. Available online: http://www.tyndall.ac.uk/events/2012/oxford-conference-2012-beyond-carbon-ensuring-justice-and-equity-redd-across-levels (accessed on 15 August 2016).

66. Carodenuto, S.; Fobissie, K. Special issue: The legal aspects of REDD+ implementation: Translating the international rules into effective national frameworks · operationalizing free, prior and informed consent (FPIC) for REDD+: Insights from the national FPIC guidelines of cameroon. *Carbon Clim. Law Rev.* **2015**, *9*, 156–167.

67. Jagger, P.; Brockhaus, M.; Duchelle, A.E.; Gebara, M.F.; Lawlor, K.; Resosudarmo, I.A.P.; Sunderlin, W.D. Multi-level policy dialogues, processes, and actions: Challenges and opportunities for national REDD+ safeguards measurement, reporting, and verification (mrv). *Forests* **2014**, *5*, 2136–2162. [CrossRef]

68. Fontana, L.B.; Grugel, J. The politics of indigenous participation through "free prior informed consent": Reflections from the bolivian case. *World Dev.* **2016**, *77*, 249–261. [CrossRef]

69. Colchester, M.; Ferrari, M.F. *Making FPIC–Free, Prior and Informed Consent–Work: Challenges and Prospects for Indigenous Peoples*; Forest Peoples Programme: Moreton-in-Marsh, UK, 2007.

70. Milne, S.; Mahanty, S. Between myth, ritual, and market value: The fetishisation of free prior and informed consent in the production of forest carbon. In Proceedings of the Association of American Geographers 58th Annual Meeting, Los Angeles, CA, USA, 12 April 2013.

71. Lounela, A. Climate change disputes and justice in central kalimantan, indonesia. *Asia Pac. Viewp.* **2015**, *56*, 62–78. [CrossRef]

72. Martin, A.; McGuire, S.; Sullivan, S. Global environmental justice and biodiversity conservation. *Geogr. J.* **2013**, *179*, 122–131. [CrossRef]

73. Hiraldo, R.; Tanner, T. Forest voices: Competing narratives over REDD+. *IDS Bull.* **2011**, *42*, 42–51. [CrossRef]

74. Hiraldo, R.; Tanner, T. *The Global Political Economy of REDD+: Engaging Social Dimensions in the Emerging Green Economy*; United Nations Research Institute for Social Development: Geneva, Switzerland, 2011.

75. Raymond, L. *Reclaiming the Atmospheric Commons: The Regional Greenhouse Gas Initiative and a New Model of Emissions Trading*; MIT Press: Cambridge, MA, USA, 2016.

76. Dryzek, J.S.; Stevenson, H. Global democracy and earth system governance. *Ecol. Econ.* **2011**, *70*, 1865–1874. [CrossRef]

77. Bäckstrand, K. Democratizing global environmental governance? Stakeholder democracy after the world summit on sustainable development. *Eur. J. Int. Relat.* **2006**, *12*, 467–498.

78. Bavikatte, K.; Robinson, D.F. Towards a people's history of the law: Biocultural jurisprudence and the nagoya protocol on access and benefit sharing. *Law Environ. Dev. J.* **2011**, *7*, 35–51.

79. Srinivas, K.R. Protecting traditional knowledge holders' interests and preventing misappropriation—Traditional knowledge commons and biocultural protocols: Necessary but not sufficient? *Int. J. Cult. Prop.* **2012**, *19*, 401–422. [CrossRef]

80. Bavikatte, K.; Jonas, H. *Bio-Cultural Community Protocols: A Community Approach to Ensuring the Integrity of Environmental Law and Policy*; Natural Justice, United Nations Environment Programme: Nairobi, Kenya, 2009. Available online: http://www.unep.org/communityprotocols/PDF/communityprotocols.pdf (accessed on 20 November 2010).

forests

MDPI

Article

Beyond Rewards and Punishments in the Brazilian Amazon: Practical Implications of the REDD+ Discourse

Maria Fernanda Gebara [1],* and Arun Agrawal [2]

[1] Development, Agriculture and Society Institute, Federal Rural University of Rio de Janeiro, Rio de Janeiro 20.071-003, Brazil
[2] School of Natural Resources and Environment, University of Michigan, Ann Arbor, MI 48109, USA; arunagra@umich.edu
* Correspondence: mfgebara@gmail.com; Tel.: +44-789-456-9543

Academic Editors: Esteve Corbera and Heike Schroeder
Received: 13 November 2016; Accepted: 27 February 2017; Published: 2 March 2017

Abstract: Through different policies and measures reducing emissions from deforestation and degradation and enhancing conservation (REDD+) has grown into a way to induce behavior change of forest managers and landowners in tropical countries. We argue that debates around REDD+ in Brazil have typically highlighted rewards and punishments, obscuring other core interventions and strategies that are also critically important to reach the goal of reducing deforestation, supporting livelihoods, and promoting conservation (i.e., technology transfer and capacity building). We adopt Foucault's concepts of governmentality and technologies of governance to provide a reading of the REDD+ discourse in Brazil and to offer an historical genealogy of the rewards and punishments approach. By analyzing practical elements from REDD+ implementation in the Brazilian Amazon, our research provides insights on the different dimensions in which smallholders react to rewards and punishments. In doing so, we add to the debate on governmentality, supplementing its focus on rationalities of governance with attention to the social practices in which such rationalities are embedded. Our research also suggests that the techniques of remuneration and coercion on which a rewards and punishments approach relies are only supporting limited behavioral changes on the ground, generating negative adaptations of deforestation practices, reducing positive feedbacks and, perhaps as importantly, producing only short-term outcomes at the expense of positive long-term land use changes. Furthermore, the approach ignores local heterogeneities and the differences between the agents engaging in forest clearing in the Amazon. The practical elements of the REDD+ discourse in Brazil suggest the rewards and punishments approach profoundly limits our understanding of human behavior by reducing the complex and multi-dimensional to a linear and rational simplicity. Such simplification leads to an underestimation of smallholders' capacity to play a key role in climate mitigation and adaptation. We conclude by highlighting the importance of looking at local heterogeneities and capacities and the need to promote trust, altruism and responsibility towards others and future generations.

Keywords: REDD+; Amazon; discourse; genealogy; rewards and punishments; behavior change

1. Introduction

During the past decade, the "carrots and sticks" approach has emerged as a combination of rewards (i.e., financial incentives) and punishments (i.e., penalties) to influence behavioral change toward reducing emissions from deforestation and degradation and enhancing conservation (REDD+) [1–7]. By undertaking a discourse analysis and drawing on practical elements of REDD+

implementation, this article explores how the rewards and punishments approach has been used to legitimize techniques of governance in the Brazilian Amazon. This analysis builds on Foucault's important contributions related to the analysis of power and knowledge [8], and is therefore critical. But rather than a complete recapitulation and elaboration of the arguments for or against rewards and punishments as the way to implement REDD+, we focus on what the current debate on REDD+ in Brazil tends to marginalize and set aside. Little if any of the work on REDD+ is attentive to insights from Foucault's work or to the kinds of analytical directions that studies related to governmentality open and make available. Our contribution, thus, is primarily a theory-building effort and it is critical in its explicit attempt to turn what is not presently conceived of as being problematic into something that is necessary to consider [9]. Our main hypothesis is that the REDD+ discourse in Brazil is making only certain things (rewards and punishments) visible, obscuring the importance of other core interventions for REDD+, such as measures focused on knowledge sharing, collective resilience, recognition of rights, technology transfer, and social and technical capacity building and development, to name a few key issues.

Since 2007, REDD+ has grown into a policy instrument to reduce deforestation in many tropical countries, with implementation occurring at local to national scales. Even as REDD+ interventions have evolved along different pathways, with many initiatives largely fulfilling roles as pilot projects, most are struggling to make a transition to sustained larger-scale success [10]. The Brazilian Amazon, which houses the largest remaining expanse of tropical forests, is a major focus of REDD+ implementation because of the country's successes in reducing deforestation. These successes can be attributed mainly to centralized command and control efforts (i.e., monitoring and law enforcement) but also to changes in relative prices of agricultural commodities [11–13]. However, recent evidence also shows command and control efforts in the Amazon have had heterogeneous effects, highlighting the need to adapt forest conservation measures to regional specificities [14]. Deforestation trends have also changed; once driven by large-scale clear-cutting, forest clearing in the Amazon now occurs mostly in small areas [15,16]. These trends have attracted the attention of the Brazilian national government to the role smallholders play in deforestation. In this paper, we focus on smallholders in particular to undertake a close reading of the rewards and punishments discourse and practice in Brazil. We also offer an historical genealogy of the approach to improve our understanding of the direction in which it is heading.

The study of policy interventions and governance tends to occur in a discursive field in which the exercise of power is "rationalized". Strategies through which such rationalization of political underpinnings and motivations occurs include the delineation of concepts, the specification of objects and borders, an emphasis on identifying models of individuals and their behaviors, and the development of arguments and justifications that naturalize observed actions and responses. Such efforts to analyze policy interventions, governance (and responses to governance interventions) portray inherently political problems as technical, and equally political actions and interventions as technical, common-sense solution strategies. They structure "a field of possibilities in which several ways of behaving, several reactions and diverse comportments, may be realized" [9]. In this structuring of the large field of possibilities, interventions and responses that do not correspond to dominant models of policy making and corresponding behavior are rendered as irrational or non-systematic and the goal of policy analysis shifts into a more normative register where the irrational and non-systematic can either be ignored or needs to be rationalized. Foucault's analyses thus provide a useful starting point and valuable insights into the different dimensions of the arts of government as ways of defining, knowing, and improving outcomes. His contributions focus on the rationality of government, meaning a way of thinking about the nature and the practice of government (who can govern; what or who is governed; how governance happens; why particular modes of governance are necessary or to be defended; and the like). We build on his work by looking into the different relationships between the governed and governing and how these relationships unfold.

Our analysis draws from the experience of six different REDD+ initiatives in Brazil. It aims at a different—compared to much of the prevailing models of REDD+ and payments for ecosystem services more generally—way of understanding the implications of categorizing REDD+ interventions into techniques of remuneration and coercion on which rewards and punishments rely. Important questions include: (a) What does the rewards and punishments discourse highlight? (b) What does this same discourse obscure from consideration? (c) What kind of subjects does it assume in practice, and through its efforts at implementation, also produce? (d) Is the assumed subject sufficiently realized for REDD+ to be effective? And, finally (e) Is the rewards and punishments discourse justifiable for REDD+? The analysis rests upon an application of Foucault's genealogical approach to some of the key underpinnings of the rewards and punishments approach (Section 2). Section 3 consists of a brief discussion of our methods and the next section hones in on the practical relevance of a Foucault-inspired approach in relation to REDD+ interventions in the Brazilian Amazon (Section 4). In Section 5 we discuss the implications of our findings for REDD+ implementation. Our analysis suggests that the techniques of remuneration (i.e., rewards) and coercion (i.e., punishments) lead to heterogeneous and at best only limited behavioral shifts on the ground, often prompting negative adaptations of deforestation practices, limiting and reducing positive feedbacks and, just as importantly, possibly producing only short-term changes at the expense of positive longer-term land use changes. In its search for a global model relevant to deforestation reduction, the approach ignores local heterogeneities and differences among agents involved in forest clearing. These findings constitute evidence that confirms our starting hypothesis: the exclusive attention to conditions conducive to the success of a rational system of rewards and punishments has led to the neglect of other forms of REDD+ interventions that may be more effective. These findings also suggest that a preoccupation with a rational rewards and punishments arrangement profoundly constrains the understanding of the multiplicity of human choices and behaviors by reducing a complex and multi-dimensional set of experiences into a linear and rational simplicity. Moreover, the approach underestimates the capacities of the very smallholders that are expected to and can play a key role in climate mitigation and adaptation. We conclude by highlighting the importance of attending to local heterogeneities, strengthening diverse capacities, and promoting trust, altruism and responsibility towards others and future generations.

2. Opening up the Rewards and Punishments Discourse

2.1. Foucault's Genealogy

In an attempt to open up the rewards and punishments discourse, we offer an historical genealogy of the approach to show how we arrived at this technique of governance to induce new behaviors, and what its practices conceal. In its ordinary sense, genealogy is synonymous with the study of one's ancestors, or in Foucault's words: "the history of the present" [8]. In a more specific sense, genealogy is a Nietzschean expression used for reconstructing a term by establishing genealogical charts of linked conceptual fragments and relationships. Examining the genealogical nature of the rewards and punishments discourse is necessary to understand the systematic ways in which colonial countries such as Brazil manage and control their forests and, in many ways, even create the knowledge and subjects of their governance in political, economic, sociological and cultural systems. The ensemble of governance practices includes a whole series of national efforts (often promoted by the discourses of Western countries and actors that are being set up on a global scale) that constitute one of the most powerful mechanisms to ensure control over the people living in forested territories. Policy instruments—such as rewards and punishments—can be viewed here as part of the evolving "set of techniques by which governmental authorities wield their power in attempting to ensure support and effect social change" [17]. Practices of government in a Foucauldian sense comprise both technologies of control and (cor)responding technologies of subject making on the part of those at whom the technologies of control are directed.

According to Foucault, the deployment of technologies of control—and subject making—rest on power which should be viewed not as something that is stratified or as an object, but as capacities diffused throughout the social body. One cannot have power; rather, power is manifested in its exercise; thus, power is everywhere, but unevenly distributed [18]. As Foucault states, "each society has its regime of truth, its "general politics" of truth: that is, the types of discourse which accepts and makes function as true; the mechanisms and instances which enable one to distinguish true and false statements, the means by which each is sanctioned; the techniques and procedures accorded value in the acquisition of truth; the status of those who are charged with saying what counts as true" [19]. The relationship between content, expression, power-relations and material practices forms discourses, which are "systems of thoughts composed of ideas, attitudes, courses of actions, beliefs and practices that systematically construct the subjects and the worlds of which they speak. He traces the role of discourses in wider social processes of legitimation and power, emphasizing the constitution of current truths, how they are maintained and what power relations they carry with them" [20]. We build on these Foucauldian insights into discursive analysis with a focus on the productive aspects of discourses, meaning "the practices they (discourses) invoke" [21].This focus on practices is important to understand what the rewards and punishments discourse is making visible, what it is obscuring, what subjects it is producing and if these subjects are sufficiently realized for REDD+ to be effective.

Two concepts used by Foucault become important in interrogating the rewards and punishments discourse: the techniques of disciplinary power and governmentality. Discipline is first and foremost a multifaceted and local technique to target the behavior of individuals in specific settings: "Discipline may be identified neither with an institution nor with an apparatus; it is a type of power, a modality for its exercise, comprising a whole set of instruments, techniques, procedures, levels of application, targets" [8]. Foucault makes explicit what discipline entails as a practice of shaping and modifying conducts to impose ends or execute overall strategies. Foucault's argument is that discipline creates "docile bodies" [8], ideal for the new economics, politics and warfare of the modern industrial age—bodies that function in factories, ordered military regiments, and school classrooms. But, to construct docile bodies, the disciplinary institutions must be able to (a) constantly observe and record the bodies they control and (b) ensure the internalization of the disciplinary individuality within the bodies being controlled. That is, discipline must come about without excessive force through careful observation, and molding of the bodies into the correct form through this observation [8].

Later, in an attempt to study the art of governing, the reasoned way of governing best and, at the same time, reflecting on the best possible ways of governing, Foucault coined a new term to capture the combination of "government" and "rationality": "governmentality." Foucault depicted "governmental rationality" through his understanding of the term "government" as having both a wide and a narrow sense. He proposed a definition of "government" in general as meaning "the conduct of conduct": that is to say, a form of activity aiming to shape, guide or affect the conduct of some persons or agents [22]. Government as an activity could concern the relations between self and self, private interpersonal relations involving some form of control or guidance, relations within social institutions and communities and, certainly, relations concerned with the exercise of political sovereignty as well. Governmentality then is also embodied in routine action and normative orientations, and is about making rational certain forms of exercise of power in preference to others. Thus, governmentality is about making things visible, in a certain, governable way; state government is not only a material structure and a mode of thinking, but also a lived and embodied experience, a mode of existence [23,24].

Smallholders have been left in a pervasive condition of vulnerability. Historically embedded processes have left lasting legacies in rural governmentality, creating what has been called "green governmentality" [25] and "environmentality" [26], whereby "the environment" is constructed in relation to the exercise of power and control. As argued by Fairhead et al. [27], these long-run historical processes provide a vital set of conditions of possibility—meaning the set of factors and relations that give discourses a sense—which those seeking to profit from the new commoditization of nature have been able to inhabit. The authors argue that "such contemporary dynamics of appropriation of nature involve new (re)configurations of actors and relationships; the reinvention of regulatory processes and the construction of novel justificatory discourses" [27].

In the Brazilian Amazon, policies historically determined by paternalistic relationships, military regiments and the substantial presence of local elites [28] strongly limit the ability of smallholders to successfully develop their own initiatives [29–31]. Forest governance in Brazil is still oriented by a development model that looks to global commodity markets and well-qualified entrepreneurs with the necessary capital for large-scale investments. This orientation continues despite a growing consensus about its negative ecological consequences, social impacts, and economic risks [32,33]. We draw upon these contextual conditions and ideas to analyze the practical elements from REDD+ implementation and how smallholders respond to rewards and punishments. By doing so, we move from the conventional focus in debates on governmentality on the rationalities of government to the social practices in which these rationalities are embedded.

2.2. The Genealogy of Rewards and Punishments

In the rewards and punishments approach, policy instruments might be formulated either in the positive sense of prescription or encouragement of actions through the creation of positive incentives (such as cash transfers), or in the negative sense of prohibition of actions through the application of penalties. These positive or negative strategies can be used to promote behaviors viewed as desirable by decision makers or inhibit behaviors deemed undesirable [34]. This distinction has its roots in conventional dual divisions between the rewards and punishments or benefits and costs. Incentives include a variety of economic and facilitative measures (not just those based on money). Penalties, on the other hand, are sanctions that involve unpleasant consequences imposed by a legally constituted authority for violation of the law [35].

However, when following Foucault's arguments, it is crucial to understand what is justifiable. We thus view both rewards and punishments as active interventions into the social field (intended to structure the fields of possible actions for determined subjects). In this sense, rewards and punishments may even be viewed as a false distinction, because both require action. Although interventions, in principle, are open to many forms of definitions and interpretations, they all rely on and constitute a particular conception of state sovereignty. The relationship works both ways. Just as interventions require sovereignty to be effective, so also sovereignty is affirmed through the implementation of interventions. This means that when leaders and decision makers justify a given intervention they are simultaneously (re)producing what is to be understood by state sovereignty. By doing so, they are granting specific authorities and competencies, while stripping away others.

2.2.1. Command and Control

The term "command and control" is based on the coercive aspects of regulation. It refers to the insurance of rules, orders, norms and provisions of an obligatory nature backed by negative sanctions or threats of negative sanctions (e.g., fines, imprisonment) by the state. The defining property of command and control is that the relationship is authoritative and based in a hierarchy system [36], meaning that the controlled persons or groups are obligated to act in the way stated by the controllers. The origins of command and control stem from the Machiavellian idea [37] of acquisition and maintenance of political power. Machiavelli saw a way to maintain monocentric political power through command and control. His ideas were further supported by Hobbes and his social contract. Both philosophers believed in the inherent selfishness of the individual. It was necessarily this belief that led them to adopt a strong central power—or the "Leviathan," in Hobbes words—as the only means of preventing the disintegration of the social order. Hardin's version of the tragedy of the commons [38] is a clear example of this Hobbesian logic playing out in the governance of natural resources.

Until recently, policy instruments for forest conservation around the world have been largely based in the rationality of command and control policies, meaning the creation of protected areas, restricting, monitoring and controlling land use and law enforcement, and enforcing compliance. The rise of command and control follows the military model of civil defense adopted by the United States after the Second World as the primary form of disaster response. In recent decades, however, many authors have pointed out fundamental limits to the regulatory system based on command and control structures. These limits become even more evident when dealing with the increasing recognition of the complexity of objects such as forest landscapes, constituted by non-linear processes, heterogeneous subjects, and inherently difficult to comprehend feedbacks related to social and biophysical interactions [39–43]. Command and control systems implicitly assume their targets to be objects that are relatively simple, clearly definable, and generally linear with respect to cause and effect. This approach has been criticized for being monocentric, authoritative, bureaucratic, unnecessarily intrusive, with high, uneven administrative costs, as well as outdated, limited innovation fraught with implementation problems [5,44,45].

Aldisert and Helms emphasize that "the imagination and interests of individuals cannot thrive in command and control structures." Instead, they argue that it is necessary to have "an environment in which all are committed to a common goal" [46]. Holling and Meffe claim that the adoption of command and control as a forest conservation measure resulted in a pathology that undermines resilience in the long run [47]. Finally, the effectiveness of command and control structures is normally associated with its legitimacy. Command and control approaches are effective in changing behavior as long as there is social consensus around the policy underlying the coercive instrument, and the capacity of the state to ensure compliance. The majority of tropical countries are still developing this ability to create consensus among different social actors, but more importantly, are also still improving their governance capacity.

2.2.2. Economic Instruments

Economic tools, in contrast, leave the subjects of governance a certain space within which to choose whether to take an action and indeed to determine the level and intensity of action taken. Or, in Foucault's words, they assume "a minimally 'free' subject" [9]. They neither prescribe nor prohibit the actions to be taken, but make them less (in the case of incentives) or more (in the case of disincentives) expensive. Here we concentrate on positive incentives, as they correspond to the ones linked to REDD+ implementation [48]. In the positive case, a material resource is handed to the agent (e.g., through cash transfers, grants, subsidies or loans), whereas in the negative case the individual or organization is deprived of some material resource (e.g., through taxes, charges or fees).

The term 'incentive' first made its appearance among American engineers in the 1880s and 1890s. It was introduced as a motivational term (or inducement), that had a strong monetary connotation. Of course, the actual use of this specific method of remuneration was not born out of a late nineteenth-century debate. Karl Marx, for one, dated the practice of paying by the piece back to the fourteenth century, serving the basic purpose of exploiting the power of labor [49]. The emergence of economic science in the eighteenth century, however, marked a turning point in the history of governmentality. Just like "the state" was central to the project of the reason of state, so "the market" became a central political and epistemological object in the formation of a distinct, liberal art of governing. A rational and self-interested individual then emerged from different economists' discourses (Francois Quesnay, Adam Smith, Robert Malthus and David Ricardo, for example) based on John Locke's Second Treatise of Government [50]. It was precisely the "alleged rationality and self-interestedness of all individuals that made them governable because it made them respond to changes in their environment in a systematic and predictable way" [22].

This rationalist mode of thought has yielded some interesting insights and results. However, it also underestimates the context in which "real individuals" come to decisions and the multitude of factors shaping decision making. It has already been argued that certain modes of valuing things can be entirely inappropriate. Anderson [51] and Sandel [52], for instance, highlight that "any attempt to commodify a good may corrupt it" and "can even make its very existence impossible." Indeed, research in areas like behavioral economics undermines the idea of a simplistic rational subject, pointing to the possibility of motivational crowding whereby material incentives for socially or environmentally desirable actions may have the effect of undermining and reducing the incidence of such actions [53]. The use of economic incentives for forest conservation has been supported since the late 1990s in a number of international and national contexts (e.g., Convention on Biological Diversity; United Nations Environment Programme; and the Sloping Lands Conservation Program in China). Since the emergence of the "ecosystem services" concept, notably used in the Millennium Ecosystem Assessment and the Economics of Ecosystems and Biodiversity processes between 2005 and 2010, economic incentives for supporting environmentally positive actions have been seen as innovative policy tools for biodiversity conservation worldwide [54]. This is part of a common contemporary shift in both ecological and economic spheres wherein "nature" is being consolidated as "natural capital" and the functions of natural systems as ecosystem services. Previously distinct domains (economics and finance vs. ecology and conservation) are coming together in an emerging "green economy" that connects actors, institutions and calculative technologies through specific discourses of rational action and behavior. These discursive shifts underpin the creation of markets for ecosystem services, including for carbon, and are critical in constituting and normalizing the logic of REDD+ [55] (see Figure 1a).

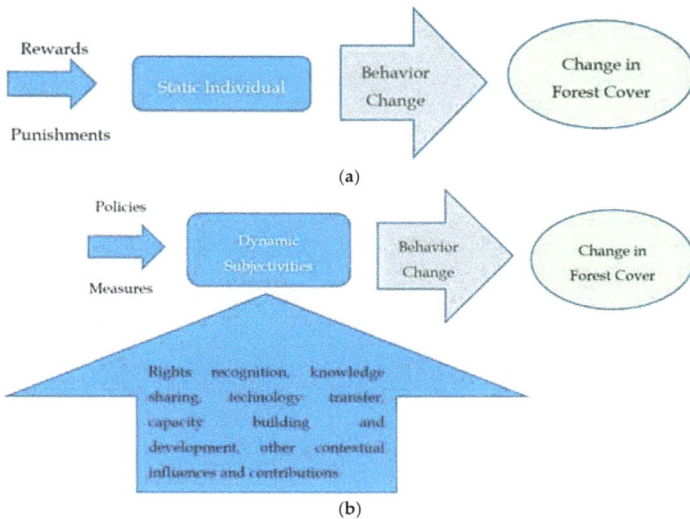

Figure 1. (**a**) Conceptual model of the rewards and punishments approach for REDD+ (reducing emissions from deforestation and degradation and enhancing conservation); (**b**) Conceptual representation of a governmentality approach for REDD+. Legend: The conventional conceptual model underpinning REDD+ assumes an individual that remains consistent in his/her subjectivity over time, responds positively (negatively) to rewards (punishments), and whose behavioral change leads to the desired outcome of increased forest cover. In contrast, as depicted in Figure 1b, a governmentality approach directs attention towards shifts in individual subjectivities, influenced in part by policies and measures that can include rewards and punishments but not necessarily in the desired direction, and with substantial uncertainties in outcomes as a result of variations in recognition of rights, shared knowledge and understandings, technical support, available capacities, and other context specific characteristics.

Although rewards can be used to support forest conservation, they come with restrictions that hinder effectiveness. These include the difficulty of measuring and valuing forest assets, institutional constraints, uncertain, ill-defined, and insecure property rights, ideological resistance, administrative complexity, and limited capacity for calculation and implementation [56–65]. The most common approach for incentives as a type of reward are payments for ecosystems services (PES). Economic arguments in favor of incentive-based policy instruments have a strong focus on cost-efficiency, normally defined as use of natural resources to optimize the production of ecosystems goods and services by equating marginal costs and benefits. The rush to confer multi-domain positive valence on such interventions by focusing on win-win or even win-win-win rhetoric can obscure negative side effects [66]. Existing research has demonstrated that financial rewards can crowd out intrinsic motivations [67,68]. Gneezy et al., for example, emphasize that the effects of extrinsic incentives depend on how they are designed, the form in which they are delivered (especially monetary or non-monetary), who receives them, knowledge about their distribution, and their interactions with intrinsic motivations [69] (see Figure 1b).

3. Methods

Our methods include the review of writings on REDD+ and forest governance to reconstruct the discursive field around deforestation and forest conservation, and empirical evidence from six REDD+ initiative sites in the Brazilian Amazon. The empirical methods used to compile this data are described in Sills et al. [70] in the evaluation of initial impacts of REDD+ that occurred under the Global Comparative Study (GCS) on REDD+, coordinated by the Center for International Forestry Research (CIFOR). The six REDD+ initiatives on which this paper focuses are: (i) the Sustainable Settlements Acre's State System of Incentives for Environmental Services (SISA); (ii) the Bolsa Floresta Program in Amazonas; (iii) the Cotriguaçu Sempre Verde in Mato Grosso; (iv) the Jari REDD+ project in Amapá; (v) the Sustainable Landscapes Pilot Program in São Félix do Xingu and (vi) the Sustainable Settlements initiative in the Transamazon (Figure 2). This research used a variety of data collection techniques, including: quantitative surveys to understand land use, measure subsistence, and learn about the tenure situation, socioeconomic context and the challenges, needs and perceptions (i.e., hopes and worries in relation to REDD+; definitions of local wellbeing) of smallholders. We define property size using the National Agrarian Institute (INCRA) classification, which classifies properties in the Amazon smaller than 300 hectares (or 1–4 fiscal modules) as "small." We also used the "Proponent Appraisal Form" to compile data on the initiatives and on the interventions being implemented. A "Survey of Project Implementation" helped gain insights into the background, history, institutional dynamics and politics surrounding the development of the initiatives. This also served to understand the challenges faced by proponents.

These six sites were selected to include a variety of areas (i.e., private and public), scale (i.e., municipality, state), proponents (i.e., non-governmental organizations (NGOs), public, private) and local actors (i.e., smallholders, large landholders, settlers). In each site, four communities were selected according to GCS methods and around 120 households in each community were randomly selected to be part of the survey. Smallholders' characteristics vary across sites in terms of socioeconomic standards, tenure, community organization, access to markets and wellbeing. More information on methods, proponents, strategies and locations can be found in Sills et al. 2014 [70] and Sunderlin et al. 2016 [71], along with extensive analysis of how REDD+ implementation affects different groups of smallholders.

We also used qualitative data collected in three of the six initiatives: SISA (2010, 2014), Bolsa Floresta Program (2009, 2011, 2015) and the Sustainable Landscapes (2010–2014). In these sites, we applied additional methods: (i) observation; (ii) open-ended questions; and (iii) focus groups. Data collection concentrated on understanding how smallholders' behavior is being shaped by the different REDD+ measures. Questions focused on smallholders' perceptions about the initiatives, including: positive and negative aspects of interventions being implemented, suggestions on how to improve them, and what would be necessary to change their practices that cause deforestation. We also conducted open-ended and semi-structured interviews with other relevant stakeholders in the sites (i.e., government actors and NGOs representatives) to capture diverse views of the initiatives and the measures being implemented to reduce deforestation. Results, therefore, are based on perceptions of REDD+ actors on ground: the recipients of interventions, but also those who were involved in REDD+ implementation (i.e., proponents and public actors). Analytical coding procedures were applied to produce a coherent analysis focused on the research questions [72]. NVivo Qualitative Content Analysis Software (QSR International Pty Ltd. Version 10, 2012) was chosen to compile, analyze, organize and reconfigure the data collected. A content analysis was then carried out, involving the use of analytical codes derived from existing theories and explanations relevant to the research focus. For a better understanding of the data collected, local perceptions were initially codified in positive and negative aspects of REDD+ measures and in challenges and needs for REDD+ implementation on the ground. After this first codification, the perceptions were codified again to understand practical aspects of governmentality, subjectivity and behavior change. Finally, we reviewed policy, programs and project documents, including the REDD+ National Strategy.

Figure 2. REDD+ initiatives analyzed.

4. Results: Rewards and Punishments in Practice

4.1. Main REDD+ Strategies in Brazil

Command and control measures have significantly curbed deforestation in Brazil in recent years. Between 2004 and 2016, the annual deforestation rate in the Brazilian Amazon fell by 71% [73]. Brazil has also taken significant steps toward setting targets and developing interventions for REDD+, including the construction of a National REDD+ Strategy [74]. There are also numerous sub-national initiatives in the Amazon, ranging from state government programs to private projects [70]. Persistent concerns, however, have been raised about the potential impacts of REDD+ on local livelihoods and socioeconomic welfare [75,76]. These impacts depend in part on the particular interventions and safeguards used to achieve REDD+. At the same time, REDD+ is a dynamic arena of negotiation with competing discourses, including those surrounding equitable benefit-sharing, financing, monitoring, indigenous peoples, landscape approaches and others (see for example [77]).

The Brazilian REDD+ national strategy has as its main guiding policies the National Policy for Climate Change and the Forest Code. It also identifies three sectorial plans as the primary channels for implementing REDD+ in Brazil: the Action Plan to Prevent and Control Deforestation in the Amazon (PPCDAm); the Action Plan to Prevent and Control Deforestation and Fire in the Brazilian Cerrado (PPCerrado) and the Plan for Low Carbon Agriculture (ABC). It supplements these plans with cross-cutting measures, including a financial architecture for REDD+ and a set of safeguards designed to ensure that REDD+ actions do not inflict social or environmental harm. Safeguards are based on social and environmental principles and criteria for REDD+ developed in 2010 by a group of civil society organizations. The REDD+ national strategy safeguard summary, however, received much criticism from the civil society and groups involved in the development of safeguards. One of the main issues is that federal safeguards only cover initiatives financed by the Amazon Fund and are related to PPCDAm implementation [78].

Additionally, the national strategy is still very unclear in terms of the types of measures it will concentrate on to implement REDD+ and to guarantee that safeguards are in place. Detailed regulation of these issues was left to the National REDD+ Entity and its Thematic Consultation Chambers, which are still in the incipient stages of designing specific measures and principles. Although participation has been historically recognized as important for strengthening forest governance in Brazil [79], the participation, consultation and representation of local populations and landowners have been highly uneven and in some cases absent [80–83].

Following the REDD+ national strategy, we define REDD+ interventions as those which are linked to the implementation of the policies and plans that guide the strategy. These involve a mix of strategies that in theory should go beyond rewards and punishments, including: tenure regularization (i.e., provision of land titles and tenure clarification), territorial management (i.e., implementation of the Rural Environmental Registry (CAR in Portuguese)), monitoring (i.e., surveillance), enforcement (i.e., application of sanctions and penalties), financial incentives for sustainable production (i.e., sustainable rural credits), PES (i.e., direct cash transfers), improvement of agricultural and livestock practices (i.e., crop-livestock-forestry integration) and environmental education (i.e., capacity building) [74]. While reviewing the initiatives, we found that in the readiness phase (between 2008 and 2015), most of them focused on (i) increasing command and control through surveillance and *in situ* monitoring and (ii) territorial management through the implementation of the Rural Environmental Registry (CAR), a tool to increase environmental compliance, and enforced nationally by the Forest Code. Other main interventions being implemented include: (iii) PES; (iv) promotion of sustainable production and better practices; and (v) increase access to financial incentives (Table 1). In this article, we concentrate on (i), (ii), (iii) and (v) since there is more evidence on their performance on the ground compared to (iv), and thus local actors have a clearer sense of their impacts on deforestation.

Table 1. Interventions of REDD+ initiatives analyzed.

Initiatives (Proponent)	Interventions
Transamazon (Amazon Environmental Research Institute—IPAM)	Household: PES *; management plans; alternative production strategies *; capacity building; technical assistance; environmental compliance; land tenure clarification through CAR * Community: deforestation surveillance and *in situ* land use monitoring *; strengthen organizations No plan to sell credits
Sustainable Landscapes (The Nature Conservancy—TNC)	Household: land tenure clarification through CAR *; livelihood alternatives *; capacity building; access to credit Community: deforestation surveillance and *in situ* land use monitoring *, sharing lessons No plan to sell credits
Cotriguaçu (Centro Vida Institute—ICV)	Household: sustainable forest management (SFM) *; alternative production strategies; land tenure clarification through CAR *; capacity building; structuring municipal environmental management; technology transfer; access to credit Community: deforestation surveillance and *in situ* land use monitoring * No plan to sell credits
SISA (State of Acre)	Household: PES *, alternative production strategies *, SFM, land tenure regularization through CAR * Community: deforestation surveillance and *in situ* land use monitoring * Selling of credits still incipient
Bolsa Floresta (Amazonas Sustainable Foundation—FAS)	Household: PES *; income diversification *, capacity building Community: deforestation surveillance and *in situ* land use monitoring *, public services * Selling of credits in some areas
Jari (Biofílica/Jari)	Household: technical assistance, sustainable production, SFM Community: deforestation surveillance and *in situ* land use monitoring * Aims at selling credits

* interventions already being implemented.

4.1.1. Monitoring and Land Clarification: Magic or Broken Sticks?

Increasing surveillance and *in situ* monitoring through command and control measures has been at the heart of REDD+ interventions on the ground. We recognize the role of command and control interventions in reducing deforestation rates in the last decade [11,12,84]. However, there is also evidence that such approaches have social implications related to distributive equity (i.e., increased social inequalities among diverse local landholders) and social welfare (i.e., unequal distribution of income losses) that need to be considered in any structure that penalizes deforestation [4,85]. PPCDAm, for example, has been criticized for limiting its implementation to command and control measures [86–88]. Assunção et al. also show that centralized measures that treat the Brazilian Amazon as a largely uniform target have heterogeneous and spillover effects (i.e., negative behavioral adaptions that drive deforestation on the ground) [89]. Command and control does not normally affect large landowners [90], who are still responsible for the bulk of forest clearance in the Brazilian Amazon [15]. Furthermore, in some cases it has also increased migration to less monitored areas [85,88], which may cause leakage.

The Brazilian Environmental Protection Agency (IBAMA) is responsible for enforcing command and control measures. Field operations are based on past deforestation patterns and on satellite imagery. Coercion measures include the collection of administrative fines and *in situ* confiscation of produce and equipment whenever illegality is obvious [91]. IBAMA has received several critiques for being ineffective, but especially for corruption scandals involving large landholders [32,91,92]. Several authors have already demonstrated that although IBAMA is responsible for issuing the largest amount of fines in Brazil, the agency is known for its inefficiency in collecting them [90,91,93,94]. To make things worse, minor offenders (often smallholders), who do not want or cannot hire lawyers to challenge or disrupt administrative proceedings, are more likely to pay fines [90].

During the field research, we observed that local practices have already been developed to sabotage satellite monitoring and IBAMA fiscalization. An important example emerging from our qualitative data is the "método quebradão" (or "broken method" in English)—adopted in the state of Pará—where deforestation occurs without the use of fire and other techniques that can be identified via satellites or even *in situ* [95]. In this method, around 20% of the forest area is kept standing and just part of the forest is deforested either manually (in the case of smallholders) or with the use of poison sprayed in the tree canopy allowing them to die in different stages (in the case of large landholders). In this way, the forest "is broken," and the forested patches are not fully deforested, making monitoring by satellite images difficult [88,96]. Another important finding regards smallholders' perceptions of IBAMA fiscalization. For them, the national government should have been able to provide environmental education (i.e., knowledge sharing on climate change and adaptation actions) and capacity building (i.e., technical assistance on sustainable land use alternatives) before increasing command and control based enforcement. Historically, they have been actively encouraged to clear the forest as proof of "productive" activity under land-titling laws and for the acquisition of rural credits. It becomes almost utopic to expect that they will change their land use practices—mainly focused on growing food (i.e., cassava, fruits) for subsistence—without any assistance. In their words: "What does the government expect me to eat? The ball?", making reference to the 2014 World Cup, year when one of the field visits took place. Many of them responded in other tropes of resistance saying: "We will show the national government and REDD+ proponents who has the power by burning the forest in the next season if technical assistance does not come."

Knowledge about the spatial boundaries of landholdings has been improving since 2006, when CAR started to be implemented by different states in the Amazon. Due to the Forest Code of 2012, CAR is mandatory for all rural properties. Smallholders and other landholders, however, were not consulted or represented during the negotiations over the instrument. When it was initially designed, CAR aimed at linking forest tenure reform efforts with environmental compliance. In practice, however, CAR has become a tool to better monitor and control forest cover changes. CAR has a declaratory and permanent nature and contains specific information about the property (limits, environmental liability, land use). It does not *per se* clarify the ownership status of rural properties, neither does it guarantee that the landowners will follow the proper steps for environmental compliance. CAR information is under the responsibility of the declarant and the instrument has already been criticized for not having the registered data verified by government agencies [97] and for being unable to guarantee compliance [88]. Other critiques include: (i) the lack of strategy and coordination of the instrument with other policies and measures; (ii) an excessive focus on registration rather than the whole process of compliance; (iii) lack of priority given to CAR by the federal government; it took two years to be regulated and is far from being comprehensive enough; and (iv) the lack of attention to overarching strategic aspects such as level of public transparency, degree of overlapping polygons and systems automation routine [88,89,97–100].

Finally, while analyzing qualitative data on local perceptions about CAR we found that they changed throughout the years from being a possible solution to the challenges of tenure clarification to becoming an undesirable tool that supports command and control measures. In addition, delays in its regulation at the national level and the recurrent prorogation of the deadline to registry properties has also meant that many landholders are starting to view CAR as possibly another policy measure that is not going to be effectively mandatory (or in their words, "is just on paper").

4.1.2. Payments for Ecosystems Services (PES) and Access to Financial Incentives: Digestive or Rotten Carrots?

Although much debated at the federal level, a national PES has not yet been established. The Forest Code authorizes the use of PES as a financial incentive, but it is not yet regulated at national level. The Bolsa Verde Program is a federal initiative that gives financial incentives to communities in conservation areas, but its main objective is to reduce poverty rather than to pay for environmental services. Famous for its visionary multi-dimensional aspects, the Bolsa Floresta Program is the main state-level PES initiative in Brazil. The program has achieved successful outcomes, such as increasing the wellbeing of households in different conservation units in the state of Amazonas [101]. But it has also suffered criticism for not allowing democratic and interactive processes of local participation [102,103]; not having different opportunity costs reflected in the payment structure [104]; possible lack of additionality [104–106]; and absence of non-compliance monitoring and conflict resolution tools [107]. Since 2010, SISA in the state of Acre has also been paying smallholders an annual bonus conditioned on the continued cessation of deforestation and burning according to the property management plan. Early evidence, however, has shown that payments are coming late or not at all, and farmers in the program areas are not receiving appropriate government support, which may result in decreased levels of participation [108]. The Transamazon initiative started implementing conditional payments in 2014, but there is no evidence yet of its outcomes. In the Bolsa Floresta Program, when asked about the direct payments being implemented, many smallholders affirmed they did not want money and instead complained: "We are prohibited from growing our own food, there is no management, authorities are distant and don't trust us. They can't just prohibit and pay; those who live in the communities need to work!".

Although recognizing that PES may play a key role in forest conservation in Brazil, Borner et al. [4,5] emphasize that PES implementation will face several limitations related to contract enforcement and unsecure and unclear tenure. The authors argue that the suspension of payments as a sanction of non-compliance becomes uncertain under imperfect enforcement. According to them, PES recipients facing a low enforcement probability may thus prefer to continue forest conversion and receive PES as an additional rent—and get away with it. It has been already acknowledged that, for various reasons, legal sanctions are seldom applied and enforcement is largely ignored because of the resistance such regulations encounter from individuals [98,109]. Indeed, as noted by different authors, the softening of restrictions on the amount of land that landholders are legally allowed to deforest under the Forest Code was a clear reflection of the lack of law enforcement in Brazil [4,89,93,110–113]. Other criticisms of PES initiatives in the Amazon include its limitations in terms of scope, outcomes and cost-effectiveness when compared to opportunity costs of actual land uses [103,114–117]. These are also reasons why some proponents of the initiatives analyzed here, such as the Sustainable Landscapes Program, opted to not include PES as a strategy for benefit-sharing.

How access to REDD+ financial incentives (i.e., ABC Program; local funds) other than PES will overcome problems of coordination with other opportunities that incentivize deforestation (i.e., rural credits) is still very unclear. At the local level, the most expressive example of a decentralized financial mechanism is the Surui Fund. Designed with high levels of local participation [118,119], the fund has been seen as an innovative way of decentralizing REDD+ fundraising and benefit-sharing while at the same time increasing autonomy [120]. The initiative, however, still presents some problems, such as absence of timely financial flows to meet the needs of the Surui tribe, principally of those who depend on productive activities, since the only currently accessible economic options in the reserve are environmentally damaging (i.e., illegal logging and cattle ranching) [121]. Divergences between members of the tribe and their different interests also generated some conflicts in relation to the fund and the REDD+ initiative as a whole [122].

At the national level, the Amazon Fund is the best example of a national results-based payment mechanism. The fund focuses on PPCDAm's main goals, such as monitoring and control, promotion of sustainable productive activities, land and territorial planning, as well as scientific and technological development. In line with federal conservation strategies, the fund has been prioritizing CAR implementation in its initiatives. It is hosted by the National Development Bank (BNDES) and proposals undergo routine procedures with bottlenecks in several stages of the process (i.e., no deadlines for analysis and disbursements). In addition, institutional characteristics of BNDES, one of the largest public investment banks funded by the Brazilian National Treasury, have generated some questions of legitimacy in the management of funds, once the bank is responsible for funding the main activities that cause the deforestation in the Amazon (i.e., large-scale hydroelectric dams; livestock and agriculture activities) [123].

Finally, outcomes of the ABC plan and its program are still very incipient. To qualify for a loan, farmers need to be registered under CAR. However, to date, the registry has faced several problems that create a barrier to financial access. Other limitations include producers' ignorance of the ABC plan, the low capacity of technology transfer, and the chronic deficiencies of the agricultural extension system in Brazil [124]. The ability of key actors to access credit is restricted by their lack of capacity and the uncertainty about the risks of new technologies. Furthermore, until recently, Banco do Brasil managers had no incentive to offer credits from the ABC plan because the provision of such credits was not factored into their performance review in the same way as the provision of other financial products, such as loans for livestock and agriculture.

4.2. Subjects and Justifications: What do Rewards and Punishments Make Possible?

This section asks two key questions: What processes of visibilization and articulation does the rewards and punishments discourse make possible? Or, in other words, what is hidden by it, what does it obscure from consideration and what kind of subject does it produce/assume? By answering these questions, we aim to shed some light on what is justifiable when it comes to REDD+ interventions and if the assumed subject is sufficiently realized for REDD+ to be effective.

The main technique of discipline and power used in the Amazon for REDD+ is spatial and, at the moment, is concentrated on increased monitoring and CAR implementation. The fact that CAR has been failing to promote environmental compliance and is serving just as another form of surveillance suggests that it has been transformed, and its old purposes are now obscured or even obliterated. Its function has become a way of exercising a more coordinated and therefore greater power that is realized at the expense of innumerable, more distributed forms of smaller powers. CAR also turns out to be a material technique to govern people on the ground. It seeks and enables control by creating the conditions for a more continuous and more comprehensive flow of information that registration under CAR makes possible. Interestingly, it requires an active acquiescence and support of the governed subjects (landowners). And, in some cases, it may even act as an incentive to increase deforestation, since registration under CAR facilitates the flow of bank loans, as also of subsidies for livestock and agriculture [121]. Recent analysis shows that at least two thirds of deforestation in the Amazon happened in areas that adhered to the CAR [125]. In other cases, CAR requires a relatively small proportion of the farm area to be maintained under forest cover, encouraging landowners with higher forest cover to undertake deforestation on their holdings for conversion towards agriculture. It shows how power works on and through the governed, as they are complicit in their "domination."

The fact that the collection of administrative fines imposed by IBAMA is ineffective and may affect only smallholders is a clear example of how power relations and hierarchies translate into heterogeneous practices of government. To understand a bit more of what was happening on the ground, we asked all participants of the qualitative research about the role of IBAMA in reducing deforestation. The majority (87%) confirmed that command and control measures undertaken by IBAMA affect more smallholders than they do other actors. Many smallholders mentioned that IBAMA agents act like "vandals" and have been inspiring a culture of fear among smallholders that in turn feel constantly threatened. In their words: "The government sent violence before education. As some parents do with children: Beatings, beatings … instead of educating."

The implicit power relations that these techniques of coercive power promote is reflected in the fears of smallholders participating in the REDD+ initiatives analyzed here. Besides the fear of IBAMA, smallholders also fear unfulfilled promises (i.e., REDD+ initiative would fail to start or fail to continue; "REDD+ proponents talk too much and act too little"); displacement (i.e., fear of being reallocated or needing to migrate to other areas because of REDD+ initiative; "will REDD+ cause smallholders to lose land?"); and decrease in wellbeing and income (i.e., REDD+ initiative would fail to provide alternative sources of income; "stopping slash and burn practices had already harmed our incomes and diets"), to name a few (Table 2). Altogether, these fears have already been shown to restrict the power of innovation and adaptation of small producers in similar contexts [126].

Table 2. Challenges, hopes and fears of REDD+ initiatives identified by local actors in the REDD+ initiatives analyzed.

Initiatives (Proponent)	Challenges	Local Perceptions
Transamazon (IPAM)	Delayed government policies that regulate CAR and REDD+; tenure regularization; modifications to the Forest Code	Fears: fail to start or continue; unfulfilled promises; changes that harm their income and diets; absence of alternative production strategies; minimal financial resources being allocated to communities Needs: improving local production; technical assistance; infrastructure; transparency; local participation; compensation for deforestation reduction
São Félix do Xingu (TNC)	Coordinating different actors; land tenure clarification; lack of policies to regulate REDD+; corruption; limited governance capacity	Fears: unfulfilled promises; decrease in wellbeing and income; starvation; dispossession; increased IBAMA monitoring Needs: unity; collectivity; innovation; technical assistance; infrastructure; education; organization; capacity building; tenure regularization
NW Mato Grosso (ICV)	Conflicts of ideologies in the state; local governance; coordinating different actors; discontinuity of policy initiatives	Fears: generation of perverse incentives Needs: infrastructure; political will
SISA (Acre)	Lack of policies to regulate REDD+; financing; support for changing BAU; isolation of communities	Fears: dispossession of smallholders and traditional peoples; reduction in other benefits and wellbeing; increased monitoring Needs: infrastructure; technical assistance
Bolsa Floresta (FAS)	Vastness and diversity of areas; transaction costs; compliance monitoring; focus on PES	Fears: decrease in HH income; lack of proper compensation Needs: capacity building; increase the value of PES; integrated development components (health, education and sustainable livelihoods); conflict-resolution tools
Jari (Biofílica/Jari)	Heavy oscillation of carbon market; adapting global standards to local level; land tenure	Fears: increased IBAMA and Jari Florestal monitoring restricts wellbeing; losing autonomy and not having assistance; benefits for proponents only; losing land Needs: tenure regularization; improving local production systems, provision of machinery and inputs, basic infrastructure, school and health center.

We observed during field research that consequences of this culture of fear include the migration of smallholders to other areas that are less controlled or to urban areas, as well as the disinterest of youth to remain in rural areas. The main actions that generate this culture reflect informal aspects of state practices that are neither accidental nor unusual, but rather correspond to what Olivier de Sardan calls "practical norms" of public agents [127]. This means that in practice there is a significant discrepancy between the law (formal rules) and the behavior of political elites and public officials (informal rules). Practical norms are part of the habits and practices of corrupt, bureaucratic and colonialist government agencies and their employees [128] and are embedded and institutionalized as social techniques of discipline, such as IBAMA's *in situ* monitoring and consequent punishments. REDD+ was supposed to provide an alternative to command and control measures and thereby overcome the limitations of such efforts. Instead, our findings show, REDD+ initiatives are incorporating past disciplinary forms of regulation, and reinscribing old practices of forest conservation (see also [31,129]). In smallholders' words: "increased monitoring just stop the smoke but do not solve the problem."

Local responses to techniques of governance, such as the "broken method," also demonstrate the complexities of behavioral changes in subjects that such techniques target. The fact that Amazon deforestation has increased in small increments, and most recently at a substantial rate [130], may be connected to such responses and behavioral adaptations and need to be considered so as to better understand the local dynamics influencing deforestation trends. In the case of the "broken method," resources, material goods and social hierarchies enabled innovation by forest managers, undermining the instrumental functionality of command and control measures. This suggests that the subjectivity and creativity of targeted smallholders and individuals are coming to surface, but with negative effects. It also shows creativity and innovation in behavioral choices, and the capacities of people to make use of the discretion that incentivization provides to subjects whose actions are to be regulated. Finding the right mix of interventions that speak to such intelligence on the ground is important to better understand the complexities of the relationships among human motivations, agent choices, and forest conservation. Expectations of strict obedience, for example, are widely recognized as lessening higher human capacities, such as independence and creativity and the ability to solve problems [131].

Most of the needs raised by local actors in the REDD+ initiatives we analyzed are related to capacity building, technical assistance and materials for alternative forms of production (Table 2). This leads to the question as to why policy discourses and practices around REDD+ have tended to focus more on rewards and punishments. The analysis suggests the discourse is both based on and is strengthening models of human and human-nature relations that prize cost-efficiency, are inattentive to complexity across targeted communities and individuals, ignore the possibility of changes in how people respond differently to material incentives, and underemphasize considerations of long-term outcomes of effectiveness and social justice. In this way, it may distract us from other measures that are equally or more effective when dealing with individual subjectivities. Table 2 demonstrates there is a diverse array of measures that address the main needs identified by smallholders—ignoring these expressions of needs raises the risks of undermining the long-term viability of REDD+. Indeed, evidence has already shown that long-term measures have effectively contributed to less damaging modifications of forest landscapes [102,132].

It is also important to recognize that PES and other rewards that frame conservation in economic terms based on techniques of remuneration may create a hierarchy within itself. From a "phatos of distance," to use Nietzsche's term, values are created by a discourse and put in to practice by a few privileged actors that are looking down on subjects and instruments [133].

This means, as Nietzsche put it, that the "pathos of distance [which] grows out of the ingrained difference between strata–when the ruling caste constantly looks afar and looks down upon subjects and instruments and just as constantly practices obedience and command, keeping down and keeping at a distance" [134]. And it is exactly this distance that reinforces the hierarchical relations between rulers and subjects. In the REDD+ carrots and sticks approach, smallholders are seen as "the poor", "the guilty" or, to better represent the discourse, "the donkey." As Pokorny et al. [31] argued, "market-oriented [initiatives] widely ignore local management practices, local ways of organizing work, and other local capacities and limitations". These approaches "demand that Amazonian rural dwellers adopt working routines and commitments which require attitudes that are essentially alien to their culture", the authors complement referring to externally defined models of development and conservation.

It has already been acknowledged that smallholders have the ability to adapt to the scarcity of natural resources in difficult conditions and without significant external support [117,119], as well as to respond to emerging production options and operate with comparatively low environmental costs [135–139]. In the rewards and punishments discourse, however, the subject is reduced to a mere instrument and has no capacity to be more connected and engaged in creative climate mitigation and adaptation solutions. At the same time, it allows certain actors (i.e., the payers, in the case of PES) to be granted an advantaged position by a way of repayment and compensation. It also assumes a stable and uniform relationship between cause and effect, which leads to an oversimplification of the complexities behind deforestation and may blur the importance of valuing individual capacities for climate adaptation and mitigation. Indeed, many smallholders in the different initiatives analyzed affirmed they do not want payments, they want to be trusted by, and work together with, governmental authorities and REDD+ proponents. In their view, they should not just be punished or rewarded; those who live in rural areas need to feel part of the process of reducing emissions by working on new solutions. This can be linked to Hegel's naturalistic approach to see human beings, in which he develops his conceptions of desire and recognition where the individual is known to use ways of intentional collaboration to be able to reach certain success for the collective, for, most of the time, this collaboration also benefits himself. The human being, in his view and for this reason, is a collective being [140].

It has already been pointed that the narrow definition of being (or "what it means to be human") as an individual guided only by self-interest and utility maximization is in conflict with the need to promote altruism and responsibility towards others and future generations. It is therefore in conflict with the notions of distributive justice [141,142]. The importance of individual and collective capacities, cultural attributes, rules and local institutions, as well as the creative and entrepreneurial capacity of the rural population have been recognized by sociologists [143,144], historians [145,146] and economists [136,147–149]. Instead of following Machiavellian and Hobbesian philosophies, perhaps is time to go back to Aristotelian ideas, in which every individual is also political and, as such, has to take responsibility for a common cause, such as fighting for our species survival.

Following the Aristotelian philosophy, happiness and wellbeing emerge through the performance of objective human activities. The latter can be seen in connection to the notion of human functioning developed by Sen and Nussbaum, both of whom stress the connection of this concept to Aristotle [150–152]. The aim is not the maximization of utility, but rather the training of dispositions that seek to find a middle ground between two extremes. In Aristotelian philosophy, virtue (or excellence) lies in achieving this middle ground. The emphasis is thus on harmony, rather than on maximization. The Aristotelian philosophy of human wellbeing points towards a conception of wellbeing in which a finite quantity of commodities does not prevent human development, since wellbeing springs from the harmonious use of existing commodities, rather than from the maximization of commodities possessed [153].

Foucault's later work on the "care of the self" constitutes a return to the Greek philosophy as he points out how one can govern oneself as an agent showing critique to be an important avenue for self-governance. Critique is then a matter of examining the *status quo* and maintaining the freedom to question it [154]. For Foucault, this freedom manifests and perpetuates itself through the Greek ancient practice called care of the self [155]. This concept remains a central theme in his analysis of the individual as subject to various power dynamics. Care of the self is lifelong work on one's body, mind, and soul, in order to better relate to other people and live an ethically driven life [155]. Foucault then, in his analysis of modes of power and care of the self, condones a full immersion into the present. In an age of mass consumerism and globalization, technological innovation and ecological consciousness, what constitutes individual identity has shifted, and discourses of power and truth have taken on new meanings [156]. Rewards and punishments approaches reflect these new meanings and place emphasis on human behavior being largely determined by incentives and penalties, sending different individuals in unanticipated directions.

There is also a contradiction in wanting to be perfectly cost-efficient in a situation whose nature is unpredictable, both because of the inherent dynamics and complexities of heterogeneous natural communities and assemblages, and related social-ecological interactions but also because of new forces and drivers such as climate change. The contradiction lies a little deeper than the mere conflict between the desire for financial efficiency and the outcome of behavior change. If we want to be cost-efficient, that is, achieving quick results with low costs, then we automatically want to be separated from the complexity of changing behavior, meaning that we forget what people are by concentrating on cheap and quick fixes to complex challenges. Yet, it is the very sense of separateness which makes us question the viability and justification of cost-efficient approaches for REDD+. To be cost-efficient means to isolate and fortify the financial aspects of REDD+, but it is exactly this isolation that distracts us from the urgent and important reasons of implementing REDD+, such as climate mitigation and adaptation. The rewards and punishments approach is also in conflict with the notions of distributive justice (i.e., focus on short-term outcomes) and it can be a very blunt and undiscriminating way of changing behavior—it can easily punish good and reward bad behavior (a critical example is the approval of the Brazilian Forest Code, which includes an amnesty for landowners who have previously deforested).

From the viewpoint of interviewed decision makers, REDD+ proponents and experts, changing the focus of interventions would not be difficult, especially under a favorable political scenario. Many answers indicated the necessity of incorporating measures focused on crop–livestock–forest integration and agroforestry systems in the REDD+ debate. These and other measures are under ABC plan goals but are still at the beginning stages of implementation. Other answers emphasized the need for an approach that engages different actors through an open-ended process, a pre-conditional state of engagement that can only exist when every actor holds a stake and has a voice and responsibilities. In this way, it is easier to address core motivations for collective behavior change [157]. A good example of interventions that encourage this collaborative state of engagement is demand-side measures, such as zero-deforestation cattle agreements signed by major meatpacking companies in the state of Pará (see [158]).

Finally, understanding persistence is increasingly important for testing policy measures and applying interventions on the ground [159,160]. Interventions have been shown "to crowd out intrinsic motivation if they are perceived as controlling—and to crowd in and thus increase intrinsic motivation if they are perceived as supportive" [64] (pp. 56–59, 71–72). Fortunately, there is a growing body of literature looking at what increases intrinsic motivation and how behavior change can be more persistent [66,161–167]. It is certainly time to strengthen the connections between the REDD+ debate and this literature.

5. Conclusions

REDD+ initiatives and national strategies can contribute to reducing deforestation and increasing the wellbeing of local people [168,169], as many of its proponents have suggested. But they can also foster social inequalities and run headlong into problems of governance and local resistances [170,171]. This article has undertaken a Foucault-inspired, close reading of the rewards and punishments discourse in relation to which REDD+ is situated to assess the nature and implications for REDD+ implementation in Brazil. We began with the fact that the prevalence of a rewards and punishments discourse as the context for REDD+ implementation necessarily normalizes and rationalizes some possibilities and forms of action, and excludes other considerations or ways of seeing and doing. The discursive dominance of rewards and punishments has real impacts in terms of the kinds of initiatives attempted, the forms of responses and adaptations to these interventions, and the outcomes that can be observed. The shortcomings and problems of interventions based on rewards and punishments approaches are particularly evident in the context of the complex, non-linear interactions of forests and people, the political dynamics of smallholders' responses, the heterogeneities and uncertainties related to targeted actors, and the evolving nature of deforestation trends in the Brazilian Amazon.

The intertwining of knowledge and power in the rewards and punishments discourse is a complex ensemble of social processes that can be better understood when broken into two main topics: (i) the limits of command and control measures for reducing deforestation, and (ii) the place of incentives as a decentralized and market-driven alternative to involve a broader range of private and public actors in the quest for low-cost climate mitigation alternatives. Flexibility and cost-efficiency are intertwined storylines articulated in climate negotiations, and the REDD+ frame as an opportunity to combine minimal-cost climate mitigation with sustainable forest management, biodiversity protection, poverty reduction and local socioeconomic development merged as a viable solution to international political negotiations. However, the local translations of international agreements around REDD+ need to come to better grips with the complexity of human selves and subjectivities at which Sen has hinted in his early criticisms of utility-maximization [172], and for which widespread evidence is now available in the work of many behavioral economists and evolutionary theorists who show that people's judgments and choices are a complex amalgam of intuitions [173], beliefs, norms, principles, dispositions, attitudes, emotions, passions and sentiments [66,174–178].

One of the central questions we aimed to answer while exploring the way in which the rewards and punishments approach has been used to legitimize techniques of governance in Brazil was: "What kind of subject does the rewards and punishments discourse produce/assume in practice?" This question is intimately linked to REDD+ because it leads directly to a consideration of the extent to which the techniques of remuneration and control, upon which rewards and punishments rely, are justifiable for REDD+. Following Foucault's ideas, it became clear that the rationalizations of state and other actors contain a particular conception of the subject that is to be governed. Each governmentality came with circumscribed expectations about the thoughts and behaviors of the incentivizable and controlled subject. The final conception of the human subject that emerges is that of the individual who acts out of self-interest, whose subjectivity is immutable over time, and whose behavior is predictably driven by techniques of coercion and remuneration. In such an approach, the malleability of subjectivity, the adaptive nature of individual and collective intelligence, and the complexity of mental and social calculations are either underplayed or ignored in favor of a simplified and simplistic model of the utility maximizer.

Our analysis provides insights about how the rewards and punishments discourse may be conditioning people into particular behavioral choices and adaptations, generating negative outcomes for deforestation, reducing positive feedbacks, and producing short-term unanticipated effects. Our results corroborate with previous research on rewards and punishments and their limits in changing behavior (see for example [173–178]). Our main argument is that exclusive attention to the conditions under which a rational system of rewards and punishments can be implemented to generate predictable forest protection outcomes in the Brazilian Amazon has led to the neglect of other ways of supporting smallholders that may in fact also help achieve the objectives of REDD+ interventions in the longer term. In this sense, one may even say that the recourse to REDD+ after a reliance on coercive techniques of forest protection is part of a process of trying to make reality match externally developed models, but its consequence is to distract attention away from trying to work with reality. Such a move profoundly constrains our understanding of human behavior and, more importantly, underestimates the capacity of smallholders to play a key role in climate mitigation and adaptation. The aptitude of smallholders in adopting a variety of practices is extraordinary. By ascribing a selfish nature to these actors, we move in the opposite direction of their ability to cooperate with each other, to create different solutions for their everyday problems, and to manage the gaps in conservation linked to remoteness and complexities of forest systems.

Our analysis and findings are less an argument in favor of doing away with rewards and punishments, and more an acknowledgement of the necessity of supplementing and moving beyond the pure calculus of efficiency and control encoded in arguments favoring rewards and punishments. Prioritizing REDD+ cost-efficiency and imposing a model based on such prioritization has the effect of ignoring the choices, selves and lives of smallholders, under-emphasizing the complexity and unpredictability of smallholder-forest-policy interactions, and leading to adverse long-term outcomes that support neither smallholder wellbeing nor climate mitigation.

It is also difficult to imagine how much of the remaining "residual" deforestation can be curbed through increased command and control and incentive-based approaches. Rather than being based on and deploying techniques of remuneration and coercion, REDD+ interventions should strive for a clearer understanding of the fundamental processes and practices that shape and drive deforestation vs conservation, such as the growing extra-local demand for forest and agricultural commodities, subsidies from outside the forest sector that encourage the production of such commodities, and the multifaceted and evolving nature of the subject targeted by incentives. Little attention focuses on the consumer, for example, in REDD+ debates. Measures associated with REDD+ interventions should reflect variations and changes in selves, social practices, and contexts. Instead of looking at smallholders as "donkeys" to be incentivized or controlled by carrots or sticks, we should embrace the idea that the current targets of forest policies are the ones most capable of creating sustainable solutions to problems of commodity provision and human survival under changing and challenging social and ecological conditions.

Acknowledgments: We are deeply grateful to all the smallholders who participated in our research for their patience and generosity in sharing their time and knowledge. REDD+ proponents and other local actors also played a key role in enabling this research. We also thank Rodrigo Calvet for his fruitful ideas. Finally, we thank the Centre for International Forestry Research for the collaboration and the opportunity of learning from REDD+ pilot initiatives.

Author Contributions: Maria Fernanda Gebara undertook the empirical and secondary data collection and analysis and coordinated the writing process. Arun Agrawal contributed to the writing by providing unique insights and by significantly helping in the alignment of theory, empirical evidence and conclusions.

Conflicts of Interest: The authors declare no conflict of interest.

References

1. Brukas, V.; Sallnäs, O. Forest management plan as a policy instrument: Carrot, stick or sermon? *Land Use Policy* **2012**, *29*, 605–613. [CrossRef]

2. Quartuch, M.R.; Beckley, T.M. Carrots and sticks: New Brunswick and Maine forest landowner perceptions toward incentives and regulations. *Environ. Manag.* **2014**, *53*, 202–218. [CrossRef] [PubMed]

3. Henderson, K.A.; Anand, M.; Bauch, C.T. Carrot or stick? Modelling how landowner behavioural responses can cause incentive-based forest governance to backfire. *PLoS ONE* **2013**, *8*, e77735. [CrossRef] [PubMed]

4. Börner, J.; Wunder, S.; Wertz-Kanounnikoff, S.; Nascimento, N. Forest law enforcement in the Brazilian Amazon: Costs and income effects. *Glob. Environ. Chang.* **2014**, *29*, 294–305. [CrossRef]

5. Börner, J.; Marinho, E.; Wunder, S. Mixing carrots and sticks to conserve forests in the Brazilian Amazon: A spatial probabilistic modeling approach. *PLoS ONE* **2015**, *10*, e0116846. [CrossRef] [PubMed]

6. Zwick, S.; Calderon, C. The Difficult Birth of Brazil's First "Green Municipality". *Ecosystem Marketplace*. 2016. Available online: http://www.ecosystemmarketplace.com/articles/paragominas-the-green-revolution-that-almost-wasnt/ (accessed on 10 November 2016).

7. Silva-Chavez, G. The Missing Link in Protecting Forests? The Private Sector. Available online: http://www.ecosystemmarketplace.com/articles/missing-link-protecting-forests-private-sector/ (accessed on 10 November 2016).

8. Foucault, M. *Discipline and Punish: The Birth of the Prison*; Vintage Books: New York, NY, USA, 1995.

9. Foucault, M. The subject and power. *Crit. Inquiry* **1982**, *8*, 777–795. [CrossRef]

10. De Sassi, C.; Sunderlin, W.D.; Sills, E.O.; Duchelle, A.E.; Ravikumar, A.; Resosudarmo, I.A.P.; Luttrell, C.; Joseph, S.; Herold, M.; Kweka, D.L.; et al. REDD+ on the ground: Global insights from local contexts. In *REDD+ on the Ground: A Casebook of REDD+ Initiatives Across the Globe*; Sills, E.O., Atmadja, S.S., De Sassi, C., Duchelle, A.E., Kweka, D.L., Resosudarmo, I.A.P., Sunderlin, W.D., Eds.; Center for International Forestry Research (CIFOR): Bogor, Indonesia, 2014.

11. Assunção, J.; Gandour, C.C.; Rocha, R. *Deforestation Slowdown in the Legal Amazon: Prices or Policies?* CPI Technical Report. NAPC/PUC-Rio: Rio de Janeiro, 2012. Available online: http://climatepolicyinitiative.org/wp-content/uploads/2012/03/Deforestation-Prices-or-Policies-Working-Paper.pdf (accessed on 10 November 2016).

12. Assunção, J.; Gandour, C.; Rocha, R. *Does Credit Affect Deforestation? Evidence from a Rural Credit Policy in the Brazilian Amazon.* CPI/NAPC Working Paper. NAPC/PUC-Rio: Rio de Janeiro, 2012. Available online: http://climatepolicyinitiative.org/wp-content/uploads/2013/01/Does-Credit-Affect-Deforestation-Evidence-from-a-Rural-Credit-Policy-in-the-Brazilian-Amazon-Technical-Paper-English.pdf (accessed on 10 November 2016).

13. Aubertin, C. Deforestation control policies in Brazil: Sovereignty versus the market. *For. Trees Livelihoods* **2015**, *24*, 147–162. [CrossRef]

14. Assunção, J.; Gandour, C.; Pessoa, P.; Rocha, R. *Deforestation Scale and Farm Size: The Need for Tailoring Policy in Brazil*; CPI Technical Report; NAPC/PUC-Rio: Rio de Janeiro, Brazil, 2015.

15. Godar, J.; Gardner, T.A.; Tizado, E.J.; Pacheco, P. Actor-specific contributions to the deforestation slowdown in the Brazilian Amazon. *Proc. Natl. Acad. Sci. USA* **2014**, *111*, 15591–15596. [CrossRef] [PubMed]

16. Bemelmans-Videc, M.L. Policy instrument choice and evaluation. In *Carrots, Sticks and Sermons: Policy Instruments and Their Evaluation*; Bemelmans-Videc, M.L., Rist, R.C., Vedung, E., Eds.; Transaction Publishers: New Brunswick, NJ, USA, 1998.

17. Foucault, M. *The Archaeology of Knowledge*; Routledge: London, UK; New York, NY, USA, 2002; p. 3.

18. Foucault, M. *The History of Sexuality Volume 1: An Introduction*; Allen Lane: London, UK, 1976.

19. Foucault, M. Truth and power. In *Power/Knowledge: Selected Interviews and Other Writings 1972–1977*; Gordon, C., Ed.; The Harvester Press: Brighton, UK, 1980; pp. 107–133.

20. Lessa, I. Discursive struggles within social welfare: Restaging teen motherhood. *Br. J. Soc. Work* **2006**, *36*, 283–298. [CrossRef]

21. Hardy, C.; Harley, B.; Phillips, N. Discourse analysis and content analysis: Two solitudes? *Int. J. Qual. Methods* **2004**, *2*, 19–22.

22. Foucault, M. *The Birth of Biopolitics: Lectures at the Collège de France, 1978–1979*; Palgrave Macmillan: Hampshire, UK, 2008; p. 269.

23. Maihofer, A. *Geschlecht als Existenzweise*; Helmer: Frankfurt, Germany, 1995.

24. Sauer, R.D. The political economy of gambling regulation. The political economy of gambling regulation. *MDE Manag. Decis. Econ.* **2001**, *22*, 1–3.

25. Luke, T. *Ecocritique: Contesting the Politics of Nature, Economy, and Culture*; University of Minnesota Press: Minneapolis, MN, USA, 1997.

26. Agrawal, A. *Environmentality: Technologies of Government and Political Subjects*; Duke University Press: Durham, NC, USA, 2005.

27. Fairhead, J.; Leach, M.; Scoones, I. Green grabbing: A new appropriation of nature? *J. Peasant. Stud.* **2012**, *39*, 253, 237–261. [CrossRef]

28. Pochmann, M. *Qual Desenvolvimento? Oportunidades e Dificuldades do Brasil Contemporâneo*; Boi Tempo: São Paulo, Brazil, 2009.

29. Cano, W. Power, Organization and Conflicts in Northern Bolivian Communities after Forest Governance Reforms. Ph.D. Thesis, Utrecht University, Utrecht, The Netherlands, 2012.

30. Medina, G.; Pokorny, B.; Campbell, B. Loggers and development agents exercising power over Amazonian villagers. *Dev. Chang. Oxf.* **2009**, *40*, 745–767. [CrossRef]

31. Pokorny, B.; Scholz, I.; de Jong, W. REDD+ for the poor or the poor for REDD+? About the limitations of environmental policies in the Amazon and the potential of achieving environmental goals through pro-poor policies. *Ecol. Soc.* **2013**, *18*, 3. [CrossRef]

32. May, P.H.; Millikan, B.; Gebara, M.F. *The Context of REDD + in Brazil: Drivers, Agents, and Institutions*, 2nd ed.; Center for International Forestry Research (CIFOR): Bogor, Indonesia, 2011.

33. Vedung, E. Policy instruments: Typologies and theories. In *Carrots, Sticks and Sermons: Policy Instruments and Their Evaluation*; Bemelmans-Videc, M.L., Rist, R.C., Vedung, E., Eds.; Transaction Publishers: New Brunswick, NJ, USA, 1998.

34. Bernard, L.L. *Social Control in Its Sociological Aspects*; Macmillan: New York, NY, USA, 1939.

35. Brigham, J.; Brown, D.W. Introduction. In *Policy Implementation: Penalties or Incentives?* Brigham, J., Brown, D.W., Eds.; Sage: Beverly Hills, CA, USA, 1980; pp. 7–77.

36. Turner, J.C.; Brown, R. Social status, cognitive alternatives and intergroup relations. In *Differentiation between Social Groups: Studies in the Social Psychology or Intergroup Relations*; Tajfel, H., Ed.; Academic Press: San Diego, CA, USA, 1978; pp. 201–234.

37. Machiavelli, N. ; Marriott, W.K., Translator; *The Prince*. 1908. Available online: https://www.gutenberg.org/files/1232/1232-h/1232-h.htm (accessed on 10 November 2016).

38. Hardin, G. The tragedy of the commons. *Science* **1968**, *162*, 1243–1248. [CrossRef] [PubMed]

39. Ackerman, B.A.; Stewart, R.B. Reforming environmental law. *Stanf. Law Rev.* **1985**, *37*, 1333–1365. [CrossRef]

40. Hahn, R.W.; Stavins, R.N. Economic incentives for environmental protection: Integrating theory and practice. *Am. Econ. Rev.* **1992**, *82*, 464–468.

41. Reitze, A.W., Jr. A century of pollution control: What's worked? What's failed? What might work? *Environ. Law* **1991**, *21*, 1549–1946.

42. Orts, E.W. Reflexive environmental law. *Northwest Univ. L. Rev.* **1995**, *89*, 1227–1340.

43. Sinclair, D. Self-regulation versus command and control? Beyond false dichotomies. *L. Policy* **1997**, *19*, 529–559. [CrossRef]

44. Gluck, P.; Rayner, J.; Cashore, B. Change in the governance of forest resources. In *Forest in the Global Balance: Changing Paradigms*; Mery, G., Alfaro, R., Kanninen, M., Labovikov, M., Eds.; World Series, 17; International Union of Forest Research Organizations: Helsinki, Finland, 2005; pp. 51–74.

45. Van Gossum, P.; Arts, B.; Verheyen, K. Smart regulation: Can policy instrument design solve forest policy aims of expansion and sustainability in Flanders and the Netherlands? *For. Policy Econ.* **2012**, *16*, 23–34. [CrossRef]

46. Aldisert, L.; Helms, E.M.M. From command and control to coaching. *IJBM* **2000**, *32*, 36.

47. Holling, C.S.; Meffe, G.K. Command and control and the pathology of natural resource management. *Conserv. Biol.* **1996**, *10*, 328–337. [CrossRef]

48. UNFCC. *Bali Action Plan.* Decision 1/CP.13, FCCC/CP/2007/6/Add.1* (14 March 2008). Available online: http://unfccc.int/files/meetings/cop_13/application/pdf/cp_bali_action.pdf (accessed on 10 November 2016).

49. Marx, K. *Capital: A Critique of Political Economy*; Penguin Books: London, UK, 1992; Volume 1.

50. Locke, J. Second Treatise of Government. Early Modern Texts. 1689. Available online: http://www.earlymoderntexts.com/assets/pdfs/locke1689a.pdf (accessed on 15 January 2017).

51. Anderson, E. *Value in Ethics and Economics*; Harvard University Press: Cambridge, MA, USA, 1993; p. 208.

52. Sandel, M.J. *What Money Can't Buy: The Moral Limits of Markets*; Farrar, Strauss and Giroux: New York, NY, USA, 2012; p. 10.

53. Agrawal, A.; Gerber, E.; Chhatre, A. Motivational crowding in sustainable development interventions: Assessing the effects of multiple treatments. *Am. Political Sci. Rev.* **2015**, *109*, 470–487. [CrossRef]

54. Clot, S.; Andriamahefazafy, F.; Grolleau, G.; Ibanez, L.; Méral, P. Compensation and Rewards for Environmental Services (CRES) and efficient design of contracts in developing countries. Behavioral insights from a natural field experiment. *Ecol. Econ.* **2015**, *113*, 85–96. [CrossRef]

55. Sullivan, S. The Natural Capital Myth; or will Accounting Save the World? Preliminary Thoughts on Nature, Finance and Values. LCSV Working Paper Series No. 3; The Leverhulme Centre for the Study of Value: Manchester, UK, 2014. Available online: http://thestudyofvalue.org/wp-content/uploads/2013/11/WP3-Sullivan-2014-Natural-Capital-Myth.pdf (accessed on 15 January 2017).

56. Markandya, A. Economic Instruments: Accelerating the Move from Concepts to Practical Application. In Fourth Expert Group Meeting on Financial Issues of Agenda 21, Santiago, Chile, 8–10 January 1997. Available online: http://repositorio.cepal.org/bitstream/handle/11362/34312/S9700533_en.pdf.txt (accessed on 10 November 2016).

57. Huber, R.; Ruitenbeeck, J.; Da Motta, R. *Market-Based Instruments for Environmental Policy Making, in Latin America and the Caribbean: Lessons from Eleven Countries*; Report; World Bank: Washington, DC, USA, 1997.

58. Borregaard, N.; Sepúlveda, C. *Introducing Economic Instruments at an Early Stage of Environmental Policy Making*; North-South Center Miami: Santiago, Chile, 1998.

59. Martinez-Alier, J. *The Environmentalism of the Poor: A Study of Ecological Conflicts and Valuation*; Edward Elgar: Cheltenham, UK, 2002.

60. Goulder, L.H.; Kennedy, D. Valuing ecosystem services: Philosophical bases and empirical methods. In *Nature's Services: Societal Dependence on Natural Ecosystems*; Daily, G.C., Ed.; Island Press: Washington, DC, USA, 1997; pp. 23–47.

61. Nunes, P.A.L.D.; van den Bergh, J.C.J.M. Economic Valuation of Biodiversity: Sense or Nonsense. *Ecol. Econ.* **2001**, *39*, 203–222. [CrossRef]

62. McCauley, D.J. Selling out on nature. *Nature* **2006**, *443*, 27–28. [CrossRef] [PubMed]

63. O'Neill, J. Managing without prices: On the monetary valuation of biodiversity. *Ambio* **1997**, *16*, 56.

64. O'Neill, J. Markets and the environment: The solution is the problem. *Econ. Political Wkly.* **2001**, *36*, 1865–1873.

65. Sagoff, M. On the economic value of ecosystem services. *Environ. Values* **2008**, *17*, 239–257. [CrossRef]
66. Neuteleers, S.; Engelen, B. Talking money: How market-based valuation can undermine environmental protection. *Ecol. Econ.* **2015**, *117*, 253–260. [CrossRef]
67. Frey, B.S.; Jegen, R. Motivation crowding theory. *J. Econ. Surv.* **2001**, *15*, 589–611. [CrossRef]
68. Bénabou, R.; Tirole, J. Incentives and prosocial behavior. *Am. Econ. Rev.* **2006**, *96*, 1652–1678. [CrossRef]
69. Gneezy, U.; Meier, S.; Rey-Biel, P. When and why incentives (don't) work to modify behavior. *J. Econ. Perspect.* **2011**, *25*, 191–210. [CrossRef]
70. Sills, E.O.; Atmadja, S.S.; de Sassi, C.; Duchelle, A.E.; Kweka, D.L.; Resosudarmo, I.A.P.; Sunderlin, W.D. (Eds.) *REDD+ on the Ground: A Casebook of REDD+ Initiatives Across the Globe*; Center for International Forestry Research (CIFOR): Bogor, Indonesia, 2014.
71. Sunderlin, W.D.; Larson, A.M.; Duchelle, A.E.; Sills, E.O.; Luttrell, C.; Jagger, P.; Pattanayak, S.; Cronkleton, P.; Ekaputri, A.; de Sassi, C.; et al. *Technical Guidelines for Research on REDD+ Subnational Initiatives*, 2nd ed.; Center for International Forestry Research (CIFOR): Bogor, Indonesia, 2014.
72. Ryan, G.W.; Bernard, H.R. Techniques to identify themes. *Field Methods* **2003**, *15*, 85–109. [CrossRef]
73. Instituto Nacional De Pesquisas Espaciais (INPE). *Monitoramento da Cobertura Florestal da Amazônia por Satélites: Sistemas PRODES, DETER, DEGRAD—2004–2015*; INPE: São José dos Campos, Brazil, 2015.
74. Ministério do Meio Ambiente (MMA). *Estrategia Nacional de REDD+; Portaria n. 370 de 2 de Dezembro de 2015*; Federal Government of Brazil: Brasília, Brazil, 2015.
75. Chhatre, A.; Lakhanpal, S.; Larson, A.M.; Nelson, F.; Ojha, H.; Rao, J. Social safeguards and co-benefits in REDD+: A review of the adjacent possible. *Curr. Opin. Environ. Sustain.* **2012**, *4*, 654–660. [CrossRef]
76. Ghazoul, J.; Butler, R.; Mateo-Vega, J.; Koh, L.P. REDD: A reckoning of environmental and development implications. *Trends Ecol. Evol.* **2010**, *25*, 396–402. [CrossRef] [PubMed]
77. Vijge, M.J.; Brockhaus, M.; Di Gregorio, M.; Muharrom, E. *Framing National REDD+ Benefits, Monitoring, Governance and Finance: A Comparative Analysis of Seven Countries*; Center for International Forestry Research (CIFOR): Bogor, Indonesia, 2016.
78. Ministério do Meio Ambiente. *Summary of Information on How the Cancun Safeguards were Addressed and Respected by Brazil throughout the Implementation of Actions to Reduce Emissions from Deforestation in the Amazon Biome between 2006 and 2010*; Ministério do Meio Ambiente: Brasilia, Brazil, 2015.
79. May, P.H.; Gebara, M.F.; de Barcellos, L.M.; Rizek, M.B.; Millikan, B. *The Context of REDD+ in Brazil: Drivers, Agents, and Institutions*, 3rd ed.; CIFOR Occasional Paper No. 160; Center for International Forestry Research (CIFOR): Bogor, Indonesia, 2016.
80. Gebara, M.F.; Thuault, A. *GHG Mitigation in Brazil's Land Use Sector: An Introduction to the Current National Policy Landscape*; Working Paper; World Resources Institute: Washington, DC, USA, 2013. Available online: http://wri.org/publication/ghg-mitigation-brazil-land-use-sector (accessed on 24 February 2017).
81. Observatório do Clima. Organizações da Sociedade Civil Voltam a Criticar Política Nacional de Mudanças Climáticas. Observatório do Clima, 2013. Available online: http://www.observatoriodoclima.eco.br/organizacoes-da-sociedade-civil-voltam-a-criticar-politica-nacional-de-mudancas-climaticas/ (accessed on 24 February 2017).
82. Observatório do Clima. Nota do Observatório do Clima sobre a Estratégia Nacional de REDD+. Observatório do Clima, 2015. Available online: http://www.observatoriodoclima.eco.br/nota-do-observatorio-do-clima-sobre-a-estrategia-nacional-de-redd/ (accessed on 24 February 2017).
83. Observatório do Clima. Especialistas do OC Questionam Decreto Que Cria Comissão Nacional de REDD+. Observatório do Clima, 2015. Available online: http://www.observatoriodoclima.eco.br/especialistas-do-oc-questionam-decreto-que-cria-comissao-nacional-de-redd/ (accessed on 24 February 2017).
84. Hargrave, J.; Kis-Katos, K. Economic causes of deforestation in the Brazilian Amazon: A panel data analysis for the 2000s. *ERE* **2013**, *54*, 471–494. [CrossRef]
85. Gebara, M.F. Sustainable landscapes program in São Felix do Xingu. In *REDD+ on the Ground: A Casebook of REDD+ Initiatives Across the Globe*; Sills, E.O., Atmadja, S.S., De Sassi, C., Duchelle, A.E., Kweka, D.L., Resosudarmo, I.A.P., Sunderlin, W.D., Eds.; Center for International Forestry Research (CIFOR): Bogor, Indonesia, 2014.
86. Abdala, G.C.; Reis Rosa, M. *PPCDam: Avaliação 2004–2007*; Revisão 2008; Ministério do Meio Ambiente: Brasília, Brazil, 2008.

87. Maia, H.; Hargrave, J.; Goméz, J.J.; Roper, M. *Avaliação do Plano de Ação Para a Prevenção e Controlo do Desmatamento da Amazônia Legal—PPCDAm—2007–2010*; Economic Commission for Latin America and the Caribbean (CEPAL), Institute of Economic and Applied Research (IPEA), German Society for International Cooperation (GIZ): Brasilia, Brazil, 2011.

88. Gebara, M.F. Governança de Paisagens Florestais: Impactos Sociais da Bricolagem Institucional em São Félix do Xingu. Ph.D. Thesis, Federal Rural University of Rio de Janeiro, 2015. Available online: http://r1.ufrrj.br/cpda/wp-content/uploads/2015/05/Governança-de-Paisagens-Florestais-Impactos-Sociais-da-Bricolagem-Institucional-em-S~ao-Félix-do-Xingu.pdf (accessed on 10 November 2016).

89. Assunção, J.; Gandour, C.; Pessoa, P.; Rocha, R. *Deforestation Scale and Farm Size: The Need for Tailoring Policy in Brazil*. CPI Technical Report. NAPC/PUC-Rio: Rio de Janeiro, 2015. Available online: https://climatepolicyinitiative.org/wp-content/uploads/2015/08/Deforestation-Scale-and-Farm-Size-the-Need-for-Tailoring-Policy-in-Brazil----Technical-Paper.pdf (accessed on 10 November 2016).

90. Telles do Valle, R.S. *Saindo do Quadrado: Propostas Para Tentar Dinamizar o Mercado de Cotas de Reserva Ambiental*; Instituto Sócio-Ambiental: Brasília, Brazil, 2013.

91. Brito, B.; Barreto, P. A eficácia da aplicação da lei de crimes ambientais pelo Ibama para proteção de florestas no Pará. *Rev. Direito Ambient.* **2006**, *43*, 35–65.

92. Moura, D.G. Mídia e Corrupção: A Operação Curupira na Amazônia. Universidade de Brasília, Centro de Desenvolvimento Sustentável: Brasilia, Brazil, 2006; p. 147. Available online: https://core.ac.uk/download/files/610/16343809.pdf (accessed on 10 November 2016).

93. Hirakuri, S.R. *Can Law Save the Forests? Lessons from Finland and Brazil*; Center for International Forestry Research (CIFOR): Bogor, Indonesia, 2003.

94. Brito, B. *Multas Pós-Operação Curupira no Mato Grosso*; Boletim "O Estado da Amazônia"; IMAZON: Belém, Brazil, 2009.

95. Barros, C.; Barcelos, L. Crime e Grilagem Com Uso do CAR. Publica, 2016. Available online: http://apublica.org/2016/08/crime-e-grilagem-com-uso-do-car (accessed on 15 January 2017).

96. Vieira, I.; Veiga, J.; Aguiar, A.; Gavina, J. *O Papel da Pecuária na Evolução da Fronteira: Relatório Geoma*; Embrapa: São Paulo, Brazil, 2010.

97. Camargo, F. *CAR Para "Inglês Ver"*; Instituto Sócio-Ambiental: Brasília, Brazil, 2014. Available online: http://www.socioambiental.org/pt-br/blog/blog-do-ppds/car-para-ingles-ver (accessed on 10 November 2016).

98. Pires, M.O.; Ortega, V.G. *O Cadastro Ambiental Rural na Amazônia*; Conservação Internacional: Brasília, Brazil, 2013.

99. Telles do Valle, R.S. *CAR Pra Q*; Instituto Sócio-Ambiental: Brasília, Brazil, 2014. Available online: http://www.socioambiental.org/pt-br/blog/blog-do-ppds/car-para-que (accessed on 10 November 2016).

100. Azevedo, A.; Rajão, L.R.; Costa, M.; Stabile, M.C.C.; Alencar, A.; Moutinho, P. Cadastro ambiental rural e sua influência na dinâmica do desmatamento na Amazônia Legal. *Bol. Amazôn. Pauta* **2014**, *3*, 3–16.

101. Börner, J.; Wunder, S.; Reimer, F.; Bakkegaard, R.K.; Viana, V.; Tezza, J.; Pinto, T.; Lima, L.; Marostica, S. *Promoting Forest Stewardship in the Bolsa Floresta Programme: Local Livelihood Strategies and Preliminary Impacts*; Center for International Forestry Research (CIFOR): Rio de Janeiro, Brazil; Fundação Amazonas Sustentável (FAS): Manaus, Brazil; Entwicklungsforschung (ZEF), University of Bonn: Zentrum für Bonn, Germany, 2013; p. 70.

102. Gebara, M.F. Importance of local participation in achieving equity in benefit-sharing mechanisms for REDD+: A case study from the Juma Sustainable Development Reserve. *IJC* **2013**, *7*, 473–497. [CrossRef]

103. Vatn, A.; Kajembe, G.; Leiva-Montoya, R.; Mosi, E.; Nantongo, M.; Silayo, D.A. *Instituting REDD+: An Analysis of the Processes and Outcomes of Two Pilot Projects in Brazil and Tanzania*; IIED: London, UK, 2013.

104. Newton, P.; Nichols, E.S.; Endo, W.; Peres, C.A. Consequences of actor level livelihood heterogeneity for additionality in a tropical forest payment for environmental services programme with an undifferentiated reward structure. *Glob. Environ. Chang.* **2012**, *1*, 127–136. [CrossRef]

105. Pereira, S.N.C. Payment for environmental services in the Amazon Forest: How can conservation and development be reconciled? *JED* **2010**, *19*, 171. [CrossRef]

106. Elfving, M. Payment for Environmental Services: A Tool for Forest Conservation and Empowerment of the Local People in the State of Amazonas, Brazil? A Case Study of Programa Bolsa Floresta. Bachelor's Thesis, Peace and Development Studies, Linnaeus University, Växjö, Sweden, 2010; p. 96.

107. Agustsson, K.; Garibjana, A.; Rojas, E.; Vatn, A. An assessment of the Forest Allowance Programme in the Juma Sustainable Development Reserve in Brazil. *Int. For. Rev.* **2014**, *16*, 87–102. [CrossRef]
108. Duchelle, A.E.; Cromberg, M.; Gebara, M.F.; Guerra, R.; Melo, T.; Larson, A.; Cronkleton, P.; Börner, J.; Sills, E.; Wunder, S.; et al. Linking forest tenure reform, environmental compliance, and incentives: Lessons from REDD+ initiatives in the Brazilian Amazon. *World Dev.* **2014**, *55*, 53–67. [CrossRef]
109. Vedung, E.; van der Doelen, C.J. The sermon: Information programs in the public policy process—Choice, effects, and evaluation. In *Carrots, Sticks and Sermons: Policy Instruments and Their Evaluation*; Bemelmans-Videc, M.L., Rist, R.C., Vedung, E., Eds.; Transaction Publishers: New Brunswick, NJ, USA, 1998; pp. 103–128.
110. Martin, S. *Potenciais Impactos das Alterações do Código Florestal Brasileiro na Meta Nacional de Redução de Emissões de Gases de Efeito Estufa*; Fundação Getulio Vargas, Observatório do Clima: São Paulo, Brazil, 2010.
111. Metzger, J.P. O código florestal tem base científica? *Nat. Conserv.* **2010**, *8*, 92–99. [CrossRef]
112. Barreto, P.; Ellinger, P. *Código Florestal: Como Sair do Impasse?* Imazon: Belém, Brazil, 2011.
113. Telles do Valle, R.S. *Lento Adeus ao Código Florestal*; Instituto Sócio-Ambiental: Brasília, Brazil, 2012. Available online: https://www.socioambiental.org/pt-br/blog/blog-do-isa/lento-adeus-ao-codigo-florestal (accessed on 10 November 2016).
114. Hall, A. Better RED than dead: Paying the people for environmental services in Amazonia. *Philos. Trans. R. Soc. Lond.* **2008**, *363*, 1925–1932. [CrossRef] [PubMed]
115. Bartels, W. *Participatory Land Use Planning in the Brazilian Amazon: Creating Learning Networks among Farmers, Non-Governmental Organizations, and Government Institutions*; School of Natural Resources and Environment, University of Florida: Gainesville, FL, USA, 2009; p. 167.
116. Börner, J.; Wunder, S.; Wertz-Kanounnikoff, S.; Tito, M.R.; Pereira, L.; Nascimento, N. Direct conservation payments in the Brazilian Amazon: Scope and equity implications. *Ecol. Econ.* **2010**, *69*, 1272–1282. [CrossRef]
117. Pokorny, B.; Jong, W.; Godar, J.; Pacheco, P.; Johnson, J. From large to small: Reorienting rural development policies in response to climate change, food security and poverty. *For. Policy Econ.* **2013**, *36*, 52–59. [CrossRef]
118. Equipe de Conservação da Amazônia (ECAM). *Free, Prior and Informed Consent. Surui Carbon Project*; ACT Brazil: Brasilia, Brazil, 2010.
119. Fundo Brasileiro para Biodiversidade (Funbio). *Fundo Paiter Surui: Manual Operacional*; Funbio: Rio de Janeiro, Brazil, 2013.
120. West, T.A.P. Indigenous community benefits from a de-centralized approach to REDD+ in Brazil. *Clim. Policy* **2015**, *1*, 1469–3062. [CrossRef]
121. Gebara, M.F.; Muccillo, L.; May, P.; Vitel, C.; Loft, L.; Santos, A. *Lessons from Local Environmental Funds for REDD+ Benefit Sharing with Indigenous People in Brazil*; Center for International Forestry Research (CIFOR): Bogor, Indonesia, 2014.
122. Lang, C. Leaders of the Paiter Suruí Ask that the Carbon Project with Natura be Terminated. REDD+ Monitor, 2015. Available online: http://www.redd-monitor.org/2015/01/13/leaders-of-the-paiter-surui-ask-that-the-carbon-project-with-natura-be-terminated/ (accessed on 10 November 2016).
123. Marcovitch, J.; Cuzziol Pinsky, V. Amazon Fund: Financing deforestation avoidance. *Rev. Adm.* **2014**, *49*, 2. [CrossRef]
124. Observatório do ABC. Propostas Para Revisão do Plano ABC. Fundação Getulio Vargas: São Paulo, Brazil, 2015. Available online: http://mediadrawer.gvces.com.br/abc/original/gv-agro_em-simples.pdf (accessed on 10 November 2016).
125. Observatório do ABC. Aumento do Desmatamento na Amazônia Legal Está Ligado à Agropecuária. Available online: http://observatorioabc.com.br/2016/12/aumento-do-desmatamento-na-amazonia-legal-esta-ligado-agropecuaria/ (accessed on 24 February 2017).
126. De Koning, J. Unpredictable outcomes in forestry—Governance institutions in practice. *Soc. Nat. Resour.* **2014**, *27*, 358–371. [CrossRef]
127. Olivier de Sardan, J.P. The delivery state in Africa: Interface bureaucrats, professional cultures and the bureaucratic mode of governance. In *States at Work: The Dynamics of African Bureaucracies*; Olivier De Sardan, J.P., Bierschenk, T., Eds.; Bill: Boston, MA, USA, 2014; pp. 399–430.
128. Bourdieu, P. *The Logic of Practice*; Stanford University Press: Stanford, CA, USA, 1990.
129. Angelsen, A.; Brockhaus, M.; Sunderlin, W.D.; Verchot, L.V. (Eds.) *Analysing REDD+: Challenges and Choices*; Center for International Forestry Research (CIFOR): Bogor, Indonesia, 2012.

130. Schiffman, R. Brazil's Deforestation Rates Are on the Rise Again. *Newsweek*. 2015. Available online: http://www.newsweek.com/2015/04/03/brazils-deforestation-rates-are-rise-again-315648.html (accessed on 15 January 2017).

131. Dix, G. Governing by Carrot and Stick. A Genealogy of the Incentive. Ph.D. Thesis, University of Amsterdam, Amsterdam, The Netherlands, 2014. Available online: http://hdl.handle.net/11245/1.417336 (accessed on 10 November 2016).

132. Kuyvenhoven, A. Creating an enabling environment: Policy conditions for less-favored areas. *Food Policy* **2004**, *29*, 407–429. [CrossRef]

133. Nietzsche, F.W. *On the Genealogy of Morals*; Vintage Books: New York, NY, USA, 1989.

134. Nietzsche, F. *Beyond Good and Evil: Prelude to a Philosophy of the Future*; Hollingdale, R.J., Ed.; Penguin Books: London, UK, 1973; p. 257.

135. Lambin, E.F.; Meyfroidt, P. Land use transitions: Socio-ecological feedback versus socio-economic change. *Land Use Policy* **2010**, *27*, 108–118. [CrossRef]

136. Biggs, S. A multiple source of innovation model of agricultural research and technology promotion. *World Dev.* **1990**, *18*, 1481–1499. [CrossRef]

137. Fearnside, P.M. Deforestation in Brazilian Amazonia: History, rates, and consequences. *Conserv. Biol.* **2005**, *19*, 680–688. [CrossRef]

138. Ostrom, E. *Governing the Commons: The Evolution of Institutions for Collective Action*; Cambridge University Press: Cambridge, UK, 1990.

139. Pacheco, P. Agrarian reform in the Brazilian Amazon: Its implications for land distribution and deforestation. *World Dev.* **2009**, *37*, 1337–1347. [CrossRef]

140. Hegel, G.W.F. *Phenomenology of Spirit*; Miller, A.V., Translator; Oxford University Press: Oxford, UK, 1977.

141. Bina, O.; La Camera, F. Promise and shortcomings of a green turn in recent policy responses to the "double crisis". *Ecol. Econ.* **2011**, *70*, 2308–2316. [CrossRef]

142. Pelletier, N. Environmental sustainability as the first principle of distributive justice: Towards an ecological communitarian normative foundation for ecological economics. *Ecol. Econ.* **2010**, *69*, 1887–1894. [CrossRef]

143. Fukuyama, F. *The Social Virtues and the Creation of Prosperity*; Penguin Books: London, UK, 1995.

144. Putman, R. *Making Democracy Work*; Princeton University Press: Princeton, NJ, USA, 1993.

145. Landes, D.S. *The Wealth and Poverty of Nations*; W.W. Norton & Co.: New York, NY, USA, 1998.

146. North, D.C. *Institutions, Institutional Change and Economic Performance*; Cambridge University Press: Cambridge, UK, 1990.

147. Arrow, K.J. *The Limits of Organization*; Norton & Co.: New York, NY, USA, 1974.

148. Guiso, L.; Sapienza, P.; Zingales, L. Does culture affect economic outcome? *J. Econ. Perspect.* **2006**, *20*, 23–48. [CrossRef]

149. Lewis, A. *The Theory of Economic Growth*; George Allen & Unwin: London, UK, 1955.

150. Sen, A.K. *Development as Freedom*; Oxford University Press: Oxford, UK, 1999.

151. Nussbaum, M.C. Human functioning and social justice: In defense of Aristotelian essentialism. *Political Theory* **1992**, *20*, 202–246. [CrossRef]

152. Nussbaum, M. *Women and Human Development: The Capabilities Approach*; Cambridge University Press: Cambridge, UK, 2000.

153. Martins, N.O. Ecosystems, strong sustainability and the classical circular economy. *Ecol. Econ.* **2016**, *129*, 32–39. [CrossRef]

154. Foucault, M. What is critique? In *The Essential Foucault*; The New Press: New York, NU, USA, 1997; pp. 263–278.

155. Foucault, M. The ethics of the concern of the self as a practice of freedom. In *Ethics: Subjectivity and Truth*; The New Press: New York, NY, USA, 1997; pp. 281–301.

156. Batters, S.M. *Care of the Self and the Will to Freedom: Michel Foucault, Critique and Ethics*; Senior Honors Projects. Paper 231; University of Rhode Island: Kingston, RI, USA, 2011.

157. Van Vugt, M. Averting the tragedy of the commons: Using social psychological science to protect the environment. *Curr. Dir. Psychol. Sci.* **2009**, *18*, 169–173. [CrossRef]

158. Gibbs, H.K.; Munger, J.; L'Roe, J.; Barreto, P.; Pereira, R.; Christie, M.; Amaral, T.; Walker, N.F. Did ranchers and slaughterhouses respond to zero-deforestation agreements in the Brazilian Amazon? *Conserv. Lett.* **2016**, *9*, 32–42. [CrossRef]

159. Thaler, R.H.; Sunstein, C. *Nudge: Improving Decisions about Health, Wealth and Happiness*; Penguin Books: New York, NY, USA, 2009.

160. Karki, S.T.; Hubacek, K. Developing a conceptual framework for the attitude–intention– behaviour links driving illegal resource extraction in Bardia National Park, Nepal. *Ecol. Econ.* **2015**, *117*, 129–139. [CrossRef]

161. Frey, E.; Rogers, T. Persistence: How Treatment Effects Persist after Interventions Stop. *Policy Insights Behav. Brain Sci.* **2014**, *1*, 172–179. [CrossRef]

162. Blasch, J.; Ohndorf, M. Altruism, moral norms and social approval: Joint determinants of individual offset behavior. *Ecol. Econ.* **2015**, *116*, 251–260. [CrossRef]

163. Kilgore, M.A.; Snyder, S.A.; Eryilmaz, D.; Markowski-Lindsay, M.A.; Butler, B.R.; Kittredge, D.B.; Catanzaro, P.F.; Hewes, J.H.; Andrejczyk, K. Assessing the relationship between different forms of landowner assistance and family forest owner behaviors and intentions. *J. For.* **2015**, *113*, 12–19. [CrossRef]

164. Bolderdijk, J.W.; Steg, L.; Geller, E.S.; Lehman, P.K.; Postmes, T. Comparing the effectiveness of monetary versus moral motives in environmental campaigning. *Nat. Clim. Chang.* **2013**, *3*, 413–416. [CrossRef]

165. Van der Linden, S. Intrinsic motivation and pro-environmental behavior. *Nat. Clim. Chang.* **2015**, *5*, 612–613. [CrossRef]

166. Clayton, S.; Devine-Wright, P.; Stern, P.C.; Whitmarsh, L.; Amanda Carrico, A.; Steg, L.; Swim, J.; Bonnes, M. Psychological research and global climate change. *Nat. Clim. Chang.* **2015**, *5*, 640–646. [CrossRef]

167. Matulis, B.S. Valuing nature: A reply to Esteve Corbera. *Ecol. Econ.* **2015**, *110*, 158–160. [CrossRef]

168. Hoang, M.; Do, T.; Pham, M.; van Noordwijk, M.; Minang, P.A. Benefit distribution across scales to reduce emissions from deforestation and forest degradation (REDD+) in Vietnam. *Land Use Policy* **2013**, *31*, 48–60. [CrossRef]

169. Maraseni, T.N.; Neupane, P.R.; Lopez-Casero, F.; Cadman, T. An Assessment of the Impacts of the REDD+ Pilot Project on Community Forests User Groups (CFUGs) and their Community Forests in Nepal. *J. Environ. Manag.* **2014**, *136*, 37–46. [CrossRef] [PubMed]

170. Dokken, T.; Putri, A.A.D.; Kweka, D.L. Making REDD+ Work for Communities and Forest Conservation in Tanzania. In *REDD+ on the Ground: A Casebook of REDD+ Initiatives Across the Globe*; Sills, E.O., Atmadja, S.S., De Sassi, C., Duchelle, A.E., Kweka, D.L., Resosudarmo, I.A.P., Sunderlin, W.D., Eds.; Center for International Forestry Research (CIFOR): Bogor, Indonesia, 2014.

171. Leggett, M.; Lovell, H. Community Perceptions of REDD+: A Case Study from Papua New Guinea. *Clim. Policy* **2012**, *12*, 115–134. [CrossRef]

172. Sen, A. Rational fools: A critique of the behavioral foundations of economic theory. *Philos. Public Aff.* **1977**, *6*, 317–344.

173. Haidt, J.; Joseph, C. Intuitive ethics: How innately prepared intuitions generate culturally variable virtues. *Daedalus* **2004**, *133*, 55–66. [CrossRef]

174. Geller, E.S. The challenge of increasing proenvironmental behavior. In *Handbook of Environmental Psychology*; Bechtel, R.B., Churchman, A., Eds.; Wiley: New York, NY, USA, 2002; pp. 525–540.

175. Geller, E.S.; Winett, R.A.; Everett, P.B. *Preserving the Environment: New Strategies for Behavior Change*; Pergamon: Elmsford, NY, USA, 1982.

176. Gärling, T.; Loukopoulos, P. Effectiveness, public acceptance, and political feasibility of coercive measures for reducing car traffic. In *Threats to the Quality of Urban Life from Car Traffic: Problems, Causes, and Solutions*; Gärling, T., Steg, L., Eds.; Elsevier: Amsterdam, The Netherlands, 2007; pp. 313–324.

177. Gärling, T.; Schuitema, G. Travel demand management targeting reduced private car use: Effectiveness, public acceptability and political feasibility. *J. Soc. Issues* **2007**, *63*, 139–153. [CrossRef]

178. Steg, L.; Dreijerink, L.; Abrahamse, W. Why are energy policies acceptable and effective? *Environ. Behav.* **2006**, *38*, 92–111. [CrossRef]

forests

MDPI

Article

Livelihoods and Land Uses in Environmental Policy Approaches: The Case of PES and REDD+ in the Lam Dong Province of Vietnam

Leif Tore Trædal * and Pal Olav Vedeld

Department of International Environment and Development Studies (Noragric), Faculty of Land and Society, Norwegian University of Life Sciences, Ås, N-1432, Norway; pal.vedeld@nmbu.no
* Correspondence: leif.tore.traedal@nmbu.no; Tel.: +47-41-188-175

Academic Editors: Esteve Corbera and Heike Schroeder
Received: 17 September 2016; Accepted: 2 February 2017; Published: 8 February 2017

Abstract: This paper explores assumptions about the drivers of forest cover change in a Payments for Environmental Services (PES) and Reduced Emissions from Deforestation and Degradation (REDD+) context in the Lam Dong Province in Vietnam. In policy discourses, deforestation is often linked to 'poor' and 'ethnic minority' households and their unsustainable practices such as the expansion of coffee production (and other agricultural activities) into forest areas. This paper applies a livelihood framework to discuss the links between livelihoods and land use amongst small-scale farmers in two communities. The findings of the livelihood survey demonstrate no clear linkages between poverty levels and unsustainable practices. In fact, the poorest segments were found to deforest the least. The ways in which current PES and REDD+ approaches are designed, do not provide appropriate solutions to address the underlying dimensions of issues at stake. The paper criticizes one-dimensional perspectives of the drivers behind deforestation and forest degradation often found in public policies and discourses. We suggest more comprehensive analyses of underlying factors encompassing the entire coffee production and land use system in this region. Addressing issues of land tenure and the scarcity of productive lands, and generating viable off-farm income alternatives seem to be crucial. Sustainable approaches for reducing deforestation and degradation could be possible through engaging with multiple stakeholders, including the business-oriented households in control of the coffee trade and of land transactions.

Keywords: Vietnam; drivers of deforestation; livelihoods; environmental policies; REDD+; PES

---

1. Introduction

Tropical deforestation has been identified as one of the main global environmental challenges, contributing to a major share of the global emissions of greenhouse gasses and to the loss of biodiversity and ecological integrity worldwide [1]. At the same time, tropical forests are key arenas for livelihoods and outcomes for indigenous people and forest communities. Over recent years, the process of establishing a global mechanism for Reduced Emissions from Deforestation and Degradation (REDD+) has increased the focus on saving tropical forests. This has prompted a number of tropical forest countries to develop policies and prepare for reducing greenhouse gas (GHG) emissions from forests. However, the lack of progress in global climate change negotiations and in establishing a global REDD+ mechanism has stalled the process and also led to a diversity of approaches and adaptations of REDD+ in many countries [2].

In this paper, we use a case study from Vietnam to explore assumed and real processes driving forest cover change, and scrutinize some of the approaches taken for REDD+. In the Central Highlands of Vietnam, small-scale coffee production has been identified as a major driver of forest cover loss, and it is also frequently associated with other negative environmental and social consequences, such as soil erosion, and economic and political marginalization of ethnic minority groups [3–5]. Various policy schemes to decrease environmental degradation and reduce poverty levels have been developed and implemented. Recently, innovative, results-based policy initiatives such as Payments for Environmental Services (PES) and REDD+ have been introduced to attempt to alleviate some of the problems. A predominant argument in forest-related policy discourses is that poverty and general low knowledge levels among ethnic groups often lead to sub-optimal and unsustainable livelihood decision making [6]. The poverty-deforestation link is not unique to Vietnam, but can be identified in readiness processes of various REDD+ countries, as documented by, for example, Dooley et al. in their review of Readiness Preparedness Proposals (R-PP) submitted to the World Bank Forest Carbon Partnership Facility (FCPF) [7].

In Vietnam, dealing with the (perceived) negative linkages between poverty levels and environmental degradation has frequently been associated with the need for awareness raising and reducing poverty rates among indigenous groups [4]. Such aims are explicitly expressed in national forest-related policy documents and discourses, such as the Lam Dong Province Provincial REDD+ Action Plan ([8], p. 8):

> 'Ethnic minority people in particular, have been carrying out deforestation and converting forest land to settlements and agricultural land to support their traditionally very large families. Awareness raising amongst both male and female members of the community is needed to reduce population growth and the deforestation associated with it.'

This paper uses an empirical case study of the interrelationships between livelihoods and land use to investigate policy assumptions about the drivers of land use change. In Vietnam, the coffee sector, and poor people in particular, are often blamed for deforestation processes, and this sector and group of households are therefore frequently targets of environmental programs, such as PES and REDD+. The study demonstrates the shortcomings of one-dimensional analyses of drivers of land use change in policies that aim to reduce deforestation and degradation. We argue that poor people do not deforest the most, and that targeting the coffee sector will be a challenging task unless the reforms are embedded in wider structures of the coffee industry. The focus on linking the direct drivers of change to small-scale coffee production by poor households seems to mask many of the underlying factors, such as longer-term 'control' and vested interests of state-owned coffee companies and the more wealthy segments of the population in maintaining a particular mode of production. Therefore, REDD+ projects are likely to fail in achieving their goals of reducing deforestation and degradation if they are targeted only at the poor, and omit the structural determinants of the coffee sector at large.

PES has been introduced in the area to cope with smallholder expansion of coffee. Based on the findings, we reflect upon the magnitude and scale of forest conversion caused by the expansion of small-scale livelihood-driven coffee production. We also investigate the social and environmental impacts of the local PES scheme. PES is also likely to become a key component of REDD+ in Lam Dong, and the paper therefore discusses the implications of the findings in terms of practical policy implementation, particularly in view of the ongoing and planned REDD+ pilot activities in the area.

More concretely, the paper explores the following research questions: (1) How do households manage and diversify assets and resources to generate livelihood outcomes? (2) What are the impacts of PES policies on livelihoods, environmental awareness, and deforestation? (3) How do the overall production structures of the coffee economy affect livelihoods and deforestation processes? (4) What are the potential implications for emerging REDD+ policies?

The paper starts by outlining the conceptual models considered relevant for the study. This is followed-up by an introduction of the methodology and case study context. Considering the high degree of market integration and the presence of various policy schemes, analyzing the overall context and dynamics of the coffee sector and policy schemes is crucial for understanding livelihoods and land use in the study area. The final section discusses the results in view of livelihood outcomes and policy implications for REDD+ and PES schemes.

2. Conceptual Framework

This paper investigates households' livelihoods and the implications for land use and deforestation in the study area. These findings are further used to analyze policy assumptions about 'poor' and 'ethnic' households and their alleged unsustainable land-use practices that tend to be prevalent in policy documents and other discourses. Perspectives from livelihood theory, political ecology, and land-use change theory consequently inspire the conceptual framework for the study.

We applied a livelihood framework (LF) to identify differences in livelihood adaptations, and the role of small-scale coffee production in local land-use changes and deforestation. According to the LF, households combine various capitals, such as natural, physical, human, social, and financial capitals, to generate livelihood outcomes in the form of agricultural and forestry outputs, and off-farm and business oriented income [9,10]. Hence, the determinant relationship between the households' asset portfolios, livelihood strategies, and outcomes is at the core of the LF. Institutional factors, such as property regimes, markets, local values, attitudes and norms, skills, and various other social institutions and decisions taken at multiple scales, also affect access to assets and livelihood decision making in households [9,11]. The clearing of forests and expansion of agricultural areas are important components in the livelihood strategies of households [12].

Defining the relevant drivers of deforestation and degradation has been identified as one of the main challenges in developing efficient and effective REDD+ strategies and policies, and is a field that requires further research [13,14]. The land-use change literature frequently differentiates between *proximate* (direct) and *underlying* (indirect) causes of environmental change [15–17]. In our case study area, smallholder coffee production could be viewed as a main *proximate* driver of deforestation. In policy discourses, this has often been linked to livelihood and poverty as *underlying* drivers [8]. Our argument is, however, that there is a need to look beyond the livelihood and poverty dimensions in order to understand the broader issues surrounding coffee production and land-use change processes in the region.

The dynamics of the drivers of deforestation have changed over time and space in different parts of the world. In a meta-analysis of place and time-specific case studies, Rudel et al. detected a general trend from small farmers as a main agent of deforestation in the pre-1990 period, towards more agribusiness related activities in the period after 1990 [18]. In Vietnam the deforestation trends have often been characterized by large-scale deforestation processes of so-called 'slash-and-burn' and state-led deforestation in the 1970s and 1980s, towards more smallholder-oriented expansions of commercial agriculture in certain areas of the country in the 1990s and 2000s [19].

Nevertheless, making wide generalizations about the drivers of environmental degradation often masks the complexities of such processes. Tim Forsyth has termed widely accepted assumptions of the drivers of environmental change as 'environmental orthodoxies' ([20], p. 52). Such orthodoxies have been predominant in Vietnam, frequently obscuring many other underlying factors of change [21]. Notions about ethnic people's unsustainable practices in Vietnam were formerly linked to 'slash-and-burn' agriculture. However, such perceptions are also found in relation to the expansion of smallholder coffee production, as demonstrated by the quotation from the Lam Dong REDD+ Action Plan above. In this sense, the way REDD+ has been conceived in Vietnam risks ignoring important dimensions and actors involved, particularly in the coffee sector and related land-use change processes. Defining forest benefits and compensations should not only relate to the 'indigenous' and the 'poor', but acknowledge the diversity of concerns about, and benefits generated from, forests [22].

3. Methodology

In order to respond to the research objectives and questions presented above, the study adopted a case study approach. The data collection instruments included a household survey and in-depth interviews with farmers, policy makers, researchers, and government officials at national, provincial, and commune levels. The findings were triangulated against other socio-economic and qualitative studies carried out in the region, such as Hoang et al. [6], Tran [23], and Ogonowski and Enright [24].

The following sections describe the case study area and key approaches used in the collection and analyses of data.

3.1. Case Study Area

The fieldwork was carried out in the Lam Dong Province (Figure 1) which has a population of 1,259,300 (2014) [25] and covers an area of 9764.8 km^2 [26]. The province is landlocked, and characterized by upland environments. The average elevation is approximately 1500 meters above sea level. Precipitation is irregular in space and time, ranging from 1600 to about 2700 mm [26]. The variations in landscapes and precipitation contribute to exceptional bio- and ecological diversity [27].

Figure 1. Study sites of Da Nhim and Da Chais, Lam Dong Province, Vietnam, 2016. Maps generated from Google Earth and Wikipedia [28].

Historically, the economy of the province has been based on agriculture. General development trends need to be understood in the light of large-scale immigrations of people in the 1970s and 1980s, both as a result of government policies to inhabit, develop, and control the Central Highlands [29], and of people from the region seeking refuge from war and unrest. More recent immigrations have taken place through people seeking new economic opportunities, mainly related to commercial agriculture, such as the production of perennial crops (mainly coffee, rubber, and tea), vegetables, and flowers. Another factor was the resettlement of various ethnic minority groups that had traditionally practiced shifting agriculture within the region. These were situated in residential areas during the government's large-scale 'fixed agriculture' schemes of the 1970s and 1980s [21]. The combination of the displacement of people, voluntary migrations, and government land confiscations caused losses of rights to ancestral lands for various ethnic groups, and increased conflict levels over access to land and natural resources [30]. In the Lam Dong Province, land-use change and deforestation have been closely associated with the marginalization of ethnic minorities, land confiscations, forced resettlement, migrations of people, and last but not least, the integration of agriculture and local communities into a market economy [27]. Recent studies have also revealed trends of displacement of shifting cultivation for annual crops into forest margins, indirectly driven by the expansion of perennial commercial crops (mainly coffee) on agricultural land, a direct manifestation of the marginalization of ethnic groups in the Central Highlands [5].

Historically, coffee was not an important agricultural crop in Vietnam. It was initially introduced by the French colonialists as a plantation crop but during history, small-scale farmers have also widely adopted it as a cash crop [31]. During the late 1980s and 1990s, coffee production exploded, and within a decade or so, Vietnam became the world's second biggest coffee producer [32]. Key reasons for this dramatic development related to a large extent to the economic and agricultural transformation processes of the market liberalization processes (often referred to as the *doi moi*) that started in the mid-1980s [33]. The shift also related to conscious government policies of improved credit facilities, research and development, general extension activities, tenure reforms, and foreign investment [34]. The government policies have been driven by a wish to commercialize and increase the number of (agricultural) export products in order to stimulate economic growth (Figure 2). Today, coffee is grown in a variety of locations and types of farmers, ranging from small-scale household producers to larger plantations [32]. The commercial coffee production has also spurred the emergence of private businesses locally, linked mainly to coffee trading, fertilizer provision, and resale and general trade in farm supplies [35]. The great influx of people to the Central Highlands to grow coffee have, however, caused tensions and conflicts over productive land between majority Viet (Kinh) people and various ethnic minorities.

Figure 2. "Grow coffee to increase the number of products for exportation"—propaganda poster for promoting coffee production in Vietnam.

The Vietnamese state has also played a very active role in the promotion of the coffee sector in the country, and continues to have an interest in its development and continued growth. In 1995, the National Coffee Corporation (VINACAFE) was established by the government under the auspices of the Ministry of Agriculture, with the mandate to organize trade and develop the coffee sector in the country [36]. Its mandate has been very diverse, ranging from implementing government policies, research, extension services, quality control, and acting as a major commercial actor controlling a major share of the coffee export [31,37].

In the late 1990s, a collapse in global coffee affected the sector negatively, including the coffee producing households and their livelihoods, demonstrating the potential vulnerability of basing livelihoods single-mindedly on producing for a global commodity market. In recent years, fluctuating global prices, high debt levels, and bankruptcies amongst banks and key investors continue to negatively impact the sector, including the economy of VINACAFE [38]. Debt levels have reportedly also increased amongst rural households in coffee-producing zones, including the area in which this study was undertaken [6]. This indicates that the integration of rural livelihoods into a global (coffee) commodity market have caused households to become more vulnerable to commodity price fluctuations. Studies have also revealed big differences between ethnic groups in terms of abilities to adapt and cope with price fluctuations, with the Kinh demonstrating higher capacities to diversify livelihoods and income sources as compared to the minority groups [39].

The agricultural system related to coffee production in Lam Dong is characterized by the *homeland* production. ('Homeland production' here does not refer to home gardens, but rather to agricultural fields close to the house. These are normally areas of land that are denominated and certified for agricultural use of coffee in combination with persimmon and vegetables), and the *hillside* production of coffee, sometimes mixed with persimmon and maize for subsistence. The homeland areas are certified agricultural land, while the hillside coffee plots are farther away and are often the result of illegal clearing of forests. Today, coffee production is considered to be a major driver of land-use change, causing the loss and degradation of large forest areas [4,5,40]. By reviewing the production statistics of Lam Dong in the 2002–2011 period, Tan [3] demonstrated that the expansion of coffee production in Lam Dong do seem to correlate with a rise in the price of coffee.

In order to reduce environmental degradation and enhance social development in the rural areas, a number of forest policy reforms and plans have been designed and implemented since the early 1990s. These include the '327' and the '661' programs (the 661 program is also known as the Five Million Hectares Reforestation Program). These programs had the twofold objectives of conserving the remaining forests and expanding tree cover on the 'barren' lands by means of large-scale replanting of trees [41]. In the Lam Dong area, the extent of forest plantations has increased from about 27,000 ha in 1999 to almost 66,000 ha in 2011 [42]. However, official data records show that in recent years in the overall region of the Central Highlands, the negative trend of deforestation has continued.

Following the Ministry of Agriculture's forest land classification system, forests are classified as *special use*, *conservation*, or *production* forests [41]. A key approach to reducing deforestation has been to establish national parks categorized as *special use* forests and large areas categorized as *protection* forests. The latter category has the purpose of conserving key environmental functions and services, such as water provision, energy production, soil protection, and protection against extreme weather events. *Production* forests are intended to be plantations that allow the production and sale of timber and non-timber forest products, in combination with environmental protection practices [43]. In Lam Dong, the relative proportion of *protection* and *special use* forests is quite high compared to *production* forests. There are two large national parks in Lam Dong (Cat Tien and Bidoup-Nui Ba), both of which are categorized as *special use* forests. The forest classification and tenure regimes provide guidelines for what types of activities are allowed on forest land, representing a rather strict system for controlling activities and resources. This, in turn, has important implications for access to land, land use, and livelihood outcomes.

Forest tenure reforms and the distribution of forest land to individual households have been carried out in Vietnam to varying extents. In Lam Dong, state entities such as the State Forest Enterprises, Forest Management Boards, and People's Committees still control most forest land. Only 1% of forest land is managed by individual households. On the other hand, many households in the area have agricultural land certificates (*red books*) to their homeland areas [4].

3.1.1. PES in Lam Dong

Since 2008, Lam Dong has been a pilot province for the national PES program [44]. Even though the prospects for successful PES implementation were described as meagre [45], the Vietnam PES 'experiment' is today often regarded as a successful case, and frequently used as a showcase for other countries in the Mekong Region and Southeast Asia (see, for example, [46]). This is mainly due to its scale and ability to generate substantial amounts of funds.

The pilot scheme for PES was established in Lam Dong and Son La provinces already in 2008 [47], and in 2010 it was scaled up as a nation-wide policy [48]. The Vietnam Forest Sector Development Fund (VNFF) is responsible for channeling, managing, and coordinating the PES funds. During the piloting phase in Lam Dong, 9870 households were included in the scheme covering nearly 210,000 ha of forest [49].

The poverty reduction objectives of PES (and REDD+) are emphasized in various policy documents, and must be understood within the general objectives of the 2006–2020 Vietnam Forestry Development Strategy [50]. This strategy identifies poverty reduction and socio-economic development as key objectives of the forest sector. On the other hand, PES in Vietnam has been criticized for elite capture of financial resources and for failing to target the underlying causes of environmental degradation, such as contested land rights and the general lack of participation and involvement in resource management [51,52].

The critique of PES also includes a general lack of clear linkages between payments and performance [4,53]. In Lam Dong, the PES setup implies a *forest leaser model* in which households, either individually or collectively, are hired by the state forest owners to look after forest land. This work is organized in so-called 'forest protection groups' that are collectively responsible for monitoring and controlling forest land. In this sense, the contracts and duties resemble labor contracts more than anything else, and disbursements are based on participation, rather than performance. The performance component of PES was reportedly rejected in Lam Dong due to resistance and lack of understanding as to why some households might receive higher payments than others [53]. With the implementation of REDD+, however, it is planned that the performance component of forest protection will be reintroduced in Lam Dong [8].

3.1.2. REDD+ in Lam Dong

Vietnam was approved as a UN REDD Programme country in 2008, and was later granted a second phase of support. In addition, the country has received readiness support from the World Bank Forest Carbon Partnership Facility (FCPF). During the first phase of REDD+, activities were limited to testing approaches for participatory carbon monitoring, and free, prior and informed consent (FPIC). In UN REDD Phase II, five more provinces were added, all of which have developed provincial REDD+ action plans (PRAPs) that detail funding streams and more concrete activities to be tested for relieving pressure on—and increasing—forest areas in Lam Dong. In total, over the five years of implementation, the PRAP has a total budget of 1,749,275 million VND, equivalent to about 83.3 million USD [8].

In anticipation of a global results-based REDD+ mechanism, the REDD+ Action Plan for Lam Dong is meant to be a coordination mechanism for various potential funding sources that are relevant in reducing GHG emissions from forests, and increasing forest carbon stocks [8]. Potential funding sources include Official Development Assistance, PES, REDD+ projects and programs (for example, by NGOs), and state-funded budgets. REDD+ will largely build on the institutional structures that were developed through PES [54]. Support for forest protection will continue to be channeled through the VNFF and carried out through the same system of forest protection groups. The results-based component that was 'lost' in PES will be reintroduced through REDD+. At the outset, the results-based component was meant to be linked to an extensive participatory carbon monitoring component. This aspect has, however, over time been toned down, and replaced by a conventional remote sensing data approach along the United Nations Framework Convention on Climate Change (UNFCCC) guidelines. This also forms the basis for Vietnam's application for support through the FCPF Carbon Fund [55]. Here the emissions reduction and removal potential for the six pilot provinces over the 2016–2020 period is estimated to be 20.66 $MtCO_2$, equivalent of more than 100 million USD taking a price of 5 USD per tCO_2 into account.

The PRAP identifies small-scale agriculture, in part linked to ethnic minorities 'unsustainable practices', as a main driver of deforestation in the Lam Dong province. On the other hand, national level studies of the drivers of deforestation and degradation tend to put more emphasis on the links between coffee expansion and the forces and elasticities of the global coffee commodity market [19]. The apparent void here between the national broad sweeping analyses of the drivers of change, and the provincial and local level approaches of livelihoods improvement and diversification is striking. The PRAP's REDD+ activities include components of training of farmers in agricultural techniques (with a focus on intensifying production in the form of more coffee per unit of land), and establishing village development funds to provide farmers with favorable loans in order to stimulate the establishment of alternative livelihoods [8]. In addition, it includes components of increasing the economic value of agriculture, cultivating multipurpose tree species that can contribute to diversifying income, planting more trees (forest rehabilitation) (similar to the 661 program approaches, except that tree species should be indigenous), and promoting livestock as a way of diversifying household economies. [8]. Nevertheless, how—and to what degree—REDD+ aims at targeting the underlying structures of the coffee sector at large remains unclear in current plans and approaches.

3.2. Study Sample

For the survey, we selected the two communes Da Nhim and Da Chais in the Lac Duong District of Lam Dong, with 2009 populations of 3347 and 1339, respectively [23]. A total of four villages were randomly selected from within the communes, and 25 households were interviewed in each village. This constituted a total sample of 100 households of the 915 households in the two communes. The ethnic composition of households is similar in the two communes (about 85% *K'ho*), and thus the *K'ho* constituted 85% of our total sample. The remaining households were *Kinh* (14%) and *Tay* (1%). With regard to socio-economic status and land use (in particular agricultural expansion), the Lac Duong District contains some of the more inaccessible areas in the province, as well as a high percentage of ethnic minorities [27]. In this sense, the communes are representative of the economically more marginal—but forest rich—areas of the province. In both of the communes, forest areas are predominantly of *protection* type (according to the Ministry of Agriculture forest classification system). In addition, in Da Nhim there are certain areas taken out of the official forest classification system that are categorized as *unclassified forests* that potentially could be used for 'planned deforestation' activities (including for hydropower, mining, ecotourism, and agriculture) [56].

The two communes were considered to be representative of the PES household population in the province, as they form key areas for PES and REDD+ policy implementation within the Da Nhim watershed. In this context, the Da Nhim hydropower station is the customer, buying environmental 'services' (water for energy production) from the forest owners (the Da Nhim Forest Management Board (DNFMB) and the Bidoup-Nui Ba National Park (BNBNP)) for managing the upstream forests. The forest owners in turn contract households to conserve and manage forests sustainably in order to provide water to the hydropower station. Households are trained and paid to carry out community patrolling of the forests. The approach and implications for the households in terms of training, duties, and remunerations are similar for the contractual arrangements with the DNFMB and the BNBNP. According to the information we received at the commune and province levels, households are selected based on ethnicity (minorities are prioritized) and income (the poor are prioritized). Seventy-nine of the households in the sample participated in PES, constituting 18% of the 450 households targeted through PES schemes in the area [6].

The REDD+ pilot activities of the Lam Dong Provincial REDD+ Action Plan started in the area only after we had conducted our survey. Our reflections around REDD+ impacts and implications are therefore based on reviews of policy documents and in-depth interviews that we carried out before the REDD+ activities were implemented in the communes and after the survey.

3.3. Data Analyses

3.3.1. Investigating Livelihoods

We investigated livelihood assets, activities, and different sources of income (including agriculture and various off-farm activities) from different households. This provided information about differences in land use. In order to further investigate the factors that determine income, we ran a multiple regression model of total income against various socio-economic assets [10]. We assumed that income is dependent on financial, physical, social, human, and cultural capitals. Households were also categorized into three equal-sized income groups ('poor', 'medium', and 'better off') based on yearly income in order to investigate differences in livelihood income sources, priorities, and decision making in relation to land use. Here we used pre-PES income levels (that is, total net income without PES) as the basis for the income categorization. This was done in order to capture the poverty dimensions predicted in policy documents regarding the manner in which households in the area were selected for PES (categorizing the households based on total net income, including PES, gave a different distribution of households as compared to without PES. When running the total income model, 16 households were categorized differently as compared to without PES). The mapping of livelihoods and land-use practices feeds into the study of impacts of PES and REDD+ policies in the area.

3.3.2. Measuring the Drivers of Changes in Forest Cover

Micro-studies of land use and decision making at the household level tend to be site specific, and it is often challenging to extrapolate results in space and over time [16,57]. However, we maintain that context-specific studies can introduce nuances in perceptions and orthodoxies about the drivers of land-use change that are widely accepted and often taken for granted, particularly in policies and measures that deal with environment and development issues.

In order to avoid potential biases of data, self-reporting on land use and land use change was consistently avoided in the survey. We did not ask people directly about whether they had cleared land or not. We calculated land-use changes on the respective households' land by combining in-field measures and observations. In total, 181 agricultural plots were walked and measured with GPS in the two communes. The data were stored as gpx-files and analyzed through various online tools, including Google Earth Pro and Landsat images developed by Hansen et al. [58] at a resolution of 30 m. This gave us an overview of changes in each measured land plot over the 2000 to 2014 period. In addition, feeding these data into GIS ArcMap base maps provided by the Lowering Emissions in Asia's Forests (LEAF) project gave us information about the types of forest categories in the various locations.

There are, of course, methodological challenges in combining time series data of this kind, with cross-sectional household income data, since household socio-economic factors change over time. We proceeded to use these data, with some caution, but we believe it is valuable in developing an indicative estimate of the average rate of forest-cover loss due to small-scale livelihood activities related to coffee production. Combining it with in-depth interview data on the coffee production system helps to contextualize policy discourses around the drivers of deforestation and environmental degradation.

3.3.3. Investigating Policy Effects and Implications

The study investigated the effects of PES payments on livelihood and land use within various household categories. We assumed that the poorest households had been prioritized for PES and that this would have contributed to reducing income inequality in the communes. We used the PES proportion of total income as a measure of its importance in terms of poverty reduction. In order to consider to what degree different land use and investment rationales might be attributed to the social and cultural status of households, we measured areas cultivated, input investments made, and yields per unit of land, and how these variables relate to income levels. The different ways in which various households have access to and manage land provided insights into the degree to which links between income levels and pressure on forests may be substantiated. This helped us to reflect upon the relevance and prospects of success for ongoing and planned REDD+ activities in the area.

Key strategies of PES and REDD+ are also to improve citizen environmental 'knowledge' and 'awareness'. Measures to accomplish this include information campaigns, stakeholder meetings, and the use of media (television, radio, newspapers) [23]. For measuring changes in environmental attitudes, we developed a composite indicator. This indicator combines six different questions regarding awareness and perceptions of the value of forests and the conservation of forests. The responses were valued along a Likert type scale, (1–5 scale, 1 implying low awareness, 5 high). The responses were merged into an attitude awareness indicator where the potential score is between 6 and 30.

3.3.4. Misrepresentations of Coffee Income—A Methodological Challenge

A particular challenge we faced was that about 25% of the households reported net negative agricultural income. This may be or lead to a source of misrepresentation in the wealth ranking of households when the estimate of annual income is based on one particular year. In many cases, households will incur high initial, investment costs in establishing new coffee plantations which do not yield any crop income for three to five years [59]. The survey results show that many households established new plantations in 2010/2011, implying high input costs in terms of fertilizers and pesticides, combined with very low income (if any), resulting in negative agricultural income in the survey year (2014). The new plantation efforts in 2010/2011 may be explained by favorable *robusta* coffee market price trends, and expansive population trends, especially in Da Chais. Similar results have been found in other studies carried out in the area, such as the one by Tran [23]. We therefore believe that the negative income can be attributed to the "decoupled" or "disjointed" nature of costs and income distribution that accrued differently for various households over the time leading up to the survey.

In order to adjust this misrepresentation to a 'smoother', more longer-term income measure, we calculated a mean income value per ha, based on the mean gross value of coffee production of the households that produce coffee on their plots. We found that the mean gross coffee production was 3664 kg per ha (on a total area of 82 ha across the sample). We also found mean net investment to be just over 17.0 million VND per year, which conforms well with the national cost figures presented by Thang et al. [59] of an average of 16.9 million VND per year (approximately 846 USD). Based on this, we calculated the average net income per ha to be about 4.8 million VND per year.

We used these figures as an indicator of the net coffee income by household, multiplied by the land available per individual household. In this paper, we refer to this measure as the 'coffee index-adjusted net income'. We then used this measure to categorize households into three, equal-sized income groups. The income adjustment caused 25 households to change income groups, in most cases moving one level either upwards or downwards. In order to maintain transparency in the analyses, the descriptive data (in Table 1) contain both the mean 'indexed' and 'actual' income values. We retain, however, the 'index adjusted income' in the statistical analyses (involving income groups), because we believe that it gives a more realistic picture of wealth and income levels amongst households in the communes in the study area.

Table 1. Socio-economic assets and income sources, Da Nhim and Da Chais communes, Lam Dong Province, Vietnam, 2014 ($N = 100$).

Variable	Mean	Std Dev	Min	Max
Total net income (1000 VND)	48,480.72	92,810.94	−24000	774,000.00
Index-adjusted net income (1000 VND)	47,684.75	91,823.96	2403.3	789,004.27
Off-farm income (1000 VND)	30,737.50	94,218.92	0	780,000.00
Indexed-adjusted agricultural income (1000 VND)	6594.40	7232.19	0	50,388.32
Livestock income (1000 VND)	221.60	2649.39	−1000.00	26,400.00

Table 1. *Cont.*

Variable	Mean	Std Dev	Min	Max
Net income: fish (1000 VND)	129.00	831.37	0	6500.00
Net income: forests (1000 VND)	1952.97	3603.71	0	25,948.00
PES income (1000 VND)	7432.00	4966.21	0	16,400.00
Other environmental schemes (1000 VND)	617.28	2746.17	0	15,600.00
Household size	5.45	2.75	2	23
Age of household head	43.66	12.71	23	76
People available to engage in adult labor	3.23	1.98	1	13
Years of education of household head	4.27	4.10	0	16
Total value of assets (1000 VND)	144,606.30	339,807.40	5000.00	3,144,000.00
Debt (1000 VND)	34,170.00	111,399.00	0	1,000,000.00
Debt:income ratio	1.23	1.25	0	15.26
Total cultivated area (ha)	1.14	0.87	0	4.51
Total area of homeland (ha)	0.35	0.43	0	1.97
Area of coffee production (ha)	0.83	0.74	0	4.22
Gross coffee production (kg)	2887.17	3911.84	0	300,000.00
Input investments per ha (1000 VND)	15,761.77	14,734.69	0	65,502.18
Coffee production (kg) per ha	3423.30	3110.30	0	12,455.52
Average distance to land plots (km)	1.68	2.17	0	10.43
Area of land cleared (ha) (2000–2014)	0.35	0.43	0	2.04
Area of uncertified land cleared (2000–2014)	0.28	0.42	0	2.04

USD1 = VND 21,000. PES, Payments for Environmental Services.

4. Results and Discussion

The objectives of this paper are threefold. First, we present general household characteristics and income levels, followed by an analysis of livelihood activities and outcomes per income group. The results mainly reflect the findings from the household survey, but in-depth interviews and secondary sources also inform the discussion—particularly in relation to understanding the production structure of the local coffee economy. The results section feeds into a discussion about how households manage resources differently in order to generate their livelihoods. Second, we measured the relative importance of PES income in household livelihoods, and the degree to which PES participation can be linked to differences between income groups, in terms of income and land use. Third, we explored the coffee economy and land use by investigating the relative importance of coffee production in household economies, and the links between household affluence and the expansion of coffee production into forest land. This discussion also reports on the general organization of the coffee economy in the area. Finally, this section ends with a discussion of future prospects for REDD+ in view of the findings of the study.

4.1. Household Characteristics and Income Levels

The average household size was 5.5 members. The average age of the household head was 43.7 years, with an average level of education of 4.3 years (see Table 1). The level of poverty in the area is high. The mean net income for all households was found to be approximately 48 million VND per annum (Table 1), equaling a mean net income per capita of 8.7 million VND per annum. This equals about 1.1 USD per person per day, i.e. a very low income level as compared with global poverty level income standards of 1 USD per person per day. Nevertheless, seen in a national context, this income level is well above the national poverty rate for rural areas of 4.8 million VND per person per year [60]. Looking at the defined income groups for the study, the income levels again appear to be very low. Both the 'poor' (2.7 million VND per person per year) and the 'medium' (4.4 million VND per person per year) are below the national poverty rate for rural areas. Nevertheless, the registered low income levels correspond with other income-related studies from the area, such as Tran [23]. There are also substantial income inequalities within the communes. The variations in net income levels demonstrate a range from about 1 to 798 million VND per household per year. The estimated Gini coefficient of the communes (0.56) shows that this is well above the national average of 0.39.

The main income source in the study area is off-farm activities (64.4%), followed by PES activities (15.6%), and agriculture (13.8%). Forestry (beyond PES) and livestock-related income are small (4.1% and 0.4%, respectively). Forest related income sources are subsistence oriented and predominantly involve collecting fuelwood. The main limiting factor for agricultural production in the area is the extent of available land [23], which partly explains the high degree of off-farm reliance amongst the sampled households. The households reported an average of 1.14 ha of cultivable land each, of which about 0.82 ha (73%) is used for coffee.

We ran an ordinary least squares (OLS) analysis to test the causal relationship between access to assets and total income ($R^2 = 0.4932$; $F = 4.8998$; $p < 0.0001$) (Table 2). We found income to be positively correlated with the level of education of the household head, distance to land plots in general (regardless of whether it was 'homeland' or 'hillside'), and the total area of certified agricultural land (homeland). We found a negative correlation between net income and the total cultivated area, indicating that the higher net income the less land households cultivate. On the other hand, and contrary to policy assumptions about the 'poor' and 'ethnic', the OLS indicates a positive correlation between income levels and clearing of land, namely that the households with higher income had cleared more land in the 2000–2014 period. Household participation in PES also correlated negatively with income. This is as expected, considering the expressed priority to include poorer households in PES.

Table 2. Total income by socio-economic assets, Dha Nhim and Da Chais communes, Lam Dong Province, Vietnam, 2014 ($N = 100$).

Term	Correlation Estimate	Std Error	t Ratio	Prob > |t|
Intercept	3221.43	40,867.85	0.08	0.9374
Village 1	−24,876.99	17,209.08	−1.45	0.1521
Village 2	−31,809.43	14,681.07	−2.17	0.0332 **
Village 3	3572.64	13,956.88	0.26	0.7986
Ethnicity (dummy = K'ho ethnicity)	8629.39	16,967.99	0.51	0.6124
Size of household	5657.58	5030.02	1.12	0.2640
Sex of household head (dummy = male)	−39,948.57	14,859.90	−2.69	0.0087 ***
Age of household head	469.95	677.98	0.69	0.4902
Adult labor	−5471.05	7347.30	−0.74	0.4586
Years of education of household head	6738.28	2456.36	2.74	0.0075 ***
Collective work Y/N (dummy = yes)	−6407.06	8595.69	−0.75	0.4582
Total value of assets (1000 VND)	0.01	0.03	0.44	0.6631
Debt (1000 VND)	−0.05	0.08	−0.64	0.5210
Total cultivated area (ha)	−39,122.85	13,779.20	−2.84	0.0057 ***
Total area of homeland (ha)	72,656.29	24,393.56	2.98	0.0038 ***
Average distance to land plots (km)	9231.01	3211.01	2.87	0.0051 ***
PES participation (Y/N) (dummy = yes)	−28,299.79	14,632.62	−1.93	0.0566 *
Area of land cleared (2000–2014)	91,889.36	27,096.12	3.39	0.0011 ***

$R^2 = 0.4979$; $F = 4.8098$; $p < 0.0001$; p-values estimate significance for differences between household assets: *** is significant at $p < 0.01$; ** is significant at $p < 0.05$; * is significant at $p < 0.1$.

Surprisingly, the OLS indicates that female-headed households generate significantly more income than male-headed ones (Table 2). However, this result is due to an outlier observation in the sample. The richest household in the sample was female headed, with about 798 million VND in total net income per annum. Omitting the outlier and re-running the test eliminated the correlation between gender and income. This also eliminated the correlation between income and location (village) (namely that location influences income levels). Since the outlier household was also ethnic minority (*K'ho*), the omission yielded a positive correlation between ethnic affiliation and income (the *Kinh* earning on average more than the *K'ho*).

In summary, the descriptive data indicate that income levels are low, with substantial differences between the wealthiest and the poorest households. The agricultural and forest-related income levels are also low. According to official statistics from 2014, the provincial mean income from agriculture, forestry, and fishery was 982,000 VND per capita per month [61]. By comparison, the survey data demonstrate an average per capita income from the agriculture and forestry of about 130,000 VND per household.

Breaking the income data down according to income levels—'poor', 'medium', and 'better off'—provides more detailed insights into livelihoods and land-use dynamics for the various income groups. The results are presented in Table 3. The 'better off' households reported significantly more assets and resources than did the 'poor' and 'medium' ones. The level of education of the household head is also on average higher for the 'better off' than for the other two income groups. The differences in average household size and the average age of the household heads were statistically insignificant.

Table 3. Socio-economic characteristics, livelihood activities, and outcomes by income groups (N = 100), Da Nhim and Da Chais Communes, Lam Dong Province, Vietnam, 2014.

Variable	Poor	Medium	Better off	p-Value
Household size	4.88	5.64	5.85	0.1276
Ethnic minorities (%)	91.20	100.00	63.60	0.0004
Age of household head	44.00	42.18	44.79	0.8030
Years of education of household head	3.53 [b]	3.15 [b]	6.15 [a]	0.0066
Total value of assets (1000 VND)	80,080.88 [b]	77,465.76 [b]	278,227.70 [a]	<0.0001 ***
Debt (1000 VND)	21,617.65	19,212.12	62,060.61	0.0866 *
Net income (coffee index-adjusted) (1000 VND)	13,081.92 [b]	24,803.05 [b]	106,217.86 [a]	<0.0001 ***
- Off-farm income (1000 VND)	333.82 [b]	4155.76 [b]	88,644.24 [a]	<0.0001 ***
- Paid work (1000 VND)	108.82 [b]	3422.42 [b]	78,352.73 [a]	<0.0001 ***
- Other business (1000 VND)	11.76 [b]	0 [b]	10,012.12 [a]	<0.0001 ***
- Income transfers (1000 VND)	213.24	733.33	279.39	0.8954
- Indexed-adjusted agricultural income (1000 VND)	2925.35 [b]	8611.93 [a]	8357.09 [a]	0.0007 ***
- Indexed-adjusted coffee income (1000 VND)	2294.47 [b]	5090.87 [a]	4501.64 [a]	0.0072 ***
- Agricultural subsistence income (1000 VND)	630.88 [b]	3521.06 [a]	3855.45 [a]	0.0153 **
- Forest related income	1296.27	2525.42	2057.13	0.3961
- PES income (1000 VND)	8529.41 [a]	9018.18 [a]	4715.15 [b]	0.0011 ***
Total cultivated area (ha)	0.70 [b]	1.47 [a]	1.27 [a]	0.0059 ***
Total area of homeland (ha)	0.26	0.35	0.44	0.0878 *
Area coffee production (ha)	0.48 [b]	1.06 [a]	0.94 [a]	0.0072 ***
Gross coffee production (kg)	1891.18 [b]	2757.58 [a,b]	4042.94 [a]	0.0123 **
Input investments per ha (1000 VND)	17,880.20	12,783.43	16,557.50	0.6793
Coffee production (kg) per ha	3836.37	2859.65	3561.37	0.7006
Average distance to land plots (km)	1.21	2.41	1.83	0.3202
Area cleared for agriculture (2000–2014)	0.23	0.42	0.40	0.0956 *
Area of uncertified land cleared (2000–2014)	0.11 [b]	0.34 [a]	0.40 [a]	0.0033 ***
PES participation (%)	88.2 [a]	100 [a]	63.6 [b]	<0.0001 ***

p-values estimate significance for differences between income groups: *** is significant at $p < 0.01$; ** is significant at $p < 0.05$; * is significant at $p < 0.1$; [a-c] Bonferroni test; groups with different letters are significantly different from each other ($p < 0.05$); USD 1 = VND 21,000.

4.2. PES Income and Land Use

There are large variations in the share of income by activity for the different income groups. The 'better off' have on average much higher off-farm income (83.0% of their total income and only 7.0% from agriculture) than the other two groups—the 'poor' and the 'medium' earn about 2.5% and 16.8%, respectively, off-farm. Off-farm income is derived from paid work, business activities, and remittances. In relation to 'paid work', the types of jobs vary—the 'poor' and 'medium' tend to be hired frequently for agricultural work on the land plots of other households' (such as for commercial flower and vegetable production, and coffee production), while the 'better off' work in the commune administration, as teachers, or in the tourism sector. Business-oriented activities are carried out almost exclusively by the 'better off' families. All business households in the sample were ethnic *Kinh*.

PES income are relatively high in the area. The findings from the survey data regarding linkages between PES and performance objectives are still mixed. In the surroundings of the Bidoup-Nui Ba National Park, anecdotal evidence shows that the number of 'violations' has decreased since the start-up of PES [21,23]. The positive correlation between income and land clearing found in the OLS may also demonstrate a link between the level of PES payments and households' expansion of agriculture into forest land. When we compared PES payments across income groups, we found that the 'poor' and 'medium' households receive the highest PES payments. This indicates that in terms of the poverty reduction objective of PES, PES has contributed to reducing income inequalities in the area. In other words, PES is relatively speaking more important for the 'poor' households than for the 'medium' or the 'better off' ones (respectively 65.2%, 38.7%, and 4.4% of total net income). Nevertheless, the 'medium' and the 'better off' households cultivate more land for coffee on average, and in this sense, PES has excluded an important segment of households that are more capable and more likely to expand coffee production.

A simple binary comparison of households in terms environmental knowledge and awareness also showed that there were insignificant differences between the PES (22.8) vs. non-PES beneficiaries (22.3) households (p = 0.5237). This partly also questions the positive effects of the knowledge and awareness related activities carried out as a part of the PES scheme.

4.3. Livelihoods, Land Use, and Coffee Production

The land-clearing data indicate that the households in Da Nhim cleared more land on average than in Da Chais in the 2000–2014 period (0.24 ha more on average). Much of the forest land clearing in Da Nhim is most likely attributable to the commune's distribution of forest land for coffee plantations. Here the authorities have deliberately chosen to distribute forest land certificates for coffee plantations to relieve the pressure for improved livelihoods, to provide more land, and to encourage the expansion of coffee production towards areas perceived as less vulnerable in terms of the provision of ecological services. If we subtract the certified legal land clearings organized by the authorities, this gives us an indicative measure of the level of what is officially perceived as illegal land clearings in the area (hereafter referred to as 'uncertified land clearing'). The average area per household across the two communes is 0.28 ha (Table 1), or 0.33 and 0.23 ha in Da Nhim and Da Chais, respectively. This may indicate that access to land is an even more pertinent political issue in Da Nhim than in Da Chais. Considering land use in relation to the various income groups, the 'medium' and the 'better off' households access and cultivate more land on average than the 'poor' (1.47 and 1.27 hectares vs. 0.70 hectares, respectively). On average, 72.8% of the cultivated land is used for coffee production. The 'medium' households cultivate the most land, both in total and for coffee production. The 'poor' households have less land than the other income groups. A modest significant difference in terms of land clearing in the 2000–2014 period was also detected between the groups (Table 3); that is, the 'poor' clear slightly less land than the other groups. Interestingly, however, if we compare income groups in terms of the level of uncertified land clearings in the 2000–2014 period, this is significantly lower for the 'poor' and highest for the 'better off' (though not significantly different from the 'medium'). No significant differences were found in input investments or in productivity between the groups. Further, in terms of output per input investment (that is kg of coffee per 1000 VND), there were no big differences between the groups (i.e., 0.21 for the 'poor', 0.22 for the 'medium', and 0.22 kg/1000 VND for the 'better off').

The data collected during the in-depth interviews indicate that illegal clearing of forest land and land transactions are more complex processes than revealed through the survey data. Land clearing and land transactions often take place in a step-wise process, where the poorer households are paid by the richer and more business-oriented households to clear new land. Thereafter, the land is sold to the other coffee producing households. This was also observed in other studies, such as Vu Tien Dien and Grais [40] and the Netherlands Development Organization (SNV) [62]. Such transactions most likely represent a substantial (illegal) off-farm income source for some of the 'business households' in the sample, a finding which was not well captured through the survey data. The numbers may have been reported as 'business activities', but they were not revealed as income from land transactions.

Business activities are carried out almost exclusively by the 'better off' households, and all of them are ethnic *Kinh*. They frequently own and run small or medium-sized businesses in the communes, typically shops where they sell food or agricultural articles, sometimes combined with cafés or restaurants. These households also provide other farmers with key agricultural inputs, and act as intermediaries for the sale of coffee. In these transactions, inputs are frequently sold on credit and the debt is then repaid in installments and interest in the form of a supply of coffee beans. In such cases, the debtors typically receive a lower price for their coffee compared to selling it directly in the markets (approximately 20% lower than the market price) [6]. This often results in the poorest households being caught in a vicious circle of debt and payments with inflated interest rates.

From a vulnerability point of view, concerns have been raised about the debt situation related to coffee production in the area [6]. The survey findings in this study demonstrate that the household debt levels are indeed high. Households have on average about 34 million VND in debt—some 70.5% of their total annual income on average (Table 3). Even though the differences in debt levels between households are statistically insignificant, the debt-to-income ratios for the 'poor' are much worse than for the 'medium' and 'better off' households (165% versus 77% and 58%, respectively). The levels of interest rates on debt repayments are similar between the different groups.

The better-off group hence has a stake in sustaining the current mode of production in the area, which may represent a major underlying factor for land-use change and forest encroachment in the area. Coffee production should hence be understood within a broader political, economic, and historic context, in which small-scale coffee producing farmers are part of a complex coffee economy, influenced locally by the business-oriented segment of the population that trades coffee, key inputs, and land. This system has been traditionally supported by the state, not only through input support and extension services, but also directly through the state-owned VINACAFE. This is the largest coffee trading and manufacturing company in Vietnam, accounting for 20% to 25% of the country's coffee exports.

4.4. Implications and Prospects for REDD+

While we agree in principle with the prominent role of the coffee sector as a driver of land-use change in the province, we challenge some of the predominant perceptions linked to the underlying poverty and livelihood-related explanation models found in key policy documents and discourses.

First, the findings question the role of smallholder coffee production in relation to deforestation in the study area. Considering the rate of land conversion evidenced through the survey gives reason to question the relative role of smallholder production in the overall picture of deforestation in the province. The land-use change data for the household plots in the 2000–2014 period indicate an increase in agricultural land of 0.35 ha (constituting about 31% of the total amount of agricultural land) (Table 1). This amounts to about 0.025 ha per household per year over this period. A simple calculation of the average total annual forest loss due to livelihood activities in the two communes yields an answer of 22.9 ha. With total forest cover of the two communes of 53,546 ha, the average loss due to household livelihood activities should be around 0.043% per year, which is well below the provincial average (about 0.5 ha per annum according to official figures from the Forest Inventory and Planning Institute [42]). These results are of course indicative figures, but considering that they are based on the current population of the communes, and since the population trend of the province has been increasing in recent years, the figures are probably not underestimating deforestation. It should also be noted that there are most likely differences in land-use practices and needs between locations in different contexts. The focus in REDD+ should hence also be on contexts where larger-scale plantation mode of coffee production is taking place.

Second, PES experiences should critically be reviewed and modified in REDD+. The way REDD+ has evolved is, to a large extent, symptomatic of how domestic REDD+ policies have evolved from global approaches and ideas, and in this case from an idea about results-based PES to more pragmatic national and local adaptations of broader policies and measures [2]. REDD+ in Vietnam in general, and Lam Dong in particular, is by-and-large viewed as a coordination mechanism for ongoing and future activities that may contribute to the reduction of GHG from forests. Hence, we foresee that the current PES setup will play a key role as a distribution mechanism for financial benefits, particularly in relation to forest conservation, thus making the experiences with PES since 2008 in Lam Dong particularly relevant for future REDD+ prospects.

When comparing PES income with coffee production per unit of land, the opportunity costs of coffee production very significantly outnumber those of PES. The net average income per hectare per year from PES is about 298,000 VND, while alternative agricultural land use would constitute about 5.7 million VND. Thus, it seems reasonable to assume that any causal links between the (low) level of PES payments and environmental performance are highly uncertain. In order for a performance-based PES or REDD+ mechanisms to succeed sustainably, the opportunity cost levels of current land uses must be addressed. If PES has had any impact, it is most likely attributable to increased control and patrolling, and information campaigns that reach all households (for example, newspapers, television, radio, etc.), rather than participation and performance-based payments (effects that were also partly questioned by comparing the environmental awareness scores between the beneficiary and the non-beneficiary households—see Section 4.3 above).

Third, the 'poor' deforest the least. Our findings demonstrate that the 'poor' households are the ones carrying out the least uncertified (illegal) land clearings, and are hence not the group of households causing most deforestation in the study area. The planned REDD+ activities seem to a larger extent to recognize some of the underlying issues of deforestation in the area than has been the case with PES. For example, the establishment of the Village Development Funds may have the potential to stimulate livelihood diversification and relieve poor farmers of debt dependency in the form of high interest rates charged by business households in the communes. Nevertheless, as with PES, the focus on the 'poor' ethnic minorities and their 'destructive' activities is also a key component of REDD+, and the lack of comprehensive analyses of the coffee sector at large is striking in plans and policies at all levels.

An underlying assumption of current REDD+ activities is also that training and capacity building for better and improved practices will increase households' agricultural investment returns, and in turn reduce the pressure on forests. Our livelihood analyses question the assumption that the production of 'poor' (and ethnic minority) households is economically and agronomically less efficient than that of wealthier households. Our findings do not reveal significant differences in levels of input investments, nor production per ha, and the data on the links between income levels and the expansion of production into forest land point to the higher income segments of the population. The effects of increasing the productivity of global commodity crops, such as coffee, are also uncertain in terms of decreasing pressure on more marginal lands, considering the high price elasticity of production levels and volumes [63].

Fourth, issues of tenure and access to productive land need to be addressed. Our findings point towards land being a major limiting factor for increased livelihood income. Issues related to sustainable access to productive land should be addressed in order to stop or constrain the illegal clearing of forest land. Improved land-use planning and securing legal and sustainable access to land for the poor and ethnic minority households should be a key component in all strategies aiming at relieving pressure on strategically important forest resources. This has been done in Da Nhim, to a certain extent, and could be a viable approach for directing coffee production strategically towards the less vulnerable areas in terms of environmental and carbon values. The interests of the business-oriented segment of wealthy households, who have a major stake in coffee-related businesses and (illegal) land transactions, seems to be a cementing factor for the current mode of production. These underlying factors need to be addressed before issues related to the expansion of coffee production into forest land can be solved in a sustainable way.

5. Concluding Remarks

The study has explored some of the predominant underlying policy 'orthodoxies' on the drivers of land-use change and forest encroachment related to small-scale coffee production amongst 'poor' and 'ethnic' households in a PES and REDD+ zone in the Lam Dong Province of Vietnam. Such perceptions are not endemic to Vietnam, but are frequently found in environmental policy schemes and discourses globally [20]. The study has explored drivers of land-use change by making use of a livelihood framework. We found that land is the main limiting factor for agricultural production and livelihoods, and most households in the study area need to supplement agricultural income with off-farm sources. Both the 'medium' and the 'better off' households have access to a more diversified set of off-farm income sources compared to the 'poor', who depend more upon agriculture for their livelihoods, both overall and in terms of coffee production. Policy discourses on the drivers of land-use change in the Central Highlands of Vietnam seem to neglect various underlying factors that drive the coffee production. The findings of this study also indicate that the coffee economy and land transactions related to the expansion of coffee production are controlled largely by the business-oriented households in the study area.

In terms of links between land-use change and poverty levels, the data showed that in absolute terms, the 'medium' and the 'better-off' households cultivate most land, both in total and for coffee production. The data also indicate that these households have cleared more forest land for agriculture over the years than the 'poor' households. Especially if we take the uncertified ('illegal') clearing into account, the 'poor' deforest the least. Thus, the focus on the linkages between poverty amongst the ethnic poor and coffee-related forest encroachment seems to be overrated in PES and REDD+ policies and discourses. The main argument of this paper is therefore that, in order to enable a more comprehensive understanding of land-use change and its management, the focus should be expanded beyond the poverty-environment nexus.

The expansion and development of the coffee sector in Vietnam must be seen in the historic and political context of marginalization of, and control over, land, resources, and people [21,30]. Historically, the general deforestation in the Lam Dong province has often been related to the large migrations of people from the north who came to populate and develop the province [27]. They represent a group of households that over the years have been encouraged by the Government of Vietnam to migrate, settle, and 'develop' the region. This study has demonstrated that business-oriented households, together with the state-owned coffee corporation (VINACAFE), are in control of the coffee sector in the region. Consequently, the current focus of PES and REDD+ on payments to 'poor' and 'ethnic' households, combined with education, information campaigns, and increased levels of forest patrolling, do not seem to be sufficient in trying to solve the underlying issues at stake.

Coffee production requires a long-term investment, and reverting to subsistence production is not a viable option for most households, considering their integration into a market and 'cash' economy. Solutions for attaining sustainable livelihoods need to address the critical lack of productive land in the area. Improved land rights and land-use planning seem to be warranted. The current development and infrastructure plans, for example, of larger-scale plantations of rubber in the Central Highlands also require more attention [64]. A focus on creating realistic and viable off-farm livelihood alternatives, other than being involved in the very dominant coffee sector, is most likely an important ingredient in any policy scheme that aims to reduce pressures on forests.

Acknowledgments: We extend special thanks to Manh Cuong Pham from the UN REDD Programme Office in Vietnam and Nam Pham Thanh from the SNV Office in Dalat, who provided invaluable support and advice during the fieldwork in Lam Dong, and comments regarding the data analyses. We thank Lam Ngoc Tuan at the University of Dalat for his comments and support during the fieldwork. We are also indebted to Professor Arild Vatn who provided valuable comments on the paper. Special thanks also go to all who assisted us during the fieldwork. We are also grateful to the two anonymous reviewers for very useful and constructive comments on this paper. This research was made possible with financial support from the Nansen Fund (managed by UNIFOR at the University of Oslo) and the Norwegian University of Life Sciences.

Contributions: Leif Tore Trædal carried out the data collection and analyses, and wrote the article. Pål Olav Vedeld provided inputs regarding the theoretical dimensions, the structure of the paper, and statistical interpretation of the data.

Conflicts of Interest: The authors declare no conflict of interest.

References

1. The Economics of Ecosystems and Biodiversity (TEEB). *The Economics of Ecosystems and Biodiversity for National and International Policy Makers—Summary: Responding to the Value of Nature*; TEEB: Geneva, Switzerland, 2009.
2. Angelsen, A. REDD+ as result-based aid: General lessons and bilateral agreements of Norway. *Rev. Dev. Econ.* **2016**, 1–28. [CrossRef]
3. Tan, N.Q. Environmental services in Vietnam: An analysis of the pilot project in Lam Dong Province. In *Forest Conservation Project, Occasional Paper*; Scheyvens, H., Ed.; Institute for Global Environmental Strategy: Kanagawa, Japan, 2011.
4. UN-REDD. *UN-REDD Viet Nam Phase II Programme: Operationalising REDD+ in Viet Nam*; UN-REDD Programme: Geneva, Switzerland, 2012.
5. Meyfroidt, P.; Vu, T.P.; Hoang, V.A. Trajectories of deforestation, coffee expansion and displacement of shifting cultivation in the Central Highlands of Vietnam. *Glob. Environ. Chang.* **2013**, *23*, 1187–1198. [CrossRef]
6. Hoang, H.C.; Giang, P.T.; Hai, B.V.; Boi, D.D.; Ha, L.V.H.; Binh, N.Q.; Phuong, D.H.; Tai, N.T.K.; Lan Phuong, N.T.; Thanh, N.D.; et al. *Participatory Rural Appraisal for the Project for Strengthening Community-Based Management Capacity of Bidoup-Nui Ba National Park*; Nong Lam University of Ho Chi Minh City, Department of Social Forestry and Agroforestry: Ho Chi Minh, Vietnam, 2011.
7. Dooley, K.; Griffiths, T.; Martone, F.; Osinga, S. *Smoke and Mirrors. A Critical Assessment of the Forest Carbon Partnership Facility*; FERN and Forest Peoples Programme: Belgium, UK, 2011.
8. DARD. *Action Plan on "Reduction of Greenhouse Gas Emissions through Efforts to Reduce Deforestation and Forest Degradation, Sustainable Management of Forest Resources, and Conservation and Enhancement of Forest Carbon Stocks" in Lam Dong Province, Period 2014–2020*; Department of Agriculture and Rural Development, Lam Dong People Committee: Da Lat, Vietnam, 2014.
9. Scoones, I. *Sustainable Rural Livelihoods: A Framework for Analysis*; IDS Working Paper; IDS: Brighton, UK, 1998.
10. Ellis, F. *Rural Livelihoods and Diversity in Developing Countries*; Oxford University Press: Oxford, UK, 2000.
11. Scoones, I. *Sustainable Livelihoods and Rural Development*; Agrarian Change and Peasant Studies Series; Fernwood Publishing: Black Point, NS, Canada, 2015; Volume 4.
12. Babigumira, R.; Angelsen, A.; Buis, M.; Bauch, S.; Sunderland, T.; Wunder, S. Forest clearing in rural livelihoods: Household-level global-comparative evidence. *World Dev.* **2014**, *64*, S67–S79. [CrossRef]
13. Hosonuma, N.; Herold, M.; De Sy, V.; De Fries, R.S.; Brockhaus, M.; Verchot, L.; Angelsen, A.; Romijn, E. An assessment of deforestation and forest degradation drivers in developing countries. *Environ. Res. Lett.* **2012**, *7*, 1–13. [CrossRef]
14. Kissinger, G.; Herold, M.; Sy, V.D. *Drivers of Deforestation and Forest Degradation: A Synthesis Report for REDD+ Policymakers*; Lexeme Consulting: Vancouver, BC, Canada, 2012.
15. Geist, H.J.; Lambin, E.F. Proximate causes and underlying driving forces of tropical deforestation. *BioScience* **2002**, *52*, 143–150. [CrossRef]
16. Hersperger, A.M.; Gennaio, M.; Verburg, P.H.; Bürgi, M. Linking land change with driving forces and actors: Four conceptual models. *Ecol. Soc.* **2010**, *15*, 1.
17. Carodenuto, S.; Merger, E.; Essomba, E.; Panev, M.; Pistorius, T.; Amougou, J. A Methodological Framework for Assessing Agents, Proximate Drivers and Underlying Causes of Deforestation: Field Test Results from Southern Cameroon. *Forests* **2015**, *6*, 203–224. [CrossRef]
18. Rudel, T.K.; Defries, R.; Asner, G.P.; Laurance, W.F. Changing drivers of deforestation and new opportunities for conservation. *Conserv. Biol.* **2009**, *23*, 1396–1405. [CrossRef] [PubMed]
19. McNally, R.; Phuong, V.T.; Chien, N.T.; Phuong, P.X.; Dung, N.V. *Analyses: Policies and Measures. Support for the Revision of Viet Nam's National REDD+ Action Programme (NRAP) (Draft Report)*; Forest Carbon: Jakarta, Indonesia; Vientiane, Laos, 2016.
20. Forsyth, T. *Critical Political Ecology: The Politics of Environmental Science*; Routledge: London, UK, 2003; p. 323.
21. McElwee, P. *Forests Are Gold. Trees, People, and Environmental Rule in Vietnam*; University of Washington Press: Seattle, WA, USA; London, UK, 2016; p. 283.
22. Forsyth, T.; Sikor, T. Forests, development and the globalisation of justice. *Geogr. J.* **2013**, *179*, 114–121. [CrossRef]

23. Tran, K.T. *Socio-Economic Survey for Assessing the VN Government Policy on Payment for Environmental Services for Lam Dong Province*; Asia Regional Biodiversity Conservation Program: Ho Chi Minh City, Vietnam, 2010.
24. Ogonowski, M.; Enright, A. *Cost Implications for Pro-Poor REDD+ in Lam Dong Province, Vietnam: Opportunity Costs and Benefit Distribution Systems*; IIED: London, UK, 2013.
25. Vietnam General Statistics Office. Population and Employment. 2014. Available online: https://www.gso.gov.vn/default_en.aspx?tabid=774 (accessed on 2 March 2016).
26. Lam Dong Portal. 2016. Available online: http://www.lamdong.gov.vn/en-us/home/Pages/default.aspx (accessed on 28 November 2016).
27. De Koninck, R. *Deforestation in Vietnam*; International Development Research Centre (IDRC): Ottawa, ON, Canada, 1999.
28. Wikipedia 2014: Location of Lam Dong within Vietnam. Available online: http://en.wikipedia.org/wiki/L%C3%A2m_%C4%90%E1%BB%93ng_Province#/media/File:Lam_Dong_in_Vietnam.svg (accessed in September 2016).
29. Déry, S. Agricultural colonisation in Lam Dong Province, Vietnam. *Asia Pac. Viewp.* **2000**, *41*, 35–49. [CrossRef]
30. Salemink, O. *The Ethnography of Vietnam's Central Highlanders: A Historical Contextualization 1850–1990*; RoutledgeCurzon: London, UK, 2003; p. 383.
31. Fortunel, F. *Le Café Au Viêtnam. De la Colonisation à L'essor d'un Grand Producteur Mondial*; L'Harmattan: Paris, France, 2000.
32. Ha, D.T.; Shively, G. Coffee Boom, Coffee Bust and Smallholder Response in Vietnam's Central Highlands. *Rev. Dev. Econ.* **2008**, *12*, 312–326. [CrossRef]
33. Doutriaux, S.; Geisler, C.; Shively, G. Competing for Coffee Space: Development-Induced Displacement in the Central Highlands of Vietnam. *Rural Sociol.* **2008**, *73*, 528–554. [CrossRef]
34. Nhan, D.T. Orientations of Vietnam coffee industry. In Proceedings of the International Coffee Conference, London, UK, 17–19 May 2001.
35. Marsh, A. Diversification by smallholder farmers: Viet Nam Robusta Coffee. In *Agricultural Management, Marketing and Finance Working Document*; FAO: Rome, Italy, 2007.
36. Stevie Ray Vaughan (SRV). *Decision No. 251-TTg on the 29th of April 1995 of the Prime Minister on the Establishment of the Vietnam National Coffee Corporation*; Socialist Republic of Vietnam: Hanoi, Vietnam, 1995.
37. Minot, N. *Competitiveness of Food Processing in Vietnam: A Study of the Rice, Coffee, Seafood and Fruit and Vegetables Subsectors*; International Food Policy Research Institute: Washington, DC, USA, 1998.
38. Petty, M. *Crippling Debts and Bankruptcies Brew Vietnam Coffee Crisis*; Reuters: London, UK, 2013.
39. Agergaard, J.; Fold, N.; Gough, K.V. Global–local interactions: socioeconomic and spatial dynamics in Vietnam's coffee frontier. *Geogr. J.* **2009**, *175*, 133–145. [CrossRef]
40. Vu, T.D.; Bay, P.N.; Stephen, P.; Van Chau, T.; Grais, A.; Petrova, S. *Land Use, Forest Cover Change and Historical GHG Emission. From 1990 to 2010. Lam Dong Province, Viet Nam*; Lowering Emissions in Asia's Forests (LEAF), Ed.; SNV, USAID: Hanoi, Vietnam, 2013.
41. Long, V.; Vu, T.P. *Report on Review of Forestry Policies in Vietnam*; Research Centre for Forest Ecology and Environment: Hanoi, Vietnam, 2011.
42. Federation of Image Professionals International (FIPI). Forest Inventory and Planning Institute. 2013. Available online: http://www.kemintran.vn (accessed in January 2015).
43. SRV. *Circular No. 34/2009/TT-BNNPTNT of June 10, 2009, on Criteria for Forest Identification and Classification*; Minsitry of Agriculture and Rural Development (MARD): Hanoi, Vietnam, 2009; p. 6.
44. SRV. *Decision No. 380 on the Pilot Policy for Payment for Forest Environmental Services*; Minister, T.P., Ed.; The Prime Minister Office: Hanoi, Vietnam, 2008.
45. Wunder, S.; The, B.D.; Ibarra, E. *Payment Is Good, Control Is Better: Why Payments for Forest Environmental Services in Vietnam Have So Far Remained Incipient*; CIFOR: Bogor, Indonesia, 2005.
46. VNS. Vietnam leads way in forest protection. *Viet Nam News*, 25 November 2014.
47. To, P.X.; Santiago, C. *Vietnam Leads Southeast Asia in Payments for Ecosystem Services*; Katoomba Group Ecosystem Marketplace: Washington, D.C. USA, 2010.
48. SRV. *Decree 99/2010/ND-CP, Payments for Forest Environment Services (PES) Policy*; Socialist Republic of Vietnam: Hanoi, Vietnam, 2010.

49. Chiramba, T.; Mogoi, S.; Martinez, I. Payment for Forest Ecosystem Services (PFES): pilot implementation in Lam Dong Province, Vietnam. In Proceedings of the UN-Water International Conference, Zaragoza, Spain, 15–17 January 2011.

50. SRV. *Viet Nam Forestry Development Strategy 2006–2020*; Socialist Republic of Vietnam: Hanoi, Vietnam, 2007; p. 53.

51. McElwee, P.D. Payments for environmental services as neoliberal market-based forest conservation in Vietnam: Panacea or problem? *Geoforum* **2012**, *43*, 412–426. [CrossRef]

52. To, P.X.; Dressler, W.H.; Mahanty, S.; Pham, T.T.; Zingerli, C. The prospects for Payment for Ecosystem Services (PES) in Vietnam: A look at three payment schemes. *Hum. Ecol.* **2012**, *40*, 237–249. [CrossRef] [PubMed]

53. Trædal, L.T.; Vedeld, P.O.; Pétursson, J.G. Analyzing the transformations of forest PES in Vietnam: Implications for REDD+. *For. Policy Econ.* **2016**, *62*, 109–117. [CrossRef]

54. UN-REDD. *Design of a REDD Compliant Benefit Distribution System for Viet Nam*; UN REDD Programme: Geneva, Switzerland, 2010.

55. SRV. *Vietnam Emission Reductions Program Idea Note (ER-PIN) for the Forest Carbon Partnership Carbon Fund*; Ministry of Rural Development and Agriculture: Hanoi, Vietnam, 2014.

56. ARBCP. *Avoided Deforestation Pilot Project in Da Nhim Watershed, Lam Dong Province, Vietnam. Project Desing Document (Working Draft)*; Asia Regional Biodiversity Conservation Programme; USAID and Winrock International: Little Rock, AR, USA, 2011.

57. Angelsen, A.; Kaimowitz, D. Rethinking the causes of deforestation: Lessons from economic models. *World Bank Res. Obs.* **1999**, *14*, 73–98. [CrossRef] [PubMed]

58. Hansen, M.C.; Potapov, P.V.; Moore, R.; Hancher, M.; Turubanova, S.A.; Tyukavina, A.; Thau, D.; Stehman, S.V.; Goetz, S.J.; Loveland, T.R.; et al. High-resolution global maps of 21st-century forest cover change. *Science* **2013**, *342*, 850–853. [CrossRef] [PubMed]

59. Thang, T.C.; Burton, M.P.; Brennan, D.C. Optimal replanting and cutting rule for coffee farmers in Vietnam. In Proceedings of the Australian Agricultural and Resource Economics Society (AARES) Annual Conference, Cairns, Australia, 11–13 February 2009; p. 34.

60. SRV. *Decision No. 09/2011/QD-TTg on Setting Norms on Poor Households and Households in Danger of Falling into Poverty for the 2011–2015 Period*; The Socialist Republic of Vietnam: Hanoi, Vietnam, 2011.

61. GSO. *General Statistics Office of Vietnam*; Government of Vietnam: Hanoi, Vietnam, 2014.

62. SNV. *Cat Tien—How Forbidden Is the "Forbidden Forest"?* SNV: Cat Tien District, Vietnam, 2010.

63. Lambin, E.F.; Meyfroidt, P. Global land use change, economic globalization, and the looming land scarcity. *Proc. Natl. Acad. Sci. USA* **2011**, *108*, 3465–3472. [CrossRef] [PubMed]

64. To, X.P.; Tran, H.N. *Rubber Expansion and Forest Protection in Vietnam*; Tropenbos International Vietnam: Hue, Vietnam, 2014.

forests

MDPI

Article

Governance Values in the Climate Change Regime: Stakeholder Perceptions of REDD+ Legitimacy at the National Level

Timothy Cadman [1,*], Tek Maraseni [2], Hugh Breakey [3], Federico López-Casero [4] and Hwan Ok Ma [5]

[1]   Research fellow, Institute for Ethics, Governance and Law, Griffith University, 170 Kessels Road, Nathan, Queensland 4111, Australia

[2]   Associate professor, Institute for Agriculture and Environment (Research), University of Southern Queensland, West Street, Toowoomba, Queensland 4350, Australia; maraseni@usq.edu.au

[3]   Research fellow, Institute for Ethics, Governance and Law, Griffith University, 170 Kessels Road, Nathan, Queensland 4111, Australia; h.breakey@griffith.edu.au

[4]   Senior research fellow, Institute for Global Environmental Strategies, 2108-11 Kamiyamaguchi, Hayama, Kanagawa 240-0115, Japan; lopezcasero@iges.or.jp

[5]   Project manager, Reforestation and Forest Management, International Tropical Timber Organization, International Organizations Center, 5th Floor Pacifico-Yokohama 1-1-1, Minato-Mirai, Nishi-ku, Yokohama 220-0012, Japan; ma@itto.int

*   Correspondence: t.cadman@griffith.edu.au; Tel.: +61-419-628-709

Academic Editors: Esteve Corbera and Heike Schroeder
Received: 10 July 2016; Accepted: 18 September 2016; Published: 23 September 2016

Abstract: This paper presents the results of two national-level studies of REDD+ governance values in Nepal and Papua New Guinea (PNG), using a hierarchical framework of principles, criteria, and indicators (PC&I), with evaluation at the indicator level. The research was conducted by means of an online survey to determine general perspectives on the governance quality of REDD+, as well as stakeholder workshops, in which participants were asked to rank indicators on the basis of perceived national significance. In the online survey, respondents in both countries identified inclusiveness and resources as the highest and lowest scoring governance values, while inclusiveness, resources, accountability, and transparency, were given priority, although their relative importance differed between countries given national circumstances. The reasons for the commonalities and differences of perceptions between these countries are discussed. The findings suggest that while a generic set of governance values may be usefully applied for determining the institutional legitimacy of REDD+, their relative importance is different. This leads to the conclusion that it may not be appropriate to use a simplified approach to REDD+ governance, focusing for example on safeguards, given different national priorities and contexts.

Keywords: governance values; legitimacy; principles, criteria, and indicators (PC&I); inclusiveness; resources; accountability; transparency; REDD+

---

1. Introduction: Governance Challenges Confronting REDD+

The United Nations Framework Convention on Climate Change (UNFCCC) mechanism 'reducing emissions from deforestation and forest degradation and the role of conservation, sustainable management of forests and enhancement of forest carbon stocks in developing countries', referred to as REDD+, has been the subject of intense academic scrutiny in recent years, notably concerning the initiative's governance at the national level. A search on Proquest using the term "reducing emissions from deforestation and forest degradation" yielded 663 peer-reviewed scholarly articles between 2008 and May 2016. Adding the term "national" generated 537 results, while inclusion of the term "governance" produced 307. This article focuses on the *types* of values that might be used to determine the institutional legitimacy of REDD+, and what governance priorities national-level stakeholders place on the mechanism. 'Governance' refers to the structures and processes used to steer and coordinate interactions [1]. It is a more useful term than 'government' for understanding the political and social relations that occur in the intergovernmental realm, and the related contexts of state, society, and the market, encompassing both state and non-state actors' relationships within contemporary systems of global policy-making, and the structures and processes which underpin them [2].

REDD+ typifies such policy-making, and is an institutional complex, made up of various inter-governmental and national elements—with varying degrees of collaboration [3] (pp. 41–47). The mechanism is still evolving, reflecting the policy machinations of the UNFCCC out of which it emerged, and is made up of a mixture of intergovernmental and national governance practices [4] (pp. 59–78). There is also a recognition that the effectiveness of REDD+ governance needs closer scrutiny, especially in relation to interest representation, accountability and transparency, and decision-making and implementation, as these all contribute to its legitimacy [5] (p. 94). This is because the mechanism is not without risks, as its governance arrangements have the potential to increase conflict between the global North and South, and marginalise local communities [6] (pp. 624–625). The current market emphasis also has implications for the types of actors involved and the degree to which the resulting structures facilitate or hinder interaction, with further implications concerning governance legitimacy [7] (p. 423).

Previous experiences with REDD+ have been largely based on 'nested' (i.e., sub-national) demonstration or pilot projects, with involvement of NGOs and other non-state actors. The trend now appears to be one of increasing centralisation and government control, with the danger that any financial benefits arising from emissions reduction payments will not flow to local actors, thereby reducing compliance. These issues, combined with inconsistent institutional approaches, as well as a high level of policy uncertainty in the wake of the Paris Agreement, are all contributing to a lack of clarity around what governance arrangements will ultimately be used to implement REDD+ on the ground [8].

At the intergovernmental policy level, there has been some recognition of the challenges confronted by national REDD+ initiatives. The Cancún Agreements of the Conference of Parties (COP) 16 in 2010 refer specifically to the need for "guidance and safeguards" including "transparent and effective national forest governance" and the "full and effective participation of relevant stakeholders, in particular indigenous peoples and local communities", but they do not stipulate how these should be implemented, referring only to "national legislation and sovereignty" [9] (p. 26). In Article 72 there is an acknowledgement of the importance of "land tenure issues, forest governance issues, [and] gender considerations", but again there is no further elaboration [9] (p. 13). Nevertheless, the recognition of the UN Declaration on the Rights of Indigenous Peoples (UNDRIP) in the Cancun Agreements means that there is an implied requirement for the free, prior, and informed consent (FPIC) of Indigenous people in REDD+ activities [10] (p. 2407). Under the Cancún Agreements, tropical forest countries involved in the various stages of REDD+ (readiness, pilot studies, and implementation) are expected to ensure that national activities do not impact negatively on the environment or people, whilst also contributing positively to governance and generating environmental and social benefits. Each country is required to demonstrate compliance by developing its own national safeguard system. Compliance is informed by domestic policies, laws, and regulations, and reporting occurs via what is referred to as a safeguard information system. UNFCCC itself does not stipulate what form safeguard systems should take, only that safeguards should be addressed [11].

The effectiveness of REDD+ has been questioned especially as national practices and related power dynamics have had a significant impact on the efficacy of REDD+ governance on the ground, notably regarding the mechanism's inability to protect biodiversity [12]. Indeed, some NGO studies have gone so far as to argue that deforestation and forest degradation are "still being compounded by poor governance, including corruption, conflicts between national and local authorities, and insufficient resources and institutional capacity" [13] (p. 2). Concerns that REDD+ might increase deforestation led to calls from the NGO community in the lead-up to COP 16 for the initiative to make sure that the safeguards it adopted were 'strong' [14].

In the wake of the Cancún Agreements, the REDD+ policy community responded by adopting a range of measures, notably around benefit sharing arrangements. The World Bank's Forest Carbon Partnership Facility (WB, FCPF) has played a significant role, which requires that allocations from its Carbon Fund occur in the context of a national benefit-sharing plan, but exact arrangements are not specified [15]. In an effort to create a broader approach to managing safeguards, the FCPF released its own Common Approach to Environmental and Social Safeguards for Multiple Delivery Partners in 2011 including guidelines as to how partners were to prepare Strategic Environmental and Social Assessment (SESA), a WB requirement for funding [16,17]. The WB revisited its policies in July 2015, and in the case of FPIC, the revisions required that consent must be demonstrated and where it could not, those project aspects relevant to Indigenous peoples could not proceed. In the case of stakeholder management, the revision required increased and ongoing levels of engagement, as well as providing new standards for international financial institutions (IFIs), leading the Bank to claim it was at the 'forefront' of safeguard policies [18].

A number of voluntary standards have arisen around benefit sharing arrangements and other safeguards, including the Climate Community and Biodiversity (CCB standards), a largely NGO-driven initiative, which requires FPIC, and the REDD+ Social and Environmental Standards, which contain provisions for the transparent, participatory, inclusive, effective, and equitable distribution of benefits [15] (pp. 271–272). It should be noted here that there are institutional linkages between the two, with the Climate, Community and Biodiversity Alliance (CARE International, Conservation International, The Nature Conservancy, Rainforest Alliance, Wildlife Conservation Society) and Care International also being involved in REDD SES. The CCB covers 'site based projects', while REDD SES covers 'government led programmes' [19]. Other organisations, including the United Nations' own Collaborative Programme on Reducing Emissions from Deforestation and Forest Degradation in Developing Countries (UN-REDD), and research organisations, such as the International Union of Forestry Research Organisations (IUFRO) and the Centre for International Tropical Forestry Research (CIFOR) and the International Union for Conservation of Nature (IUCN) have also developed their own "guiding principles", but in the absence of formal rules, countries ultimately determine their arrangements [15] (p. 271).

In the light of these issues, this article explores the country-level expressions of REDD+ governance in two countries, Nepal and Papua New Guinea (PNG). REDD+ activities started in Nepal and PNG in 2008. In Nepal they focused largely on capacity building around carbon accounting and arrangements for sharing the benefits arising from carbon payments. Various pilot projects were conducted by a range of international NGOs and local partners in Nepal's well-developed system of community forests, which are managed by community forest user groups (CFUGs), while remaining under state control [20]. The experience of REDD+ on the ground to date has been both positive and negative. On the one hand the financial and social capacity of the CFUG network has improved, but this has been at the expense of autonomous decision-making and customary rights to forest access [21] (p. 39). Despite Nepal's history of inclusive forest management, the technical orientation of REDD+, it has been suggested, has contributed to inhibiting local participation [8] (p. 66). In PNG, the implementation of pilot projects has been somewhat sporadic, due in part to an unstable political climate and forest governance issues. Although land tenure resides with customary landowners, projects have been largely directed by government agencies with only a secondary role for NGOs [22]. Local communities have had no formally specified role in either REDD+ design, or monitoring reporting and verification (MRV) [8]. As a result, REDD+ projects have been largely unable to secure social license for their activities in PNG (particularly in relation to FPIC), not the least because of the difficulties of registering landowner groups—and thereby demonstrating formal land tenure—which is a precondition for participation in REDD+ projects [23] (p. 152).

Scholarly analysis suggests that while safeguards have the potential to balance the over-simplification of social-ecological systems that such globally driven and technically oriented initiatives as REDD+ can lead to, this is dependent on the design of standards under which safeguards operate. The standards that have arisen in the context of REDD+ safeguards recognise FPIC, but interpretation can vary, understanding 'consent' as 'consultation' and leaving the ultimate determination in the hands of auditors, and can be largely procedural, merely ensuring that duties and responsibilities are fulfilled when rights are ignored [24] (pp. 3350–3354). This can be due to political sensitivities in developing countries around issues of democracy and human rights, resulting in the application of national approaches, rather than international norms [25] (p. 2406). This has led to contestation around issues of legitimacy, notably regarding which approach to environmental governance should be adopted. As a result, non-state actors have begun to develop standards themselves, based on their own conceptual frameworks [24] (p. 3355). Consequently, in the world of REDD+ (as in other market-based instruments) there are inconsistencies between standards and the values that inform them [26] (pp. 3–4).

In order to determine REDD+ legitimacy and effectiveness in these countries, and draw some broader conclusions, the article uses a comprehensive framework of governance values, which bear directly on the question of how the initiative is steered and its collective action coordinated. The findings suggest that a generic set of governance values may be usefully applied for determining the institutional legitimacy of REDD+, but the emphasis placed on values is different across the two case study countries, with implications for national REDD+ initiatives more broadly.

2. Materials and Methods: Identifying and Assessing REDD+ Governance Values

Before outlining the methods adopted in this study, a few preparatory comments regarding the approach towards governance and legitimacy is required. It is beyond the scope of an article of this length to go into any great detail, as both subjects have been submitted to exhaustive exploration in recent years, especially amongst scholars of international relations and public policy. There was something of a turn in regime theory in the 1990s and early 2000s, which emphasized a shift away from the notion of 'government' to 'governance', as the latter more accurately described the nature of the structures and processes underlying social-political interactions between state and non-state actors within institutions of global governance [2,27–29].

A parallel debate centered upon the notion of institutional legitimacy. The orthodox view was that the state was the sole repository of political power [30] (p. 187) [31] (pp. 3–4). But the view emerged that legitimacy resided within institutions, state or non-state. Two distinct schools of thought emerged: one emphasizing the importance of democratic processes and rules of procedure, referred to as input legitimacy (the means justify the ends); the other focusing on results, referred to as output legitimacy (the ends justify the means) [32] (pp. 152–155), [33] (p. 12), [34] (p. 45). This approach to understanding legitimacy has been subsequently expanded to encompass 'throughput legitimacy', which focuses on the institution's internal organizational arrangements—both in terms of who is represented, and how deliberation occurs [35] (pp. 2–5).

Given the institutional emphasis in this discussion, governance can therefore be understood as a value-neutral term, referring to the way in which an institution engages with its stakeholders, makes decisions, and accounts for itself to its constituents and the public at large [36]. 'Governance arrangements' refer to the structures and processes an institution uses to facilitate steering and coordination [37] (footnote 13, p. 24). Governance values in turn, as explained below, delineate the specific attributes necessary to inform a global public standard for the normative legitimacy of global governance institutions [38] (p. 405).

In order to critically interrogate the governance underpinning REDD+, it is necessary to make some methodological decisions about what types of values will or will not be included in the analysis. The method includes consideration and analysis of eleven different governance values (at the level of 'indicators', which can be grouped into the larger categories of 'criteria' and 'principles'): accountability; transparency; inclusiveness; resources; equality; democracy; agreement; dispute settlement; behavioural change; problem solving; and durability. While these governance values exist in many explorations of governance and legitimacy in the literature [39] (pp. 12–15), the use of this specific list nevertheless requires justification in two directions. On the one hand, it must be shown why the research method limited itself to these, rather than including further potentially attractive values. On the other hand, it must be shown why the use of all these eleven values was required, rather than just considering a narrower focus on, for example, transparency and accountability. First, these are of course not the only values that could conceivably bear on an institution's legitimacy. Buchanan and Keohane's analysis of the legitimacy of global governance institutions also includes substantive values, such as respect for human rights [38] (p. 419). A similar list by Sampford also includes human rights, and adds the substantive values of fraternity and environmental value [40]. The term 'substantive values' connotes specific goals the institution should seek to further, or moral constraints it should recognize. As such, rather than evaluating *who* was involved, and *how* the process occurred, substantive values make up considerations that should inform the institution's executive decision-makers. While these substantive values are no doubt crucial to an all-things-considered appraisal of an institution's legitimacy, the research method here takes a more focused approach by considering only the values that bear on the nature of the structures and processes of governance themselves. Leo Huberts marks this distinction by separating the *ethics* of governance from the *results* of its process [41]. In other words, because 'governance' refers to the structures and processes used to steer and coordinate interactions, 'governance values' should refer to the qualities and characteristics of these structures and processes—rather than more substantive moral goals and purposes an institution might pursue.

This research project employs this exclusive focus on qualities of governance (rather than wider substantive values) for several reasons. First, as seen earlier in Section 1, a consistent charge laid against REDD+ has been on the basis of its quality of governance, in particular in the form of local stakeholder input. This particular instance follows a larger pattern of local communities challenging the legitimacy of global governance [42]. Without prejudicing the validity of other moral concerns, this specific charge about governance failures warrants a dedicated treatment, which this research aims to provide—at least as it applies to the two case-study areas analyzed. Second, REDD+ already has its substantive values laid down by the UNFCCC, and as expressed institutionally via UN-REDD, FCPF, national level agencies, and so forth (reducing emissions, conservation and sustainable management, and enhancement carbon stocks, and so forth). Achieving these will be captured by the aforementioned governance values of behavioural change, problem solving, and durability. To include in the analysis additional substantive values outside the program's mandated goals would court controversy that a strict focus on structures and processes may avoid. Third, and perhaps most important, a strong performance on the governance values presented here will tend to facilitate the types of substantive goals lauded by Buchanan, Keohane, and Sampford. For example, improved inclusiveness and equality, as well as deliberation and democracy will necessarily further the human rights requirement of giving subjects (including local stakeholders) involvement in decision-making, and input into the decisions of the political authorities that affect them, as well as helping ensure the respect of their other human rights (such as their property or cultural rights pertaining to local environments). Improving these governance values therefore advances the human right to take part in government enshrined in Article 21 of the Universal Declaration of Human Rights [43], UNDRIP [10], and the social safeguards laid down in Cancun [9], and helps facilitate the specific rights of property and culture pursuant to Articles 17, 22, and 27 of the Universal Declaration.

What then of the second concern, which in coming from an opposite direction queries why so many governance values are employed, rather than just cleaving to a narrow focus on usual suspects like transparency and accountability? Again, there are several reasons for this methodological choice. Centrally, it is plausible to think that transparency and accountability are not valuable in themselves: they are valuable insofar as they contribute positively to an institution's performance and legitimacy [44]. If an institution has transparency, but stakeholders possess no way to impact upon its working or decision-making, then the transparency will fail to improve the institution's quality or its legitimacy. Similarly, accountability itself does not guarantee good outcomes—it just ensures that someone can be held responsible for rule-breaches that lead to bad outcomes, which is a much narrower quality [45]. These two points hint at the deeper reason why it is necessary to employ a comprehensive array of governance values, namely, governance values inter-link with each other. Each governance value's full worth remains dependent on the presence and quality of the others. For example, the benefits of having inclusive practices of allowing many stakeholders a seat at the decision-making table are lost if there is no equality in power relations (or at least efforts made to address imbalances), or access to resources, so that poor, local stakeholders can afford to take the seats allocated to them. A deliberately comprehensive array of governance values can therefore identify an institution's weaknesses that prevent it from achieving its goals and responding to stakeholders. Finally, considering a wide array of governance values allows the project to test which methods the stakeholders prioritize themselves. This helps the institution to respond to the concerns of specific cultural groups, and helps ensure that the research remains sensitive to the priorities of local and national stakeholders. As Section 4 shows, the cultures in the two case-study areas did indeed prioritize different qualities of the governance system differently, suggesting the merit of the methodological approach employed.

In order to evaluate REDD+ institutional expression at the national level, a conceptual continuum of values is presented here. The tripartite typology of governance value sets presented here moves from 'thin' to 'thickish' to 'thick'. Each set differs from the others in terms of the richness of the governance values it includes, with 'thick' governance values incorporating all of the qualities earmarked under the 'thickish' category—and both of these including the sparse requirements laid down by 'thin' governance values.

Thick governance values capture the full gamut of social and moral qualities that can be demanded of an institution's mechanisms of steering and coordinating. If an institution possesses all of these qualities, then (while this may not cover everything that morality might demand of the institution) the institution will possess legitimacy in the ways it goes about making, authorizing, implementing, and appraising its decisions. Thick governance values thus constitute a plausible answer to the question of what sorts of internal organizational arrangements would legitimize institutions and institutional complexes [38].

Institutions and networks gain in legitimacy the more that their actions are transparent and accountable, the more they employ constructive deliberation, the more they provide affected stakeholders with the capacity to provide input into their workings, and the more they produce effective results [39] (pp. 15–18). Legitimacy in an institutional context is therefore closely linked to governance quality and also applies to the structures and processes that steer an institution, in the theoretical terms of input-, throughput- and output legitimacy, as discussed above. Figure 1 below is an integrated model that captures these three previously discrete theories [32,33,35].

In order to evaluate the case studies selected, Table 1 below contains a comprehensive set of governance values using a hierarchical framework of principles, criteria, and indicators (PC&I), derived from a review of contemporary governance literature [12] (pp. 12–18).

PC&I have become a common method of assessment for sustainability, and sustainable forest management (SFM) in particular, popularised as a consequence of the 1992 'Rio' Earth Summit, and reflected in its foundational document Agenda 21 [46] (p. 61). Frameworks such as these have been developed to ensure the consistency of evaluation, by placing each element in its appropriate location, from principle to criterion and thence to the relevant indicator, to avoid duplication or overlap. A principle represents a fundamental rule or value to be determined, and which is categorised into criteria, which in turn are broken down into indicators, or parameters, for assessment purposes.

Assessment itself occurs at the indicator level, as both principles and criteria are ideational in nature, and not directly measurable [47] (pp. 5–35).

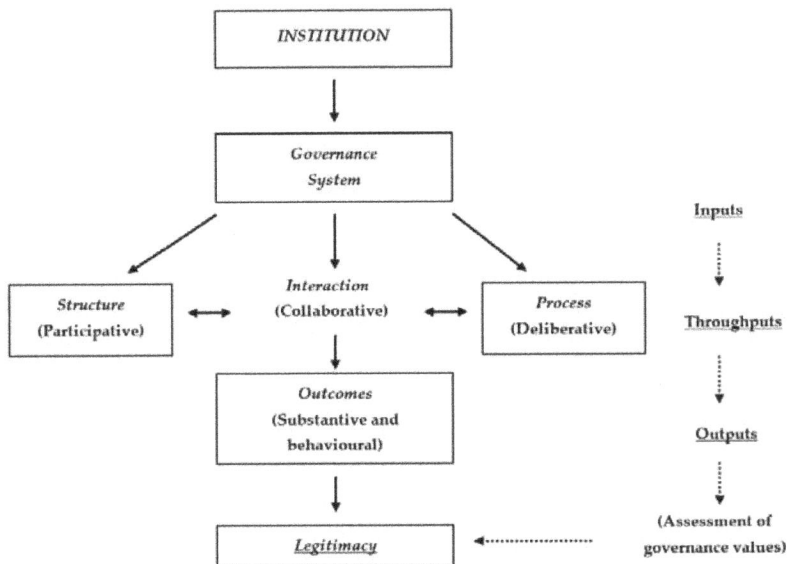

Figure 1. Theoretical model for the evaluation of contemporary global governance [39] (p. 5, adapted; courtesy of Palgrave Macmillan).

Table 1. Hierarchical framework of governance values. [1]

Principle	Criterion	Indicator	Thickness
"Meaningful participation"	Organisational responsibility	Accountability	Thin
		Transparency	Thin
	Interest representation	Inclusiveness	Thickish
		Resources	Thickish
		Equality	Thick
"Productive deliberation"	Decision making	Democracy	Thickish
		Agreement	Thickish
		Dispute settlement	Thick
	Implementation	Behavioural change	Thick
		Problem solving	Thick
		Durability	Thick

[1] Reproduced by courtesy of Palgrave Macmillan (adapted).

In terms of governance, 'values' refer to specific aspects of the structures and processes that determine how things are done within an institution. 'Thin' values tend to be the focus of staple governance mechanisms around ensuring organisational responsibility to other participants and the public at large, such as accountability and transparency. 'Thickish' values add more elements beyond the 'usual suspects' of accountability and transparency, incorporating qualities around interest representation, notably the degree of inclusiveness of stakeholders (access), and the extent to which an institution resources those who would otherwise not be able to participate meaningfully. They also contain some recognition of the need for democratic practices for reaching agreement (such as consensus or voting). However, viewed from the perspective of assessing legitimacy, thickish values remain incomplete. By contrast, 'thick' governance encompasses the existing suite of values, but is concerned with ensuring equality of relations between stakeholder interests as well as 'deeper' procedural values around how disputes arising from decisions are settled, and how decisions are implemented. Implementation in this thick context looks beyond outputs, such as rules of procedure, or standards, etc., to see if these products result in substantive outcomes that all stakeholders can claim ownership of and support, and which result in lasting behavioural change and solutions to the problem that the institution was established to address (in the case of REDD+, greenhouse gasses arising as a consequence of deforestation and forest degradation) [48].

The two studies were conducted in Nepal over the period from July–December 2011, and in PNG from April–May 2015, following the methodology developed by López-Casero et al. [49] (pp. 12–13). The research design was intended to foster collaboration with REDD+ participants in country, using action research methods, whereby an atmosphere is created that allows stakeholders to develop their own solutions to the problems they are tackling, and solve them through their own efforts [50].

At the commencement of the research, participants were invited to express their opinions regarding the structures and processes of governance relevant to forest management and REDD+, and were enlisted from the economic, social, and environmental sectors. A purposive sampling method was used for the selection of participants for the survey because only a subset of the national population was familiar with REDD+ [51–53]. Their contact details were sourced from publically available online documents using the search terms 'REDD+', 'participants' list', 'Nepal' and 'Papua New Guinea', as appropriate. In the case of PNG, these contacts were supplemented by a further database of stakeholders provided by the research project's national contacts. They included non-state actors active in the forest sector, i.e., non-governmental organisations/civil society organisation (NGOs/CSOs), as well as state interests, i.e., governmental agencies.

Given the specific national context of each country, it should be noted that the relevant stakeholder sectors included in the survey varied slightly: in Nepal, Indigenous people play a role in REDD+ as a specific interest grouping; in PNG, the entire national citizenry sees itself as indigenous, making such a designation less relevant. Care was therefore taken in ensuring the relevant stakeholder sectors were included in each country survey.

The views of stakeholders were collected by use of on online survey tool. The survey was in English, and translated into one of each country's national language (Nepali and Tok Pisin). Participants were contacted by email, while the survey itself was conducted anonymously to encourage participation. Respondents were asked to identify their sector, as well as country status (national citizens or international) and to provide a rating for their perceptions regarding the governance of REDD+ using a Likert scale of 1–5 ('very low' to 'very high') using the indicators of Table 1 above (resulting in a possible minimum and maximum total score for the 11 indicators of 11 and 55, respectively). Opportunity for comment was provided under each of the indicators. A selection of relevant comments is included in the analysis below.

Some caveats to the research must be made here. The surveys were deployed in Nepal in July 2011 and in PNG in April 2015, and represent perceptions regarding REDD+ prior to any direct interaction with the researchers. While this reduced the impacts of researcher-subject influence, the surveys were conducted in the two countries over different time periods. Comparison is therefore not entirely like-for-like, and respondents' perceptions were probably influenced by developments at the point in time in which the surveys were conducted. Furthermore, the mechanism is also still undergoing development, and it is likely that perceptions will continue to change as REDD+ processes further evolve in the two countries.

Table 2 below breaks down the survey respondents by country, stakeholder sectors, nationality, and number of surveys commenced and completed.

The survey was followed by a two-day national workshop in each country; in Nepal forty-three multi-stakeholders were present, in PNG thirty-three. Emphasis was again placed on ensuring a diverse representation of interests, targeting those groups least represented in the survey (e.g., women, minority groups). This led to a mixed group of participants in both workshops: some were recruited from the online survey, while others were identified over the course of the research. During the course of a two day workshop. Participants were provided with further information on the indicators and survey on the basis of the results in their respective countries. This provided them with more detailed knowledge to assist in the ranking of the indicators. At the conclusion of the workshop, participants were asked to rank all indicators on a 1–10 scale on the basis of their importance in their respective countries (1 being the least important and 10 being the most important). In the case of the workshops it should be noted that the prioritisation exercise happened after interaction with the researchers, and after the online survey (December 2011 in Nepal, and May 2015 in PNG). While the survey may have influenced some (but not all) of the workshop participants, the objective was to identify national-level priorities amongst the workshop participants in countries. Unlike the survey, the prioritisation exercise reflected future aspirations (i.e., what workshop participants thought was significant for REDD+ moving forward). In both countries, the total number of participants in the workshops was over 30 and the respondents were completely different from each other in each workshop. On that basis and in order to determine whether the mean perceptions of Nepalese and PNG stakeholders for the 11 indicators were significantly different statistically, the independent *t*-test (or student-*t* test) was applied [54] (p. 158). Given the variations in the size of each sample, the limitations of such a test should be recognised, but it should also be acknowledged that the overall sample size is large enough to warrant such an approach [55].

Table 2. Summary of REDD+ survey participants in Nepal and PNG.

Participants	Nepal 2011		PNG 2015	
Initial cohort	300		380	
	Aid programme	7	Aid programme	7
			Community Based Organisation (CBO)	2
	Community forest users	11	Community forest users	1
			Cooperative Societies	2
	Dalit	2		
			Faith Based Organisation (FBO)	1
Stakeholder sectors	Finance	3	Finance	0
	Forest-based industry	3	Forest-based industry	3
	Government	23	Government	24
			Incorporated Land Group (ILG)	1
	Indigenous peoples' organisation	3		
			International NGO	7
			Landowner group	0
			Local NGO	11
	Madhesi	3		
	NGO	49		
	Womens' organisation	1	Womens' organisation	1
	Other	26	Other	14
National citizen	121		59	
International	6		15	
Attempted surveys	131		74	
Completed surveys	66		45	

3. Results and Discussion

3.1. Online Survey

The results of the online survey are reproduced in Table 3 below.

The overall survey scores for REDD+ governance in both countries were relatively similar (34.85 out of 55 in Nepal cf. 32.73 in PNG) (see Table 3 above). These equate to what the researchers refer to as a 'legitimacy rating' of 63% and 60%, respectively (with rounding up)—satisfactory performances, but not overwhelming. Secondly, the highest and lowest performing indicators at the national level were inclusiveness (3.83 out of 5 for Nepal cf. 3.49 for PNG), and resources (2.3 and 2.42). In the case of resources, the ratings provided by respondents were below the threshold 'pass' of 2.5 out of 5. In both Nepal and PNG, accountability and transparency were the second and fourth lowest indicators in both countries (in Nepal 3.01 and 3.08 respectively, while democracy was the third lowest indicator at 3.02; in PNG 2.71 and 2.78, with dispute settlement as the third lowest indicator at 2.64).

Table 3. Nepalese (2011) and PNG (2015) respondents' online survey rating of REDD+.

Indicator	Thickness	Nepal (66)	PNG (45)	Significance
Accountability	Thin	3.01	2.71	$t = 1.56, p = 0.12$
Transparency	Thin	3.08	2.78	$t = 1.40, p = 0.16$
Inclusiveness	Thickish	3.83	3.49	$t = 1.61, p = 0.11$
Resources	Thickish	2.3	2.42	$t = -0.44, p = 0.66$
Equality	Thick	3.15	2.84	$t = 1.55, p = 0.22$
Democracy	Thickish	3.02	3.04	$t = -0.07, p = 0.95$
Agreement	Thickish	3.31	2.84	$t = 2.20, p = 0.03$
Dispute settlement	Thick	3.17	2.64	$t = 2.54, p = 0.01$
Behavioural change	Thick	3.62	3.36	$t = 1.14, p = 0.26$
Problem solving	Thick	3.23	3.27	$t = -0.24, p = 0.81$
Durability	Thick	3.13	3.33	$t = -0.97, p = 0.34$
Total (out of 55)		34.85	32.73	N/A

Despite the differences in sample size, the similarity of the results in both countries led the researchers to conclude that they could be interrogated by means of an independent *t*-test to determine if there were any statistically significant differences between the two sets of results at the indicator level. The analysis of the online surveys revealed that the perceptions of Nepalese and PNG stakeholders for nine of the 11 indicators were not significantly different statistically (i.e., that they rated them equally, at least statistically). In the case of the two statistically significant indicators (*p*-value $\leq$ 0.05), the mean value of perceptions for respondents from PNG was lower for agreement (2.84 in PNG and 3.31 in Nepal) and dispute settlement (2.64 in PNG and 3.17 in PNG) than for those from Nepal.

It should be noted here that adding or subtracting ordinal values is not common when using Likert scales to determine respondent perceptions. However, it is not uncommon in political science for indicator-based assessments to convert verbal values (low, medium, high) to numerical scores, although ordinal scales are usually applied to measure the extent of variability rather than quantifiable levels of difference [56] (p. 373; footnote 7). Certification bodies routinely apply numerical values to verbal assessments. For example, in the case of Forest Stewardship Council-accredited certifier, Smartwood, compliance is evaluated at the criterion level using a five-point score (extremely weak, weak, satisfactory, favourable, outstanding), based on the averaging of related indicators, which in turn are assessed using a high-to-low scale [57] (p. 33). The point being made here is that the quantification of indicators is not a precise science, and a range of methods for aggregating verbal assessments and determining pass/fail thresholds are used in the field of SFM [58].

3.2. Prioritisation Execrcise

The results of the prioritisation exercise are presented in Figure 2 below.

The four most significant governance values at the indicator level for both countries were inclusiveness, resources, accountability, and transparency, but in different orders of priority. In Nepal, stakeholders ranked them in the order of transparency (8.6 out of 10), inclusiveness (8.4), accountability (7.8), and resources (7) (TIAR), and in PNG, inclusiveness (7.4), accountability (7.3), resources (7.2) and transparency (7.1) (IART). The lowest prioritised indicators were also the same for both countries—agreement, dispute settlement, and problem solving, but again in different orders;

in Nepal, problem solving (3), dispute settlement (3.3), and agreement (3.6), while in PNG they were dispute settlement (3.3), problem solving (4.6), and agreement (5).

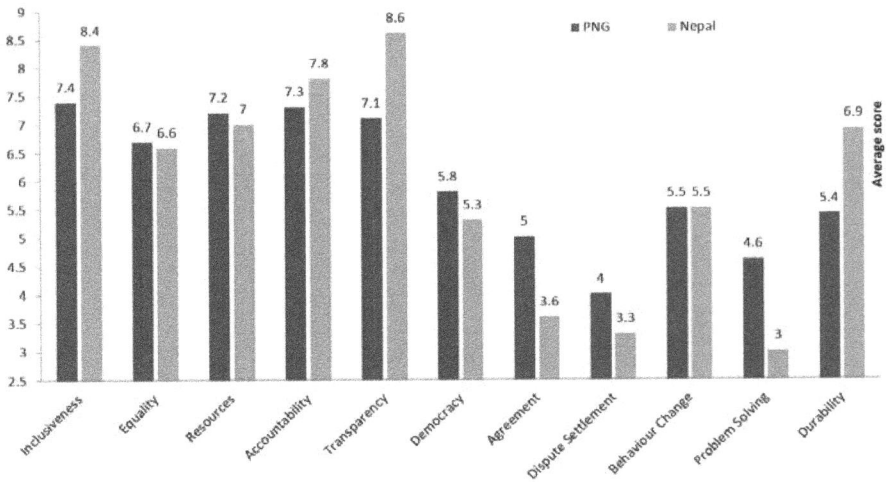

Figure 2. Prioritisation at the indicator level of 11 different governance values in Nepal and PNG.

The analysis of ranking by Nepalese and PNG stakeholders reveals that for eight of the 11 indicators there was no significant difference statistically. In the case of the three statistically significant indicators (at a 95% confidence level), the mean value of rankings for respondents from PNG was higher for agreement (5 in PNG and 3.6 in Nepal) and problem solving (4.6 in PNG and 3 in Nepal) than for those from Nepal, but the mean value of ranking for respondents from PNG for transparency was lower than that of Nepal (7.1 in PNG and 8.6 in Nepal).

4. Discussion and Comments

The results of the two governance surveys reveal more than just which indicators were the best and worst performers. While it is commendable that both countries' participating stakeholders thought REDD+ was inclusive, the low score for resources is a cause for some concern, as organisations lacking the capacity (technical, financial, institutional) to represent their interests cannot participate meaningfully in making and implementing decisions. A review of the comments of the survey participants reveals why resources may have been rated so low. In the Nepal survey one respondent ('other' who identified as a PhD student) commented that: "The resources available for the participation of different interest groups is currently limited to certain elite groups who are misleading the representation of that particular group". Another respondent (also 'other', with experience in a donor agency) observed that resources tended to stay within the REDD+ working group, while "outside access is rare". An 'NGO' respondent made the case that "resources should be transparent and shared among the stakeholders". In PNG, one local NGO was of the view that it was "absolutely important that communities in our areas are assisted to be proactive", while another international NGO (INGO) indicated that they had been able to "meet local communities, local landowner group representatives and district as well as provincial authorities" in pursuit of their own REDD+ activities, but stressed that: "more resources should be out into building their capacity on what roles they play in reducing emissions, in whatever development activities taking place in their areas". A further INGO noted that while the PNG authorities had been "doing a lot in the REDD domain", "more of what they have done" needed to be "in the public domain". Seeing a relationship between resources and transparency, another government respondent also agreed, that "prior information on REDD+ resource provision [would] be appreciated". One community forest user saw a "need to consult more land owners and take their options" regarding resources, while a third INGO was concerned that "information provided" was not compatible "to the Resource owners' understanding" of REDD+.

Comments relating to inclusiveness were also revealing. In both countries, respondents were keen to ensure that in addition to government, donors, and implementing agencies, that local communities were included in REDD+ project activities. In Nepal, emphasis was given to Dalit, Indigenous people, and—in the words of one student, who had previously worked with an international NGO—"different interest groups based on gender, caste, class, ethnicity and geographic origin". In PNG, it was not surprising to see several respondents stress the importance of including local landowner groups, given their significance in relation to forest management. In the words of one government respondent it was important because:

> The people own the land, where REDD+ Projects would most likely be implemented (not government). Therefore we have to put more focus on the landowners and how the practical benefits (not in monetary terms) will help the local people on the ground.

Respondents also made linkages between inclusiveness and transparency. In PNG, one community based organisation representative who lived in a village within a REDD+ project area was of the view that "we should be updated on the progress of this project". Another national respondent who identified as being with a "global NGO/CSO" made the point that "implementing agencies need more transparency, and to involve all stakeholders within the scope of REDD+". As one 'research-based' NGO from Nepal explained, inclusiveness could only happen:

> By engaging the diverse stakeholders in REDD decision forums; existing forums are dominated by government officials and very few, elite, members from indigenous communities and community networks.

Respondents made a range of pertinent observations regarding accountability. In Nepal, one respondent who identified as 'NGO' felt that "the powerful can influence and the powerless may lose". They saw a connection between accountability and implementation, and wanted to see measures that would lead to "accountable service providers". In PNG, one 'government' respondent believed that there were "no transparency or governance mechanisms in place to account for such projects". They thought that "until such measures" were in place that it could not be said that REDD+ projects were "acting in the accountable manner". Another PNG national who identified as 'INGO' pointed to "the irresponsible nature of the process as it is happening in PNG", putting this down to "the regulatory system, complex landowners issues, and the inability of those responsible to be held accountable". This meant that the "REDD+ process is bound to be entangled in all this". This led one Women's organisation respondent to conclude that: "some systems need to be put in place in order to clearly see that it is accountable".

Problem solving and agreement were rated comparatively lower in PNG than Nepal. In relation problem solving, one PNG academic/government respondent was of the view that "OCCD [Office of Climate Change Development; now the Climate Change Development Authority] lacks technical leadership". Another local NGO thought that: "we need political will and commitment from the government of the day to address governance issues and corruption in the country's forestry sector". Concerning agreement, one INGO respondent noted that landowners were not really involved, observing that: "most time it is only the policy makers, and the very owners of the forest are not included." The academic/government respondent commented further that:

> The Agreements in REDD+ are not that effective because nationals with technical expertise are not often involved while most technical expertise [is] from outside (overseas) to cover up for those at OCCD. There should be a core team of technical expertise including foreigners to be involved in any agreements related to REDD+ in the country.

Both agreement and dispute settlement were perceived not to be such significant issues in Nepal as in PNG. This may be because REDD+ in Nepal has focused exclusively on the country's well-established community forestry programme, which has its own policies, rules, and legislation [59], even if social and environmental tensions exist with REDD+ [60]. In PNG, REDD+ is occurring in a larger and more complex forest governance environment as discussed above. Notably, there are likely to be interactions between the reaching of agreements and resolution of conflicts (or failure to do so). As the local NGO respondent who previously commented on agreement noted, "it is often difficult to resolve conflicts over natural resources in PNG". Another INGO respondent explained that it was hard for local stakeholders to meet with those responsible for managing REDD+, as they were based in the urban centres, which meant that:

> People don't have access to them in discussions to address any issues related to REDD+. It would be better to have offices dealing with these kinds of things placed in the Districts or Provinces so people have access to them, and discuss and settle issues. Sometimes people just discuss amongst themselves knowing that issues will never get resolved because no one is listening and taking their concerns in, so they are just wasting their time.

Clearly given the low score for resources, it is not especially surprising that in both workshops, it was a priority for stakeholders. In the case of inclusiveness, it is possible that despite the high score in the survey, stakeholders were concerned to make REDD+ more inclusive, or to make sure it remained so. In the case of transparency, and in the light of comments from the survey respondents, workshop participants may also have made a link between resources and their allocation in an open and visible manner. In this case, transparency was perceived as a bigger issue in Nepal than that of PNG. This may be on account of the fact that at the time of the ranking exercise, Nepal was fully engaged in a series of REDD+ pilot projects, and research participants may have been concerned about fund details, and the allocation of project resources to the different levels of government (national, sub-national, and local).

5. Conclusions

In view of the complex relations between the different governance values discussed above, the researchers observe that the legitimacy of governance systems employing only thin values may be limited. The variations between the two cases are also noteworthy, such as the different perceptions of governance quality regarding dispute settlement. The relationship between different governance values is also interesting—for example, the linkages identified by respondents between accountability and implementation. An analysis measuring only transparency and accountability for example would have awarded REDD+ a 'pass', remarking only on the mild difference between the two countries. However, in drilling down to a more granular level, and by including a broader suite of governance values, it is possible to identify areas of shared satisfaction and dissatisfaction. In this regard, the researchers especially note the scores for inclusiveness and resources, as these national level results accord with a range of global surveys of REDD+ governance they have undertaken over the past five years [26,61–65]. It might be easy to dismiss the low score for resources as being a reflection that there is never enough money, or capacity building, to satisfy stakeholders' needs. However, the comments made by respondents, and the linkages they made to issues around accountability and transparency of resource allocation, lead to a further conclusion that this may be a systemic problem throughout the REDD+ programme. This may be further reflected by the prioritisation of inclusiveness, resources, accountability, and transparency amongst research participants at the national level in Nepal and PNG. This might demonstrate that these governance values are important to REDD+ stakeholders in developing countries regardless of their location and socio-political circumstances. In measuring the impacts and effectiveness of REDD+ policies, an emphasis on broader governance values may be merited. This leads to the conclusion that a 'safeguards' based approach, which stresses only a limited set of values, may not be sufficient to carry the whole burden of responsibility for ensuring the 'good' governance of REDD+ at the national level.

Acknowledgments: The authors acknowledge the Institute for Global Environmental Strategies and the Ministry of Environment, Japan, for their financial support in carrying out research in Nepal, and the International Tropical Timber Organization for its financial support for the PNG research.

Author Contributions: Timothy Cadman, main author; Tek Maraseni, data analysis; Hugh Breakey on governance values; Federico López-Casero, for field research assistance in Nepal; Hwan Ok Ma for field research assistance in PNG.

Conflicts of Interest: The authors declare no conflict of interest.

References

1. Pierre, J.; Guy Peters, B. *Governance, Politics and the State*; Macmillan: Basingstoke, UK, 2000.
2. Kooiman, J. *Societal Governance: Levels, Models, and Orders of Social-Political Interaction, in Debating Governance: Authority, Steering and Democracy*; Pierre, J., Ed.; Oxford University Press: Oxford, UK, 2000; pp. 138–166.
3. Lubell, M. Collaborative Partnerships in Complex Institutional Systems. *Curr. Opin. Environ. Sustain.* **2015**, *12*, 41–47. [CrossRef]
4. Palmujoki, E.; Virtanen, P. Global, National, or Market? Emerging REDD+ Governance Practices in Mozambique and Tanzania. *Glob. Environ. Polit.* **2016**, *16*, 59–78. [CrossRef]
5. Corbera, E.; Schroeder, H. Governing and implementing REDD+. *Environ. Sci. Policy* **2011**, *14*, 89–99. [CrossRef]
6. Gupta, J. Glocal forest and REDD+ governance: Win–win or lose–lose? *Curr. Opin. Environ. Sustain.* **2012**, *4*, 620–627. [CrossRef]
7. Vatn, A.; Vedeld, P.O. National governance structures for REDD. *Glob. Environ. Chang.* **2013**, *23*, 422–432. [CrossRef]
8. Vijge, M.J.; Brockhaus, M.; Di Gregorio, M.; Muharrom, E. Framing national REDD+ benefits, monitoring, governance and finance: A comparative analysis of seven countries. *Glob. Environ. Chang.* **2016**, *39*, 57–68. [CrossRef]
9. UNFCCC. *Outcome of the Work of the Ad Hoc Working Group on Long-Term Cooperative Action under the Convention (Cancun Agreements)*; UNFCCC: Bonn, Germany, 2011.
10. Ruggie, J. Report of the Special Representative of the Secretary—General on the issue of human rights and transnational corporations and other business enterprises. *Neth. Q. Hum. Rts.* **2011**, *29*, 224.
11. ICIMOD. Regional Workshop on 'REDD+ Safeguards for Himalayas'. Available online: http://www.icimod.org/?q=23022 (accessed on 30 August 2016).
12. Hill, R.; Miller, C.; Newell, B.; Dunlop, M.; Gordon, I.J. Why biodiversity declines as protected areas increase: The effect of the power of governance regimes on sustainable landscapes. *Sustain. Sci.* **2015**, *10*, 357–369. [CrossRef]
13. Hall, R. *REDD+ and the Underlying Causes of Deforestation and Forest Degradation*; Global Forest Coalition: Asunción, Paraguay, 2013.
14. WWF. NGOs Call for Strong Safeguards in Efforts to Halt Deforestation to Help Address Climate Change. Available online: http://wwf.panda.org/?193441/NGOs-call-for-strong-safeguards-in-efforts-to-halt-deforestation-to-help-address-climate-change (accessed on 29 August 2016).
15. Chapman, S.; Wilder, M.; Millar, I. Defining the Legal Elements of Benefit Sharing in the Context of REDD. *Carbon Clim. Law Rev.* **2014**, *8*, 270–281.
16. FCPF. *Forest Carbon Partnership Facility (FCPF) Readiness Fund Incorporating Environmental and Social Considerations into the Process of Getting Ready for REDD+*; FCPF: Washington, DC, USA, 2010; Available online: http://www.forestcarbonpartnership.org/sites/forestcarbonpartnership.org/files/Documents/PDF/Jun2010/2g_FCPF_FMT_Note_2010_16_SESA_Mainstreaming.pdf (accessed on 29 August 2016).

17. FCPF. *Forest Carbon Partnership Facility Common Approach to Environmental and Social Safeguards for Multiple Delivery Partners*; FCPF: Washington, DC, USA, 2011.

18. Bank, W. World Bank Board Committee Authorizes Release of Revised Draft Environmental and Social Framework. 2015. Available online: http://www.worldbank.org/en/news/press-release/2015/08/04/world-bank-board-committee-authorizes-release-of-revised-environmental-and-social-framework (accessed on 30 August 2016).

19. CCBA. Home. 2016. Available online: http://www.climate-standards.org (accessed on 30 August 2016).

20. Paudel, N.S.; Karki, R. *The Context of REDD+ in Nepal: Drivers, Agents and Institutions*; CIFOR: Bogor, Indonesia, 2013; Volume 81.

21. Poudel, M.; Thwaites, R.; Race, D.; Dahal, G.R. REDD+ and community forestry: Implications for local communities and forest management—A case study from Nepal. *Int. For. Rev.* **2014**, *16*, 39–54. [CrossRef]

22. Babon, A.; Gowae, G.Y. *The Context of REDD+ in Papua New Guinea: Drivers, Agents, and Institutions*; CIFOR: Bogor, Indonesia, 2013; Volume 89.

23. Venuti, S. REDD+ in Papua New Guinea and the protection of the REDD+ safeguard to ensure the full and effective participation of indigenous peoples and local communities. *Asia Pac. J. Environ. Law* **2014**, *17*, 131–153.

24. Aicher, C. Discourse practices in environmental governance: Social and ecological safeguards of REDD. *Biodivers. Conserv.* **2014**, *23*, 3543–3560. [CrossRef]

25. Pham, T.T.; Castella, J.-C.; Lestrelin, G.; Mertz, O.; Dung, N.L.; Moeliono, M.; Tan, Q.N.; Hien, T.V.; Tien, D.N. Adapting Free, Prior, and Informed Consent (FPIC) to Local Contexts in REDD+: Lessons from Three Experiments in Vietnam. *Forests* **2015**, *6*, 2405–2423. [CrossRef]

26. Cadman, T.; Maraseni, T.; Ma, H.O.; Lopez-Casero, F. Five years of REDD+ governance: The use of market mechanisms as a response to anthropogenic climate change. *For. Policy Econ.* **2016**, in press. [CrossRef]

27. Rosenau, J. *Change, Complexity and Governance in a Globalising Space, in Debating Governance: Authority, Steering and Democracy*; Pierre, J., Ed.; Oxford University Press: New York, NY, USA, 2000; pp. 167–200.

28. Haas, P.M. UN conferences and constructivist governance of the environment. *Glob. Gov.* **2002**, *8*, 73–91.

29. Perrons, D. *Globalization and Social Change: People and Places in a Divided World*; Routledge: London, UK, 2004.

30. Lawson, S. Conceptual issues in the comparative study of regime change and democratization. *Comp. Polit.* **1993**, *25*, 183–205. [CrossRef]

31. Young, G. *Introduction, in Legitimation and the State*; Maddox, G.Y.A.G., Ed.; Kardororair Press: Armidale, Australia, 2005; pp. 1–14.

32. Scharpf, F.W. *Games Real Actors Play: Actor-Centered Institutionalism in Policy Research*; Westview Press: Boulder, CO, USA, 1997.

33. Kjaer, A.M. *Governance*; Polity Press: Cambridge, UK, 2004.

34. Cadman, T. *The Legitimacy of ESG Standards as an Analytical Framework for Responsible Investment, in Responsible Investment in Times of Turmoil*; Springer: Dordrecht, The Netherlands, 2012; pp. 35–53.

35. Schmidt, V.A. Democracy and Legitimacy in the European Union Revisited: Input, Output and 'Throughput'. *Polit. Stud.* **2013**, *61*, 2–22. [CrossRef]

36. Breakey, H.; Cadman, T.; Sampford, C. *Conceptualizing Personal and Institutional Integrity: The Comprehensive Integrity Framework, in The Ethical Contribution of Organizations to Society*; Emerald Group Publishing Limited: Bingley, UK, 2015; pp. 1–40.

37. Koenig-Archibugi, M. *Introduction: Institutional Diversity in Global Governance, in New Modes of Governance in the Global System: Exploring Publicness, Delegation and Inclusiveness*; Koenig-Archibugi, M., Zurn, M., Eds.; Palgrave Macmillan: Basingstoke, UK, 2006; pp. 1–30.

38. Buchanan, A.; Keohane, R.O. The legitimacy of global governance institutions. *Ethics Int. Aff.* **2006**, *20*, 405–437. [CrossRef]

39. Cadman, T. *Quality and Legitimacy of Global Governance: Case Lessons from Forestry*; Palgrave Macmillan: Basingstoke, UK, 2011.

40. Sampford, C.J. *Challenges to the Concepts of 'Sovereignty' and 'Intervention', in Human Rights in Philosophy and Practice (Applied Legal Philosophy)*; Tom, B.M.L., Campbell, D., Eds.; Ashgate: London, UK, 2001; pp. 335–392.

41. Huberts, L. *The Integrity of Governance: What It Is, What We Know, What Is Done and Where to Go*; Springer: Dordrecht, The Netherlands, 2014.

42. Bodansky, D. The legitimacy of international governance: A coming challenge for international environmental law? *Am. J. Int. Law* **1999**, *93*, 596–624. [CrossRef]

43. UN General Assembly. Universal Declaration of Human Rights. 1948. Available online: http://www.un.org/en/udhrbook/pdf/udhr_booklet_en_web.pdf (accessed on 20 September 2016).

44. Curtin, D.; Meijer, A.J. Does transparency strengthen legitimacy? *Inf. Polity* **2006**, *11*, 109–122.

45. Grace, D.; Cohen, S. *Business Ethics*, 5th ed.; Oxford University Press: Oxford, UK, 2013.

46. Rametsteiner, E.; Pülzl, H.; Alkan-Olsson, J.; Frederiksen, P. Sustainability indicator development—Science or political negotiation? *Ecol. Indic.* **2009**, *11*, 61–70. [CrossRef]

47. Lammerts van Beuren, E.M.; Blom, E.M. *Hierarchical Framework for the Formulation of Sustainable Forest Management Standards*; The Tropenbos Foundation: Leiden, The Netherlands, 1997.

48. Breakey, H.; Cadman, T.; Sampford, C. Governance Values and Institutional Integrity. In *Governing the Climate Regime: Instituional Integrity and Integrity Systems*; Cadman, T., Maguire, R., Sampford, C., Eds.; Routlege: London, UK.

49. López-Casero, F.; Cadman, T.; Maraseni, T. *Quality-of-Governance Standards for Forest Management and Emissions Reduction: Developing Community Forestry and REDD+ Governance through a Multi-Stage, Multi-Level and Multi-Stakeholder Approach—2016 Update—2016*; Institute for Global Environmental Strategies (IGES): Hayama, Japan, 2016; p. 49.

50. Hall, B.L. Knowledge as a Commodity and Participatory Research. *Prospects Q. Rev. Educ.* **1979**, *9*, 393–408. [CrossRef]

51. Bernard, H.R. *Research Methods in Anthropology: Qualitative and Quantitative Approaches*; Rowman Altamira: Walnut Creek, CA, USA, 2011.

52. Blythe, J.L. Social-ecological analysis of integrated agriculture-aquaculture systems in Dedza, Malawi. *Environ. Dev. Sustain.* **2013**, *15*, 1143–1155. [CrossRef]

53. Oteros-Rozas, E.; González, J.A.; Martín-López, B.; López, C.A.; Montes, C. Ecosystem Services and Social-Ecological Resilience in Transhumance Cultural Landscapes: Learning from the Past, Looking for a Future. In *Resilience and the Cultural Landscape. Understanding and Managing Change in Human-shaped Environments'*; Plieninger, T., Bieling, C., Eds.; Cambridge University Press: Cambridge, UK, 2012; pp. 242–260.

54. Mankiewicz, R. *The Story of Mathematics*; Princeton University Press: Princeton, NJ, USA, 2000.

55. Ruxton, G.D. The unequal variance *t*-test is an underused alternative to Student's *t*-test and the Mann–Whitney U test. *Behav. Ecol.* **2006**, *17*, 688–690. [CrossRef]

56. Nanz, P.; Steffek, J. Assessing the democratic quality of deliberation in international governance: Criteria and research strategies. *Acta Politica* **2005**, *40*, 368–383. [CrossRef]

57. Smartwood, SmartWood Certification Assessment Report for Alberta-Pacific Forest Industries Inc. 2004. Available online: https://alpac.ca/application/files/1614/1876/0528/2005_Final_Report.pdf (accessed on 20 September 2016).

58. Mendoza, G.A.; Prabhu, R. Fuzzy methods for assessing criteria and indicators of sustainable forest management. *Ecol. Indic.* **2004**, *3*, 227–236. [CrossRef]

59. Acharya, K.P. Twenty-four years of community forestry in Nepal. *Int. For. Rev.* **2002**, *4*, 149–156. [CrossRef]

60. Saito-Jensen, M.; Rutt, R.L.; Chhetri, B.B.K. Social and Environmental Tensions: Affirmative Measures under REDD+ Carbon Payment Initiatives in Nepal. *Hum. Ecol.* **2014**, *42*, 683–694. [CrossRef]

61. Maraseni, T.N.; Cadman, T. A comparative analysis of global stakeholders' perceptions of the governance quality of the clean development mechanism (CDM) and reducing emissions from deforestation and forest degradation (REDD+). *Int. J. Environ. Stud.* **2015**, *72*, 288–304. [CrossRef]

62. Cadman, T.; Maraseni, T. More equal than others? A comparative analysis of state and non-state perceptions of interest representation and decision-making in REDD+ negotiations. *Innovation* **2013**, *26*, 214–230. [CrossRef]

63. Cadman, T.; Maraseni, T. The governance of REDD+: An institutional analysis in the Asia Pacific region and beyond. *J. Environ. Plan. Manag.* **2012**, *55*, 617–619. [CrossRef]

64. Cadman, T.; Maraseni, T.; Blazey, P. Perspectives on the quality of global environmental governance: An evaluation of NGO participation in global climate negotiations in the Asia-Pacific and beyond. *Third Sect. Rev.* **2012**, *18*, 145–169.

65. Cadman, T.; Maraseni, T. The Governance of Climate Change: Evaluating the Governance Quality and Legitimacy of the United Nations' REDD-plus Programme. *Int. J. Clim. Chang. Impacts Responses* **2011**, *2*, 103–124.

forests

MDPI

Article

Early REDD+ Implementation: The Journey of an Indigenous Community in Eastern Panama

Ignacia Holmes [1,*,†], Catherine Potvin [1,2,*,†] and Oliver T. Coomes [3]

1   Department of Biology, McGill University, 1205 Dr Penfield, Montreal, QC H3A 1B1, Canada
2   Smithsonian Tropical Research Institute, Panama 0843-03092, Panama
3   Department of Geography, McGill University, 805 Sherbrooke St. West, Montreal, QC H3A 0B9, Canada;
    oliver.coomes@mcgill.ca
*   Correspondence: ignacia.holmes@mail.mcgill.ca (I.H.); catherine.potvin@mcgill.ca (C.P.);
    Tel.: +1-514-398-6726 (C.P.)
†   These authors contributed equally to this work.

Academic Editors: Esteve Corbera and Heike Schroeder
Received: 7 November 2016; Accepted: 24 February 2017; Published: 3 March 2017

Abstract: Reducing Emissions from Deforestation and Forest Degradation (REDD+) offers developing countries an opportunity to engage in global climate change mitigation through the sale of carbon credits for reforestation, avoided deforestation and forest conservation projects. Funding for REDD+ projects has increased in recent years and REDD+ projects have proliferated, but relatively few studies have, as yet, examined their implementation. Here, we present a synthesis of the challenges and lessons learned while implementing a REDD+ project in an Emberá community in Panama. Our case study, documented in four cycles of collaborative action research over 11 years, examines how local communities sought to reduce emissions from deforestation and benefit from carbon offset trading while improving local livelihoods. Through semi-structured interviews and participatory methods, we found that success with REDD+ hinges on broader issues than those widely discussed in the literature and in policy circles. Though economic incentives for participants and the equitable distribution of benefits remain important to project participants, our study finds that, in adapting REDD+ strategies to best suit community needs, the role of a support system for implementation ("bridging institutions") and REDD+'s potential as a conflict resolution mechanism for tenure issues deserve more attention as key factors that contribute to meaningful participation in REDD+.

Keywords: REDD+; community REDD+; carbon-offset projects; land tenure; deforestation; climate change mitigation

---

1. Introduction

Reducing Emissions from Deforestation and Forest Degradation (REDD+) is now an accepted climate change mechanism under the United Nations Framework Convention on Climate Change (UNFCCC). It allows developing countries to contribute to mitigation by undertaking one or more of five activities: reducing emissions from deforestation, reducing emissions from degradation, conserving carbon stocks, managing forest sustainably and enhancing forest carbon stocks [1]. REDD+ is mobilizing significant financial resources from the international community as well as the private sector [2]. To date, funding from public sources dominates contributions, including bilateral agreements (i.e., the Norway–Indonesia REDD+, US$1 billion), and multilateral funds such as the Forest Carbon Partnership Facility (FCPF, US$385 million) [3] and the United Nations REDD Programme (UN-REDD, US$227.28 million) [4]. Much of such funding is aimed at REDD+ readiness, assisting countries in preparing to develop and implement carbon mitigation measures. In contrast, funding for early pilot initiatives comes mostly from the private sector through voluntary carbon funds [5,6]. Investment in REDD+ greatly exceeds historical spending in conservation. The total budget for protected areas in developing countries during the 1990s was estimated at US$ 0.70 billion [7], a fraction of REDD+ investment. Such high levels of investment have raised hopes for enhanced forest conservation in parallel with effective carbon management [8].

REDD+ provides an opportunity for developing countries to engage in a process of change regarding forest resources, to address the causes of deforestation [6] and to create incentives for farmers and communities that influence local land use [9]. Many countries are currently hosting REDD+ pilot projects and their numbers are growing rapidly. In 2013, 300 REDD+ pilot projects were reported around the world [10], most of which were in the development phase and for which experience with implementation was relatively limited [11]. In 2012, we found information on only 20 REDD+ projects under implementation in Latin America [12], while Panfil and Harvey [13] in 2013 analysed 80 Project Design Documents for REDD+ projects worldwide under the Climate Community Biodiversity Alliance (CCBA) standards, only 15 of which filed an Interim progress report. The CCBA data however currently report 142 verified projects for Latin America [14].

REDD+ faces both opportunities and challenges that are contingent on local contexts and actors [12,15,16]. CIFOR's Global Comparative Study on REDD+ [17] recently yielded a number of important analyses of early REDD+ implementation in Brazil, Cameroon, Tanzania, Indonesia and Vietnam [18,19], recognizing that challenges were rooted deep in history and likely to require landscape-wide reforms. Their conclusion was echoed by Corbera et al. [20], who analysed tenure rights in Brazil, Mexico and Costa Rica, highlighting different 'bundles of rights' amongst countries. The authors stressed the importance of effective, equitable and legitimate policies and measures addressing not only tenure insecurity but also the ability of communities to exercise their rights to manage the forests. In fact, Chhatre et al. [21] identified tenure security and effective participation as the principal safeguards for REDD+ implementation. In Latin America, community forest management is particularly advanced, emerging as a solution from grassroots pressures to retain control over forests [22]. If REDD+ is implemented centrally, however, some authors fear a reversal of the "democratization" of forest governance that evolved over the last 20 years [23]. Although nested approaches to REDD+ implementation offer, in theory, the possibility for jurisdictional- or project-level REDD+ initiatives to inform national policy, a review of multilevel governance in Brazil, Peru, Indonesia, Tanzania, Cameroon and Vietnam suggest that positive feedback between levels is the exception rather than the norm [18].

The literature on early REDD+ implementation suggests that strategies for implementation of community-based REDD+ projects tend to be similar around the globe as rural and indigenous communities try to address common challenges they face [24,25]. For example, the majority (91%) of the projects reviewed in the literature address small-scale drivers of land-use change mostly related to agriculture and cropping systems [24,25]. Most of the projects (80%) include afforestation/reforestation strategies with both timber and agroforestry. As noted by Lawlor [24], the incentives most commonly used in projects are payments for ecosystem services (39%) and integrated conservation and development (29%). Securing land tenure is also an important motivation for community participation; Sills et al. [25] found that most projects support communities in their claims to secure tenure. The need to obtain land tenure also holds true for a project with 14 communities in Colombia [26]. Several projects have focused on land invasions and supporting communities in enforcing their land demarcations [27].

The present study contributes to this growing literature by providing empirical insights from a long-term REDD+ participatory action research initiative in an indigenous Emberá community in eastern Panama. The initiative, which spanned 11 years (2002–2013), was originally conceived in the context of the voluntary carbon market that developed in parallel with the Clean Development Mechanism. The initiative evolved to include avoided deforestation, livelihood enhancement and conflict resolution as part of a voluntary REDD+ pilot project. We present here the lessons learned from the project, identifying successes and barriers, and providing recommendations for future REDD+ implementation, while adopting a people-centred approach. We position the villagers' reality as a starting point and examine the options offered by REDD+ for capability, equity and sustainability as necessary conditions to enhance livelihoods.

2. Methods

2.1. Introducing the Case Study

The long-term participatory action research initiative reported here took place in an indigenous Emberá community of eastern Panama. The region receives an annual average of 2500 mm of rainfall, with a marked dry season between December and April; the mean annual temperature is 26 °C [28] and the primary vegetation is tropical moist forest (Holdridge Life Zones system). Migration and settlement patterns of the Emberá population in eastern Panama's Bayano region are well-described in the literature [29,30]. The Emberá people first settled in the area during the 1950s when they migrated from the Darien, Panama's easternmost province. In the mid-1970s, a hydroelectric dam was constructed in the Bayano River, creating Lake Bayano and displacing some 400 Emberá, 1500 Guna and 2500 colonist farmers [30]. The Majecito Agreement signed between President Omar Torrijos and the Emberá community in 1975 entitled the Emberá to new land without granting full legal land rights. Forty years after the resettlement, in 2015, the Panamanian government officially recognized the territory as a Collective Land under Law 72 following ruling of the Inter-American Court of Human Rights in favor of the Indigenous groups of the Bayano watershed [31].

The community is governed by a political body, the *dirigencia*, led by a community chief (*noko*) chosen by a community assembly. The community is also home to a community-based non-governmental organization (NGO) established in 1998. Its mission is to promote conservation and sustainable development as well as to preserve the culture and traditions of the Emberá people. In 2004, when interest in a carbon project first emerged, the community consisted of 550 people with most of the 71 families residing in a nucleated settlement [32]. The community's territory encompasses 3145 ha, with the land divided into plots ranging in size from 1 ha to 100 ha. Plots are allocated by the chief to individual households and decisions on land-use management are made at the household level. Community regulations prohibit households from selling their plots. Land cover in 2004 included forest (46%), pasture (18%), tall fallow fields (19%), short fallow fields (8%) and cropland (9%) [33]. Households rely, for their livelihoods, on subsistence cultivation, cattle ranching, day labour and handicraft production. Timber, beef and manioc (*Manihot esculenta* Crantz) are the principal market goods.

The community initiated a long-term collaboration with the Neotropical Ecology Laboratory of McGill University in 1996. Since the onset, an action research approach was employed, in which research is conducted jointly between researchers and local participants to empower villagers in identifying issues that affect their lives and promote social change [34]. Action research is undertaken in cycles initiated by implementing an action and reflecting upon the experience to develop a new research cycle, enabling both participants and researchers to "learn by doing" [34]. Lessons from the earlier phase allow the formulation of new research questions and actions for the next cycle, in a progressive learning path [35,36]. Multiple methods are used in action research, and the participatory nature and full involvement of local participants are of central importance.

The first studies in the community, in the late 1990s, examined the conservation status of traditional plant resources [37,38]. Research showed that about a third of the studied species were considered by villagers to be threatened or potentially threatened [38]. These results stimulated community interest in exploring land-use alternatives that would allow for forest restoration and conservation, particularly reforestation. A former community chief mentioned that: *"Through this study, we realized that if we continued cutting trees we were going to end up with no forest; we needed to do something about it."* (*Former community chief in the 1990s, interview conducted in 2009*).

In 2002, the community began exploring the feasibility of implementing a reforestation/afforestation initiative for carbon storage, species conservation and enhancing local livelihoods [33]. Earlier studies suggested that carbon stocks in the community were likely to decline by more than 50% between 2004 and 2024 (from 301,859 t C to 155,730 t C) due to a projected increase in pasture area and an expected reduction in fallow cycle duration on established croplands [33]. This led the community to engage, in 2008, in a voluntary carbon-offset project with a Panama-based research institute (hereafter the client) that engaged to purchase a total of 6900 t CO_2e over 25 years. The contract aimed at enhancing forest carbon stocks through reforestation (3600 t CO_2e equivalent to 10 ha) and avoiding deforestation (3300 t CO_2e equivalent to 24 ha in three years, 8 ha per year). The client committed to providing the seedlings for the first planting season (2008) and to assisting the community-based NGO in establishing a local nursery for future plantations. In this paper, we describe the research cycles that unfolded during this carbon-offset project implementation and draw the lessons learned from the community's experience with REDD+.

The overarching research question that guided the participatory action research between 2009 and 2013 was "How can forest carbon offset initiatives benefit local communities, reconciling emission reduction and local livelihoods?" To answer this question, the project engaged in three cycles of research action. The first research cycle documented here addresses climate change mitigation, focusing on the carbon contract's participatory design, including the internal process undertaken by community leaders to engage local villagers, find a carbon buyer and negotiate the carbon price. The specific question asked during this research cycle was "From participants' viewpoints, what are the best combinations of land-use alternatives to increase carbon stocks and what is an appropriate benefit-sharing design?" The second research cycle focused on early implementation of the carbon contract, asking "What barriers were faced during implementation on the ground?"; "How were they overcome?"; and "What is the participants' perception of the project?" The third research cycle broadened the scope of analysis, aiming to understand interactions with other stakeholders, in particular *colonos* (non-indigenous farmers of Latino origin) living adjacent to Emberá land. This research cycle explored possible ways forward.

2.2. Participatory Design and Implementation of a Carbon Project

In 2004, villagers developed a participatory map of current land uses on their territory that served as a basis to determine a carbon baseline [33]. Members of the community then began discussing what future high-carbon landscapes could look like. They crystallized their desire to implement a carbon project by developing a Project Design Document (PDD) for submission to ANAM, Panama's National Authority for the Environment, which was acting then as the clearing-house for carbon projects in Panama. Interviews conducted in 2005 with 60 of the 70 community landowners helped them determine (1) how much land could be allocated to a carbon project; and (2) which new land uses would be preferred [33].

Between 2009 and 2013, 108 semi-structured interviews were conducted with community leaders, project participants and other stakeholders (Table 1). Interviews were designed to answer questions from each research cycle and to shed light on diverse aspects of project design and implementation. Interviews were conducted with the help of McGill's Panama Field Study Semester (PFSS) students and community research assistants; the latter ensured full comprehension of questions and answers and translated between Spanish and Emberá as needed. The semi-structured interviews were coded qualitatively and key concepts were identified and categorized into interview themes [39]. Multiple coding was used to cross-check the data and interpretation by different members of the research team [40]. For example, data gathered and analyzed by students were always validated in supervisory meetings with I.H. and C.P. Further, member checking—a qualitative validation method for verifying research findings with participants [41]—was used to validate and modify preliminary interpretations as needed. Participant observation, using field notes, was used throughout the research cycles to record community meetings [42,43]. The study followed McGill University's Policy on the Ethical Conduct of Research Involving Human Subjects, the Neotropical Environment Option Protocol for conducting research with Indigenous Peoples in Panama as well as the agreement and rules set forth in Resolution 2 of 16 March 2008 by the *Congreso Local de la Comunidad*. For more details about interviews, including type of interviewees and interview guides and initial coding and analysis of interviews, see [44–47].

Table 1. Total number of interviews conducted throughout the research cycles (2008–2012).

Interviewee	2008	2009	2010	2011	2012	Total
Former community chiefs	1	1				2
Local NGO * representative	1	1	1	2	2	7
National NGO representative		1				1
Representative of the client	1	1				2
ANAM ** representatives		1				1
Project participants						
Participants in avoided deforestation ***		3	3			6
Participants in reforestation ***			7	8	20	35
Interested potential participants					54	54
Total	3	8	11	10	76	108

* Non-Government Organization (NGO); ** National Authority for the Environment (ANAM); *** Some participants were interviewed in multiple years; total participants in avoided deforestation 3 and in reforestation 22.

2.3. Land Invasion and REDD+ Implementation

One year into implementation, in 2009, participants reported that the land set aside for REDD+ was invaded by *colonos*. To document what was happening and determine the rate of deforestation in this area due to *colono* invasion, the area dedicated to REDD+ was visited on three separate occasions—4–14 February and 28 March 2009 (see details in [47]). During each visit, GPS coordinates were systematically taken at points of interest and at minimum intervals of every 200 m (using both Garmin GPS III Plus™ and Legend HCx to improve the precision of locations). At each point, the land use type was noted and points of interests, including houses and working huts, were recorded. The land uses noted followed those used in the 2004 map that served to develop the carbon baseline [33]: pasture, short fallow (1–4 year), tall fallow (>5 year), primary forest, secondary forest and new deforestation. The latter category included freshly cut trees, burned fields and sites in which the understory was entirely cut. As necessary, 360-degree pictures were taken to give a sense of surroundings. Sketches relying on estimating relative distances walked and orientation with a compass were drawn in the field to help subsequent mapping. GPS coordinates and field notes were imported into ArcGIS. A satellite image from Google Earth was then overlaid in ArcGIS using landmarks, such as rivers, mountain ridges and roads. After developing the map, notes for each GPS point were reviewed along with its corresponding photographs to delineate the current land-use areas on the map. Polygons were created to represent the distribution of each land use in the area. The exact extent of the polygons relied heavily on photographs and field sketches. Pictures from different field days from the same area were also compared to verify accuracy; however, distances were often estimated in the field or from photographs. ArcGIS was used to determine polygon areas using geographic coordinates [47].

3. Results

3.1. Research Cycle 1: Participatory Design of a Forest-Carbon Project: What Are the Best Combinations of Land-Use Alternatives to Increase Carbon Stocks and What Is an Appropriate Benefit-Sharing Design?

As the community began to discuss a carbon-offset initiative, there was much interest expressed by local leadership and households. Coffee plantation under forest trees, silvopastoral systems and living fences were identified as attractive novel land uses to increase carbon stocks. Only eight of the 60 land owners initially interviewed in 2005 were not interested in investing part of their land in a carbon project. The vast majority of landowners proposed to convert half of their lands into tree plantations, due to the high market value of timber, while the other half showed interest in fruit tree plantations for their market and subsistence value. In addition, some villagers showed interest in silvipastoral practices (the combination of livestock and trees in the same land); others were interested in putting living fences around their land. The 2005 Project Design Document indicated that 52 owners would participate, dedicating 180 ha of pasture, 66 ha of low fallow, 268 ha of tall fallow and 65 ha of mechanized rice for a total of 579 ha of land. The novel land uses would consist of 290 ha of native timber plantations, 155 ha of agroforestry, 140 ha of silvopastoral and ~7500 m of living fences.

Despite initial enthusiasm, it took four years for community members to initiate the carbon project. In 2006, a working group established by the *cacique* (regional chief) assessed options to go forward with the project and decided to begin the project as a pilot initiative. Recruiting community members to participate in the carbon project took considerable time. The project started in 2008 with seven participants. After three years of implementation, there were 22 participating households. During a community meeting undertaken in 2012, four years after project implementation, 54 people expressed an interest in engaging in a second carbon-offset project, suggesting that they were less reticent to participate after seeing results of others that engaged first. Still, the 22 participating households were more cautious when it came to project establishment then they had been in the PDD development. Their initial pledge in 2005 during the PDD development was on average 30% of their total land. In reality, participants allocated an aggregate of 19 ha to the carbon project, representing 1.3% of their total land. Allocation to novel land uses was consistent with their initial intentions but at a much-reduced scale, and participants' interest shifted over time from timber to agroforestry (Figure 1).

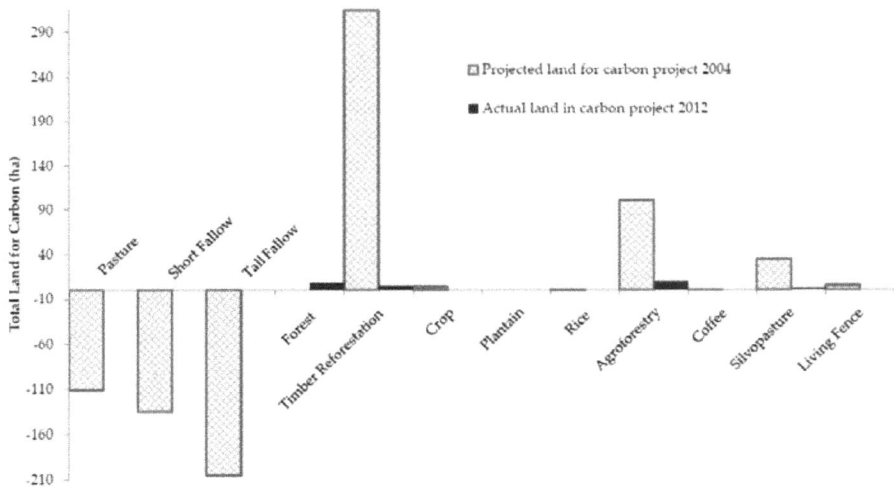

Figure 1. Initial projection and actual land (in hectares) allocated by participants to the carbon-offset project in 2004 and in 2012.

Interviews with both local leaders and participants revealed tensions within the community regarding the notion and design of a benefit-sharing mechanism, suggesting that this tension might have been holding back participation. There was friction particularly over whether the revenues should be for community development or for participating households. The former *cacique* mentioned that, in her view, a carbon project should aim towards the collective development of the community, an opinion shared by the President of the community-based NGO. The local authority wanted to establish a community development fund for improving local infrastructure and access to services such as education and health.

> *Families who wanted to participate sought to receive the entire portion of the carbon funds; while our vision (as leaders) was that part of the money should be for a collective fund to benefit the entire community—for us, this is development—and another portion to families so they could*

> *buy their personal stuff. This internal difference blocked the idea of a carbon project until 2007.*
> *(Former community chief between 2002 and 2004, interview conducted in 2008)*

This view, however, was not shared among interested community members and potential participants.

> *I am the one that works all day under the sun weeding my plot, so why should the carbon project*
> *renovate and buy seats for the communal meeting house? I prefer that this money is in my pocket.*
> *(Project participant (male), reforestation with agroforestry, interview conducted in 2012)*

In the end, 80% of payments were to be made directly to participants and 20% pooled into a community development fund managed by the community-based organization [45]. Per the contract, direct payments would be made bi-annually to participants if they complied with their contract obligation. These commitments included ensuring the preservation of standing forest (avoiding deforestation) and, for the reforestation component of the carbon project, weeding and replacing dead trees.

Results of semi-structured interviews, carried out in 2012, revealed different views as to how financial incentives for participation in a carbon project should be calculated [46]. Some landowners thought the project should compensate participants for the potential gains they would have made in the various land uses that they instead chose to allocate to the project; to this end, villagers estimated land use opportunity costs. Others were comfortable with evidence-based payments, i.e., payments based on carbon sequestered. Finally, some felt that the project should ensure participants' costs of living, providing them with desired social benefits in exchange for their role in maintaining an important ecosystem service. The proponents of this "quality of life approach" estimated that a family income of US$700 per ha per year per family was needed to ensure access to education and community development. Ultimately, the carbon contract used both reforestation and avoided deforestation. The contract allocated US$4500/ha (US$10.22/t CO_2e) for reforestation, with funds to be disbursed over the first eight years. For avoiding deforestation, the client agreed to pay US$100/ha per year based on estimated net returns for cattle ranching over a 25-year period [48].

Interviews held in 2012 with 54 potential new participants revealed that receiving monetary compensation was the main attraction for participating in REDD+ (72% of interviewees) [46]. Thirteen percent of interviewees stated a better quality of life as a reason for participating in a future REDD+ project. Facilitating access to education was the second most common reason (61% of interviewees), with 42.6% of interviewees specifying that they wanted money to send their children to school. Community development was cited by 39% of interviewees. Of these, 57% mentioned improving community infrastructure, while 28.5% specified community development through capacity-building. Further, 26% of interviewees listed improving the environment—in particular, forest preservation and climate change mitigation—as their reason for wanting to participate. Insufficient monetary incentive was a concern for 24% of interviewees. Price volatility and the general increase in cost of living were emphasized, leading interviewees to stress the need for monetary incentives being flexible over time and adjustable according to the rising cost of living [46].

3.2. Research Cycle 2: Early Implementation of the Forest-Carbon Contract: What Barriers Were Faced during Implementation on the Ground? How Were These Overcome? What Is the Participants' Perception of the Project Strategy?

The initial carbon-offset contract included reforestation of existing pasture and short fallow areas. Although reforestation was initially planned with native timber species only (on 10 ha), in 2009, project participants explained that they did not have enough land to devote exclusively to timber and asked to include agroforestry systems that combine fruit and native timber trees as a reforestation option. Project participants argued that exclusive allocation of land to timber could compromise household food security and livelihoods, and they were concerned about the long maturation time and lag for receiving benefits from timber. An interviewee noted that:

"*It is important to have options so reforestation could ensure that I could continue using my land to grow cassava, plantains and the products I need to feed my family*". (*Project participant, reforestation with agroforestry, interview conducted in 2009*)

The client agreed to adjust the reforestation strategy, and agroforestry became a reforestation option in 2009. This change was accepted because the client's interest lay in the total amount of carbon sequestered, regardless of the manner by which it was achieved. The client's representative mentioned that: "*This is not different from buying pineapples; you buy a number of them and want to have them in hand when you pay*". Agroforestry systems were chosen as the reforestation strategy by 55% of project participants (12 households), and only one new project participant chose to reforest exclusively with timber species [45].

Interviews conducted with project participants in 2012 reveal concern regarding the disbursement mechanism of the project. Participants and community leaders alike expressed reservations, specifically regarding the Communal Fund. The idea behind the Communal Fund was to support community development initiatives so project benefits could be broadly shared. However, there was a lack of agreement on the fund's objectives as most project participants (62.5%) did not agree that they should "donate" money for community development from the money they generate through individual efforts on the carbon project. Project participants and community leaders so far have been unable to agree on how the money should be disbursed. Consequently, the Communal Fund has accumulated as clarity on how it should be spent is pending [46].

The interviews in 2012 highlight another concern about the disbursement mechanism, namely lack of cash flow. In all participating families, men—who are mostly in charge of agricultural work—receive daily or monthly payment for daily labour (i.e., as agricultural worker-peons, loggers, teak plantation workers, security guards, etc.). Seventy-five percent (15/20) of participants mentioned hiring agricultural workers (*peones*) for weeding and planting on their reforestation plots [45]. These project participants asked the client to pay them a bi-annual upfront payment rather than paying them after plantations were cleaned, to allow them to hire labour. This, however, has not been accepted by the client, who indicated that such a payment method would entail an investment risk (Field notes from meeting, December 2010).

The initial interviews held in 2010 to inform project design revealed leadership and fund management as an *a priori* concern for 24% of interviewees, who noted the lack of transparency in fund management by existing leaders [44]. This concern emerged as a central theme in the interviews carried out in 2012 after years of implementation [46]. Project participants (reforestation) overwhelmingly (88%) felt that the project lacked the leadership needed to support, motivate and galvanize the project. A key factor that might explain the lack of strong local leadership throughout the project life is that, in 2008, a generational change of leaders occurred and the capacities of elder leaders were not passed on to the new generation. For the new, young leaders of the community who faced the demands of having young children and of building an economic future for their families, time committed to project coordinating activities without compensation was costly and sometimes counterproductive. The carbon-offset contract contemplated the need to support project implementation and develop leadership capacity. A national NGO was invited by the client to monitor the reforestation plots (verification) and assist participants and the community-based NGO in building project implementation capacity. The national NGO initially played a very positive role in the community, bringing additional corporate social responsibility funds from a national enterprise. In the end, the national NGO could not fulfil its role because of lack of funding to support its continued engagement. Its lack of experience in supporting carbon sequestration projects in indigenous communities might have also played a role, since the project was the first of its kind in Panama. "*This is the first experience we have [as organization] with projects related to carbon sequestration and so far we are finding the implementation process slower than expected*" (*National NGO representative, interview conducted in 2009*).

3.3. Research Cycle 3: Unresolved Foreseen Complications

Land tenure was a concern voiced by 16.7% of interviewees during the project design phase in 2008. Community members feared that *colonos* might invade areas dedicated to the carbon project, threatening the success of REDD+. Furthermore, some mentioned that Panama's Constitution indicates that if land is not used, it is not considered as being owned. They were thus concerned that the Panamanian Government could take back their land if they were to "stop using" it in order to conserve the forest under REDD+. At the project onset, only three landowners chose to engage in the avoided deforestation component of the project. These landowners held parcels far away from the village and without direct access by road. As the project unfolded, it became evident that these parcels had historically faced intense land conflicts between the community and local *colonos*, and that the landowners saw in REDD+ an opportunity to resolve the situation. Since 1992, *colono* families that lived on land near the community began clearing forest for agriculture on Emberá community land. To enforce REDD+ on their land, the participants' action plan included: (i) posting signs to delineate the avoided deforestation parcels; (ii) training a community-based patrol to ensure compliance; and (iii) establishing a reforestation border in the conflict area.

Within six months of action plan implementation, it became clear that the land conflict between *colonos* and the community was deeper and more complex than originally feared. Despite sign postage, colonists continued to invade community land. Site visits showed that 36 ha of community land had been cleared by colonists between February and March 2009, including short fallow fields (31 ha), tall fallow fields (4 ha) and primary forest (1 ha) [47]. Tensions increased as clearing continued and threats of violence by colonists rose. Mapping revealed the presence of *colono* houses established for over a decade both along the river and deep inside the Emberá territory. We also documented the presence of pastures used either by *colono* farmers living outside of or *colonos* established within the Emberá land [49].

Interviews with a representative from ANAM indicated that, in theory, *colono* farmers could be forced to pay a fine between $1000 and $1500, and cease any "environmentally detrimental" activities. The official, however, noted that this would not be easy as the *colono* farmers have lived in the area for some time, and it is therefore unclear whether the Emberá have legal title to their land [49]. To do this investigation, they would need to review laws and legal documents to verify that the land is indeed Emberá-owned. ANAM's representative also noted that ANAM alone cannot solve this problem, as other authorities—such as those responsible for ownership rights or the agricultural sector—also must be involved in inherently complex cases of land invasions [49].

Community leaders decided to initiate political actions aimed at resolving the land conflict through dialogue and state representation. Their complaints, however, were lost in the bureaucracy; none of the formal notes sent were acknowledged nor could they be found in the government record or archives. Interviews held in 2009 with representatives of two key agencies—ANAM and Política Indigenista—suggest that the unresolved land tenure status of the community prevented the agencies from enforcing any action on the ground to resolve the land conflict. Further, the agencies lacked clarity in their mandate to address land and land-use conflicts where tenure status was unclear; interviewees argued that their agencies had no legal mandate per se to resolve conflicts.

To support dialogue between local stakeholders with competing interests for the land and the government and identify possible ways forward, researchers from the Neotropical Ecology Laboratory invited local and national stakeholders to form a working group, The Advisory Council on Conflict Resolution and REDD+ (hereafter The Council). Between May and December 2011, The Council brought together 68 participants representing 34 organizations including representatives from government agencies (14), indigenous peoples (5), colonists (5) and NGOs (4) to establish a successful intercultural collaborative dialogue on the contentious issue of territorial conflicts among different sectors [50].

The Council developed a series of recommendations shedding light on the considerable confusion about the roles and responsibilities of government agencies in land tenure and invasion law enforcement (Table 2) [50]. Nonetheless, despite the efforts to build capacity for land conflict resolution and new government laws since The Council meetings, much confusion remains about the roles and responsibilities of government agencies in land law enforcement.

Table 2. Recommendations issued by The Advisory Council on Conflict Resolution and REDD+ to the National Land Council Panamanian Government [50]. ANATI: National Authority for Land, the government agency responsible for land titling.

1.	The National Land Council should lead territorial conflict resolution following ANATI's advice.
2.	It is essential to precisely define territorial limits as proof of judicial processes that will determine relocation and/or eviction decisions. The national Geographic Institute should be responsible for field analyses relying on geographical information, in coordination with the national Limits Commission.
3.	Where they do not exist, Municipal Courts should be created to manage territorial conflicts and mandate executions in coordination with responsible authorities.
4.	It is imperative to harmonize the work of different government entities and clarify the legal context. Legal gaps and overlapping/conflicting legislation need to be identified and corrected; clear rules defining institutional responsibilities that apply to land conflicts are a must.
5.	ANATI should implement an extensive and in-depth divulgation campaign to present and clearly explain hierarchical order and institutional Government mandates pertaining to territorial conflict resolution, as well as the corresponding processes to be followed.
6.	The National Land Council should create a Follow-up Agreement Commission.
7.	The Advisory Council offers that its members be integrated into the Follow-up Agreement Commission as they are personally acquainted with territorial conflicts, have received training and tools for their resolution through dialogue and have demonstrated genuine interest in reaching consensual and beneficial solutions for all parties involved.

4. Discussion

The socio-economic and cultural features of communities in which REDD+ is being implemented vary widely around the world. Lawlor et al. [24] reviewed 41 projects across 22 countries—from Latin America, Asia and Africa—and Sills et al. [25] reviewed another 23 projects in six countries. These projects take place in a variety of ecosystems, from drylands to tropical rainforests, with project size ranging from 42 to 642,184 ha and local populations varying from 1025 to 250,000 people [24,25]. The Panamanian project analyzed here, with 19 ha of reforestation and 22 participants, is smaller than most other REDD+ case studies reviewed. However, it began in 2002 when the community sought financial support to estimate the carbon stocks of its territory and is therefore one of the longest ongoing such initiatives in Latin America, together with Noel Kempf Mercado in Bolivia and Scolel-Té in Mexico, initiated in 1996 and 1997, respectively.

Overall, the community's efforts to engage in REDD+ documented here show that implementing REDD+ may be particularly challenging for poor communities. REDD+ was initially proposed as a "cost-effective strategy" for climate change mitigation [51]. Attention has since been drawn to the fact that cost-effectiveness is context-dependent and conditional on the remoteness of areas, the nature of social, cultural and environmental landscapes, governance and tenure regimes, and more [52]. Indeed, the general claim of cost-effectiveness has been called into question [9,53]. In addition, REDD+, as an initial "carbon-market" vision, was conceived as a financial rather than governance mechanism [54] but implementation of a people-centred approach requires that REDD+ be understood as a new environmental governance structure [55].

Our long-term study showed that, to embrace livelihood enhancement in rural communities, REDD+ design and implementation must consider the underpinning complex of social, cultural, political and economic factors operating at different levels (Figure 2). The journey we described in this article is rich in lessons learned and we use them to guide the discussion of key observed implementation barriers and attempts to overcome them.

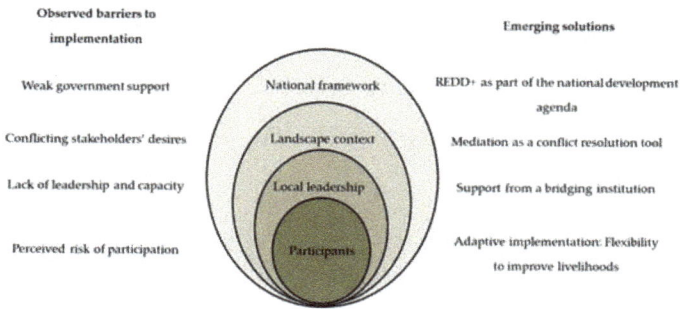

Figure 2. Concept map showing REDD+ implementation considering a landscape approach and summary of observed barriers to REDD+ implementation and emerging solutions.

4.1. Lesson 1: REDD+ Participation Entails a Slow Process of Engagement

A key lesson learned is that engagement in REDD+, in the context of poor communities, is a slow process. The concept of carbon credits or selling carbon is apparently not easily understood by local communities and there is considerable uncertainty about how REDD+ will affect existing tenure and use rights (i.e., changes in carbon tenure) [52]. For REDD+ to be fully appreciated by communities, sufficient time must be available to negotiate conditions under which communities will feel motivated to participate [52,56]. Consent must be understood as a process and negotiated between affected parties throughout all stages of a project, which inevitably takes time [57]. Three years into project implementation, we felt that residents were still risk-averse but willing to take a chance after seeing the benefit/risk ratio of early participants. This "demonstration effect" is consistent with reports that pilot initiatives can act as a catalyst for others to engage and scale up [15,58]. Local willingness to participate and engage in new strategies can require long timeframes which need to be considered in the project design phase [59]. Cronkleton et al. [22] forewarned REDD+ planners that they should expect a long period of learning and adaptations and therefore find "patient" investors.

Previous studies have emphasized the need to provide a suite of activities from which participants can choose to ensure broad participation and shape the project strategy to local livelihood needs [8,33,60]. A review of 40 Payment for Ecosystem Service projects noted that initiatives that improved livelihoods and community development were more successful than those that did not [61]. In the present case study, the flexibility of the client who accepted a modification in the original plantation design to allow the establishment of agroforestry systems played a key role. The tangible livelihood benefits incurred by establishing agroforestry plots facilitated adoption by reducing the perceived risk of engaging in REDD+. We therefore call for adaptive implementation allowing to reorganize project activities during implementation to align with the evolving needs and constraints of participants.

4.2. Lesson 2: Strong Sustained Local Leadership and Nested Bridging Institutions Are Critical for Successful REDD+ Implementation

A second lesson drawn from the community's journey is that the "willingness" to manage the forest sustainably is insufficient, alone, to ensure success. REDD+ is a complex mechanism and implementation in indigenous and small rural communities poses particular challenges. The effectiveness of REDD+ projects rests largely on the ability and interest of local communities to manage their forests sustainably while enhancing forest carbon stocks [55]. Participants in the project identified project leadership as a major barrier in implementation. Evaluating the use of payment for ecosystem services to foster community development and forest conservation in Latin America, Couto Pereira [62] noted that discussions around capability, equity and sustainability were often absent from ongoing dialogues.

The framework developed by Cerbu et al. [5] to understand the capacity barriers of communities seeking to implement REDD+ is highly useful. The authors considered four categories of capacity: (1) technical; (2) managerial; (3) organizational; and (4) related to dealing with risk and potential REDD+/livelihood development conflicts. In the Panama project, we noted strong capacity for both the technical and managerial aspects of resource use [33,45]. The leadership issues noted as a major impediment were organizational in nature.

We propose that, for small communities, REDD+ projects should be considered as emerging sustainable social-ecological systems, *sensu* Ostrom [63]. In such a context, bridging institutions could assist communities in building their capacity for successful REDD+ implementation [12]. Bridging institutions, a key element in Ostrom's [63] framework, as Berkes ([64], p. 1692) notes, "bridging organizations provide a forum for the interaction of these different kinds of knowledge, and the coordination of other tasks that enable co-operation: accessing resources, bringing together different actors, building trust, resolving conflict, and networking". Recognition of the need for capacity-building and the role of bridging institutions should be considered in designing payments. Payment designed to foster stewardship or to compensate for lost opportunities might be more appropriate for grassroots communities than payments targeting ecosystem services [55]. The Juma Sustainable Development Reserve in Brazil, with its four components of supporting families—sustainable income generation, local organization strengthening and participation and social development—provides an interesting example of payments to support forest stewardship [65].

4.3. Lesson 3: Multi-Actor Land Conflicts Can Undermine REDD+

A third lesson is that inability to engage with multiple agents responsible for deforestation can imperil communities' efforts to implement REDD+. This study was carried out in an intercultural context where interest groups collide forcefully over the fate of the forest. Land allocation policies in developing countries often consider intact forests as "unproductive" resources [66]. One way of "improving" land is through deforestation [66,67] as cleared land is more valuable than standing forest [68,69] and, in countries such as Panama, the Constitution and Agrarian Code indicate that deforestation signals land ownership. As pressure on land resources becomes more acute, pressure on forests can mount and excluding outsiders becomes even more challenging [19]. It has long been recognized that uncertain tenure can facilitate deforestation [67,70]. Corbera et al. [20], however, cautioned against using REDD+ as an instrument for gaining land title and tenure security over their lands, suggesting that income would be insufficient to enable the full enforcement of forest rights.

Recently, a landscape approach that integrates agriculture, conservation and other land uses is gaining momentum in REDD+ implementation [71]. This approach acknowledges the decisions made by multiple actors and cultures at the landscape level [72] and enables the search for solutions that broadly address conservation and development challenges [71]. Undertaking REDD+ implementation based on a landscape approach would allow us to look beyond the forest, acknowledging competing land uses as well as the complete array of actors and institutions that shape land use beyond local communities [21].

In the Panama project, the community chose to engage in a process involving conflict resolution methods such as mediation [73] to try to find a way forward. The Council that served as a multi-stakeholder platform for discussing the challenges of REDD+ implementation was successful in that its analysis shed clarity on the responsibilities of the many institutions involved in land tenure and management. However, The Council failed to attract the attention of higher levels of government, despite being composed of representatives from five governmental agencies.

4.4. Lesson 4: The Self-Interest of the State

Possibly the greatest impediment faced during the Panama carbon project's development was the lack of support from the State. This was first evident in 2005 when the community sent their PDD to ANAM. Despite numerous efforts to follow-up, the community never heard back from the government. This absence of interest prevented the community from garnering support to identify potential carbon buyers internationally. A similar failure to support the community was observed when government authorities were presented with both the evidence of *colono* invasion of the communal land and The Council's proposal for mediation. The lack of cooperation and assistance from the State is consistent with reports of poor communication between the government and indigenous peoples [74].

Policy options that could allow for reducing deforestation in Latin America include the need to stop illegal expansion of the agricultural frontier [75]. An essential question therefore is "Why would this be in the interest of the State?" Forests have long been perceived as a supply of land in Latin America [23]. In this context, "predatory forest exploitation" is often financially more attractive than careful management [75]. As noted by Chhatre et al. [21], some of the most forested regions in the world are characterized by weak rule of law and a low level of public accountability. In a comparison of REDD+ policies in Brazil, Cameroon, Tanzania, Indonesia and Vietnam, Ravikumar et al. [18] pointed to the absence of harmonization between national and local jurisdictional policies. Brazil is apparently the only country in which land tenure regularization by proponents is more closely aligned with national policies and in which a series of measures adopted by the State have allowed successful reduction of deforestation in the Amazon [19,23].

In a review of 23 REDD+ projects at the jurisdiction and project level, Nasi et al. [75] trace the willingness of States to manage forest as emerging from social and political values influenced by the positioning of extractive industries and society. Scholars have emphasized the need to trigger a values shift in response to climate change—a task that requires action on the part of governments, private sector leaders, and civil society—and to create a vision of the future that is both desirable and feasible [76]. Imaginative, novel values could possibly emerge from a national REDD+ agenda that proposes another way to "improve" land, such as partnering with conservation organizations [77]. As noted by Sjaastad and Bromley ([78], p. 553) "although insecurity of tenure is a disincentive to invest, it is—paradoxically—often also an incentive because investment will itself increase security". REDD+ could therefore offer a way for local communities to prove that the land is being used "productively" (i.e., with respect to carbon), showing investment in and improvements to their land and increasing the value of standing forests.

At the time of our case study, combating deforestation has not been a national priority for Panama. Indeed, current as well as past priorities centered around poverty elimination, security and health [79]. At the Conference of Parties of the UN Framework Convention on Climate Change in Paris in 2015, Panama's President launched a national reforestation campaign pledging to reforest one million hectares over 20 years [80]. It will be interesting to see if this new initiative will finally provide the community with the support needed to advance their goal to maintain forest cover in their territory.

5. Conclusions

Whether the Panama project was a success or failure depends on which objectives are taken as paramount. From the strict viewpoint of carbon sequestration, the Panama project could be interpreted as a failure. Project participants were unable to implement the avoided deforestation component of the original carbon contract, equivalent to 3300 t CO_2e, and the reforestation component fell short of reaching its carbon sequestration objective. It was estimated that agroforestry systems would sequester 71 ± 2.5 t C ha^{-1} but projections showed that, due to tree mortality, carbon sequestration would reach only 59 ($\pm$ 16) t C ha^{-1} [45]. This outcome contrasts strongly with the trajectory of the Scolel-Té project of Chiapas Mexico [81], a project that started with 43 participants who reforested 77.5 ha and 10 years later, in 2015, reached 2437 participants for 9645 ha. A possible explanation for the marked difference in trajectory between the two projects is that, while both were initiated as "ways to advance community development and provide environmental services" [82], Scolel-Té evolved in response to new possibilities offered by the carbon market whereas the Emberá project remained a community-development strategy.

From the perspective of community development, the Panama project was a success. Three years following project implementation, half of the project participants were receiving income from sales of early-maturing fruit, and project participant interviews indicated that they felt their livelihoods had improved [45]. Furthermore, indigenous authorities and villagers from nearby Emberá territories strive to emulate the community in which the carbon project took place, perceiving it as successful in avoiding environmental degradation [83]. Thus, from a people-centered approach or from a community development viewpoint, the Panama carbon project did achieve favorable outcomes.

We propose that successful implementation of REDD+ projects in small rural or indigenous communities demands a shift in paradigm, away from "evidence-based payment" and towards an integrated development approach. The key lesson we learned is that REDD+ takes much time to implement because it involves redefining livelihood strategies. It is indeed unlikely that, at the current market price for carbon, evidence-based payments could support the transformative change that is required at the community level, e.g., development of enhanced leadership capacity of local organizations or of bridging institutions. Alternate models of support for community-based REDD+ initiatives are needed, such as the Juma project in Brazil [65] or the N'hambita community carbon project in Mozambique [15] that are combining evidence-based payments with development and poverty reduction, and apparently are better suited to support community development needs and challenges in the context of REDD+.

Acknowledgments: We thank all PFSS students that assisted in data collection and early analysis throughout the various research cycles including: Phillippe Brunet, Etienne Lafortune, Alexandre Duchesne, Maxime Lemoyne, Jena Whitson, Danylo Bobyk, Kirsten Wiens, Fréderic Lebel, Kiley Remiszewski, Mya Sherman and Marie Verrot. Special thanks to the entire Emberá community and the carbon-offset project participants in this study. Thanks to my host family in the community, Olga Dogírama and Rigoberto Caisamo, from whom I learned a great deal. Thanks also to the community leaders, especially Bonarge Pacheco and Alba Caisamo. Special thanks to Lady Mancilla and José Monteza for their constant support during fieldwork and beyond. Thanks to Laura Donohue who reviewed early versions of this paper and assisted in formatting and preparation of maps and figures. Thanks to CONICYT for the scholarship provided to Ignacia Holmes. We are very grateful for the two anonymous reviewers for their comprehensive reading and remarks that allow to greatly improved this manuscript.

Author Contributions: Ignacia Holmes conceived and designed the research, and conducted a large part of data collection and analysis. Catherine Potvin and Ignacia Holmes wrote the manuscript with contributions from Oliver Coomes. Catherine Potvin supervised the research.

Conflicts of Interest: The authors declare no conflict of interest.

References

1. United Nations Framework Convention on Climate Change. *Proceedings of UNFCCC 2007, Decision 2/CP.13.* Available online: http://unfccc.int/resource/docs/2007/cop13/eng/06a01.pdf#page=8 (accessed on 26 February 2017).
2. Angelsen, A.; McNeill, D. The evolution of REDD+. In *Analysing REDD+ Challenges and Choices*; Angelsen, A., Ed.; Center for International Forestry Research: Bogor, Indonesia, 2012; pp. 31–51.
3. FCPF. About FCPF. Available online: https://www.forestcarbonpartnership.org/about-fcpf-0 (accessed on 10 October 2015).
4. UN-REDD. UN REDD Programme Fund. Available online: http://mptf.undp.org/factsheet/fund/CCF00 (accessed on 10 October 2015).
5. Cerbu, G.A.; Swallow, B.M.; Thompson, D.Y. Locating REDD: A global survey and analysis of REDD readiness and demonstration activities. *Environ. Sci. Policy* **2011**, *14*, 168–180. [CrossRef]
6. Streck, C.; Parker, C. Financing REDD+. In *Analysing REDD+ Challenges and Choices*; Angelsen, A., Ed.; Center for International Forestry Research: Bogor, Indonesia, 2012; pp. 91–111.
7. James, A.N.; Green, M.; Paine, J. *A Global Review of Protected Area Budgets and Staff*; World Conservation Monitoring Centre: Cambridge, UK, 1999.
8. Venter, O.; Koh, L.P. Reducing emissions from deforestation and forest degradation (REDD+): Game changer or just another quick fix? *Ann. N. Y. Acad. Sci.* **2012**, *1249*, 137–150. [CrossRef] [PubMed]

9. Minang, P.A.; van Noordwijk, M. Design challenges for achieving reduced emissions from deforestation and forest degradation through conservation: Leveraging multiple paradigms at the tropical forest margins. *Land Use Policy* **2013**, *31*, 61–70. [CrossRef]
10. Kshatriya, M.; Sills, E.O.; Lin, L. Global Database of REDD+ and Other Forest Carbon Projects. Interactive Map. Available online: http://www.forestsclimatechange.org/redd-map (accessed on 28 January 2017).
11. Danielsen, F.; Adrian, T.; Brofeldt, S.; van Noordwijk, M.; Poulsen, M.K.; Rahayu, S.; Burgess, N. Community monitoring for REDD+: International promises and field realities. *Ecol. Soc.* **2013**, *18*, 41. [CrossRef]
12. Holmes, I.; Potvin, C. Avoiding re-inventing the wheel in a people-centered approach to REDD+. *Conserv. Biol.* **2014**, *28*, 1380–1393. [CrossRef] [PubMed]
13. Panfil, S.N.; Harvey, C.A. REDD+ and biodiversity conservation: A review of the biodiversity goals, monitoring methods, and impacts of 80 REDD+ projects. *Conserv. Lett.* **2015**, *9*, 143–150. [CrossRef]
14. VCS. VCS Project Database: About the CCB Project Database. Available online: http://vcsprojectdatabase.org/#/ccb (accessed on 28 January 2017).
15. Jindal, R.; Kerr, J.M.; Carter, S. Reducing poverty through carbon forestry? Impacts of the N'hambita community carbon project in Mozambique. *World Dev.* **2012**, *40*, 2123–2135. [CrossRef]
16. Maraseni, T.N.; Neupane, P.R.; Lopez-Casero, F.; Cadman, T. An assessment of the impacts of the REDD+ pilot project on community forests user groups and their community forests in Nepal. *J. Environ. Manag.* **2014**, *136*, 37–46. [CrossRef] [PubMed]
17. CIFOR. Global Comparative Study on REDD+. Available online: http://www.cifor.org/gcs/ (accessed on 28 January 2017).
18. Ravikumar, A.; Larson, A.M.; Tovar, J.G.; Duchelle, A.E.; Myers, R. Multilevel governance challenges in transitioning towards a national approach for REDD+: Evidence from 23 subnational REDD+ initiatives. *Int. J. Commons* **2015**, *9*, 909–931. [CrossRef]
19. Sunderlin, W.D.; Larson, A.M.; Duchelle, A.E.; Resosudarmo, I.A.P.; Huynh, T.B.; Awono, A.; Dokken, T. How are REDD+ proponents addressing tenure problems? Evidence from Brazil, Cameroon, Tanzania, Indonesia, and Vietnam. *World Dev.* **2014**, *55*, 37–52. [CrossRef]
20. Corbera, E.; Estrada, M.; May, P.; Navarro, G.; Pacheco, P. Rights to land, forests and carbon in REDD+: Insights from Mexico, Brazil and Costa Rica. *Forests* **2011**, *2*, 301–342. [CrossRef]
21. Chhatre, A.; Lakhanpal, S.; Larson, A.M.; Nelson, F.; Ojha, H.; Rao, J. Social safeguards and co-benefits in REDD+: A review of the adjacent possible. *Curr. Opin. Environ. Sustain.* **2012**, *4*, 654–660. [CrossRef]
22. Cronkleton, P.; Bray, D.B.; Medina, G. Community forest management and the emergence of multi-scale governance institutions: Lessons for REDD+ development from Mexico, Brazil and Bolivia. *Forests* **2011**, *2*, 451–473. [CrossRef]
23. Petkova, E.; Larson, A.; Pacheco, P. Forest governance, decentralization and REDD+ in Latin America. *Forests* **2010**, *1*, 250–254. [CrossRef]
24. Lawlor, K.; Madeira, E.; Blockhus, J.; Ganz, D. Community participation and benefits in REDD+: A review of initial outcomes and lessons. *Forests* **2013**, *4*, 296–318. [CrossRef]
25. Sills, E.O.; Atmadja, S.; de Sassi, C.; Duchelle, A.E.; Kweka, D.; Resosudarmo, I.A.P.; Sunderlin, W.D. *REDD+ on the Ground: A Case Book of Subnational Initiatives Across the Globe*; Center for International Forestry Research (CIFOR): Bogor, Indonesia, 2014.
26. Castro-Nunez, A.; Mertz, O.; Quintero, M. Propensity of farmers to conserve forest within REDD+ projects in areas affected by armed-conflict. *For. Policy Econ.* **2016**, *66*, 22–30. [CrossRef]
27. Hayes, T.M. A challenge for environmental governance: Institutional change in a traditional common-property forest system. *Policy Sci.* **2010**, *43*, 27–48. [CrossRef]
28. Instituto Geográfico Nacional. *Atlas Nacional de la República de Panamá*; Instituto Geográfico Nacional Tommy Guardia: Panama city, Panama, 1998.
29. Wali, A. *Kilowatts and Crisis: Hydroelectric Power and Social Dislocation in Eastern Panama*; Westview Press: Boulder, CO, USA, 1989.
30. Wali, A. The transformation of a frontier: State and regional relationships in Panama, 1972–1990. *Hum. Organ.* **1993**, *52*, 115–129. [CrossRef]
31. Corte Interamericana de Derechos Humanos. Caso de Los Pueblos Indígenas Kuna de Madigandí y Emberá de Bayano y sus miembros vs. Panamá: Sentencia de 14 de Octubre de 2014. Available online: http://www.corteidh.or.cr/docs/casos/articulos/seriec_284_esp.pdf (accessed on 28 January 2017).

32. Tschakert, P.; Coomes, O.T.; Potvin, C. Indigenous livelihoods, slash-and-burn agriculture, and carbon stocks in eastern Panama. *Ecol. Econ.* **2007**, *60*, 807–820. [CrossRef]
33. Potvin, C.; Tschakert, P.; Lebel, F.; Kirby, K.; Barrios, H.; Bocariza, J.; Caisamo, J.; Caisamo, L.; Cansari, C.; Casama, J.; et al. A participatory approach to the establishment of a baseline scenario for a reforestation clean development mechanism project. *Mitig. Adapt. Strateg. Glob. Chang.* **2007**, *12*, 1341–1362. [CrossRef]
34. Greenwood, D.J.; Levin, M. *Introduction to Action Research*, 2nd ed.; Sage Publications, Inc.: Thousand Oaks, CA, USA, 2007.
35. Elliott, P.W. *Participatory Action Research: Challenges, Complications, and Opportunities*; Centre for the Study of Co-Operatives; University of Saskatchewan: Saskatoon, SK, Canada, 2013.
36. Kapoor, D.; Jordan, S. Introduction. In *Education, Participatory Action Research and Social Change: International Perspectives*; Kapoor, D., Jordan, S., Eds.; Palgrave Macmillan: New York, NY, USA, 2009.
37. Dalle, S.P.; López, H.; Díaz, D.; Legendre, P.; Potvin, C. Spatial distribution and habitats of useful plants: An initial assessment for conservation on an indigenous territory, Panama. *Biodivers. Conserv.* **2002**, *11*, 637–667. [CrossRef]
38. Dalle, S.P.; Potvin, C. Conservation of useful plants: An evaluation of local priorities from two indigenous communities in eastern Panama. *Econ. Bot.* **2004**, *58*, 38–57. [CrossRef]
39. Auerbach, C.F.; Silverstein, L.B. *Qualitative Data: An Introduction to Coding and Analysis*; New York University Press: New York, NY, USA, 2003.
40. Barbour, R.S. Checklists for improving rigour in qualitative research: A case of the tail wagging the dog? *Br. Med. J.* **2001**, *322*, 1115–1117. [CrossRef]
41. Morse, J.M.; Barrett, M.; Mayan, M.; Olson, K.; Spiers, J. Verification strategies for establishing reliability and validity in qualitative research. *Int. J. Qual. Methods* **2002**, *1*, 13–22. [CrossRef]
42. Cook, I. Participant observation. In *Methods in Human Geography: A Guide for Students Doing a Research Project*; Flowerdew, R., Martin, D., Eds.; Pearson Education: Upper Saddle River, NJ, USA, 2005.
43. DeWalt, K.M.; DeWalt, B.R. *Participant Observation: A Guide for Fieldworkers*; Rowman & Littlefield: Lanham, MD, USA, 2011.
44. Brunet, P.; Lafortune, E. *PFSS Internship Report: Evaluating the Constraints, Opportunities and Challenges to the Adoption of a Reforestation-Based Carbon Sequestration Project as a Means of Conservation and Economic Development in an Indigenous Community of Eastern Panama*; McGill University: Montreal, QC, Canada, 2010.
45. Holmes, I.; Kirby, K.R.; Potvin, C. Agroforestry within REDD+: Experiences of an indigenous Emberá community in Panama. *Agrofor. Syst.* **2016**, 1–17. [CrossRef]
46. Remiszewski, K.; Sherman, M.; Verrot, M. *PFSS Internship Report: An exploration of the Ipetí-Emberá Community's Vision for a Possible REDD+ Project*; McGill University: Montreal, QC, Canada, 2012.
47. Whitson, J.; Bobyk, D. *PFSS Internship Report: Land Use Change and the Agricultural Frontier in Ipetí-Emberá*; McGill University: Montreal, QC, Canada, 2009.
48. Coomes, O.T.; Grimard, F.; Potvin, C.; Sima, P. The fate of the tropical forest: Carbon or cattle? *Ecol. Econ.* **2008**, *65*, 207–212. [CrossRef]
49. Duchesne, A.; Lemoyne, M. *PFSS Internship Report: Living on a Modern Colonization Frontier: An Assessment of Colono Necessities and Livelihood Strategies in the Buffer Zone of Ipetí*; McGill University: Montreal, QC, Canada, 2009.
50. Smithsonian Tropical Research Institute. *Recomendaciones del Consejo Consultivo en Resolución de Conflictos en REDD+*; Smithsonian Tropical Research Institute: Panama city, Panama, 2012.
51. Stern, N. *The Economics of Climate Change: The Stern Review*; Cambridge University Press: Cambridge, UK, 2006.
52. Poudel, D.P. REDD+ comes with money, not with development: An analysis of post-pilot project scenarios from the community forestry of Nepal Himalaya. *Int. J. Sustain. Dev. World Ecol.* **2014**, *21*, 552–562. [CrossRef]
53. Eliasch, J. *Climate Change: Financing Global Forests*; The Eliasch Review; Earthscan: London, UK, 2008.
54. Larson, A.M.; Petkova, E. An introduction to forest governance, people and REDD+ in Latin America: Obstacles and opportunities. *Forests* **2011**, *2*, 86–111. [CrossRef]
55. Bayrak, M.M.; Marafa, L.M. Ten years of REDD+: A critical review of the impact of REDD+ on forest-dependent communities. *Sustainability* **2016**, *8*, 680. [CrossRef]
56. Tiani, A.M.; Bele, M.Y.; Sonwa, D.J. What are we talking about? The state of perceptions and knowledge on REDD+ and adaptation to climate change in central Africa. *Clim. Dev.* **2015**, *7*, 310–321. [CrossRef]

57. Salim, E. *Striking a Better Balance: The Final Report of the Extractive Industries Review*; The World Bank and Extractive Industries: Washington, DC, USA, 2004; Volume 1.

58. Poffenberger, M. Restoring and conserving Khasi forests: A community-based REDD strategy from northeast India. *Forests* **2015**, *6*, 4477–4494. [CrossRef]

59. Mayrand, K.; Paquin, M. *Payments for Environmental Services: A Survey and Assessment of Current Schemes*; Unisfera International Center: Montreal, QC, Canada, 2004.

60. Tomich, T.P.; van Noordwijk, M.; Vosti, S.A.; Witcover, J. Agricultural development with rainforest conservation: Methods for seeking best bet alternatives to slash-and-burn, with applications to Brazil and Indonesia. *Agric. Econ.* **1998**, *19*, 159–174. [CrossRef]

61. Grima, N.; Singh, S.J.; Smetschka, B.; Ringhofer, L. Payment for Ecosystem Services (PES) in Latin America: Analysing the performance of 40 case studies. *Ecosyst. Serv.* **2016**, *17*, 24–32. [CrossRef]

62. Couto Pereira, S. Payment for environmental services in the Amazon forest: How can conservation and development be reconciled? *J. Environ. Dev.* **2010**, *19*, 171–190. [CrossRef]

63. Ostrom, E. A general framework for analyzing sustainability of social-ecological systems. *Science* **2009**, *325*, 419–422. [CrossRef] [PubMed]

64. Berkes, F. Evolution of co-management: Role of knowledge generation, bridging organizations and social learning. *J. Environ. Manag.* **2009**, *90*, 1692–1702. [CrossRef] [PubMed]

65. Agustsson, K.; Garibjana, A.; Rojas, E.; Vatn, A. An assessment of the forest allowance programme in the Juma Sustainable Development Reserve in Brazil. *Int. For. Rev.* **2014**, *16*, 87–102. [CrossRef]

66. Rudel, T.K. *Tropical Forests: Regional Paths of Destruction and Regeneration in the Late Twentieth Century*; Columbia University Press: New York, NY, USA, 2005.

67. Geist, H.J.; Lambin, E. *What Drives Tropical Deforestation? A Meta-Analysis of Proximate and Underlying Causes of Deforestation Based on Subnational Case Study Evidence*; LUCC International Project Office, University of Louvain: Louvain, Begium, 2001.

68. Holland, T.G.; Coomes, O.T.; Robinson, B.E. Evolving frontier land markets and the opportunity cost of sparing forests in western Amazonia. *Land Use Policy* **2016**, *58*, 456–471. [CrossRef]

69. López, R.; Valdés, A. Fighting rural poverty in latin america: New evidence and policy. In *Rural Poverty in Latin America*; Lopez, R., Valdes, A., Eds.; Palgrave Macmillan: London, UK, 2000.

70. Skutsch, M.; Vickers, B.; Georgiadou, Y.; McCall, M. Alternative models for carbon payments to communities under REDD+: A comparison using the polis model of actor inducements. *Environ. Scie. Policy* **2011**, *14*, 140–151. [CrossRef]

71. Sayer, J.; Sunderland, T.; Ghazoul, J.; Pfund, J.; Sheil, D.; Meijaard, E.; Venter, M.; Boedhihartono, A.K.; Day, M.; Garcia, C. Ten principles for a landscape approach to reconciling agriculture, conservation, and other competing land uses. *Proc. Natl. Acad. Sci. USA* **2013**, *110*, 8349–8356. [CrossRef] [PubMed]

72. Pfund, J.-L.; Watts, J.D.; Boissiere, M.; Boucard, A.; Bullock, R.M.; Ekadinata, A.; Dewi, S.; Feintrenie, L.; Levang, P.; Rantala, S. Understanding and integrating local perceptions of trees and forests into incentives for sustainable landscape management. *Environ. Manag.* **2011**, *48*, 334–349. [CrossRef] [PubMed]

73. Amado, A.; St-Laurent, P.; Potvin, C.; Llapur, R. Conflictor territoriales—Modelo para su resolución en preparación para la protección de bosques. In *Experiencias Latinoamericanas en el Abordaje de Conflictos*; Pfund, A., Ed.; Universidad Para la Paz: San José, Costa Rica, 2015; pp. 81–99.

74. Potvin, C.; Mateo-Vega, J. Panama: Curb indigenous fears of REDD+. *Nature* **2013**, *500*, 400. [CrossRef] [PubMed]

75. Nasi, R.; Putz, F.; Pacheco, P.; Wunder, S.; Anta, S. Sustainable forest management and carbon in tropical Latin America: The case for REDD+. *Forests* **2011**, *2*, 200–217. [CrossRef]

76. Burch, S.; Shaw, A.; Dale, A.; Robinson, J. Triggering transformative change: A development path approach to climate change response in communities. *Clim. Policy* **2014**, *14*, 467–487. [CrossRef]

77. Schwartzman, S.; Zimmerman, B. Conservation alliances with indigenous peoples of the Amazon. *Conserv. Biol.* **2005**, *19*, 721–727. [CrossRef]

78. Sjaastad, E.; Bromley, D.W. Indigenous land rights in sub-Saharan Africa: Appropriation, security and investment demand. *World Dev.* **1997**, *25*, 549–562. [CrossRef]

79. Gobierno de la Republica de Panamá. INICIO. Available online: https://www.presidencia.gob.pa/Inicio (accessed on 28 January 2017).

80. ANCON. Alianza por el Millón de Hectáreas. Available online: http://www.ancon.org/alianza-por-el-millon-de-hectareas/ (accessed on 28 January 2017).
81. Hendrickson, C.Y.; Corbera, E. Participation dynamics and institutional change in the Scolel-Té carbon forestry project, Chiapas, Mexico. *Geoforum* **2015**, *59*, 63–72. [CrossRef]
82. Nelson, K.C.; de Jong, B.H.J. Making global initiatives local realities: Carbon mitigation projects in Chiapas, Mexico. *Glob. Environ. Chang.* **2003**, *13*, 19–30. [CrossRef]
83. Sharma, D.; Holmes, I.; Vergara-Asenjo, G.; Miller, W.N.; Cunampio, M.; Cunampio, R.B.; Cunampio, M.B.; Potvin, C. A comparison of influences on the landscape of two social-ecological systems. *Land Use Policy* **2016**, *57*, 499–513. [CrossRef]

![forests logo]

forests

MDPI

Article

Forest Islands and Castaway Communities: REDD+ and Forest Restoration in Prey Lang Forest

Courtney Work

International Institute of Social Studies, Erasmus University, 2502 LT The Hague, Netherlands; courtneykwork@gmail.com; Tel.: +855-66-799-559

Academic Editors: Esteve Corbera and Heike Schroeder
Received: 31 October 2016; Accepted: 2 February 2017; Published: 17 February 2017

Abstract: Climate Change policies are playing an ever-increasing role in global development strategies and their implementation gives rise to often-unforeseen social conflicts and environmental degradations. A landscape approach to analyzing forest-based Climate Change Mitigation policies (CCM) and land grabs in the Prey Lang Forest landscape, Cambodia revealed two Korea-Cambodia partnership projects designed to increase forest cover that are juxtaposed in this paper. Case study data revealed a REDD+ project with little negative impact or social conflict in the project area and an Afforestation/Reforestation (A/R) project that created both social and ecological conflicts. The study concludes that forest-based CCM policies can reduce conflict through efforts at minimal transformation of local livelihoods, maximal attention to the tenure rights, responsibilities, and authority of citizens, and by improving, not degrading, the project landscapes. The paper presents the circumstances under which these guidelines are sidestepped by the A/R project, and importantly reveals that dramatic forest and livelihood transformation had already affected the community and environment in the REDD+ project site. There are deep contradictions at the heart of climate change policies toward which attention must be given, lest we leave our future generations with nothing but forest islands and castaway communities.

Keywords: Cambodia; climate change; landscape; REDD+

---

1. Introduction

On opposite sides of Prey Lang forest in central Cambodia are two Korea-Cambodia partnership projects that are part of a larger Memorandum of Understanding between Cambodia's Forest Administration and the Korean Forest Service to invest in Cambodia's forests and provide administrative assistance toward reducing climate change (Interview "NGO Forum" 15 February 2015; Interview Think Biotech 11 January 2016). The most recent is the Korea-Cambodia Tumring REDD+ Project (T-REDD), established in July 2015 between Cambodia's Forest Administration (FA) and the Korean Forest Service (KFS). The second is an afforestation/reforestation (A/R) project in Kratie Province, established in December 2010, between the Forest Administration and the Think Biotech Cambodia, co. ltd. Phnom Penh, Cambodia (TB), a subsidiary of Korea's Hanwha Corporation. Their juxtaposition as forest-based climate change mitigation (CCM) projects enacted in the same landscape highlights some of the contradictions through which such policies are enacted.

This paper engages in the emerging global discourse in forest-based CCM initiatives that facilitate international claims to forests in Cambodia. The green grabs described in this paper are an unapologetic initiative in "securing overseas forest resources" by the Korean government [1]. Issues explored through this investigation speak to a growing body of literature concerned with the structures of governance that control exclusion and access [2,3] in the wake of land grabs in both their conservation-oriented green varieties and their more traditional economically-oriented earth toned versions [4,5]. This study situates conflict at the level of political, economic, and ecological policies and practices that privilege one type of land use over another and divest certain communities of their use rights.

A landscape approach to analyzing CCM and land grabs revealed the productive juxtaposition of two case studies enacted on opposite sides of the same vital protected area, which may have been analyzed in isolation using a project-based approach. Data were collected using a method that co-produced knowledge with citizen researchers, revealing the forest as a common pool resource that was and could continue to be sustainably managed by communities interconnected through it. Further, this research method supports actions for change, which generated alternative scenarios for project implementation discussed at the end of the paper. I argue that new tools are necessary for grappling with the profound contradictions of our coming green economy and that the three-part method of landscape perspective, local collaborations, and actions for change applied in this research can make space for fresh perspectives.

The data presented here contributes to critiques of market-based approaches to climate change policies [6,7] and discourses of sustainable forestry [8,9]. The forest restoration project is enacted as a for-profit business venture that can be green-washed because it claims to enhance forest carbon sequestration to mitigate global climate change [10]. The carbon utility of plantation forests is under debate [11], and TB has negative effects on the communities and natural resources in its area. The T-REDD project circumvents some of the pitfalls encountered by other REDD+ projects [12,13] for two reasons. The first is timing. The project area was already transformed and people already divested of their access to forest resources before the enactment of the project. The second is market orientation. T-REDD is a publicly funded project not directed toward supplying carbon credits for sale on the open market; it is designed to provide direct offsets for carbon-intensive Korean business practices inside Cambodia (Interview FA 5 September 2016).

I will demonstrate in this paper two distinct discourses engendered by climate change policies. The first involves the cynical green-washing of intensive extraction and the second shows measured attempts to grapple with the shortcomings of REDD+ governance structures in order to retain threatened forest areas. The TB project explicitly devalues traditional, sustainable livelihoods and citizen authority over land use in favor of planned reforestation and biodiversity conservation with corporate authority over land use. The discourse of the T-REDD project explicitly values citizen authority, sustainable livelihoods, and forest retention within the project boundaries. What is provocative about this juxtaposition, is that the landscape of the T-REDD project is the least conducive for meeting many of the project's objectives, it is already very degraded, is easily accessible by roads, and is under great pressure from a densely populated area. The landscape of the TB project, on the other hand, retains a great deal of biodiversity, has limited pressure for land conversion, and is in a very remote part of the forest. Both areas have a strong community, engaged in sustainable forest use, but in Kampong Thom they are pressured by incoming migrants. The evidence I will present suggests that switching project landscapes, or re-adjusting boundaries, could reach the full potential of each project's objectives with maximum benefit for the social and ecological communities in each location.

To support this claim, I first introduce the project landscape and the particular methodology deployed for this study. A brief review of the relevant literature on forest plantations and on REDD+, as carbon-reducing enterprises is followed by a discussion of those initiatives in Cambodia and descriptions of both projects from the ground, informed by the work of citizen researchers. The policy documents for each project are then evaluated considering Cambodia's national policies and the social and environmental objectives for each. The closing discussion describes the altered landscape possible through flexible planning that attempts maximum benefit for local communities.

2. The Landscape: Prey Lang

Prey Lang remains one of the largest contiguous lowland forests in Southeast Asia, ranging from 300,000 to 600,000 ha, depending on who reports the figures. The newly designated Prey Lang Wildlife

Sanctuary (PLWS), is 431,683 hectares and captures most, but not all, of the forest recognized by long-term residents of the area (Figure 1).

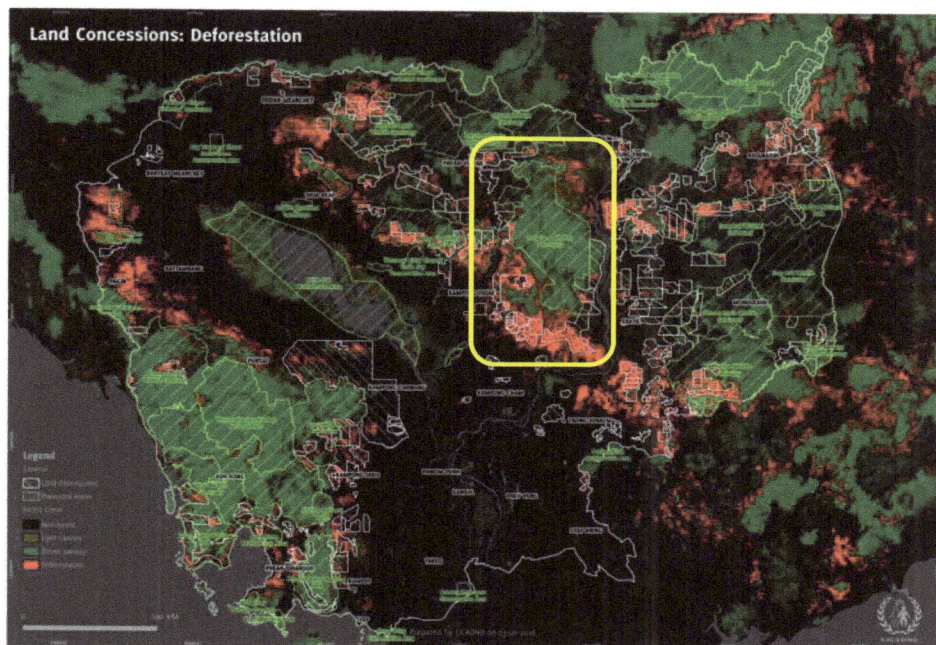

Figure 1. LICADHO (Cambodian League for the promotion and defense of human rights) forest cover map with Economic Land Concession (ELC) and protected area (PA) boundaries, 2016. Forest cover is green, deforestation is red. Green outlines are Protected Areas and white outlines are ELC [14].

The boundaries of the forest intersect four provinces: Preah Vihear, Steung Treng, Kampong Thom, and Kratie and sit between the Mekong and the Sen Rivers (Figure 2). Prey Lang is an important watershed for the Mekong and the Tonle Sap lake and the region is the site of a drawn-out struggle for forest resources that began with Cambodia's transition to a market economy in the early 1990s [15,16]. The degrading effects of this market transition on the landscape and lives of the people in Prey Lang makes the forest a target area for forest-based CCM policies [6]. The projects highlighted here, T-REDD and TB, are located in Kampong Thom and Kratie/Steung Treng, respectively (Figure 3) and each will be discussed in their affected areas and in relation to the larger landscape.

T-REDD is located at the south-western edge of Prey Lang forest in Kampong Thom province. It is an easy to access area of intense land conversion that began in the early 1990s with Forest Concessions (FCs) for industrial logging and continued through Economic Land Concessions (ELCs), a land titling campaign conducted by Prime Minister Hun Sen in 2012, and a Social Land Concession (SLC) in 2013. After Forest Concessions were cancelled in 2001 and before the land titling project in 2012, the national government awarded 16 ELCs (outlined in white) and 23 community forests (CFs) in Sandan District, Tumring, and Mien Rith communes (for more on the CF initiatives in Cambodia, please see [17–19]). Today, some stand as islands, cut off from Prey Lang forest by clear-cutting and ELCs, while others reach out from Prey Lang forest like archipelagos in a sea of cassava and rubber plantations visible inside the circle of Figure 3. They testify to the dense forest that once stood in this place and to the strength of the community that fought for them and protects them today. The T-REDD project encompasses 14 of these CFs in an 88,444 ha carbon offset initiative (Figure 4).

Figure 2. Prey Lang Wildlife Sanctuary (PLWS) in relation to provincial boundaries. Map from GoPhnomPenh http://gophnompenh.com/phnom-penh-overview/.

Figure 3. Tumring REDD+ Project (T-REDD) circled (project area is green); Think Biotech Cambodia, co. ltd. (TB, Phnom Penh, Cambodia) squared.

The TB project site is at the eastern boundary of Prey Lang forest, which ends at the Mekong River in Kratie and Steung Treng provinces. This is an isolated area, only accessible by boat or by crossing the forest. The area was also affected by FC, but the customary livelihood strategies of wet rice agriculture, shifting cultivation, forest product collection, resin tapping, and fishing remained strong. After FCs were curtailed in 2001, logging still continued illegally in the area, but ELCs did not immediately follow. In this case, livelihoods were altered, but customary practices remained the primary source of livelihoods. After the FCs, people also planted plantation crops and engaged in freelance logging (I use the term freelance logging to refer to for-profit logging done by local villagers. There are middle men who hire locals to work for them. It is a well-paying job and often the only job available. This is especially poignant in areas where companies impacted traditional livelihoods and left few avenues for people to buy what they used to gather from the forest. When villagers do contract logging for outsiders I argue that this is not illegal logging. It is freelance work. This is distinctly different from clearing land for market crops, which is also often an outcropping of plantation encroachment but cannot be considered freelance). Five CFs were established between 2010 and 2012 and people report an abundant lifestyle until 2012, when two companies began clearing the forest (Group Discussion 13–14 February 2016). One of these was TB, the 34,007 hectare forest restoration project that began operations in inhabited areas and encompassing four of the CFs in the areas of Kampong Cham and Beung Char commune, Sambor district, Kratie province (Figure 5), the omitted company, the Chhun Hong Rubber Better, Co. ltd. (Phnom Penh, Cambodia), will not be discussed here.

These two CCM projects involve the same governing bodies and are part of the same Memorandum of Understanding between the Korean Forest Service (KFS) and Cambodia's FA, designed to enhance forest resources, increase biodiversity, and enhance livelihoods. REDD+ does this by controlling cutting in natural forest; TB does this by removing natural forest and replacing it with, controlled plantation forest that will decrease reliance on the natural forest. The TB project is clear-cutting natural forest for a tree plantation on the most isolated edge of Prey Lang where customary cultivation practices were the norm and pressure on the forest was light. The company's dramatic land conversions have caused both environmental and social conflicts within and without its project boundaries. The T-REDD project, on the other hand, encompassed existing CFs along with remaining forested areas of state land. It was implemented on the most vulnerable boundary of Prey Lang in an area where customary livelihoods were already transformed and the forest was heavily degraded. This project has created few reported conflicts with local communities.

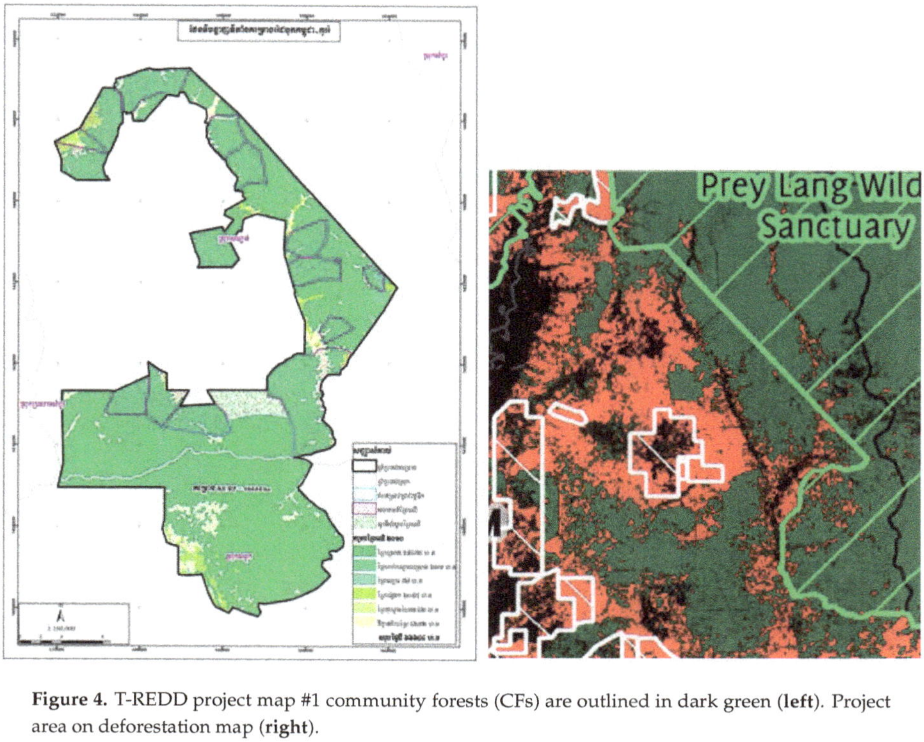

Figure 4. T-REDD project map #1 community forests (CFs) are outlined in dark green (**left**). Project area on deforestation map (**right**).

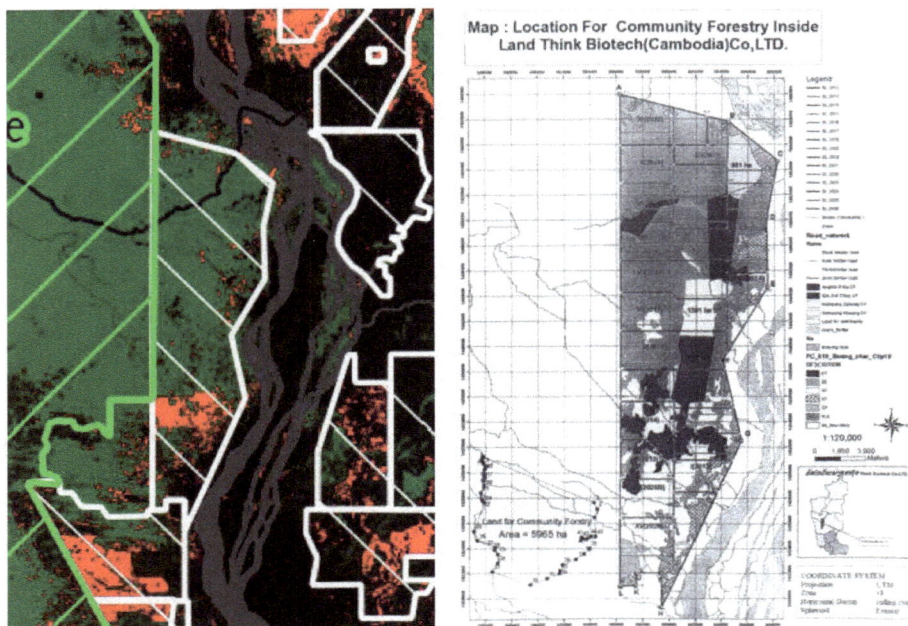

Figure 5. Think Biotech (TB) forest cover map (**left**). Think Biotech project map with four CFs labeled (**right**).

3. Methods

This paper is part of a broader research project, called "Mosaic: Climate change mitigation policies, land grabbing and conflict in fragile states: understanding intersections, exploring transformations in Myanmar and Cambodia". Mosaic develops a new research agenda that explicitly studies the interactions between climate change mitigation initiatives, land grabs, and resulting patterns of conflict [10]. The methodological approach has three parts. The first is to take a landscape perspective, the second is to co-produce knowledge between academic researchers and affected communities, and the third is to support actions for change. The landscape approach is outlined in the framework paper for this project [8], where the authors "conceptualize landscape as a 'place' where physical and socio-cultural elements occur in localized, spatially specific combinations and where human actors dynamically interact. Thus, a landscape is both ecologically and socially fluid and changeable, but also holds continuities . . . . A landscape is thus a space larger than a farm but smaller than a region, in which physical, ecological, and human dimensions co-exist as a product of socio-ecological and cultural co-evolution . . . . There is no single formula for determining where a landscape 'ends'." For this project, the landscape is based on the interconnected ecosystems of the large forest, and on the social relationships among the Prey Lang Community Network (PLCN), collective forest activists from all four provinces since the early days of exploitation [9].

In this research project, the co-production of knowledge emerges through three main activities: training, research, and information sharing. First, Mosaic researchers trained the PLCN in research methods and forest-based climate change mitigation policies. Second, Mosaic researchers and the PLCN conduct research related to Mosaic objectives together. The PLCN also conducts research in their areas according to their needs, for example, researching the activities of new mining operations, elite capture of forest land, or the exploitation of CFs. Data and insights from this research overlaps with Mosaic concerns and are shared and discussed with Mosaic researchers. Third, the PLCN shares data with Mosaic about forest crimes and the activities of companies, government officials, and conservation organizations. In turn, Mosaic researchers share information from their desk and urban research by translating and explaining new policy and land use initiatives that may affect the PLCN and by sharing new information learned through interviews with donors or companies. Community and academic researchers are in regular contact and alert each other to emerging concerns.

The third part of the Mosaic research initiative, supporting action for change, entails bringing what academics know about the concerns and situations of affected communities to other stakeholders informing and influencing policy and projects. It also involves helping to bridge the divide between local authorities and the PLCN.

This method brings to light the spillover effects of climate change policies like REDD+ or A/R, which include the disjuncture between project documents and project activities, the privileging of monitory benefits over biodiversity or other benefits, and the leakage of deforestation outside the project area. These policies and projects also spillover and intersect through social and ecological feedbacks that shape each other within a given landscape. When that landscape lacks strong governance and is being rapidly converted toward commercial purposes, the dynamics of land use, conflict, and cooperation are instructive for understanding the political-institutional conditions in which just and inclusive solutions might be created.

Research was conducted between January 2015 and September 2016 (with the assistance of RONG Vannrith, THUON Ratha, SONG, Danik, and SEAKCHHY Monyrath, each connected with Mosaic partner organizations in Cambodia: Equitable Cambodia and the Cambodian Peace Building Network—data was importantly supplemented with reports from the PLCN), combining participant observation methods with formal and informal interviews, group discussions, and secondary literature reviews. All interviews were conducted in Khmer or English. Grassroots-level data were collected in cooperation with local research teams. Mosaic researchers conducted research during four forest patrol activities, each consisting of five days and four nights in the forest with PLCN members and in 18 meetings with government officials, non-governmental organizations (NGOs) and international non-governmental organizations (INGOs) and community representatives related to forest governance, climate change, and conservation initiatives. These activities provide important context for the interview data collected from the four communes affected by the two projects of this paper: eight interviews with local authorities, focus group discussions with 11 of the 14 CFs in the REDD+ project (researchers were unable to reach the last three CFs and since the data received from all the others was remarkably similar and corroborated with that of other participants during forest patrol and meetings this was not pursued further), focus group discussions with seven villages affected by TB, and individual interviews with seven key informants at the community level. Additionally, interviews were conducted with eight NGOs working in the area, and six interviews with national Forest Administration officials, two interviews with TB company representatives, and one with the company that performed the Environmental Impact Assessment for TB. National authorities and company voices are underrepresented in this study, despite efforts to reach them. Especially missing are the voices of the Korean Forest Service, with whom every attempt at contact failed. Nonetheless, I feel confident that the data presented supports the claims of this paper.

The benefits of gathering data using the approach here described is that it situates two CCM projects, REDD+ and A/R, together in the same social and physical landscape where research can draw out the historical, bureaucratic, and socio-economic intersections of each project. The input from citizen researchers brings important information to bear on the land-uses and use potentials of the respective project sites, as well as understandings of community capacity for conservation initiatives. This information was instrumental in considering how changes to project implementation and governance strategies could decrease conflicts while achieving project goals. Analyzed separately, these two projects show the characteristics of conflict-light or conflict-heavy implementations of CCM policies. Thinking about them together in the same physical, social, historical, and policy landscape, however, provides space to move beyond the silos of activities within individual projects and begin to address the implications of the system as a whole.

4. Forest-Centered CCM: REDD+ and Forest Plantations

REDD+ is an international instrument designed to preserve forest resources by creating a market for the carbon that they store. The mechanisms through which carbon is captured, measured, bought, and sold vary considerably, but the use of REDD+ to offset other carbon-intensive activities continues to create inequitable outcomes by privileging forest management programs under international frameworks over more traditional methods of sustainable forest management [10]. Globally, REDD+ exhibits multiple types of implementation "running ahead of policy processes and state-driven decisions" [12] and also ahead of any internationally agreed upon process [13]. The discourse and practice is increasingly dominated by donor-driven policy narratives and technological interventions that depoliticize climate mitigation, maintaining current structures [20] and obscuring the discourses and values that underlie them. By focusing on the value of carbon and future funds from donor organizations or rich countries, the issue of deforestation and forest protection becomes "a mere footnote" [21]; diminishing the value of biodiversity [22], and the value of healthy human communities in forests [23]. Additionally, situating shifting cultivation or charcoal production as the main drivers of deforestation in REDD+ project areas, devalues the benefits of shifting cultivation, and leaves the industrial drivers of forest loss unmentioned, unquestioned, and unaffected by REDD+ [24].

As forest-based processes, the main difference between REDD+ and A/R is their approach: REDD+ is designed to keep forests standing, while A/R creates plantation forests where there are no trees. A/R projects attempt to increase industrial forest conversion through the Clean Development Mechanisms that promote managed forest cultivation and increase leafy green canopy above the bare earth for carbon capture [25,26]. In doing so, tree plantation developers that are able to invest in large-scale forestry can provide sustainable forest management and climate change mitigation [27]. In many cases, however, A/R projects are implemented in forested areas and follow the patterns of the global land grab crisis [5,28], in which small holders and indigenous people are violently dispossessed of their lands and livelihoods [29]. Case studies suggest that industrial forest plantations are neither more profitable [30], nor sequester more carbon than REDD+ or business as usual timber harvesting [31].

Both REDD+ and A/R projects settle in landscapes with existing political and economic activities [32,33] and are affected differently by various tenure and documentation issues. Communal tenure in a Cambodian REDD+ site resulted in restricted access to forest resources and resin trees [34], while this case study shows how clarifying CF tenure made protecting forests easier (*cf.* [35]), but issuing individual land titles for market crops in the same landscape put significant pressure on the forest (*cf.* [36]). REDD+ is continually confronted by its dual focus at once enhancing forest protection and increasing commercial activities [37,38]. Forest plantations tend to have less varied and more destructive effects on the social and ecological environments of their enactment [39].

The TB project documents state that the project will "stop forest clearance . . . and reduce the utilization of natural forest" [40] (translation by author from original Khmer). This effectively excludes forest dependent people from using the natural forest for subsistence, and increases timber exploitation outside project boundaries because there is legal timber exported from the region (Interview community rep 15 February 2016). Additionally, it opens the region with roads and infrastructure that facilitates cutting high-value timber for elite markets [41], makes space for local elites to profit from the timber trade and also convert land to grow market crops, and it pushes marginal villagers to clear forest to replace land lost to the company and to replace subsistence economies with market-based crops. Plantation jobs are limited and the salary is incomparable to traditional livelihoods (*cf.* [42]). What is available is freelance logging and plantation crops, both of which put pressure on forest land and increase market-based economies.

REDD+, on the other hand, is not market based due to the limited market for carbon [13]. The process of offsetting was designed to function under free market principals, but after many years of attempts, REDD+ looks more like direct patronage. For example, a in transaction orchestrated by the Wildlife Conservation Society (WCS), Disney corporation purchased 14 million metric tons of carbon dioxide equivalent emissions for $2.6 million. Funds from this deal are "earmarked to help the government protect the Keo Seima Wildlife Sanctuary and will be handled through the Ministry of Environment" [43]. The influx of capital from the private sector and other governments flows into the newly created bureaucratic structures for managing climate finance in fragile host governments [44,45]. The emerging effects of this are ministries competing for environmental funding sources [46] and continued forest loss. The logics of a market economy seem to be unequal to the challenges of deforestation and of mitigating its climate changing effects.

In this environment of under-funded and donor-dependent state systems, the two cooperative projects between Cambodia and the Korean government are both framed using the discourse of climate mitigation and clean development. The next section will describe the state of carbon capture activities in which the projects under discussion were implemented.

5. Cambodia and Carbon Capture

Cambodia came early to internationally conceived climate mitigation initiatives. The first REDD+ project started in 2008. Cambodia's Forest Administration in partnership with the International Non-governmental Organization (INGO) PACT administered the Oddar Meanchey REDD project [47,48] (PACT was not the original INGO on this project and they pulled out due to complications related to the actual sale of carbon (see, [48] for more details of this case)). The United Nations Development Program (UNDP) started REDD readiness preparations in Cambodia in 2009 [49], supporting the National REDD+ Program and REDD+ readiness activities (Interview UNDP 16 March 2015). Six years later, elaborate bureaucratic structures exist and four REDD+ projects are operating with another on the way: The Oddar Meanchey project, the WCS sponsored project in the Seima Protected Area (see, [34,50]); the International Tropical Timber Association (ITTO) project "SFM management through REDD+ mechanisms in Kampong Thom province, Cambodia" (PD 740/14 Rev.2 (F)); the Tumring REDD+ project described in this paper, and a new Japanese project in Prey Lang is entering a three year planning phase. The ITTO project objective is to get experience to do REDD on the ground. It is an office-centered capacity building exercise—not designed to sell carbon (Interview FA 9 May 2016). Even with these active projects, Cambodia is still transitioning out of the "readiness" phase and preparing to move into the implementation phase of the process [51]. Tree plantations for carbon capture have not had the same uptake, but were expected to contribute to the 60% forest cover by 2015 promised in Cambodia's Millennium Development Goals (MDG) [52], and continue to play a role in achieving the 60% forest cover goal not yet reached [53].

The most successful commercial tree plantation industry in Cambodia is rubber, accounting for over 900,000 hectares in 2013–2014 [54]. This does add green leafy canopy in a manner that satisfies the Food and Agriculture Organization (FAO) guidelines for forests, but they were planted on forested land and are thus not a vehicle for carbon capture [11]. Timber plantations in Cambodia reported by FAO in 2010 covered 70,000 ha [55]. Many of these were discontinued due to the destruction of local livelihoods, alienation of local communities, negative impacts on the environment, decreased biodiversity, and reduced water quality [56,57]. Satellite data from 2013–2014 [58] shows less than 1000 ha dedicated to timber production activities in Cambodia, and TB is one of two public-private industrial tree plantation initiatives currently active in Cambodia (the Oji Paper Co., Ltd (Tokyo, Japan) will not be discussed here). These new attempts at timber plantations also have negative impacts for society and the environment, but when confronted with the destructive activities of TB by a news reporter, Cambodia's environment minister, Say Samal, conceded a loss of biodiversity, but maintained that development was more important. "We have to be realistic, we want to build our economy, we want to create jobs for our people so we have to balance that out" [54].

According to a recent study conducted for the Technical Working Group on Forest Reform in Cambodia, the implementation of industrial tree plantations lacks strategic planning, adequate forestry skills, and would benefit from more intensive and thorough study [59]. REDD+ in Cambodia suffers similar effects from rapid implementation in advance of policies or frameworks to support it, and the Disney purchase was the country's first official sale of carbon since 2008 (policies were not in place with the Ministry of Finance to effectively manage a carbon sale in Oddar Meanchey, which remains stalled at the time of this writing). In the current environment of increasing climate financing, government officials are under pressure to adopt REDD+ schemes both from the ministry's need and desire for funding, and from international discourses of climate change mitigation. The role of donor organizations and private companies pushing carbon capture schemes was critiqued by both UNDP and Forest Administration representatives in Phnom Penh. They are "approaching the government trying to push carbon capture . . . [they put] the sale of carbon before the community and before there are legal systems in place" (Interview UNDP 16 March 2015).

Not only are the legal systems to manage REDD+ still emerging, Cambodia currently has multiple systems for managing forests. This is changing, and in April 2016 almost 100,000 hectares of forest for conservation were transferred to the Ministry of Environment (MoE), while the Ministry of Agriculture, Forestry, and Fisheries (MAFF) lost control of those same areas for conversion purposes. This is an emerging and interesting conflict in the arena of climate change policies in Cambodia, as the land conflict between MAFF and MoE precedes climate funding. This environment of shifting jurisdictions amid flows of money and policy initiatives exacerbates competition among ministries for climate finance (see also, [45]), thereby obscuring the rampant deforestation.

"We have done enough damage with our money," says UNDP representative Napoleon Navarro. Working on both sides on this project, trying to facilitate government capacity building and community awareness development, Navarro contends that "you can't organize anyone around the promise of money" (Interview 16 March 2015). This sentiment was contextualized by Chhun Delux, administrator of forest finance for the Forest Administration. "Among the government and business people, there is this push for money—they are less concerned with the ecological aspects, it is a new business incentive . . . Everyone is working for their own benefit and no one looks at the big picture." He mapped out the donor tree and the breakup of the various NGO activities "...this one biodiversity, this one community forests, this one capacity building . . . see? islands. But the whole landscape [Prey Lang] is under served, which fact goes unrecognized by the officials and donors...." (Interview FA 13 July 2015).

The speculative and non-transparent nature of land conversion in Cambodia in the context of weak state institutional systems is not ideal for sustainable forest financing. This issue is exacerbated by the problem of forest financing itself, which has yet to produce a systematic framework for this endeavor. The situation in Cambodia consists of bureaucratic structures staffed by inexperienced officials, large expenditures by donor organizations, a focus on carbon over forest and community health, and continued pressure on forest resources from market-based activities like industrial plantations, selective logging, and market-crop conversion. The next section provides a closer inspection of the T-REDD project area, which has overcome some of the pitfalls related here, but remains mired in others.

6. The T-REDD Project Area

Multiple land claims, including ELC, FC, SLC, private homestead land, CF and permanent forest estate influence the T-REDD project. The history of war, forest exploitation, and marketization remain embedded in resource claims at this site. After 1993, in the post-election "transition period", abundant forests were identified by the World Bank as key export commodities [60,61]. By 1998 all forests outside of protected areas were granted as forest concessions to well-connected governors and Southeast Asian corporations [62]. This initiative provided none of the expected state-building revenue-generating effects; it was de-emphasized by 2001 [63] and replaced with ELCs.

The T-REDD site wraps around the first ELC in Prey Lang. A 6200-ha "state-led development project", the Tumring Rubber Plantation [64] was issued well inside the locally conceived boundary of the forest. People previously accustomed to shifting agriculture and collecting forest products (almost 60% of reported income, 31% and 28%, respectively) were divested of forest resources and shifting cultivation lands and asked to grow rubber [65].

In the opening ceremony for the plantation, Prime Minister Hun Sen explicitly encouraged citizens to " ... change from collecting resin, tapping resin ... to tapping rubber" (GW 2007, 30). This was a dramatic revaluation of the land and people echoed in the TB sub-decree described below. Effects of these land-use policies are palpable for local communities, "Before the companies we rarely got sick, and if we did we could go find medicines from the forest. Now we look [for medicine] in the CF but don't find it and the people are more and more sick" (Group Discussion KT 28 February 2016). Effects are also visible in the landscape, and by 2012 the forest had given way to all the concessions outlined in white in Figure 4.

The CF initiatives started in response to the forest concessions, but due to ministerial back logs and inattention, few were in place before the implementation of ELCs. Between 2001 and 2012 a nation-wide initiative to establish community forests ensued, and communities around Prey Lang worked to establish CFs and defend the remaining unprotected forest. This was, and remains a disputed initiative in Prey Lang, as it limits community use to small designated areas. For good or for ill, by 2010, Kampong Thom had 46 CFs totaling over 35,000 ha, more than any other province [66]. Local participants say that protecting their CF "was so difficult, even the police were involved in cutting the CF" (Group Discussion KT 27 February 2016), after official ministry recognition, this task became much easier. These CFs are visible in Figure 4 as the green patches that spread from the forest, like island archipelagos.

The red deforested area surrounding the CFs and the Tumring Rubber Plantation in Figure 4 represents small-holder cassava plantations. These areas were converted in increments and sold to migrants by local elites until the roll out of Hun Sen's "Order 01 on Measures for Strengthening and Increasing the Effectiveness of the Management of Economic Land Concessions" (Order 01) in 2012. Instigated to address growing unrest over the effects of ELCs across the country, Order 01 morphed to include issuing land titles for citizens, which drove elite land capture in advance of titling (this was a terribly complicated and controversial initiative that affected people differently across the country—for a comprehensive discussion of this complex process, see [36,67,68]). Another effect of Order 01 is the replacement of ELCs with SLCs for economic endeavors. In 2013, an SLC was awarded right behind one of the T-REDD community forests. There are no landless settlers in this Social Land Concession, which is for "poor citizens who are landless or land poor" [69]. There are no landless settlers inside the SLC area, only sawmills operated by one of the same well-connected Cambodian businessmen involved in creating the Tumring Rubber Plantation [64,70].

All the land use strategies described above occurred before T-REDD and were known at the time the documents were signed; the sub-decree was not acknowledged in the first map (Figure 4), but the map was revised (Figure 6).

The once forest subsistent people now administering T-REDD were divested of their modes of production and forced into the market economy 13 years before the project. After two meetings with the Forest Administration in June and July 2015, REDD+ CF communities had low, but positive expectations. They receive patrolling equipment and 50 USD/month "if we keep the forest they will help us—if not, they won't" (Group Discussion KT 28 February 2016).

Community forest committees directly involved in the project reported moderate added benefits. "The FA trained us about REDD+ and now they support us with equipment and money for patrols" (Group Discussion KT 28 March 2016). "If we can get help from the government to protect [the forest] it is better for us" (Group Discussion KT 30 March 2016). In terms of the community's ability to protect the CF area, "REDD+ does not change much for us, only that we get training, equipment, and patrol money, not enough money though. It doesn't replace money we can make doing other things, so it's hard to get people to patrol. It only pays for our gas" (Group Discussion 29 February 2016). When asked directly about conflicts involving REDD+, people reported conflicts with illegal loggers, companies, elites, NGOs, or neighbors, not with REDD+, except that it did not do enough. It did not solve their problems with forest protection or getting support from the Forest Administration to arrest loggers or evict encroachers, "we call them [the FA]. Sometimes they come, but mostly not" (Group Discussion KT 29 March 2016). Neither did it capture all the CFs in the area, and the excluded CF committees wished they too would receive the benefits (Meeting, Prey Lang Working Group, 16 August 2016).

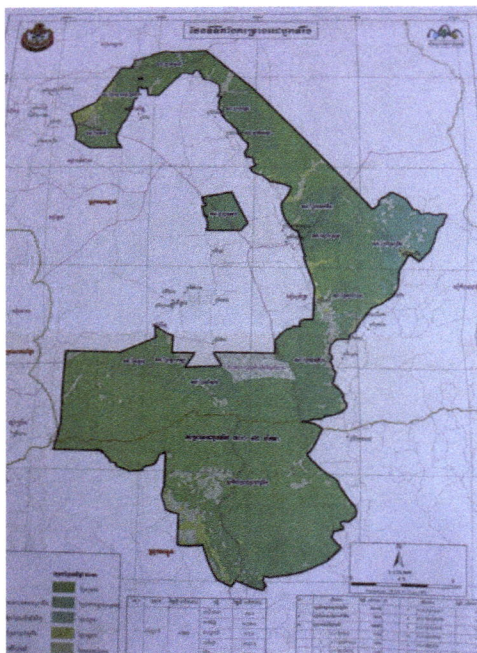

Figure 6. Photo of T-REDD Map 2 taken by author.

This site was purposefully chosen for REDD+ because of its high deforestation rate [71]. The project uses a simple system for measuring carbon and deforestation through a projection scenario based on historical deforestation data. Success demonstrates less deforestation than projected (Interview FA 5 September 2016). The T-REDD project documents give thoughtful attention to the challenges of REDD+ and attempt to foster community empowerment. Household uses of the forest, like shifting cultivation or charcoal production, are explicitly not considered part of the deforestation problem and community capacity to "keep the forest" is honored. T-REDD aims to, "assist community forests to scale up their forestland management area to cover the remaining permanent forest estate ... legalized as parts of the provincial and commune land-use planning framework" [72].

As the law stands at the time of this writing, however, citizens do not have the authority to patrol the larger PLWS protected area. "They give us the CF to protect while they cut the Prey Lang as they like ... even the police go too" (Group Discussion KT 28 February 2016); "they take trucks every day along the new road ... into the forest" (Group Discussion KT 29 February 2016). "If they give us authority, we can protect it [PLWS]," people say, "it has use for us—this is our resin forest" (Group Discussion KT 28 February 2016). The T-REDD project description cannot contain all the contradictions and pitfalls of REDD+ projects described above, but those it can transform suggest that that climate change policies may benefit from non-market approaches, and that effective REDD+ implementation should not degrade and at least marginally benefit local economies.

7. Think Biotech Cambodia, co. ltd.

Think Biotech was established as a public-private partnership between the Forest Administration of Cambodia and the Think Biotech Cambodia, co. ltd. in December 2010. Think Biotech is a subsidiary of the Hanwha corporation that specializes in the manufacture of explosives and military equipment. Before 2010, the Think Biotech Cambodia, co. ltd. was not listed in Hanwha's annual report and Hanwha had no experience in forestry initiatives [72].

Nonetheless, in Article 2 of the sub-decree signed by MAFF, it states that Think Biotech will establish an A/R project that will "improve soil fertility through reforestation and biodiversity conservation ... [and will be] part of Clean Development Mechanisms or other mechanisms that contribute to the reductions of greenhouse gas emissions and climate change mitigation" [40]. The conflict drivers in this case are the other objectives in Article 2 that include, "to stop slash and burn activities, and ... illegal claims to trees ... " [40]. The project deliberately alters the local economy in ways that also increase pressure on the PLWS outside the project boundary by pushing subsistence and market activities deeper into the forest.

In June 2012, the company began operations in Kampong Cham commune, Sombor District, Kratie province. Unlike in Kampong Thom, there was a period of 10 years between the slowdown in Forest Concession activity and the rise of ELCs, during which time freelance logging rose considerably. Even so, when TB came in 2012, traditional livelihoods of shifting cultivation, wet-rice agriculture, and resin collecting were the primary economic activities (Group Discussion KR 13 February 2015). As is typical of ELCs in Cambodia, TB started operations with no community consultation. At first people thought the excavators were for road development, it then became clear that they were making a business and some people took small jobs creating the tree nursery or other buildings (Interview Mr. Som No 15 February 2016).

The director of TB stated that he knew there were people living inside the concession boundaries (Interview TB 1 November 2016). The commune chief confirmed this as well, but suggested that the company cleared community land because they did not recognize it; community land for shifting cultivation looked like degraded forest (Interview KR 14 February 2015). The company signed an agreement with MAFF to begin operations in 2012 and hired CES co. ltd. (Phnom Penh, Cambodia) to conduct an environmental impact assessment (EIA). This assessment was completed in February 2013 and the project was found to have "a lot of problems" and "would be bad for the community" (Interview CES 8 November 2016). This information was shared with the MoE and with the company. According to an unpublished report by NGO Forum, the project would affect approximately 1900 families, 4412 hectares of rice fields, 3534 hectares of plantation land, 5970 hectares of community forests, 5 hectares of spirit forests, and 5 hectares of burial grounds ("NGO Forum" notes 17 November 2015).

After the negative EIA, in May of 2013, company bulldozers began clearing community lands along the old road. During this clearing, 178 households lost nearly 1000 hectares of farm and shifting cultivation land. Strong community protests kept the company from clearing more land (Interview community rep. 14 February 2015), but there remain 400 hectares of disputed land inside the company boundaries (Interview commune chief KR 3 February 2015). In addition to community rice fields and farm lands, the company cleared nearly 5000 ha of forest land in the southern end of their project next to the affected villages (Figure 5, red deforested area). In so doing, they cleared the forest right up to the banks of the streams. The effects of this on rain-fed streams are visible (Figure 7). The sun exposure kills fish eggs in the streams and causes them to dry completely when the rains stop (Field Notes, personal communication, Mr. Som No 6 August 2016).

Figure 7. Stream *O Sro Lork* in the rainy season, 9.7.2016. Photo by Seay Monyrath.

At peak production, the company employed approximately 800 workers earning between 150–180 USD per month (Interview KR 16 February 2016), which is well below the 250–300 USD people report from tapping resin and selling market crops (Field Notes Steung Treng 19 December 2016, see also [73]). The commune chief sees the benefit of the jobs (Interview 16 February 2015), but was not pleased that the promised road was never built (Field Notes, personal communication Mr. Nak Virak 9 February 2016). The only roads built by the company were for company use and community members had to protest to gain access (Group Discussion KR 14 February 2016). Since February 2016 the company has been quiet. They laid off most of their workers and the saw mill stopped operations. In September 2016 there were about 200 workers planting saplings, many of whom are migrants (FN KR 7 September 2016).

While in the south the company seems quiet, in June of 2016 they conducted a public consultation in Steung Treng with about 100 members of the soon to be affected community. A local researcher accompanied company representatives and provincial and national level ministry officials on their mission to mark the TB boundary. At the most isolated edge of the project bordering the PLWS they placed company boundary markers and drew a map. While mapping this territory, the researcher informed the officials they will need a map of community holdings to conform to Order 01 guidelines that require companies to develop around, and not through, community lands. The Forest Administration representative told him, "we don't have to follow the rules of Order 01, this is not an ELC. It's a government partnership" (Action Research Interview ST 18 August 2016). Since the boundary mapping, community researchers have mapped 15 of the 19 resin forests in use at the present time (Figure 8).

That the company moved to map and begin developing the northernmost region of their project could be because they have cleared the area in the south right up to the boundary of the southernmost CF that is inside the concession (Figure 9). In many ways the company is frustrated with this project as well. The CFs inside project boundaries were a surprise according to the company director, and his face was visibly concerned when the author showed him the mapped boundaries of resin forests inside "his" company boundaries (Interview TB 1 November 2016).

The situations on the ground at the TB project site differ widely from those at the T-REDD site. Most importantly, the evacuation of local subsistence practices was explicitly part of the project's goal. There are many other factors that contributed to the conflict-heavy implementation of this project, these include the lack of community consultation, ignoring EIA recommendations, clearing community land holdings, providing undesirable jobs, and restricting access to company roads. The company is actively avoiding the CF area inside their project, which is commendable, but are surprised and dismayed by the presence of community holdings inside their northern boundary. This issue is currently under negotiation.

Figure 8. Community resin forest shapefiles. Red is project boundary, white N-S line is PLWS boundary, white E-W line is provincial boundary (Steung Treng to the north, Kratie to the south). Blue are mapped forests and yellow are remaining areas to be mapped. GoogleEarth screen shot by author.

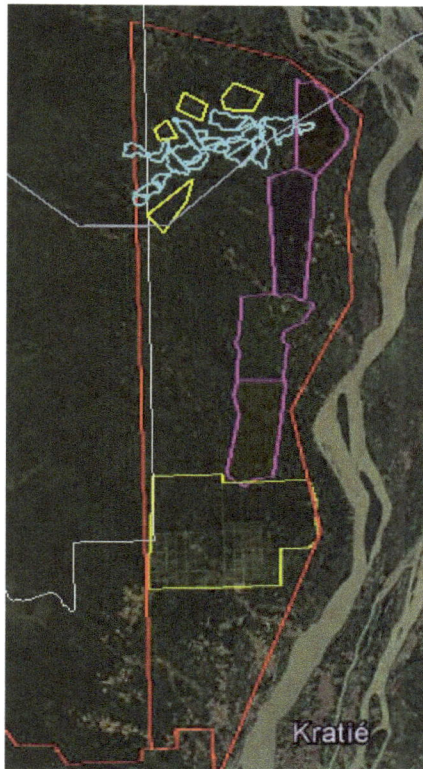

Figure 9. Google Earth screen shot with shape files. Pink are CFs and yellow is cleared company area as of September 2016. Affected area boundaries created using GPS data points collected by PLCN and author in September 2016. CF shape files are from Cambodia's Forest Administration (FA).

8. Discussion

The evidence presented in this study shows two governance strategies for CCM projects as well as the social, economic, geographic, and historical circumstances of the people and the regions in which each operates. Through this analysis, focused on the potential for decreasing the incidence of conflict in CCM policies, some fundamental differences between the projects are visible. First, in the project with little conflict, both project documents and implementation practice were focused on granting tenure rights and management responsibility to participants. By contrast, in the conflict-heavy project, the language of the project documents, the implementation of project activities, and the voices of company representatives were all directed at divesting communities of their rights and responsibilities. Second, the conflict-light scenario is not focused on monitory trade or profit, a source of regular conflict in REDD+ case studies [74]. Communities know exactly what to expect from this initiative. It is not a lot and the $50/month does not compensate them for their time, but it covers the costs of patrolling. The TB project gives rise to multiple financial conflicts beyond clearing village resources for company profits: the wages are low, salaries are often late, large lay-offs have occurred, and locals accuse the company of selectively logging high-value timber from the deep forest.

These data suggest that conflict-light CCM projects can be implemented with close attention to the relationship between selective project outcomes and the physical, social, and economic landscape in which it is implemented. The evidence from these two case studies suggests that the TB restoration project is not executed in the best landscape for an industrial tree plantation and that both the social and physical landscape is better suited for a REDD+ project under the Korea-Cambodia MoU for enhancing forest resources and mitigating climate change. First, it is in an area populated with people actively protecting the forest and using the forest for subsistence. Second, it is in a remote area with very little infrastructure, which adds to transport costs for industrial production and limits encroachment, enhancing conservation. Third, it is covered with forest that in many parts of the project area is quite dense and diverse.

There is very little support for creating industrial timber plantations in already forested landscapes. Guidelines insist that A/R activities should be on degraded forest land [26], but the practice of labeling rich forest as degraded forest is widespread [75], with profoundly negative effects on social and ecological communities [28,39]. Increased deforestation beyond project boundaries is a recorded effect of forest-conversion for development. The map of the Tumring area shows clearly what this looks like (Figure 4). The TB project is in a very remote area, but is currently building roads to transport timber and equipment. Not only does this add costs to their operation, but satellite imagery shows increased forest conversion all along these roads and researchers have mapped these roads to points well inside the PLWS.

The final, and most important reason that suggests a successful REDD+ project on this site is that the community is both capable and willing to engage in forest protection activities. This most important fact, coupled with the very progressive and flexible project design of the T-REDD initiative could go a long way to protecting valuable and biodiverse forests in an isolated and naturally protected forest landscape.

Korea is innovative and is engaged in a number of development experiments; juxtaposing REDD+ and A/R is not new for them [76]. The KFS currently partners with 12 countries in forestry initiatives and is well-known for public private partnership funding arrangements [77]. The implementation of T-REDD was unique, according to the technical partner in the Korea-Cambodia T-REDD project (Meeting FA 13 July 2015), who claimed that waiting for the legal frameworks was too slow and government to government initiatives do not need them. Local sources in the community and in the Forest Administration, suggest that the government has more power to implement a REDD+ project, "the NGO has no power and local authorities have no reason to do what they ask" (Interview FA 17 December 2015); "this project is strong because it is with the government, they have to help us" (Group Discussion 29 March 2016). It is worth discussing if a Korean-led publicly engaged REDD+

development project could be instituted at the TB site and if the already cleared SLC in Tumring could be reforested for commercial timber production.

What I suggest here is a radical, perhaps impossible, undertaking of project manipulation. However, had the governments and companies involved adopted an un-cynical approach to their projects and attended to larger project landscapes for maximum project/public/forest benefit, it could have produced conflict-light CCM policy enactments.

9. Conclusions

The case studies presented here demonstrate how a particular method to analyze the relationship between CCM and conflict exposed the social and ecological impacts of two related projects. The negative impacts of the TB project hit at the intersecting agendas of a loosely accountable government and a for-profit business investment, both focused on monetary benefits T-REDD project effects cannot really be called positive, but by all accounts from participants they are not negative, and the project design makes explicit attempts to mitigate the known challenges to REDD+ implementation. These case studies also reveal a possible pathway to avoided conflict. The T-REDD and TB data suggest that forest-based CCM policies can reduce conflict through efforts at minimal transformation of local livelihoods, maximal attention to the tenure rights, responsibilities, and authority of citizens, and by improving, not degrading, the environment in project landscapes.

In the A/R study cited here, each of these conflict-reducing practices was violated, resulting in a conflict-heavy CCM enactment. TB was typical of other Cambodian economic land grabs, but was also explicitly designed to relieve citizens of their livelihoods with no considerations of tenure rights or respect for community authority. In addition, the environmental impact of large-scale clearing of forested landscapes spreads the conflict into the environment both within and beyond project boundaries. On the other hand, T-REDD did enact the conflict reducing practices. The explicit attempt to put communities at the front of managing state forest reserves marks a turning point for Cambodian forestry policy. It is important to note, however, that the project was conflict-light primarily because livelihoods had already been transformed through ELCs and other tenure schemes and its lack of reported conflict from project participants should not overshadow the "perverse logic" [78] under which it works. For $50/month the T-REDD+ project uses "if not blatantly exploits" those least responsible for causing climate change to do the "messy, time-consuming, labor-intensive and dangerous work of protecting forests, which are of global benefit" (ibid). They are put to this work to offset the carbon-intensive activities of Korean companies in Cambodia.

It is important at this historical juncture to make explicit political claims about the effects of pretending that preserving forest in one place can mitigate the large-scale emission of carbon in another place [24,37,79]. Climate change is a complex issue and each attempt to mitigate it through bureaucratic and/or market-based endeavors exposes a dangerous ambivalence. This study suggests that conflict is at the heart of CCM policies, connected as they are to the resource capture necessary for economic growth. This is a fundamental contradiction with which we must all grapple, lest we are left with nothing but forest islands and castaway communities.

Acknowledgments: Funded by the Dutch National Science Foundation (NWO), and the Department for International Development (DFID) through the CoCooN—Conflict and Cooperation in the Management of Climate Change—Integrated Project. See: http://www.iss.nl/mosaic.

Conflicts of Interest: The author declares no conflict of interest.

References

1. Jae-un, L. Korea Begins Preserving Tropical Forests in Cambodia. Korea.Net. Available online: http://www.korea.net/NewsFocus/Policies/view?articleId=127298 (accessed on 7 May 2015).
2. Ribot, J.C.; Peluso, N.L. A Theory of Access*. *Rural. Sociol.* **2009**, *68*, 153–181. [CrossRef]
3. Hall, D.; Hirsch, P.; Li, T. *Powers of Exclusion: Land Dilemmas in Southeast Asia*; University of Hawaii: Honolulu, HI, USA, 2011; in press.

4. James, F.; Leach, M.; Scoones, I. Green Grabbing: A new appropriation of nature? *J. Peasant. Stud.* **2012**, *39*, 237–261.
5. Borras, S.; Franco, J. Global Land Grabbing and Trajectories of Agrarian Change: A preliminary analysis. *J. Agrar. Chang.* **2012**, *12*, 34–59. [CrossRef]
6. Theilade, I.; Schmidt, L. *REDD + and Conservation of Prey Long Forest, Cambodia Summary of Scientific Findings 2007–2010*; No. 66; Forest and Landscape: Cornelius, OR, USA, 2011; Volume 66.
7. Touch, S.; Neef, A. Local Responses to Land Grabbing and Displacement in Rural Cambodia. In *Global Implications of Development, Climate Change and Disasters: Responses to Displacement from Asia–Pacific*; Price, S., Singer, J., Eds.; Routledge/Earthscan: London, UK; New York, NY, USA, 2015; pp. 124–141.
8. Hunsberger, C.; Corbera, E.; Borras, S.M., Jr.; Rosa, R.; de Eang, V.; Franco, J.C.; Herre, R.; Kham, S.S.; Park, C.; Pred, D.; et al. Land-Based Climate Change Mitigation, Land Grabbing, and Conflict: Towards a landscape-based and collaborative research agenda. *Can. J. Dev. Stud. Rev.* **2015**, in press.
9. Parnell, T. Story-Telling and social change: A case study of the Prey Lang Community Network. In *Conservation and Development in Cambodia: Exploring Frontiers of Change in Nature, State and Society*; Milne, S., Mahanty, S., Eds.; Routledge: London, UK; New York, NY, USA, 2015; pp. 258–279.
10. Fox, J.; Castella, J.-C.; Ziegler, A.D. Swidden, rubber and carbon: Can REDD+ work for people and the environment in Montane Mainland Southeast Asia? *Glob. Environ. Chang.* **2014**, *29*, 318–326. [CrossRef]
11. Ziegler, A.D.; Phelps, J.; Yuen, J.Q.; Webb, E.L.; Lawrence, D.; Fox, J.M.; Bruun, T.B.; Leisz, S.J.; Ryan, C.M.; Dressler, W.; et al. Carbon outcomes of major land-cover transitions in SE Asia: Great uncertainties and REDD+ policy implications. *Glob. Chang. Biol.* **2012**, *18*, 3087–3099. [CrossRef]
12. Corbera, E.; Schroeder, H. Governing and implementing REDD+. *Environ. Sci. Policy* **2011**, *14*, 89–99. [CrossRef]
13. Savaresi, A. A Glimpse into the Future of the Climate Regime: Lessons from the REDD+ Architecture. *Rev. Eur. Comp. Int. Environ. Law* **2016**, *25*, 186–196. [CrossRef]
14. Hansen, M.C.; Potapov, P.V.; Moore, R.; Hancher, M.; Turubanova, S.A.; Tyukavina, A. High-resolution global maps of forest cover change. *Science* **2013**, *342*, 850–853. [CrossRef] [PubMed]
15. Le Billon, P. The Political Ecology of Transition in Cambodia 1989–1999: War, Peace and Forest Exploitation. *Dev. Chang.* **2000**, *31*, 785–805. [CrossRef]
16. Le Billon, P.; Springer, S. Between war and peace: Violence and accommodation in the Cambodian logging sector. In *Extreme Conflict and Tropical Forests*; Springer: Dordrecht, The Netherlands, 2007; pp. 17–36.
17. Biddulph, R. In whose name and in whose interests? An actor-oriented analysis of community forestry in Bey, a Khmer village in Northeast Cambodia. In *Conservation and Development in Cambodia: Exploring Frontiers of Change in Nature, State and Society*; Milne, S., Mahanty, S., Eds.; Routledge: London, UK; New York, NY, USA, 2015; pp. 160–176.
18. Pasgaard, M.; Chea, L. Double Inequity? The Social Dimensions of Deforestation and Forest Protection in Local Communities in Northern Cambodia. *Austrian J. South East Asian Stud.* **2013**, *6*, 330–355.
19. Royal Government of Cambodia (RGC). *Community Forest Governance*; Ministry of Agriculture, Foretry, and Fisheries: Phnom Penh, Cambodia, 2003.
20. Käkönen, M.; Lebel, L.; Karhunmaa, K.; Dany, V.; Try, T. Rendering Climate Change Governable in the Least-Developed Countries: Policy Narratives and Expert Technologies in Cambodia. *Forum Dev. Stud.* **2014**, *4*, 351–376. [CrossRef]
21. Eilenberg, M. Shades of green and REDD: Local and global contestations over the value of forest versus plantation development on the Indonesian forest frontier. *Asia Pac. Viewp.* **2015**, *56*, 48–61. [CrossRef]
22. Hinsley, A.; Entwistle, A.; Pio, D.V. Does the long-term success of REDD+ also depend on biodiversity? *Oryx* **2014**, *49*, 1–6. [CrossRef]
23. Beyene, A.D.; Bluffstone, R.; Mekonnen, A. Community Controlled Forests, Carbon Sequestration and REDD+: Some Evidence from Ethiopia. *Environ. Dev. Econ.* **2016**, *21*, 249–272. [CrossRef]
24. Ingalls, M.L.; Dwyer, M.B. Missing the forest for the trees? Navigating the trade-offs between mitigation and adaptation under REDD. *Clim. Chang.* **2016**, *136*, 353–366. [CrossRef]
25. Angelsen, A.; Wertz-Kanounnikoff, S. *Moving Ahead with REDD Issues, Options and Implications*; Center for International Forestry Research: Kota Bogo, Indonesia, 2008; Volume 6.
26. United Nations Framework Convention on Climate Change (UNFCC). *Afforestation and Reforestation Projects under the Clean Development Mechanism*; UNFCC Secretariat: Bonn, Germany, 2013.

27. Cyranoski, D. Biodiversity: Logging: The new conservation. *Nature* **2007**, *446*, 608–610. [CrossRef] [PubMed]
28. Lyons, K.; Westoby, P. Carbon colonialism and the new land grab: Plantation forestry in Uganda and its livelihood impacts. *J. Rural Stud.* **2014**, *36*, 13–21. [CrossRef]
29. Neef, A.; Touch, S.; Chiengthong, J. The Politics and Ethics of Land Concessions in Rural Cambodia. *J. Agric. Environ. Eth.* **2013**, *26*, 1085–1103. [CrossRef]
30. Sasaki, N.; Yoshimoto, A. Benefits of tropical forest management under the new climate change agreement-a case study in Cambodia. *Environ. Sci. Policy* **2010**, *13*, 384–392. [CrossRef]
31. Khun, V.; Sasaki, N. Cumulative Carbon Fluxes Due to Selective Logging in Southeast Asia. *Low Carbon Econ.* **2014**, *5*, 180–191. [CrossRef]
32. Sikor, T.; Cam, H. REDD+ on the rocks? Conflict over Forest and Politics of Justice in Vietnam. *Hum. Ecol.* **2016**, *44*, 217–227. [CrossRef] [PubMed]
33. Scheidel, A.; Work, C. Large-scale forest plantations for climate change mitigation? New frontiers of deforestation and land grabbing in Cambodia. In *Global Governance/Politics, Climate Justice and Agrarian/Social Justice: Linkages and Challenges*; ICAS Colloquium: Hague, The Netherlands, 2016; p. 11.
34. Baird, I.G. "Indigenous Peoples" and land: Comparing communal land titling and its implications in Cambodia and Laos. *Asia Pac. Viewp.* **2013**, *543*, 269–281. [CrossRef]
35. Thuon, R. *REDD+ and Tenure Security of Community Forestry: A Case Study of Oddar Meanchey Community Forestry REDD+ Project*; University of Tokyo: Tokyo, Japan, 2013.
36. Milne, S. Under the leopard's skin: Land commodification and the dilemmas of Indigenous communal title in upland Cambodia. *Asia Pac. Viewp.* **2013**, *54*, 323–339. [CrossRef]
37. Vongvisouk, T.; Broegaard, R.B.; Mertz, O.; Thongmanivong, S. Rush for cash crops and forest protection: Neither land sparing nor land sharing. *Land Use Policy* **2016**, *55*, 182–192. [CrossRef]
38. Kurashima, T.; Matsuura, T.; Miyamoto, A.; Sano, M.; Tith, B.; Chann, S. Changes in Income Structure in Frontier Villages and Implications for REDD+ Benefit Sharing. *Forests* **2014**, *5*, 2865–2881. [CrossRef]
39. Olwig, M.F.; Noe, C.; Kangalawe, R.; Luoga, E. Inverting the moral economy: The case of land acquisitions for forest plantations in Tanzania Inverting the moral economy: The case of land acquisitions for forest plantations in Tanzania. *Third World Q.* **2016**, *6597*, 2316–2336.
40. Ministry of Agriculture, Forestry, and Fisheries (MAFF). *Prakas: Declare the Establishment of Bung Cha Camp for Promotion and Forest Restoration*; MAFF: Prakas, Cambodia, 2010.
41. Global Withess. *The Cost of Luxury*; Global Witness: London, UK, 2015.
42. Michaud, A. *Prey Lang Development Case: Do People Benefit from Its Development?* NGO Forum: Phnom Penh, Cambodia, 2013.
43. Kotoski, K. Disney carbon deal sets stage for more partnerships. Available online: http://www.phnompenhpost.com/business/disney-carbon-deal-sets-stage-more-partnerships (accessed on 12 August 2016).
44. Food and Agriculture Organization; United Nations Development Programme; United Nations Environment Programme. Cambodia REDD+ Roadmap. In Proceedings of UN-REDD programme 5th policy board meeting, Washington, DC, USA, 4–5 November 2010; UN-REDD: Washington, DC, USA, 2010; pp. 1–54.
45. The United Nations Programme on Reducing Emissions from Deforestation and Forest Degradation. *Myanmar REDD+ Readiness Roadmap*; The Government of the Union of Myanmar: Yangoon, Myanmar, 2013.
46. Atela, J.O.; Quinn, C.H.; Minang, P.A.; Duguma, L.A.; Houdet, J.A. Implementing REDD+ at the national level: Stakeholder engagement and policy coherences between REDD+ rules and Kenya's sectoral policies. *For. Policy Econ.* **2016**, *65*, 37–46. [CrossRef]
47. Lang, C. *Konflikte um Rohstoffe in Asien REDDheads: The People Behind REDD and the Climate Scam in Southeast Asia*; Stiftung Asienhaus: Cologne, Germany, 2016.
48. Yeang, D.; Brewster, J. *REDD+ Demonstration Activities in Cambodia: The Case of the Oddar Meanchey Community Forestry REDD+ Project*; PACT Cambodia: Phnom Penh, Cambodia, 2012.
49. The United Nations Programme on Reducing Emissions from Deforestation and Forest Degradation. UN-REDD Programme Welcomes Five New Countries. Press Release. Available online: http://www.yeangdonal.net/2009/11/un-redd-programme-welcomes-five-new.html (accessed on 7 March 2015).
50. Lynham, T.; Evans, T.; Pet, P.; Phien, S. *Monitoring Systems for Illegal Land Encroachment at the Seima Protection Forest REDD+ Demonstration Site*; Wildlife Conservation Society and the Forestry Administration of the Ministry of Agriculture, Forestry, and Fisheries: Phnom Penh, Cambodia, 2014.

51. Delux, C.; Nguon, P. *National REDD Strategy, 4th Draft*; Forest Administration of the Ministry of Agriculture, Forestry, and Fisheries: Phnom Penh, Cambodia, 2015.

52. Royal Government of Cambodia (RGC). *Cambodia Millennium Development Goals Report*; Ministry of Planning: Phnom Penh, Cambodia, 2003.

53. Royal Government of Cambodia (RGC). *National Strategic Development Plan, 2014–2018*; Ministry of Planning: Phnom Penh, Cambodia, 2014.

54. Board, J. Cambodian villagers fear for future amid forest burning dispute. Available online: http://www.channelnewsasia.com/news/asiapacific/cambodian-villagers-fear/3020786.html (accessed on 11 August 2016).

55. Forest Administration of Cambodia. *Cambodia Forestry Outlook Study*; No. Working Paper No. APFSOS II/WP/2010/32; Ministry of Agriculture, Forestry, and Fisheries: Phnom Penh, Cambodia, 2010.

56. Lang, C. The Expansion of Industrial Tree Plantations in Cambodia and Laos. *Focus Asien Schr. Asienhauses* **2006**, *29*, 24–27.

57. Middleton, C.; Sokleap, H. *Fast-Wood Plantations, Economic Concessions and Local Livelihoods in Cambodia.* 2007. Available online: http://wrm.org.uy/oldsite/countries/Cambodia/EFCT_Plantations_Report.pdf (accessed on 15 March 2015).

58. Petersen, R.; Aksenov, D.; Esipova, E.; Goldman, E.; Harris, N.; Kuksina, N.; Sargent, S.; Manisha, A.; Loboda, T.; Shevade, V. *Mapping Tree Plantations with Multispectral Imagery: Permiminary Results for Seven Tropical Countries*; Technical Note; World Resources Institute: Washington, DC, USA, 2016.

59. Ra, K.; Kimsun, T. Financial Viability of Plantations of Fast- Growing Tree Species in Cambodia. Phnom Penh, Cambodia, 2012. Available online: http://twgfr.org/download/StudyReports(2)/13-Financialviabilityofplantation-fast-growingtreespeciesinCambodia-2012.pdf (accessed on 30 January 2017).

60. World Bank. *Cambodia Agenda for Rehabilitation and Reconstruction*; World Bank: Washington, DC, USA, 1992.

61. Hughes, C. *The Political Economy of Cambodia's Transition, 1991–2001*; Routledge/Curzon: London, UK; New York, NY, USA, 2003.

62. Le Billon, P. Logging in Muddy Waters: The Politics of Forest Exploitation in Cambodia. *Crit. Asian Stud.* **2002**, *34*, 563–586. [CrossRef]

63. Milne, S. Cambodia's Unofficial Regime of Extraction: Illicit Logging in the Shadow of Transnational Governance and Investment. *Crit. Asian Stud.* **2015**, *47*, 200–228. [CrossRef]

64. Global Witness. *Cambodia 's Family Trees: Illegal Logging and the Stripping of Public Assets*; Global Witness: Phnom Penh, Cambodia, 2007.

65. Dararath, Y.; Top, N.; Lic, V. *Rubber Plantation Development in Cambodia: At What Cost?* The Economy and Environment Program for Southeast Asia (EEPSEA): Phnom Penh, Cambodia, 2011.

66. Blomley, T.; Tola, P.; Kosal, M.; Dyna, E.; Dubois, M. *Review of Community Forestry and Community Fisheries in Cambodia; Report Prepared for the Natural Resource Management and Livelihoods Programme*; Natural Resource Management and Livelihoods Programme: Phno Penh, Cambodia, 2010.

67. Grimsditch, M.; Schoenberger, L. *New Actions and Existing Policies: The Implementation and Impacts of Order 01*; NGO Forum: Phnom Penh, Cambodia, 2015.

68. Focus on the Global South. *Study on the Impacts of the Implementation of Order 01BB on Selected Communities in Rural Cambodia*; Focus on the Global South: Phnom Penh, Cambodia, 2013. Available online: http://focusweb.org/sites/www.focusweb.org/files/StudentVolunteersReport_ENG.pdf (accessed on 12 December 2012).

69. Royal Government of Cambodia. *Sub-Decree for the Transfer of 2133.99 ha of State Public Land in Samaoch Village, Mien Rith Commune, Sandan District, Kampong Thom Province, for a Social Land Concession to Give to Landless and Land Poor Citizens, Cambodia*; Royal Government of Cambodia: Phnom Penh, Cambodia, 2013.

70. Titthara, M. Deforestation Continues. Availiable online: http://www.khmertimeskh.com/news/28863/the-deforestation-continues/ (accessed on 24 August 2016).

71. Chheng, K.; Bun, R.; Williams, B. *Tumring REDD+ Project: A Joint Korea-Cambodia Project*; Ministry of Agriculture Forestry and Fisheries: Phnom Penh, Cambodia, 2015.

72. PWC, P. *Hanwha Corporation Non-Consolidated Financial Statements*; Samil Price Waterhouse Coopers: Seoul, Korea, 2010.

73. Jiao, X.; Smith-Hall, C.; Theilade, I. Rural household incomes and land grabbing in Cambodia. *Land Use Policy* **2015**, *48*, 317–328. [CrossRef]

74. Milne, S.; Adams, B. Market Masquerades: Uncovering the Politics of Community-level Payments for Environmental Services in Cambodia. *Dev. Chang.* **2012**, *43*, 133–158. [CrossRef]
75. Harms, E.; Baird, I.G. Wastelands, degraded lands and forests, and the class (ification) struggle: Three critical perspectives from mainland Southeast Asia. *Singap. J. Trop. Geogra.* **2014**, *35*, 289–294. [CrossRef]
76. REDD Desk. *Korea Indonesia Joint Project for Adaptation and Mitigation of Climate Change in Forestry*; Global Canopy Programme (GCP): Oxford, UK, 2016. Availiable online: http://theredddesk.org/countries/initiatives/korea-indonesia-joint-project-adaptation-and-mitigation-climate-change (accessed on 5 September 2016).
77. Kim, J.; Kim, J.; Shin, S.; Lee, S. *Public-Private Partnership Infrastructure Projects: Case Studies from the Republic of Korea*; Asian Development Bank: Mandaluyong, Philippines, 2011; Volume 1.
78. Frewer, T. Code REDD+ in Cambodia. Availiable online: http://asiapacific.anu.edu.au/newmandala/2015/08/11/code-redd-in-cambodia/ (accessed on 12 December 2015).
79. Käkönen, M.; Karhunmaa, K.; Bruun, O.; Kaisti, H.; Tuominen, V.; Thuon, T.; Luukkanen, J. *Climate Mitigation in the Least Carbon Emitting Countries—Dilemmas of Co-Benefits in Cambodia and Laos*; University of Turku: Turku, Finland, 2013.

forests

MDPI

Article

Evolving Protected-Area Impacts in Mexico: Political Shifts as Suggested by Impact Evaluations

Alexander Pfaff [1],*,†, Francisco Santiago-Ávila [2],† and Lucas Joppa [3]

[1]   Duke University, 302 Towerview Drive, Durham NC 27708, USA
[2]   University of Wisconsin, 1220 Linden Drive, Madison, WI 53706, USA; santiagoavil@wisc.edu
[3]   Microsoft Research, One Microsoft Way, Redmond, WA 98052, USA; lujoppa@microsoft.com
*   Correspondence: alex.pfaff@duke.edu; Tel.: +1-919-613-9240
†   Co-lead authors.

Academic Editors: Esteve Corbera and Heike Schroeder
Received: 8 October 2016; Accepted: 22 December 2016; Published: 29 December 2016

Abstract: For protected areas (PAs), variation in forest impacts over space—including types of PA—are increasingly well documented, while shifts in impacts over time receive less attention. For Mexico, in the 1990s, PAs effectively were 'paper parks'. Thus, achieving impacts on the forest would require shifts over time in the politics of PA siting and PA implementation. We rigorously analyze the impacts of Mexican PAs on 2000–2005 loss of natural land cover, using matching to reduce location bias caused by typical land-use economics and politics. We find a 3.2% lower loss, on average, due to PAs. Since politics often vary by type of PA, we also show that in Mexico stricter PAs are closer to cities and have greater impact than mixed-use PAs. These shifts in impacts suggest some potential for PAs to conserve forests.

Keywords: Mexico; deforestation; conservation; protected areas; impact evaluation; matching

1. Introduction

In recent decades, national and international efforts to reduce forest loss have had some impact but have not substantially slowed tropical forest loss. Adding climate change to concern about species, such forest losses now account for about one-sixth of anthropogenic greenhouse gas emissions. Climate-related incentives could enhance various programs and policies that affect deforestation, most likely by emphasizing performance with monitoring, reporting, and verification of outcomes. Such emphasis, with incentives, could increase policies' deforestation impacts [1].

Understanding of past policy impact also supports greater future impacts, through better design. It is crucial for countries to know what has (not) worked to date in reducing forest loss, and why. Here, we follow a recent explosion of literature that has provided improved impact evaluations for conservation policies, such as protected areas (PAs—impacts reviewed in [2]). Improved methods often reduce estimates of average PA impacts, while also clearly demonstrating variation in impacts across space and PA types [3,4].

First, to highlight the actors and incentives relevant for variation in impacts of PAs upon forests, we present conceptual models of individual land-use choices and the political economy of PA siting. They suggest key influences on PAs' impacts. Our results shed empirical light on those influences, suggesting that the politics of PA implementation in Mexico clearly have shifted over time.

We start with a classic land-use model, taking as given both the PAs' locations and their enforcement. The clear prediction that typical landscapes vary in their baseline deforestation rates is the motivation for our efforts to control for characteristics of sites that affect individual decisions that yield deforestation. Only by removing the influence of such characteristics, which might well be expected to differ for protected versus unprotected locations [5], can we infer the impact of the PAs.

As to why PA locations might differ from unprotected sites in relevant characteristics, we highlight political and economic tradeoffs. Those tradeoffs predict variations in both locations and enforcement across PA types and, thus, also varied impacts [6]. In sum, these models suggest that, depending on the political economic context, either strict or mixed-use PAs can have more impact.

For Mexico, an important basis for comparison is recent work showing no PA impacts in the 1990s ([7] on 'paper parks'). Thus, we consider PA impacts on more recent changes in land cover (2000–2005) after a shift in the politics of conservation in Mexico, with a change in administration in 2000 to President Fox, under whom Ernesto Enkerlin shifted management of PAs. In addition, GEF provided stable funds for some PAs (https://www.thegef.org/country/mexico). The time period we study, coming just after the 1990s, offers a useful window on a possible sharp shift in politics. We estimate the average PA impact and then differences in impacts between strict and mixed-use PAs.

Our application of matching to improve estimates of PA impact—following both of our models—reduces estimated impacts, as hypothesized, relative to when ignoring key location characteristics. Nonetheless, improved estimates show PAs in Mexico did lower land-cover change. Specifically, PAs across Mexico reduced the 2000–2005 loss of land cover by 3.2%, on average.

In comparing PA types, following the political economy model, we find that strict PAs avoid more land-cover change (5.2%) than mixed-use PAs (2.7%). That is intuitive, if enforcement is tougher. However, we show that greater impact from strict PAs is influenced by the relative locations of the PA types. That may seem counterintuitive but it is consistent with the PAs' sites having been chosen at a time when PA enforcement was low.

The paper proceeds as follows. Section 2 provides background for Mexico and prior literature evaluating policy impacts. Section 3 presents both of the conceptual models just discussed above. Section 4 then describes our data and methods, Section 5 conveys results, and Section 6 concludes.

2. Background and Related Literature

Mexican forests cover 67 million hectares, about one-third of the country (198 million hectares). Along with agriculture and fishing, forestry accounted for about 5% of the country's GDP in 2006. Agriculture and forestry are areas where greenhouse gas (GHG) emission can be reduced, having generated about 135 MtCO$_2$e or 21% of Mexican emissions in 2002 [8], with two-thirds from the forest sector. Proximate causes of both deforestation and degradation are conversion to grassland, slash-and-burn agriculture, illegal logging, and fire. Underlying forces include a lack of investment in the forestry sector, low income from forest activities, multiple agricultural and livestock activities, uncertainty related to use rights, poverty, and a general lack of opportunities for forest owners [9]. The drivers are complex and they vary between regions.

It is widely acknowledged that successful forest interventions could not only reduce emissions but also generate ecosystem services, income, and employment—among other co-benefits [8]. Interventions such as reforestation and commercial plantations are what account for 85% of the state's proposed mitigation in agriculture and forestry. Success would depend on institutional changes, better public financing, and sustainable forest product markets [8].

This direction for policy appears to have some support in Mexico, despite a large illegal timber market, a lack of financial and human resources for operational capability, and (drug-related) insecurity. Mexico currently is emphasizing a national, multi-functional and multi-scale mechanism for monitoring, reporting and verification (MRV) based on remote sensing and ground-based forest inventory methodologies [9]. It could include early detection for changes in land cover and land use [10]. Such a mechanism could provide the relevant authorities with more precise measures to address concern about detection of small-scale land-cover changes.

2.1. Evaluating PA Impacts

Joppa and Pfaff [2] review the PA literature—as do Naughton-Treves [11], Nagendra [12], and Campbell et al. [13]—emphasizing obstacles to inferring PA impacts on forest. This follows global documentation [5] that, at least on average, the locations of PAs across national landscapes are significantly biased in ways relevant to deforestation rates. They stress that observing fully forested PAs (e.g., [14,15]) falls short of observing impacts, as impacts require a comparison to what would have happened without PAs. Given that it is literally impossible to observe such 'counterfactuals'—scenarios without PAs—one must estimate what would have happened on PA land without protection, using outcomes from unprotected sites. Siting bias means the average outcome for all unprotected lands (as in [16–18]) is a poor counterfactual estimate for PAs, since sites are different on average. Comparing to areas around PAs [15,19–23] can help. Yet, without explicitly comparing land characteristics, it is hard to know how similar they are. At least as troubling, those nearby lands could be affected by the establishment of the PA if that generates local spatial spillovers such as 'leakage'.

Matching methods explicitly compare land characteristics between sites, aiming to increase the similarity of controls to PA observations. Andam et al. [24]'s application of matching to Costa Rican PAs reduces the estimated impact from about 44%, over decades, to about 11% of the protected area. Joppa and Pfaff [3] provide analogous results for each of over 100 countries around the globe. Given those average impacts, Pfaff et al. [4] revisit Costa Rican PAs in subsets, using matching, to confirm predicted variation in PAs' impacts over the landscape. Impacts are higher near roads and cities and on flat land. Shah and Baylis [25] also show heterogeneous PA impacts on forest, for Indonesia, while Joppa and Pfaff [3] confirm such predictable variation in impacts in their global study of PAs.

2.2. Evaluating Conservation Impacts in Mexico

Early conservation evaluations for Mexico included payments by a federal agency to upstream land, which can affect water quality. Munoz [26] finds significant location bias in payments contracts, lowering impact. Alix-Garcia et al. [27] extend this with similar results while Alix-Garcia et al. [28], using additional data, consider both forest and poverty impacts of these hydro-services payments. Recently, Pfaff et al. [29] and Kaczan et al. [30] consider institutional adjustments in such payments.

Concerning PAs, Blackman et al. [7] is analogous to our study except for the time period, i.e., the 1990s, when it was widely asserted that PAs were under-resourced 'paper parks' (while Sims and Alix-Garcia 2016's analogous ongoing work extends further forward in time and compares outcomes of PAs to PES [31]). Blackman et al. [7] find no PA effect. Figueroa and Sánchez-Cordero [32] also find limited effect for the 1990s. Low intention or ability to enforce can affect not only PA outcomes given PA sites but also PAs' sites because, without enforcement, there is little reason for locals to push back. Mas [33] and Durán-Medina et al. [34] find some impact for particular PA sites, as do Honey-Roses et al. [35,36], for a site where PES and a PA are combined, while Miteva et al. [37] examine tenure.

3. Modeling PA Impact

3.1. PA Impact by Location

From von Thunen [38] to the 'monocentric model', landscape analysts assert that the pressure to clear forest falls as we move outward from the market center (in Figure 1, a city where the axes cross). Transport costs imply falling profits from agricultural products to be sold in the city, as we move out. If all land is originally forested and only transport matters, forests will remain farther from markets (in Figure 1, forest will remain to the right of where the 'Expected Profits' line crosses the '0'axis).

Of course, factors other than transport also affect relative profits from agriculture versus forests: e.g., high slopes near markets may stay forested; and good soils far from market may be cleared. From an analyst's point of view, some of these factors are observed while others are unobserved, as there are limits on all datasets. The empirical analyses we review include observable factors; however, Figure 1 does not explicitly depict the observables, focusing on representing unobserved factors in the form of a distribution, or varying density, of land-parcel profits around the Expected Profit line.

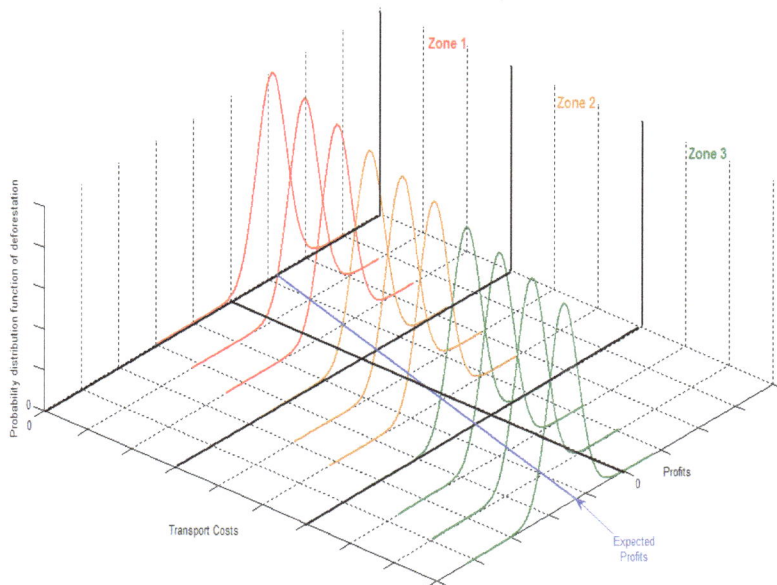

Figure 1. Expected Profits Across Landscape with Implications for PA Impact & Siting.

Conservation policies keep forests standing, using enforcement if there is pressure to clear forest. Without pressure, a fully forested PA does not imply impact, i.e., avoidance of baseline deforestation. Thus, if private pressure would not have led to clearing of the forest anyway, the policy did not have impact. More generally, a conservation policy's impact equals the private or 'baseline' deforestation rate that would have arisen without the policy minus the deforestation rate observed with the policy. Within Figure 1, if transport cost is a significant factor in the private (or 'baseline') rate of clearing, then a PA that is far to the right may not have had much impact on the forest, even if fully forested.

3.2. PA Impact by Type

Given the underlying logic above, we can see that PA sites determine average PA impacts. Those sites will be affected by the political economy of development: PAs block production; thus, lobbying by producers tends to push PAs to low-profit locations. Where agricultural profit is high, lands are expensive to buy for conservation, plus lobbying against allocating public lands to conservation will be significant. Formally, a simple theory of PA location would be that for a PA to be established its cost—including local profit in Figure 1 that is foregone—must be below benefit (perceptions of which vary by decision maker, as World Bank [39] shows for Brazil). For any given level of PA benefit, this theory predicts that PA siting may avoid costs by avoiding pressures (as confirmed by, e.g., [5]).

The political economy can vary with the PA type, as well as by country and across time periods. Development tradeoffs for 'strict' PAs differ from those for mixed-use PAs that permit some local extraction and, therefore, induce less lobbying against the establishment of the PA. Thus, PA types can end up with different locations. For instance, if differences in PA regulations by type are actually enforced, we might well expect to see stricter PAs further to the right in Figure 1 [40,41].

Yet where public actors locate new PAs depends on whether they intend to enforce them. For instance, if locals believed that PAs in Mexico would not be enforced, i.e., expected the 1990s 'paper parks' outcome in [7], then they would have had little reason to lobby against a PA. Even strict PAs might be permitted near cities. Once PAs are sited, enforcement may occur but may fall with isolation [42,43]. Thus, interactions with pressure could produce varied patterns across space of PAs' impacts.

4. Data and Matching Methods

4.1. Land Cover Data

Our data are from a global data set with land cover for 2000 [44] and land cover for 2005 (ESA) [45]. The 2000 data, GLC2000, have 23 classifications of land cover. They were reclassified to 'natural' or 'human modified', the latter including categories 16 (cultivated and managed areas), 17 (mosaic of cropland with tree cover or other natural vegetation), 18 (mosaics of cropland, with shrubs or grass cover), 19 (bare areas), and 22 (artificial surfaces and associated areas). The same was done for 2005 GLOBCOVER300 data. Again, multiple categories are placed into "natural" and "modified", the latter including categories 11 (irrigated croplands), 14 (rainfed croplands), 20 (mosaic cropland (50%–70%)), 30 (mosaic cropland (20%–50%)), and 190 (urban areas >50%). These datasets were not constructed for precise intertemporal comparison, yet this transformation allows for a reasonable comparison of years. In each dataset, the spatial scale of land observations is 1-km^2 polygons.

The change between the datasets—which we call 'vegchange'—was calculated after that process. With the transformation, the data in principle track change from a 'natural' to a 'human modified' landscape and this 2000–2005 change is the dependent variable upon which we focus. We want to study recent changes that can be affected by the presence of PAs (though they are net changes, i.e., from 'natural' to 'human modified' and vice versa, which matters for indicators of species habitat). Given any issues with comparing datasets, we also analyze spatial patterns in the 2005 land cover.

4.2. Land Characteristics Data

Table 1 provides descriptive statistics for land characteristics in our analysis. Elevation (m) is from the Shuttle Radar Topography Mission [46], with slope in degrees from horizontal. The roads and urban areas used to compute distances (km) are from VMAP0 Roads of the World [47] and the Global Rural Urban Extent data [48]. While not the best quality, these VMAP0 data are all that is freely accessible to define the global road network.

Table 1. Land characteristics and land cover (including PA type).

	Unprotected	Protected	Protected–Stricter	Protected–Mixed-Use
Protected, Stricter Subset (1/0)	0	0.21	1	0
Protected, Mixed-use Subset (1/0)	0	0.79	0	1
IUCN Category (1–6)	0	5.0	1.3	6.0
Distance Outside PA Edge (km)	55.8	−14.0	−10.5	−15.0
Urban Distance (km)	35.6	95.7	58.4	105.3
Road Distance (km)	8.5	12.6	14.3	12.2
Elevation (m)	1081	622	885	554
Slope (degrees)	2.81	2.85	3.42	2.70
Agricultural Suitability (0–9)	6.17	6.81	6.57	6.88
Relatively Low Agric. Suitability (1/0)	0.39	0.56	0.38	0.60
Relatively High Agric. Suitability (1/0)	0.37	0.21	0.22	0.21
Fires (number detected in 2000–2006)	0.005	0.003	0.004	0.003
Dummy for Whether Any Fires (1/0)	0.005	0.003	0.004	0.003
Pine Oak Forest Dummy (1/0)	0.115	0.003	0.005	0.003
Chihuahua Desert Dummy (1/0)	0.163	0.064	0.012	0.078
GLC 2000 Natural Land Cover (1/0)	0.840	0.920	0.912	0.921
Globcover 2005 Natural Land Cover (1/0)	0.862	0.957	0.970	0.954
2000–2005 Loss of Natural Cover (1/0)	0.092	0.029	0.022	0.031
# observations	1,801,935	121,847	25,096	96,751

PAs (protected areas) are from World Database on Protected Areas [49]. In these data, Categories I–II allow less human intervention, while Categories III–VI tend to be less protected, allowing for multiple uses. For any overlaps of categories, the polygon was categorized into the stricter category. We created three dummy variables: 'protected', for any protection (regardless of IUCN category); 'mixed-use PA' for IUCN categories III–VI; and 'strict PA', if the IUCN category is I or II. In order to check robustness, we also compared categories I–IV versus V–VI (following Joppa and Pfaff [3], as well as Nelson and Chomitz [40]). Also, for our time period, we used only the PAs created by 2000.

The World Wildlife Fund classified ecoregions [50]. Unclassified ecoregions were dropped (n = 1747) and we created dummy variables for the two ecoregions with the highest frequencies: 'pineoakdum' (10.7% of total) if the ecoregion is Sierra Madre Occidental Pine Oak Forest; and 'chidesertdum' (15.6% of total) if the ecoregion is Chihuahuan desert. The agricultural suitability is from International Institute for Applied Systems Analysis' Global Agro-Ecological Zones data [51]. It uses climate, soil type, land cover, and slope to assign a value to each polygon, ranging from 0 (meaning no constraints) to 9 (severe constraints). We created two dummy variables: 'low' suitability for more agricultural constraints (i.e., agricultural suitability score of 8 or higher); and 'high' for situations with fewer agricultural constraints, (i.e., agricultural suitability score of 5 or lower). Additionally, as a robustness check, we use the full variable (which does not affect any of the results).

The fire variable, as a continuous metric, simply captures the number of fires from 2001 to 2006. The 'firedum' dummy we created takes a value of '1' if that polygon experienced any fires ($\geq$1). Finally, the distance-to-edge variable measures the distance to the edge of a protected area (km) from the polygon in question. A negative distance value indicates that the observation is inside of the PA.

These variables are not expected to fully explain land-cover change or the location of protection. Yet as all influence profit, they predict deforestation as well as the local resistance to protected areas. It is a combination of relevance to protection and deforestation that makes them useful for matching. The data contain approximately 1,935,301 observations (1-km^2 polygons of land) and 13 variables. We have run many analyses with random 10% samples to confirm our results are robust to sampling. As noted, Table 1 presents summary statistics for all the aforementioned variables. For our results, we explore their impacts upon land-cover change as well as their significance for locations of PAs.

4.3. Matching Methods

If PAs were sited randomly, PA impacts would be easy to estimate using the differences between deforestation inside and outside of PAs. Deforestation outside would be an unbiased estimate of what would have been occurred inside PAs had there been no protection, as other factors would cancel out. Yet neither PAs in general, nor any PA type, seems to be distributed as if at random. Further, that non-randomness often appears to be along dimensions that can affect deforestation. To isolate PA impact by removing the influences of these differences, we use 'matching' methods to improve controls. One 'matches' each protected polygon with the most similar unprotected polygon(s) to get as close as possible to 'apples-to-apples' comparisons. Thus, PAs are compared not to all unprotected land but, instead, to the most similar unprotected land. Here, we apply propensity-score matching in our many initial analyses and then, for our final results, confirm robustness to covariate matching.

Propensity-score matching assesses the 'similarity' of sites using the predicted probability of a polygon being in a PA. PA polygons are then compared to unprotected polygons with similar enough characteristics to yield a similar probability of protection. Probabilities are generated by a probit model using factors in protection and deforestation to explain where protection occurred [52]. More weight is given, then, to the variables that are important determinants of protection. In covariate matching, similarity is assessed using the distance between polygons in the covariate space. For either method, we matched each treated polygon with the single most similar unprotected polygon.

However, selecting the most similar polygon does not guarantee that controls are, in fact, similar. Thus, we also check explicitly whether the selected unprotected polygons are similar to the protected.

We examine balance, i.e., if the characteristics' values are distinguishable between the protected and matched unprotected observations. Ideally, they should not be. Assuming large differences to begin with, we would expect at least a significant reduction in differences between groups, due to matching.

Given balanced characteristics, the deforestation in matched unprotected sites is an improved estimate of the deforestation that would have occurred at PA locations without the protection. PA impact is calculated as that counterfactual rate minus the observed deforestation (given protection). However, still there will be differences between these groups, in terms of characteristics relevant for deforestation. Thus, our preferred matching estimates involve first matching and then regressions. We refer to the latter as 'bias adjustment', as this addresses remaining differences in characteristics.

If the unobservable or omitted factors are correlated with "treatment", i.e., with where PAs are, that could bias estimates of impact. Matching can control only for the included observable factors. For instance, we do not know anything about the populations in the PAs versus unprotected sites. We suspect that factors we do observe, however, such as the road and city distances, correlate with the unobservables. Thus, given our observables, we cannot be sure of the sign of any residual biases. As a robustness check, we compute 'Rosenbaum bounds' to estimate how sensitive are these results.

5. Results

5.1. Descriptive Statistics

Table 1 provides means for our outcomes variables as well as our metrics for drivers, including Mexico's protection interventions—both in general and by PA type. Note that 79% of protection is a 'mixed-use' type, with an average IUCN category of 6, while 21% of the PA observations are stricter protection, having an average IUCN category of 1.3.

In terms of drivers of deforestation, Table 1 leads with a very important driver, urban distance. It is the driver that differs most in these data between PAs and unprotected polygons, and between the two types of PAs. Protected polygons are roughly three times as far as unprotected. Mixed-use PAs are roughly twice as far from cities as strict PAs—key for different PA impacts across types. We note that these relative locations of PA types in Mexico differ from the Brazilian Amazon [41] and from the averages across the globe (see Nelson and Chomitz [40] concerning PAs' impacts on fires).

The relative location of PAs on average fits our political economy model in which, for a given societal benefit, opportunity costs drive PAs to relatively unprofitable sites. Table 1 shows this for urban distance and, to a lesser extent, for road distance, as well as for the fractions of high and low quality soils. Without considering such differences in location, Table 1's last row provides a form of estimate of PAs' average impact: there is a 2.9% loss of natural land cover in PAs, lower than the 9.2% loss for unprotected observations. Further, Table 1 suggests that there was actually some enforcement difference across stricter and mixed-use PAs during our time period (after alleged political shifts). The land-cover loss of 2.2% is lower for strict PAs, despite the fact that the only major difference in characteristics is that strict are *closer* to market (recall, siting occurred before our time period). However, actually estimating PA impact requires revisiting with control for all land characteristics.

5.2. Drivers of Protection

Table 2 uses regression to consider which locations are protected. Considering urban distance, PAs are further from urban areas even controlling for other influences. As expected from Table 1, the coefficient on urban distance for mixed-use PAs is larger than for strict, i.e., mixed-use are farther. For road distance, the smaller coefficients also support Table 1, i.e., the mixed-use PAs are closer (PAs may be near rural *ejidos* with road access, as 70% of land with forest is owned by *ejidos*, as Mexico's 1992 Agrarian Law, Article 1 gave these communities legal status and land ownership and they can participate in conservation programs such as Mexico's PES (The REDD Desk, 2012).) The other findings in Table 2, e.g., for agricultural suitability, also reflect the differences seen in Table 1.

Table 2. Drivers of protection (including by type).

	Protected	Protected–Stricter Subset	Protected–Mixed-Use Subset
Urban Distance	0.00251 ***	0.00032 ***	0.00248 ***
	(0.000)	(0.000)	(0.000)
Road Distance	0.00031 *	0.00066 ***	−0.00035 ***
	(0.000)	(0.000)	(0.000)
Elevation	−0.00001 ***	0.00000 ***	−0.00001 ***
	(0.000)	(0.000)	(0.000)
Slope	0.00120 ***	0.00048 ***	−0.00044 ***
	(0.000)	(0.000)	(0.000)
High Agricultural Suitability	−0.01533 ***	−0.01053 ***	−0.00564 ***
	(0.000)	(0.000)	(0.000)
Pine Oak Forest	−0.10585 ***	−0.03090 ***	−0.08272 ***
	(0.000)	(0.000)	(0.000)
Chihuahuan Desert	−0.10554 ***	−0.03271 ***	−0.08405 ***
	(0.000)	(0.000)	(0.000)
constant	−0.00390 ***	−0.00566 ***	0.00461 ***
	(0.000)	(0.000)	(0.000)
# obs	1,923,782	1,827,031	1,898,686
R^2	0.182	0.021	0.196
Adjusted R^2	0.182	0.021	0.196

Robust standard errors in parentheses: *** $p < 0.001$; ** $p < 0.01$; * $p < 0.05$.

5.3. Drivers of Land-Cover Change

Before estimating impacts of PAs on land cover, with controls for effects of other determinants, we would like to empirically confirm for our data what was driving the loss of natural land cover in Mexico in this time period. To do this, we examine rates of loss in just the unprotected locations. Table 3 presents OLS regressions of both land-cover loss and 2005 land cover on characteristics (note that we would expect the coefficients in these two outcomes columns to have the opposite signs).

Table 3. Drivers of natural land cover, explaining both loss & final amount (outside PAs).

	Natural Land Cover **Loss 2000–2005**	Natural Land Cover **Amount 2005**
Urban Distance	−0.00101 ***	0.00137 ***
	(0.000)	(0.000)
Road Distance	−0.00074 ***	0.00064 ***
	(0.000)	(0.000)
Elevation	−0.00004 ***	0.00006 ***
	(0.000)	(0.000)
Slope	−0.00445 ***	0.00572 ***
	(0.000)	(0.000)
High Agricultural Suitability	0.03979 ***	−0.06338 ***
	(0.001)	(0.001)
Pine Oak Forest	0.06457 ***	−0.06964 ***
	(0.001)	(0.001)
Chihuahuan Desert	−0.02074 ***	0.01979 ***
	(0.001)	(0.001)
Constant	0.19813 ***	0.75283 ***
	(0.001)	(0.001)
# obs	1,514,249	1,801,935
R^2	0.03	0.06
Adjusted R^2	0.03	0.06

Robust standard errors in parentheses: *** $p < 0.001$, ** $p < 0.01$, * $p < 0.05$.

In Table 3, as expected, urban distance has a significant effect on the profitability of land-cover loss. Also expected, road distance matters, even controlling for urban distance. Transport is critical. Higher elevation and slope also discourage whatever land uses involve the loss of natural cover, again as expected. Further, also expected, better conditions for agriculture increase the loss of cover. Both the Sierra Madre pine oak forest and the Chihuahuan desert are statistically significant controls.

5.4. Matching on Drivers of Land-Cover Change

Table 4 communicates whether our matching approach has actually made controls more similar. Again we start with the urban distance. For all PA polygons, or the strict and mixed-use subsets, Table 4 shows over 95% reductions in the differences between the PAs and unprotected controls. As the other differences noted in Table 1 were for road distance and high agricultural suitability, Table 4 shows similar reductions for them, as is the case for the significant and large ecoregions. These reductions in group differences are, we feel, a solid basis to claim improved estimates of PA impact.

Table 4. Matching balance.

Propensity Score Matching	Protected–Unprotected Characteristics Differences	Protected–Stricter Subset	Protected–Mixed-Use Subset
Urban Distance			
Pre-match difference	60.064 ***	19.273 ***	69.409 ***
Post-match difference	−2.580 ***	−0.817	−2.759 ***
% residual difference	−4.3%	−4.2%	−4.0%
Road Distance			
Pre-match difference	4.128 ***	5.584 ***	3.622 ***
Post-match difference	0.223 ***	0.174	−0.029
% residual difference	5.4%	3.1%	−0.8%
Elevation			
Pre-match difference	−459.298 ***	−169.814 ***	−524.714 ***
Post-match difference	−126.659 ***	26.358 **	−61.156 ***
% residual difference	27.6%	−15.5%	11.7%
Slope			
Pre-match difference	0.039 ***	0.617 ***	−0.118 ***
Post-match difference	−0.684 ***	−0.133 **	−0.003
% residual difference	†	−21.6%	2.5%
High Agric. Suitability			
Pre-match difference	−0.163 ***	−0.139 ***	−0.165 ***
Post-match difference	−0.049 ***	0.013 **	−0.002
% residual difference	30%	−9.4%	1.2%
Pine Oak Forest			
Pre-match difference	−0.111 ***	−0.105 ***	−0.110 ***
Post-match difference	−0.007 ***	−0.002 **	−0.005 ***
% residual difference	6.3%	1.9%	4.5%
Chihuahuan Desert			
Pre-match difference	−0.099 ***	−0.147 ***	−0.084 ***
Post-match difference	−0.021 ***	−0.001	−0.025 ***
% residual difference	21.2%	0.7%	29.8%

† increased small difference; *** $p < 0.001$; ** $p < 0.01$; * $p < 0.05$.

That said, matching always leaves residual characteristics differences between treated and control, noting that we also tested matching 'calipers' of 0.01 and 0.001, yet that did not improve balances. Within Table 4, many differences are statistically significant, in part due to the very large sample (a reason we checked for robustness to random 10% samples (Supplemental Materials: Tables S1 and S2). Thus, the post-matching regressions that we have employed are well motivated in Table 4—but also we simply flag these residual differences (another reason we test robustness to covariate matching). Finally, we note that unobservable differences could also exist—always a key issue for matching.

5.5. Average PA Impact

For loss of natural land cover and the 2005 amount of land cover, Table 5 presents PAs' impacts. The table's first column presents average impact, i.e., the interventions are any kind of protection. We discuss here the 2000–2005 loss of natural land cover (as impacts upon 2005 cover are analogous).

Table 5. PA impact estimates.

	Protected	Protected–Strict Subset	Protected–Mixed-Use Subset
2000–2005 Natural Land Cover Loss			
Pre-match difference in means	−6.3 % ***	−7.0 % ***	−6.1%***
# *observations*	1,923,782	1,827,031	1,898,686
Pre-match regression (all the data)	−3.5 % ***	−5.8 % ***	−2.6 % ***
SE	(0.001)	(0.001)	(0.001)
# *observations*	1,626,295	1,537,139	1,603,405
Post-match regression (most similar)	−3.2 % ***	−5.2 % ***	−2.7 % ***
SE	(0.002)	(0.003)	(0.001)
# *observations*	181,844	38,633	143,318
2005 Natural Land Cover			
Pre-match difference in means	9.5 % ***	10.8 % ***	9.2 % ***
# *observations*	1,923,782	1,827,031	1,898,686
Pre-match regression (all the data)	3.7 % ***	6.8 % ***	2.5 % ***
SE	(0.001)	(0.001)	(0.001)
# *observations*	1,923,782	1,827,031	1,898,686
Post-match regression (most similar)	3.5 % ***	6.3 % ***	2.9 % ***
SE	(0.002)	(0.003)	(0.001)
# *observations*	201,569	43,655	157,826

*** $p < 0.001$; ** $p < 0.01$; * $p < 0.05$.

Table 5's first row shows means differences, analogous to the impact estimate implied by Table 1 (a little different because, as seen in Table 3's first column, this outcome has fewer observations). Its second row introduces controls for the effects of other determinants of natural land-cover loss using regression analysis with all observations. This lower the estimated impact by almost half (as in Andam et al. [24] for Costa Rica and globally, for 141 countries, within Joppa and Pfaff [3]). The matching estimate in Table 5's third row confirms robustness to matching's use of observables, finding a 3.2% reduction in the 2000–2005 loss of land cover, using just the matched observations.

5.6. Impacts by PA Type

Table 5's second and third columns separate the PA treatment into the strict versus mixed-use PAs. We note that enforcement choices are illuminated by simple differences in means, in the top row, where the loss rate in each PA type is compared to the loss rate within all unprotected polygons. Despite the fact that mixed-use PAs are further from cities, which should lower clearing pressures, more loss of natural land cover is occurring within mixed-use PAs. This simplest 'impact estimate' is lower for mixed-use PAs (6.1%) than for strict PAs (7.0%). Enforcement seems to occur on average and, it seems here, enforcement choices differed by PA type in that clearing differs across PA types.

The differences between Table 5's first row and its second and third rows, which employ controls, in turn illuminate the different location choices for the two types of PAs. The more biased are PAs' locations towards low pressure—as we predict given apparent enforcement on average for this time period (again, per Blackman et al. [7], such enforcement did not always occur)—the greater the reduction in estimated PA impact we should see when controlling for all the land characteristics.

When we control using observables, impact estimates for strict and mixed-use PAs both fall, consistent with the effect in the column for all of the PAs. The estimate for the mixed-use PAs falls by more, however, consistent with greater urban distance for mixed-use reducing the true impact, given the apparent impact. Given enforcement and locations, stricter PAs have greater impact (5.2% versus 2.7%, noting that this difference in estimated impacts across types is statistically significant). These results may be explained by siting in the 'paper park' era, before increases in enforcement.

These results are supported by our robustness checks making use of 10% samples, which also addresses potential issues with spatial autocorrelation. Further, calculation of Rosenbaum bounds, given that there could be important unobserved factors for which matching clearly cannot control, shows that to eliminate the significance of strict impacts requires unobserved factors to make non-PA sites almost three times as likely to be cleared–versus only twice as likely for impacts of all PAs.

6. Conclusions

With two goals in mind, we estimated a suite of PA impacts upon land-cover changes in Mexico. One motivation was to study a period after the 1990s, as conservation politics were alleged to have shifted and prior work had demonstrated that, during the 1990s, PAs functioned as 'paper parks'. Another was to demonstrate the need to address how land-use decisions imply the possibility of bias in PA impact estimates and how the political economy of public PA choices implies its likelihood.

PAs did reduce losses of natural land cover within their boundaries during 2000–2005. Thus, it would appear that conservation politics shifted, at least as revealed by our impact assessment. Further, it was important in estimating the impacts to check for, and then control for, differences in characteristics of protected (versus unprotected) sites, as they directly influence deforestation. Controlling for them lowers the estimated average impact of PAs by about half. Further, controls generate different adjustments to the estimated impacts for the strict versus the mixed-use PAs.

Across PA types, a combination of siting and enforcement differences yielded different impacts, with strict PAs generating greater reductions in loss within their boundaries than mixed-use PAs. These results may reflect shifts over time in conservation politics, as PA locations seem consistent with a lack of intention to enforce during the 1990s that, as our results suggest, later was reversed.

Extensions could improve on these analyses. Our data for land cover are not ideal and Mexico's new MRV mechanism may offer options. For moving forward to future analyses, e.g., INEGI has invested in data on roads and vegetation, for example, and the Nature Conservancy has invested in tracking sites and types of protection. Finally, more study of differences in dynamics across space could further contribute to an understanding of the conditions that drive the impacts of protection.

Supplementary Materials: The following are available online at www.mdpi.com/1999-4907/8/1/17/s1, Table S1: Rosenbaum Bounds, Table S2: 10% Samples.

Acknowledgments: We are grateful for helpful suggestions from Maria Carnovale and Elizabeth Shapiro. This paper derives from Francisco J. Santiago-Ávila's master's project (or MP) at Duke University, concluded in 2013.

Author Contributions: Alexander Pfaff conceived of the paper; all the authors designed analyses; Lucas Joppa organized most of the data; Francisco J. Santiago-Ávila performed the analyses with guidance from Alexander Pfaff and Lucas Joppa; and all of the authors wrote the paper.

Conflicts of Interest: The authors declare no conflict of interest.

References

1.   Pfaff, A.; Amacher, G.S.; Sills, E.O. Realistic REDD: Improving the forest impacts of domestic policies in different settings. *Rev. Envir. Eco. Policy* **2013**, *7*, 114–135. [CrossRef]

2.   Joppa, L.; Pfaff, A. Re-assessing the forest impacts of protection: The challenge of non-random protection & a corrective method. *Annal. NY Acad. Sci* **2010**, *1185*, 135–149.

3.   Joppa, L.; Pfaff, A. Global Park Impacts. *Proc. Royal Society B.* **2010**. [CrossRef]

4.   Pfaff, A.; Robalino, J.A.; Sanchez-Azofeifa, G.A.; Andam, K.; Ferraro, P. Park Location Affects Forest Protection: Land Characteristics Cause Differences in Park Impacts across Costa Rica. *BE J. Econ. Anal. Poli.* 2009. Available online: http://www.bepress.com/bejeap/vol9/iss2/art5 (accessed on 24 December 2016).

5.   Joppa, L.; Pfaff, A. High & Far: Biases in the location of protected areas. *PLoS ONE* **2009**, *4*, e8273. [CrossRef]
6.   Pfaff, A.; Robalino, J. Protecting Forests, Biodiversity and the Climate: Predicting policy impact to improve policy choice. *Oxford Rev. Econ. Policy* **2012**, *28*, 164–179. [CrossRef]
7.   Blackman, A.; Pfaff, A.; Robalino, J. Paper Park Performance: Mexico's natural protected areas in the 1990s. *Glob. Environ. Chang.* **2015**, *31*, 50–61. [CrossRef]
8.   Johnson, T.; Alatorre, C.; Romo, Z.; Feng, L. *Low-Carbon Development for Mexico*; The World Bank: Washington, DC, USA, 2010.
9.   The REDD Desk. REDD in Mexico. 2011. Available online: www.theredddesk.org/countries/mexico/readiness_overview (accessed on 3 February 2013).
10.   The Forest Carbon Partnership. REDD Readiness Progress Fact Sheet. 2012. Available online: www.forestcarbonpartnership.org/ (accessed on 3 February 2013).
11.   Naughton-Treves, L.; Holland, M.B.; Brandon, K. The role of protected areas in conserving biodiversity and sustaining local livelihoods. *Annual Rev. Environ. Resour.* **2005**, *30*, 219–252. [CrossRef]
12.   Nagendra, H. Do Parks Work? Impact of Protected Areas on Land Cover Clearing. *Ambio* **2008**, *37*, 330–337. [CrossRef] [PubMed]
13.   Campbell, A.; Clark, S.; Coad, L.; Miles, L.; Bolt, K.; Roe, D. Protecting the future: Carbon, forests, protected areas and local livelihoods. *Biodiversity* **2008**, *9*, 117–122. [CrossRef]
14.   Fuller, D.; Jessup, T.; Salim, A. Loss of forest cover in Kalimantan, Indonesia, since the 1997–1998 El Nino. *Conserv. Biol.* **2004**, *18*, 249–254. [CrossRef]
15.   Sanchez-Azofeifa, G.A.; Quesada-Mateo, C.; Gonzalez-Quesada, P.; Dayanandan, S.; Bawa, K.S. Protected areas and conservation of biodiversity in the tropics. *Conserv. Biol.* **1999**, *13*, 407–411. [CrossRef]
16.   DeFries, R.; Hansen, A.; Newton, A.C.; Hansen, M.C. Increasing isolation of protected areas in tropical forests over the past twenty years. *Ecol. Appl.* **2005**, *15*, 19–26. [CrossRef]
17.   Gaveau, D.L.A.; Wandono, H.; Setiabudi, F. Three decades of deforestation in southwest Sumatra: Have protected areas halted forest loss and logging, and promoted re-growth? *Biol. Conserv.* **2007**, *134*, 495–504. [CrossRef]
18.   Messina, J.P.; Walsh, S.J.; Mena, C.F.; Delamater, P.L. Land tenure and deforestation patterns in the Ecuadorian Amazon: Conflicts in land conservation in frontier settings. *Appl. Geogra.* **2006**, *26*, 113–128. [CrossRef]
19.   Bruner, A.; Gullison, R.E.; Rice, R.E.; Da Fonseca, G.A. Effectiveness of parks in protecting tropical biodiversity. *Science* **2001**, *291*, 125–128. [CrossRef] [PubMed]
20.   Curran, L.; Trigg, S.N.; McDonald, A.K.; Astiani, D.; Hardiono, Y.M.; Siregar, P.; Caniago, I.; Kasischke, E. Lowland forest loss in protected areas of Indonesian Borneo. *Science* **2004**, *303*, 1000–1003. [CrossRef] [PubMed]
21.   Kinnaird, M.F.; Sanderson, E.W.; O'Brien, T.G.; Wibisono, H.T.; Woolmer, G. Deforestation trends in a tropical landscape and implications for endangered large mammals. *Conserv. Biol.* **2003**, *17*, 245–257. [CrossRef]
22.   Liu, J.G.; Linderman, M.; Ouyang, Z.; An, L.; Yang, J.; Zhang, H. Ecological degradation in protected areas: The case of Wolong Nature Reserve for giant pandas. *Science* **2001**, *292*, 98–101. [CrossRef] [PubMed]
23.   Sader, S.; Hayes, D.J.; Hepinstall, J.A.; Coan, M.; Soza, C. Forest change monitoring of a remote biosphere reserve. *Int. J. Remote Sens.* **2001**, *22*, 1937–1950. [CrossRef]
24.   Andam, K.; Ferraro, P.J.; Pfaff, A.; Sanchez-Azofeifa, G.A.; Robalino, J.A. Measuring the effectiveness of protected area networks in reducing deforestation. *Proc. Natl. Acad. Sci. USA* **2008**, *105*, 16089–16094. [CrossRef] [PubMed]
25.   Shah, P.; Baylis, K. Evaluating Heterogeneous Conservation Effects of Forest Protection in Indonesia. *PLoS ONE* **2015**, *10*, e0124872. [CrossRef] [PubMed]
26.   Munoz-Pina, C.; Guevara, A.; Torres, J.M.; Brana, J. Paying for the hydrological services of Mexico's forests: Analysis, negotiations and results. *Ecol. Econ.* **2008**, *65*, 725–736. [CrossRef]
27.   Alix-Garcia, J.M.; Shapiro, E.N.; Sims, K.R.E. Forest Conservation and Slippage: Evidence from Mexico's National Payments for Ecosystem Services Program. *Land Econ.* **2012**, *88*, 613–638. [CrossRef]
28.   Alix-Garcia, J.M.; Sims, K.R.E.; Yanez-Pagans, P. Only One Tree from Each Seed? Environmental effectiveness and poverty alleviation in Mexico's Payments for Ecosystem Services Program. *Am. Econ. J. Econ. Policy* **2015**, *7*, 1–40. [CrossRef]

29. Rodriguez, L.A. On the Regulation of Small Actors: Three Experimental Essays about Policies based on Voluntary Compliance and Decentralized Monitoring. Ph.D. Thesis, University Program in Environmental Policy. Duke University, Durham, NC, USA, 2016.

30. Kaczan, D.; Pfaff, A.; Rodriguez, L.; Shapiro, E.N. Increasing the Impact of Collective Incentives: Conditionality on additionality within PES in Mexico. Unpublished work, 2016.

31. Sims, K.R.E.; Alix-Garcia, J.M. Parks versus PES: Evaluating direct and incentive-based land conservation in Mexico. *J. Environ. Econ. Manag.* **2016**. [CrossRef]

32. Figueroa, F.; Sánchez-Cordero, V. Effectiveness of Natural Protected Areas to Prevent Land Use and Land Cover Change in Mexico. *Biodivers. Cons.* **2008**, *17*, 3223–3240. [CrossRef]

33. Mas, J.F. Assessing Protected Areas Effectiveness Using Surrounding (Buffer) Areas Environmentally Similar to the Target Area. *Envir. Monit. Assess.* **2005**, *105*, 69–80. [CrossRef]

34. Durán, E.; Mas, J.F.; Velázquez, A. Land Use/Cover Change in Community-Based Forest Management Regions and Protected Areas in Mexico. In *The Community Forests of Mexico*; Bray, D.B., Merino-Pérez, L., Barry, D., Eds.; University of Texas Press: Austin, TX, USA, 2005; pp. 215–238.

35. Honey-Roses, J.; Lopez-Garcia, J.; Rendon-Salinas, E.; Peralta-Higuera, A.; Galindo-Leal, C. To pay or not to pay? Monitoring performance and enforcing conditionality when paying for forest conservation in Mexico. *Environ. Conserv.* **2009**, *36*, 120–128. [CrossRef]

36. Honey-Roses, J.; Baylis, K.; Ramirez, M.I. A Spatially Explicit Estimate of Avoided Forest Loss. *Conserv. Biol.* **2011**, *25*, 1032–1043. [CrossRef] [PubMed]

37. Miteva, D.; Ellis, E.; Ellis, P.; Griscom, B. The role of property rights in resisting forest loss in the Yucatan Peninsula. Unpublished work. 2016.

38. Von Thünen, J.H. Der Isolierte Staat in Beziehung der Landwirtschaft und Nationalökonomie (1996, trans.). In *von Thünen's The Isolated State*; Hall, P., Ed.; Pergamon Press: Oxford, United Kingdom, 1826.

39. World Bank. *Policies and Deforestation in the Brazilian Amazon: Roads, Protected Areas, Their Interactions, and Their Impacts*; Technical Paper; World Bank: Washington DC, USA, 2013.

40. Nelson, A.; Chomitz, K. Effectiveness of Strict vs. Multiple Use Protected Areas in Reducing Tropical Forest Fires: A Global Analysis Using Matching Methods. *PLoS ONE* **2011**, *6*, 322722. [CrossRef] [PubMed]

41. Pfaff, A.; Robalino, J.; Lima, E.; Sandoval, C.; Herrera, L.D. Governance, Location and Avoided Deforestation from Protected Areas: Greater restrictions can have lower impact, due to differences in location. *World Develop.* 2014. Available online: http://dx.doi.org/10.1016/j.worlddev.2013.01.011 (accessed on 24 December 2016).

42. Albers, H.J. Spatial modeling of extraction and enforcement in developing country protected areas. *Resour. Energy Econ.* **2010**, *32*, 165–179. [CrossRef]

43. Borner, J.; Kis-Katos, K.; Hargrave, J.; Konig, K. Post-Crackdown Effectiveness of Field-Based Forest Law Enforcement in the Brazilian Amazon. *PLoS ONE* **2015**, *10*, e0121544. [CrossRef] [PubMed]

44. Bartholome, E.; Belward, A. GLC2000: A new approach to global land cover mapping from Earth observation data. *Int. J. Remote Sens.* **2005**, *26*, 1959–1977. [CrossRef]

45. European Space Agency (ESA); ESA GlobeCover Project lead by MEDIAS-France. Ionia GlobCover. Available online: http://due.esrin.esa.int/page_globcover.php (accessed on 24 December 2009).

46. United States Geological Survey. Shuttle Radar Topography Mission, 30 Arc Second scene SRTM_GTOPO_u30, Mosaic. 2006. Available online: http://www2.jpl.nasa.gov/srtm/ (accessed on 24 December 2008).

47. National Imagery and Mapping Agency (NIMA). Vector Map Level. Available online: http://egsc.usgs.gov/nimamaps/ (accessed on 24 December 2010).

48. United Nations Environment Program-Center for International Earth Science Information Network (UNEP-CIESEN). Global Rural-Urban Mapping Project (GRUMP), Alpha Version: Urban Extent. 2006. Available online: http://sedac.ciesin.columbia.edu/gpw/ancillaryfigures.jsp (accessed on 24 December 2010).

49. World Conservation Monitoring Center. World Database on Protected Areas (WDPA). World Conservation Union (IUCN) and UNEP-World Conservation Monitoring Center Cambridge, UK. 2007. Available online: http://www.wdpa.org/ (accessed on 24 December 2010).

50. Olson, D.; Dinerstein, E.; Wikramanayake, E.D.; Burgess, N.D.; Powell, G.V.; Underwood, E.C.; D'Amico, J.A.; Itoua, I.; Strand, H.E.; Morrison, J.C.; et al. Terrestrial ecoregions of the world: A new map of life on Earth. *BioScience* **2001**, *51*, 933–938. [CrossRef]
51. Fischer, G.; van Velthuizen, H.; Nachtergaele, F.; Medow, S. Global Agro-Ecological Zones. 2002. Available online: http://www.fao.org/nr/gaez/en/ (accessed on 24 December 2010).
52. Rosenbaum, P.R.; Rubin, D.B. The central role of the propensity score in observational studies for causal effects. *Biometrika* **1983**, *70*, 41–55. [CrossRef]

forests

MDPI

Article

Does REDD+ Ensure Sectoral Coordination and Stakeholder Participation? A Comparative Analysis of REDD+ National Governance Structures in Countries of Asia-Pacific Region

Taiji Fujisaki [1,*], Kimihiko Hyakumura [2], Henry Scheyvens [3] and Tim Cadman [4,5]

[1] Graduate School of Integrated Sciences for Global Society, Kyushu University, 744, Motooka, Nishi-ku, Fukuoka City 819-0395, Japan
[2] Institute of Tropical Agriculture, Kyushu University, 6-10-1 Hakozaki, Higashi-ku, Fukuoka City 812-8581, Japan; hyaku@agr.kyushu-u.ac.jp
[3] Institute for Global Environmental Strategies (IGES), 2108-11 Kamiyamaguchi, Hayama, Kanagawa 240-0115, Japan; scheyvens@iges.or.jp
[4] Institute for Ethics, Governance and Law, Griffith University, Macrossan Building (N16), Nathan QLD 4111, Australia; t.cadman@griffith.edu.au
[5] University of Southern Queensland, West Street Toowoomba QLD 4350, Australia
* Correspondence: taiji.fujisaki@gmail.com; Tel.: +81-92-623-074

Academic Editors: Timothy A. Martin, Esteve Corbera and Heike Schroeder
Received: 22 June 2016; Accepted: 27 August 2016; Published: 31 August 2016

Abstract: Reducing emissions from deforestation and forest degradation in developing countries (REDD+) requires harmonizing different policy sectors and interests that have impacts on forests. However, these elements have not been well-operationalized in environmental policy-making processes of most developing countries. Drawing on five cases—Cambodia, Indonesia, Lao PDR, Papua New Guinea, and Vietnam, this article aims to determine whether emerging governance arrangements help REDD+ development by delivering participatory mechanisms for policy coordination. Building upon literature on environmental governance and stakeholder participation, the article examines national governance structures for REDD+ and identifies who participates where, and what decision-making powers they have. Despite structural differences between the countries, our analysis illustrates that REDD+ potentially encourages a new form of environmental governance promoting a cross-sectoral approach and stakeholder participation. Cohesiveness of the structures within a broader governance system is key to defining the capacity of REDD+ governance. The result also poses a question as to the inclusiveness of the state actors involved in order to tackle the different pressure on forests. Considering structural inequalities, the analysis further suggests a need of policy support for those who are affected by REDD+ to ensure that their voices could be heard in decision-making processes.

Keywords: REDD+; national governance structures; cross-sectoral coordination; stakeholder participation; Cambodia; Indonesia; Lao PDR; Papua New Guinea and Vietnam

---

1. Introduction

Reducing emissions from deforestation and forest degradation in developing countries (REDD+) is an emerging policy instrument in climate negotiations, and represents a next-generation mechanism in encouraging the sustainable management of forests as a means of reducing greenhouse gas emissions. In the Asia-Pacific region, 20 countries have been engaged in preparing to be ready for implementing REDD+ at country-level (referred to REDD+ Readiness), with support from multilateral initiatives, including the Forest Carbon Partnership Facility (FCPF) of the World Bank and/or UN-REDD Programme [1,2]. To move REDD+ Readiness forward, these countries have designed and developed institutional frameworks for REDD+, whose objectives include ensuring overall responsibilities and coordination for REDD+, developing strategy, action plan and programs, channeling international funding, and monitoring and reporting REDD+ actions [3].

Because such framework introduced for REDD+ Readiness defines policy actors, their roles and capacity, as well as interactions among them [3–5], it has critical significance in decision-making processes, accommodating different interests and ideas, and operationalizing policy decisions on managing forest resources and combatting climate change. Accordingly, REDD+ can be considered as a new form of environmental governance that frames problems of forest management and land use in the context of climate change and sets up new rules and norms aligning specific stakeholders and vision, while marginalizing others [4,6]. This notion of governance poses important concerns about the practice of institutional arrangements for REDD+ at country-level: What do they look like, who participates where, and what decision-making powers they have (or do not have) within the frameworks?

A number of publications have reviewed the progress of national REDD+ Readiness of particular countries, covering diverse aspects of REDD+ [7–18]. In addition, Minang et al. [19] present an assessment framework for REDD+ Readiness to compare countries. Yet, little has been done to analyze structural settings to explain the ways that governments address ambitious objectives of REDD+. More attention is needed on the structural challenges for REDD+ institutional arrangements [8]. In particular, challenges associated with sector coordination and participation of stakeholders have been among the most prominent governance issues shared by countries developing REDD+, and evidenced in their national strategic documents for preparing REDD+, such as Readiness Preparation Proposals (R-PP) submitted to the FCPF, and have received particular attention in the literature [3,7,19–24].

Drawing on five cases—Cambodia, Indonesia, Lao PDR, Papua New Guinea, and Vietnam, this article aims to determine whether and how emerging institution arrangements help or hinder REDD+ Readiness, delivering participatory mechanisms to harmonize different policy sectors and interests that have an impact on the forests. The multiple cases of these countries provide structural elements to understand the nature of REDD+ governance, especially how issues of sectoral coordination and stakeholder participation are addressed. The article is organized as follows: The first part provides the analytical approach adopted and the research methods used. The second part describes a theoretical focus on governance and stakeholder participation that guides the analysis of REDD+ frameworks. The third section describes the case studies selected, followed by discussion on the identified structures and actors involved in the institutional arrangements. Finally, the article comments on the findings, and identifies the structural challenges confronting REDD+ as a policy instrument for combatting climate change into the future.

2. Method of the Study

Using an analysis based on the evaluation of the institutional arrangements, what follows is a comparative study of the five countries chosen to identify the structural and institutional elements necessary for explaining governance arrangements for developing REDD+.

These five countries in the Asia-Pacific Region were selected because of their policy commitment and creation of institutional frameworks for REDD+ Readiness as shown in Figure 1 and Table 1, as well as their presentation of official policy documents to REDD+ design, which have enabled the assessment in this study. Engaged with the international initiatives to support developing countries, namely FCPF and the UN-REDD Programme, the five countries have formulated the key national strategic documents for preparing REDD+, such as R-PP for FCPF and Joint Programme Document (JPD) for UN-REDD Programme, where they present the design of institutional frameworks to move REDD+ Readiness forward. In addition, these countries have issued legal decisions, such as ministerial decisions and presidential decrees, to establish and implement such frameworks. Analysis was based on surveying those policy documents and other relevant to national REDD+ strategies, combined with key informant interviews with government officials, donor agencies, and non-governmental organizations (NGOs) working on national level REDD+ in each country in order to confirm the frameworks and understand interactions among actors. The government officials interviewed were from the main management body of REDD+ (Table 1), and ministry departments and/or agencies responsible for forestry, environment and climate change in each country (24 in total: 5 in Cambodia, 7 in Indonesia, 4 in Lao PDR, 4 in Papua New Guinea, and 4 in Vietnam).

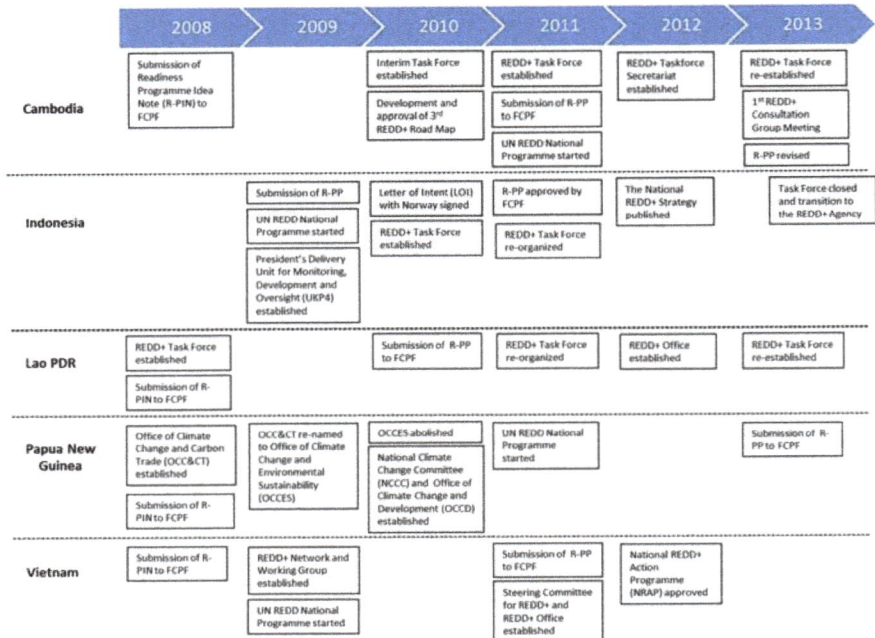

Figure 1. REDD+ timeline in the five countries 2008–2013 (Institutional frameworks and strategic documents). Sources: [11,12,14–17,25–30].

Table 1. Institutional arrangements for REDD+ Readiness in the five countries in 2013.

Country	Cambodia	Indonesia	Lao PDR	Papua New Guinea	Vietnam
Higher political level	National Climate Change Committee (2006) [30]	Delivery Unit of the President (UKP4) (2009–2014) [11,31]	National Environmental Council (NEC) (2002) [27,32]	National Climate Change Committee (NCCC) (2011) [9,28]	Steering Committee for REDD+ (2011) [18,29,33]
Main management body	REDD+ Taskforce (2010, 2013) [30,34]	National REDD+ Task Force (2010–2013) [7,11,31]	REDD+ Task Force (2008, 2011, 2013) [10,14,25,27]	Office of Climate Change and Development (OCCD) (2010). The REDD+ and Mitigation Division within OCCD [9,28]	Steering Committee for REDD+ (2011) and REDD+ Network (2009) [18,29,33]
Administrative matter	Taskforce Secretariat (2012) [30,34]		REDD+ Office (2012) [14,25]		REDD+ Office (2011) [29,33]
REDD+ Technical matters	Technical Teams [30,34]	Working Groups under the Task Force [11,35]	Working Groups [14,25,27]	REDD+ Technical Working Group [9,17,28]	Technical Working Groups [29,33]

(year) is when the organization was established.

It is important to note that REDD+ institutions seem to be evolving rapidly at different stages in the five countries. However, as the analysis focuses on the Readiness phase, in which national governments prepare for REDD+, the paper has selected the period of 2013 for comparison purposes.

3. Theoretical Focus of the Study: Sector Coordination and Stakeholder Participation

As widely recognized, deforestation and forest degradation cannot be framed as simply forestry problems [5,36]. In addition to commercial timber exploitation, industrial agricultural development, shifting cultivation, and infrastructure expansion all play a significant role in current levels of deforestation and forest degradation [23,37]. Furthermore, it is not possible to discuss improved forest and land use without addressing social, environmental and economic aspects, such as rural poverty, land tenure, environmental services, and financial and market issues. Consequently, REDD+ requires a cross-sectoral response and the consideration of all relevant factors inside and outside forests. From this viewpoint, representation and interaction of actors across sector boundaries and diverse knowledge and values are necessary features for REDD+ institutional arrangements to respond to the wider scope of the problems to be solved.

3.1. Cross-Sectoral Coordination

Drawing on Cadman [20] and Hogl [38], it is understood that cross-sectoral coordination is enabled through collaboration of different policy actors, notably via their participation in decision-making processes. Sectoral interests are represented by not only state government bodies, but also by a range of non-state actors and networks. However, it is the responsibility of states and their agencies to prepare and implement sectoral policies at national scale by exercising authority and actual implementation practice. Within this perspective, it should be also understood that the state government is conceived not as a single-unity but as a network of different ministries and agencies [39], which have different objectives and roles with different areas of knowledge, resources and constituencies.

Representation of state actors from different policy sectors alone does not necessarily ensure improved policy coordination as embedded political interests, power relations, and historical institutional path dependencies could undermine the effectiveness of institutional arrangements in policy outcomes [40,41]. Well-facilitated participatory processes that cross policy sector boundaries support communication and ensure that decisions are accountable, are necessary characteristics to promote integration across policy sectors [41,42]. Particular attention should be paid to ensuring that state actors from different policy sectors are involved in REDD+ institutional arrangements. Drawing on the quality of the policy coordination proposed by Metcalfe [39], the interaction of ministries and state agencies can be understood by identifying how and where they are engaged, and their roles and responsibilities in decision-making processes. Following the typologies based on different degrees of participation [43–46], the engagement of policy sectors are organized as follows: leading role (i.e., lead decision-making and coordination); interactive participation (i.e., involvement in analysis and development of strategies and plans); functional participation (i.e., provision of technical knowledge and guidance); and consultative function (i.e., provision of information and comments upon request).

3.2. Stakeholder Participation

Significant roles for stakeholder participation are increasingly recognized in REDD+ debates. But the term "REDD+ stakeholder" is too generally used to be easily understood. While a range of typologies have been developed to understand stakeholder participation in different contexts [43,46], the common meaning of "stakeholder" refers to those who are affected by or can affect a decision [47]. Bearing this in mind, this paper focuses on non-state actors affected by and/or directly involved in REDD+ and forest management, following the categories identified by Brockhaus et al. (2014) [40]: multilateral and/or bilateral donor; international NGOs, academic and/or research entity; civil organization and/or domestic NGO; private sector (business); and indigenous group and/or local community.

In the literature on contemporary environmental governance, there are mainly two views about the purpose and role of stakeholder participation. First, looking the matter through the perspectives of justice and equity, participation is understood as procedural justice [48]. Since indigenous peoples and forest dependent communities are often vulnerable to external political and economic interests [49] and are considered as most affected by the implementation of REDD+ [5], their participation needs specific attention. In addition, the *pragmatic value* of stakeholder participation has also been stressed. By building consensus among different groups and having more comprehensive information inputs through stakeholder participation, the quality and implementation of decisions are likely to be greater [50–52].

Consequently, an investigation into which actors are/are not involved in REDD+, the roles they are/are not afforded in related processes, can help to determine the nature of stakeholder participation in the institutional frameworks, as explored below.

4. Results

4.1. Overview of Institutional Arrangements for REDD+ Readiness in the Five Countries

The five countries share similarities in terms of hierarchy structures composed of basically three or four strata in the institutional arrangements for REDD+ Readiness. Table 1 gives a profile of how national REDD+ framework are organized in terms of administrative levels, functions and roles.

As a higher political forum for guiding strategic decision for REDD+ Readiness and promoting sectoral coordination, each country has formed or appointed a governmental body at the uppermost level of the arrangement. In the cases of Cambodia, Lao PDR, Papua New Guinea, and Vietnam, this body takes the form of a committee or council, chaired by the ministerial or secretary general level, and involving representatives from other relevant ministries. In the case of Indonesia, along with the co-operation with the Norway Government (Letter of Intent: LOI), overall responsibility for developing REDD+ Readiness was mandated under the authority of the President's Delivery Unit for Monitoring, Development and Oversight (UKP4) (while the analysis for Indonesia's REDD+ focuses on the REDD+ governance arrangement under the UKP4, it should be noted that the UKP4 was dissolved in December 2014 according to the government's reform led by Indonesia's new president, President Joko Widodo. Since 2015, the newly established Ministry of Environment and Forestry is taking a lead in REDD+ by Presidential Decree No. 16, 2015) established within the President Office in 2009, the intention of which is to improve inter-ministerial coordination [53,54].

As the main management and coordination body to develop REDD+ Readiness, a special government unit or management office was set up in each of the five countries: REDD+ Taskforce in Cambodia (2010–now), National REDD+ Task Force in Indonesia (2010–2013), REDD+ Task Force in Lao PDR (2008–now), and the Office of Climate Change and Development (OCCD) in Papua New Guinea (2010–now). In Vietnam, the REDD+ Steering Committee (2011–now) supported by the National Network (2009–now) is viewed equivalent to this operating level (detailed discussion is given below). These organizations are considered as the main government instrument for REDD+ Readiness in each country, as they are directly responsible for the formulation of national strategy, managing overall development of REDD+ activities, and taking the lead in coordinating measures over stakeholders. To provide support in terms of administrative operations, Cambodia, Lao PDR and Vietnam set up secretarial offices. National REDD+ Task Force in Indonesia and the OCCD in Papua New Guinea are provided with the necessary administrative capacities.

In order to address technical issues related to REDD+, such as monitoring, reporting and verification (MRV) system, forest reference emission levels (RELs), and social and environmental safeguards, several technical team/working groups are under development within the REDD+ institutional frameworks in each country.

4.2. REDD+ Intuitional Arrangements in Relation with Existing Forest Institutions

Each of the five countries has established governance structures for REDD+ Readiness, which may provide vertical linkages from higher political level to administrative and technical matters associated with REDD+. While the frameworks present common features as shown in Table 1, two types of governance arrangement are found in relation to existing forest institutions: based on the present forest administration (the arrangements of Cambodia, Lao PDR, and Vietnam); and creation of new institutions for REDD+ Readiness instead of the present government body for forest administration (the arrangements of Indonesia and Papua New Guinea).

In the former type of arrangement, the forest administration body officially plays a central role in leading REDD+ Readiness and serves as the national focal point. For example, in Cambodia, the Forestry Administration (FA) under the Ministry of Agriculture, Forestry and Fisheries (MAFF) and the Ministry of Environment (MOE) chair REDD+ Taskforce and take the lead on REDD+ [30]. In Vietnam, the Vietnam Administration of Forestry (VNFOREST) under the Ministry of Agriculture and Rural Development (MARD) acts as a national focal point, hosting the REDD+ Office to support REDD+ Steering Committee and organize REDD+ Network activities [29].

In contrast, new organizations were appointed or created with authority to lead REDD+ Readiness in Indonesia and Papua New Guinea outside the existing forest institution. In Indonesia, even though the preparation for REDD+ had been originally conducted by the Ministry of Forestry with the collaboration of BAPPENAS (The Ministry of National Development and Planning) [11], coordination in formulating the REDD+ National Strategy and lead preparation for REDD+ shifted to the National REDD+ Task Force, which was independently established under the authority of the UKP4. In Papua New Guinea, the OCCD was created by National Executive Council (NEC) decision 54/2010 and appointed to develop and manage REDD+ Readiness, while the Papua New Guinea Forest Authority (PNGFA) is a government body with regulatory and administrative responsibility for managing the forest sector [55,56].

4.3. Involvement of Ministries and Agencies in the Inner-Ministerial Coordination Body in the Arrangements

In the five countries, main coordinating structures between different policy sectors have been established at the management level of the governance arrangements for REDD+ Readiness. Table 2 shows policy sectors involved in the inter-ministerial body through participation of responsible ministries and agencies of the five countries.

Table 2. Cross-sectoral dimension in key REDD+ management bodies in the five countries in 2013.

Country	Cambodia	Indonesia	Lao PDR	Papua New Guinea	Vietnam
Cross-sectoral coordination structure	REDD+ Taskforce	REDD+ Task Force	REDD+ Taskforce	Technical Working Group	REDD+ Steering Committee
Forestry	†	***	***	*	†
Environment/ Climate change	***	***	†	*	***
Agriculture	-	***	***	*	-
Finance	***	***	***	*	***
Land use planning	***	***	***	*	***
Energy and mining	***	***	***	*	-
Indigenous rights and social welfare	§	-	§	§	***
Other	***	***	***	*	***

†, leading role; ***, interactive participation; **, functional participation; *, consultative function; §, not clear; -, not involved. Source: [7,11,16–18,25–29,33–35,57].

In Cambodia, the need for sectoral coordination for REDD+ was primarily answered by the creation of REDD+ Task Force (the inter-ministry REDD+ Taskforce was first established in January 2010 as an interim measure and then re-established in February 2013 by Decision No. 87 of Ministry of Agriculture, Forestry and Fisheries on Establishment of Cambodian REDD+ Taskforce), which is the main REDD+ management body with mandates to oversee and coordinate processes to formulate national REDD+ strategy and relevant actions to REDD+. This arrangement was created in 2010, and since then, the number of state actors participating in the Task Force has increased to ten members, consisting of senior officials from the forest, environment, land use planning, finance, rural development, and mining ministries. The Task Force was given the role of high-level cross sectoral coordination and policy guidance from the National Climate Change Committee (NCCC), which is chaired by the Prime Minister and involves representatives from 20 ministries and three government agencies [58]. However, since the Committee addresses all aspects of climate change including both mitigation and adaptation, the Task Force serves as the most powerful instrument to provide sectoral coordination on REDD+ decision-making in Cambodia.

A similar mechanism is found in Lao PDR, where sectoral coordination is primarily introduced at the same level of government organization as Cambodia's REDD+ Task Force. The inter-ministerial REDD+ Task Force was originally established by the Ministry of Agriculture and Forestry (MAF) in 2008 via Ministerial Decree No. 1313 to oversee and to provide coordination over REDD+ [27]. Following the establishment of the new ministry (Ministry of Natural Resources and Environment: MONRE), the responsibility of REDD+ was transfer from the MAF and the Task Force was re-established by Department of Forest Resources Management (DFRM) under MONRE in October 2013 via MONRE decision No. 7176. Following this reform, the number of members comprising the Task Force was increased from 15 to 24. The Task Force comprises members at director general/deputy director general level from ministries and other agencies related to forestry and agriculture, energy and mining, development planning and industry sectors. In addition, the R-PP of Lao PDR suggests high-level cross sectoral coordination to be provided from the National Environment Council (NEC), which consists of ministers and vice ministers from 14 key agencies [27]. The Council was established in 2002 to provide advice to the government regarding environmental issues [32], however it has not functioned effectively and its linkage with the Task Force remains unclear.

In Indonesia, REDD+ Task Force (2010–2013) served as the main instrument to implement the LOI with Norway's Government and the most powerful to develop Indonesia's REDD+ Readiness. Its tasks included to create an institutional and legal framework for REDD+ in Indonesia, including formulation of National REDD+ Strategy and establishment of the REDD+ Agency (in 2014, the REDD+ Agency started its operation as a successor to National REDD+ Task Force in order to move REDD+ toward implementation phase. However, REDD+ Agency was closed down in 30 January 2015 following the government's reform led by Indonesia's new president, President Joko Widodo and the Ministry of Environment and Forestry is currently taking a lead in REDD+). Established through the Presidential Decree No. 19/2010, and chaired by the head of UKP4, the Task Force was theoretically able to lead the activities of all other line ministries and agencies regarding REDD+ [53]. In addition, the Task Force involves members of senior officials of relevant ministries and other agencies for "forestry", "environment", "land use", "planning", and "finance" [31]. During the first period of the Task Force (2010–2011), "agriculture" and "mining" sectors were not formally been represented in the REDD+ institutional arrangements, however, in its second period (2011–2013), these two sectoral ministries participated as members, appointed in the Presidential Decree No. 25/2011. Under the Task Force, ten thematic working groups were established, each headed by senior government officials and experts in their field.

In Papua New Guinea, the OCCD acts as the responsible government body for REDD+, while REDD+ sector programs will be implemented by other government departments and agencies in accordance with their mandates [28]. Under the arrangement, a cross-departmental working group, namely Technical Working Group for REDD+, was established in 2010. Chaired by the OCCD, the Working Group has diverse membership, comprising 21 members from departments of the government, non-government organizations, donors and industry associations [28]. Its main duty is to provide technical guidance for the implementation of REDD+ activities, yet the Working Group seems to function primarily as consulting groups that provide comments on information from the OCCD [17].

In Vietnam, the National REDD+ Steering Committee was created in early 2011 via MARD Decision No.39/QD-BNN-TCCB to facilitate cross-sectoral coordination among government agencies. The main functions of the Steering Committee is to propose relevant policies relating to REDD+, and direct the formulation of the National REDD+ Programme (NRAP). Chaired by the Minister of MARD, the Committee's members include representatives from seven departments of MARD as well as other ministries and agencies, which are responsible for "environment and climate change", "planning", "finance", "foreign affairs", and "ethnic issue" [29]. In addition, the government established the National REDD Network via MARD Decision No.2614/QD-BNN-LN, together with the National REDD Working Group including several Sub-Technical Working Groups, which actually serve as forums for information sharing and discussions about the design and implementation of all elements of national REDD+ system. Several donor organizations, international NGOs, and many of the forestry institutions of the MARD are engaged as members in the Network and the Working Group [18]. However the participation of other ministries and government agencies is limited at these technical and operational levels. The sectoral coordination for REDD+ is attempted primarily through the National REDD+ Steering Committee.

4.4. Participation of Non-State Actors in the Institutional Arrangements for REDD+ Readiness in the Five Countries

The government documents for designing REDD+, such as R-PP, of the five countries identify a range of non-state actors who would be affected by or would have an impact on the design and implementation of REDD+, including multilateral and bilateral donor organizations, international NGOs, academic and research organizations, civil organizations and domestic NGOs, private sector, indigenous groups, and local communities [26–30]. However, the five countries demonstrate a great variation in engagement and coordination with stakeholders. Table 3 summarizes the non-state actors involved and their given role in the institutional arrangements for REDD+ Readiness of the five countries.

Table 3. Participation of non-state actors in REDD+ institutional arrangements in the five countries in 2013.

Country	Cambodia	Indonesia	Lao PDR	Papua New Guinea	Vietnam
Structure for stakeholder participation	REDD+ Consultation Group	REDD+ Working Group	REDD+ Taskforce	Technical Working Group	REDD+ Network and Working Group
Multilateral and/or Bilateral donor	-	***	-	*	**
International NGO	*	***	-	*	**
Academic and/or research entity	*	***	**	*	**
Civil organization and/or domestic NGO	*	***	§	*	-
Private sector	*	***	-	*	-
Indigenous group and/or local community	*	***	-	-	-

***, interactive participation; **, functional participation; *, consultative function; §, not clear; -, not involved.
Source: [7,9,11,15,17,25,27–30,57,59,60].

In recognition of the importance of stakeholder participation and consultation in developing national REDD+ system, the Royal Government of Cambodia decided to set up the REDD+ Consultation Group within the REDD+ institutional framework. Involving international organizations, academic institutions, civil society, indigenous peoples, forest dependent communities, NGOs, and the private sector, the Consultation Group aims to support the REDD+ Taskforce by providing comments and recommendations, and/or express concerns especially on issues associated to stakeholder engagement and REDD+ safeguards [59,60]. Representing diverse groups of stakeholders, its fundamental role is to introduce an effective consultation process in developing REDD+ in Cambodia.

In contrast to Cambodia's case, stakeholder participation is not easily grasped in the REDD+ framework of Lao PDR. The framework constitutes mostly by the government organizations in accordance with their mandates. As a member of the REDD+ Taskforce, the National University of Laos plays a think-tank role in designing REDD+ elements. A few non-government entities are also involved in the Taskforce, namely Lao Front for National Construction (LFNC), Lao Women's Union (LWU) and Lao National Chamber of Commerce and Industry, which are considered to represent views of civil society and particular groups [25]. However, these organizations are considered as inherently political, under the ruling Lao People's Revolutionary Party. The space for non-state actors to engage with REDD+ decision-making is likely to be limited due to the political history and culture of the country.

The REDD+ frameworks of Indonesia and Vietnam provide non-state actor with opportunities to participate actively in discussion and analysis for developing REDD+. In Indonesia, there were ten working groups under the REDD+ Task Force. The members of the groups were not only from the governmental organizations, but also included experts from multilateral organizations, NGOs, civil society, private sector and indigenous groups. Each working group, consisting of 4–7 members was responsible for developing elements of REDD+, such as a national strategy, the MRV institution, safeguards system and a financial mechanism. However, participation is not on a voluntary basis; rather it is based on individual expertise to complete the responsibilities of the Task Force.

In Vietnam, non-state actors are allowed to participate in REDD+ institutional framework through the REDD+ Network and a number of Sub-Working Groups. Together with government agencies, ten non-state organizations are listed as members of these groups [57]. However civil society, private sector, indigenous people groups, and local communities are not involved as the members, even though the arrangement takes a form of open-ended participation. Including academic institutions, international development partners, and NGOs, the arrangement is viewed to contribute to generating knowledge for REDD+ and encouraging communication between state and non-state actors. There have been a number of outputs from the Sub-Working Groups on design of REDD+ elements (e.g., MRV framework document, a design document for benefit distribution). However, Stewart et al. [18] pose a question about the actual impacts of the mechanism to facilitate policy inputs from non-state actors, as the government has no obligation to adopt these technical recommendations.

In Papua New Guinea, the Technical Working Group for REDD+ under the OCCD has diverse members including 15 non-government organizations such as NGOs, donors and industry associations [28]. Such a multiple-actors framework is designed to provide overall technical knowledge and guidance for the implementation of REDD+ activities; and to support the OCCD to develop constructive relationships with a variety of different stakeholders [9]. However, the non-state members listed in the R-PP are international and national actors, while local actors such as indigenous communities are not represented. In addition, as pointed out to date, mainly the function has been focused on consultation to provide comments on information from the OCCD and it is less clear how and whether comments would be incorporated or not.

5. Discussion

5.1. Two Types of Governance Arrangement for REDD+ Readiness

Because REDD+ essentially copes with forests and would be largely consistent with governments' efforts to improve forestry sector such as forest conservation, sustainable management of forests, community forestry and certification of forest products, the forestry administration-led governance arrangement might be the most appropriate (found in the frameworks of Cambodia, Lao PDR and Vietnam). This arrangement appears to allow governments to take fullest advantage of existing institutions, resources, capacities and networks of forestry sector. However, the arrangement may limit scope and capacity of REDD+, as evidenced in the Cambodia's R-PP [30], which places the development of the REDD+ strategy within the existing forest policy framework and strategies.

To enable REDD+ to address drivers of deforestation from outside the forestry sector for climate change mitigation, design and implementation of REDD+ would be beyond the mandates of forest institutions. However, the ability of the forest administration body to coordinate with and have influence on decisions of other ministries could be inherently limited due to its capacity articulated within the border governance system. The ministry or agency responsible for managing state's forests is essentially technical administration, and inter-sectoral coordination roles are often given to another ministry or government agency dealing with planning or financial issues. As an operational matter, the compartmentalized administrative structure of forest administration is likely to require complex communication to coordinate with other sectors. In addition, as Luttrell et al. [15] discuss, there remains a concern about embedded economic interests and politically connected networks in the natural resources and forest management in many countries that are engaging in REDD+ Readiness. The weak forest governance may undermine efforts to reduce emissions from forest, and presents a great risk for accountability and transparency in REDD+ decision-making process and for the rights of indigenous groups and local communicates.

To respond to the need to tackle these weaknesses and challenges, where reformation of existing forest institutions would be required, new agencies have been created or appointed in Indonesia and Papua New Guinea. Setting up new institutions for REDD+ separately from the present forest administration should be able to frame and deal with deforestation problems in a broader perspective towards the climate change mitigation objective, and bring about sector coordination. However, new institutions for REDD+ appear to cause political tensions on account of their overlapping roles with the present forest administration bodies and other line ministries, resulting in uncertainty over leadership and the legal force of agreed proposals from the new institution.

For instance, Resosudarmo et al. [53] point out the limited influence of Indonesia's REDD+ Task Force on other ministries over its decisions due to weak institutional capacity. Even though its tasks were clear, the Task Force was on an ad hoc basis taking roles between 2010 and 2013 and not institutionally well-established in relation to the cabinet ministries, which led to confusion and tension in relation with the broader governance system [15]. Its political instability is evidenced in legal recognition of the National REDD+ Strategy finalized by the REDD+ Task Force in 2012. Without obtaining strong support from line ministries, the Strategy was issued through the Chairman's Decree of the REDD+ Task Force without a relatively stable center of authority, such as presidential recognition. Accordingly, there remains an uncertainty regarding the legal binding of the strategy and its influence on the sectoral policies. Political confusion is also found around REDD+ in Papua New Guinea, where the OCCD has struggled to show strong leadership in developing national REDD+ Readiness [9]. Tension was especially found between the OCCD and the Forestry Authority [17]. Even though new institutions are built to introduce cross-sectoral approach for REDD+ Readiness, their effectiveness remains questionable due to institutional uncertainty and/or rival relations generated with regard to the present government administrations.

However, the overlap and/or tension within the government structure over forests and REDD+ are also found in Cambodia, Lao PDR and Vietnam, where REDD+ institutional arrangements are based on the present forest administration. In Cambodia and Lao PDR, responsibility over forest resources is not mandated to a single government agency, but distributed to plural organizations at ministerial level. For instance, most state forest areas in Cambodia are managed by the Forestry Administration (FA) and the Ministry of Environment (MOE) according to the management objectives of forests [61]. There exists competition over mandates and rival relations in managing the natural resources [62]. In Lao PDR, the government created the Ministry of Natural Resources and Environment (MONRE) in 2011, which took overall responsibility for conservation and protection of forests from the Ministry of Agriculture and Forestry (MAF), which has led to uncertainty and confusion over leadership and coordination for REDD+ [14].

Even though forestry institutions generally control all state forest areas, other institutions, especially those responsible for environmental issue also deal with forest resources in certain respects [36]. In Vietnam, the Ministry of Agriculture and Rural Development (MARD) has overall responsibility for the forest lands, while the environmental ministry has a mandate for management of state lands. This overlap mandate is often perceived as a drawback in the administrative process for decision-making over forest land in Vietnam [12,16,63].

The five countries illustrate two types of arrangements to govern REDD+ Readiness, which highlight differences in what each country considers to develop REDD+. However, both arrangements also demonstrate challenges to ensure the adequate capacity of REDD+ institutions to achieve the sectoral coordination.

5.2. Understanding Cross-Sectoral Coordination Mechanisms in REDD+ Governance Strucurres

In both arrangements, the attempt for cross-sectoral coordination is found through the creation of inter-ministerial bodies where ministries and agencies formally participate as members. Participating ministries and agencies are embedded in the formal decision-making processes for REDD+, which theoretically stimulates inter-ministerial communication and allows sectoral perspectives and interests to be integrated into REDD+ direction.

The policy sectors commonly involved in the five countries are "forestry", "environment", "finance", and "land use planning". However, the "agriculture" sector, which has a critical influence on land use change, is missing in Cambodia and Vietnam. The difference is also found in the participation of the responsible ministry or agency regarding the rights of indigenous people and social welfare. In Vietnam, the REDD+ Steering Committee involves the Committee on Ethnic Minority Affairs, which is a ministry-level agency that exercises the functions of state management of ethnic minority affairs nationwide [64]. In Cambodia, however, the Department of Ethnic Minorities Development under the Ministry of Rural Development is not involved in the REDD+ Task Force, while the Department of Rural Water Supply from the same ministry, which is tasked to develop water supplies for rural areas [65] and Local Administration of the Ministry of Interior are listed in the REDD+ Taskforce [34]. In Indonesia, the Ministry of Social Affairs, which deals with Indigenous Peoples' issues [66] and the Ministry of Law and Human Rights were not involved in the Task Force. The cases of Lao PDR and Papua New Guinea present less certainty in terms of representativeness of this sector within the government. Notably, issues related to rights, rural development and social welfare are in general not coordinated by a single ministry; rather different ministries and agencies are working on the issues from different aspects such as land rights, health care, infrastructure and education.

It is also important to understand what roles and responsibilities are given to the participating ministries and agencies within the coordination body. In four of the five countries, the member ministries and agencies are given a position to participate in the discussion and analysis for design, implement and monitor REDD+ development. With such interactive participation of different policy sectors, REDD+ decision-making processes may lead towards a better recognition of diverse sectoral interests and their collaboration. However, Papua New Guinea shows less significance of participation of line ministries and agencies, where they participate through the REDD+ Technical Working Group. While 19 government departments and agencies are involved in the Group [28], their function seems primarily to be as consultative groups to provide comments and/or technical recommendation to the OCCD's ideas and plans for REDD+ [17].

REDD+ institutional arrangements in the five countries demonstrate that the range of state actors participating in REDD+ decision-making processes has gone beyond the forestry sector with inter-ministerial participation. Notably, the number of state actors participating in the inter-ministerial bodies has increased in the case of Cambodia, Indonesia and Lao PDR, which implies growing attention from different policy sectors to REDD+ as a potential to achieve their own policy goals. On the other hand, the countries also show variations especially in terms of the policy sectors involved. The result poses a question as to whether the state actors involved have enough authority and roles to tackle the different pressure on forests from, for example, the agriculture sector. It also suggests a challenge to coordinate the issues related to indigenous groups, local communities and peoples' welfares from a structural point of view. These are the cases that need to be analyzed country by country. Nevertheless, the REDD+ strategic documents such as R-PP are unlikely to explain the criteria to be used in selecting government bodies to be involved in the structure for REDD+ decision making.

5.3. Different Approaches to Stakeholder Participation

There are significant differences in terms of non-state actors involved and ways of their engagement across the five countries, but overall opportunities to participate in REDD+ decision making structures can be viewed as unevenly distributed between different types of stakeholders. Most of the non-state actors involved in the structure in the five countries are multilateral and/or bilateral donor organizations and international environmental NGOs, which have expertise on elements of REDD+, or resources to support government efforts. On the other hand, civil society, private sector, indigenous groups and forest dependent communities are likely to be provided less opportunity to participate in REDD+ decision-making structures. This implication is critical especially for indigenous groups and forest dependent communities as they would be mostly affected by the implementation of REDD, but their social values and interests associated with forests are not always the main concern of national administrations [67–69].

In addition to the types of stakeholder, it is also crucial to identify the purposes and functions of stakeholder participation. In Indonesia, the non-government actors were involved in the decision-making process to design and develop elements of REDD+. However, it should be noted that the participation is not open-ended, but based on expertise of each actor. Accordingly, it is viewed that the participation is pragmatic rather than normative following Reed (2008) [46], as it centers on technical concerns rather than representing the full range of relevant stakeholders.

The arrangement in Vietnam shows a different approach to stakeholder participation, which is characterized as being organizational and voluntary. The purpose of stakeholder participation is to provide technical knowledge and guidance for developing elements of REDD+. Importantly, it can be understood that the opportunity to participate is likely to be given through existing actor networks, since the participants are international organizations those who can affect rather than those who are affected. The participation is, therefore, viewed as political, technical and resource-oriented. This is also found in the case of Papua New Guinea, although its function is more consultative-oriented. The stakeholder participation in both countries raises a question about criteria and transparency in the process to decide who are involved and who are not.

Cambodia's arrangement demonstrates another approach. With the aim of ensuring stakeholder engagement, the selection process for membership in REDD+ Consultation Group was carefully designed with clear criteria, and a broader range of stakeholders are involved including indigenous group and local community. Compared with other countries, however, the role of non-stakeholders in the group is relatively limited, which is viewed as "consultative" rather than "functional" following Farrington [44].

As one of the agreements at the sixteenth session of the Conference of the Parties (COP 16) to United Nations Framework Convention on Climate Change (UNFCCC), countries developing

REDD+ are requested to ensure "the full and effective participation of relevant stakeholders, inter alia indigenous peoples and local communities" [70]. Because values of forest vary depending on stakeholders [67], in theory, involving stakeholders could contribute to higher quality and durability of decisions [36,51]. Stakeholder participation is also discussed in view of representing the full range of groups that would be affected by REDD+ (e.g., Guidelines on Stakeholder Engagement in REDD+ Readiness [71]). However, the institutional frameworks of the five countries demonstrate differences in governments' interpretations and responses. In addition to viewing this only from the representativeness of different groups, stakeholder participation in REDD+ as an institutional arrangement is likely to be understood through broader operational objectives for which participation and the capacity of stakeholders is used to meet such objectives.

6. Conclusions

REDD+ requires national governments to address cross-sectoral coordination and ensure inclusive and meaningful participation of stakeholders. However, these elements have not been well-operationalized in the forest and environmental policy-making processes of most developing countries [36,67].

By examining the institutional arrangements for REDD+ Readiness in the five countries, this empirical study has identified the approaches taken to organize the structures (i.e., either organizing it based on an existing forest administration, or creating a new one) and demonstrated that REDD+ potentially encourages new forms of environmental governance that seeks a cross-sectoral approach and the participation of non-state actors. Despite variations found between the countries, state actors from different policy sectors directly participate in the decision-making processes to develop REDD+ including formulation of national strategy. It should be also highlighted that the institutional arrangements provide formal spaces for non-state actors to engage in developing national REDD+ systems.

However, the legitimacy and ability of the institutional arrangements are being influenced by existing governance arrangements and inter-ministerial relationships over forest resources. REDD+ governance needs to be institutionally well-established within the broader national governance system. In addition, it is crucial not to overlook the questions of appropriate government bodies and their roles structured in the frameworks. Certainly, further analysis is required for each of the countries studied here in order to fully understand the complexity of deforestation drivers as well as social, environmental and economic aspects of REDD+ inside outside forests. Yet, from a structural view, our study has the questioned inclusiveness and has assumed the absence of criteria to select state actors in order to address diverse drivers of deforestation and relevant factors to REDD+, especially issues of indigenous rights and social welfare. Furthermore, the study of these five countries has demonstrated that the argument on sectoral coordination is not straightforward. In policy practice, there may be competence and tension among ministries and agencies for influence over REDD+. The incumbent government bodies, particularly those dealing with forests and climate change are likely aiming to use REDD+ to advance their own interests and policy agenda. There might also be an institutional risk as more state actors are involved from different policy sectors of conflicting political interests with the ability to decide; REDD+ decision-making would require more complex communication processes and policy deliberations.

Participation of non-state actors is generally found in the institutional structures in the five countries, whereas the approaches differ by country. The result suggests that stakeholder participation is objectively defined, addressing not only representation of interests. Rather, consultative function and provision of technical knowledge seem to be the main purpose of involving non-state actors. Hence, opportunities for participation are distributed according to the capacity of actors to meet such objectives and supported by the political networks and resources that they have. Certainly, it is not only formal governance structures, which provide participation, but non-state actors are likely to be engaged in policy dialogues at multiple levels in both formal and non-formal ways [72,73]. Nevertheless, considering such structural inequalities, our study assumes that the powers and responsibilities of the local stakeholders who are and will be affected by REDD+ are relatively less significant in REDD+ formal decision-making processes. It also suggests a need for specific policy support, resources and participatory processes for these local actors to ensure that their voices would be heard in REDD+ decision-making processes.

The status of REDD+ is moving forward in these five countries and their institutional arrangements are highly dynamic. In particular the arrangements in Indonesia and Lao PDR have drastically changed following the government's reform. Yet, it is a fundamental feature of REDD+ decision-making processes to deal with a growing set of policy sectors and diverse interests. Focusing on the Readiness phase, in which the state governments seek to build a basis for REDD+, the analysis of the five countries has proved that REDD+ has succeeded in introducing participatory mechanisms to support sectoral coordination and dialogue with stakeholders into a national policy arena over forests for combatting climate change. It is a new but also challenging form of environmental decision-making process. It is thus critical to explore why such mechanisms either work or do not work to achieve policy coordination and strengthen REDD+ governance.

Acknowledgments: The initial country surveys were conducted under IGES REDD+ research project funded by the Ministry of Environment, Japan.

Author Contributions: Taiji Fujisaki designed and wrote the article in collaboration with Kimihiko Hyakumura; Henry Scheyvens initially designed the country study and collected information in Papua New Guinea; Tim Cadman provided suggestions regarding the theoretical and practical aspects of stakeholder participation in REDD+ and forest management.

Conflicts of Interest: The authors declare no conflict of interest.

References

1. Forest Carbon Partnership Facility (FCPF) REDD+ Countries. Available online: https://www.forestcarbonpartnership.org/redd-countries-1 (accessed on 20 April 2016).
2. UN-REDD Programme Regions and Countries Overview. Available online: http://www.unredd.net/index.php?option=com_unregions&view=overview&Itemid=495 (accessed on 20 April 2016).
3. Vatn, A.; Angelse, A. Options for a national REDD+ architecture. In *Realising REDD+: National Strategy and Policy Options*; Angelsen, A., Brockhaus, M., Kanninen, M., Sills, E., Sunderlin, W.D., Wertz-Kanounnikoff, S., Eds.; Center for International Forestry Research (CIFOR): Bogor, Indonesia, 2009; pp. 57–74.
4. Thompson, M.C.; Baruah, M.; Carr, E.R. Seeing REDD+ as a project of environmental governance. *Environ. Sci. Policy* **2011**, *14*, 100–110. [CrossRef]
5. Schroeder, H. Agency in international climate negotiations: The case of indigenous peoples and avoided deforestation. *Int. Environ. Agreem. Polit. Law Econ.* **2010**, *10*, 317–332. [CrossRef]
6. Corbera, E.; Schroeder, H. Governing and implementing REDD+. *Environ. Sci. Policy* **2011**, *14*, 89–99. [CrossRef]
7. Situmorang, A.W.; Nababan, A.; Kartodihardjo, H.; Jossi, K.; Achmad, S.; Safitri, M.; Soeprihanto, P.; Effendi, S. *Participatory Governance Assessment: The 2012 Indonesia Forest, Land and REDD+ Governance Index*; UNDP Indonesia: Jakarta, Indonesia, 2012.
8. Aquino, A.; Guay, B. Implementing REDD+ in the Democratic Republic of Congo: An analysis of the emerging national REDD+ governance structure. *For. Policy Econ.* **2013**, *36*, 71–79. [CrossRef]
9. Babon, A.; Gowae, G.Y. *The Context of REDD+ in Papua New Guinea*; Center for International Forestry Research (CIFOR): Bogor, Indonesia, 2013.
10. Dwyer, M.B.; Ingalls, M. *REDD + at the Crossroads: Choices and Tradeoffs for 2015—2020 in Laos*; Center for International Forestry Research (CIFOR): Bogor, Indonesia, 2015.
11. Ibarra-Gene, E. *Indonesia REDD+ Readiness—State of Play*; Institute for Global Environmental Strategies (IGES): Hayama, Japan, 2012.
12. International Development Law Organization (IDLO); Food and Agriculture Organization of the United Nations (FAO). *Legal Preparedness for REDD+ in Vietnam: Country study*; UN-REDD Programme; International Development Law Organization (IDLO): Rome, Italy, 2011.

13. Larson, A.M.; Petkova, E. An introduction to forest governance, people and REDD+ in Latin America: Obstacles and opportunities. *Forests* **2011**, *2*, 86–111. [CrossRef]
14. Lestrelin, G.; Trockenbrodt, M.; Phanvilay, K.; Thongmanivong, S.; Vongvisouk, T.; Pham Thu, T.; Castella, J.C. *The Context of REDD+ in the Lao People's Democratic Republic: Drivers, Agents and Institutions*; Center for International Forestry Research (CIFOR): Bogor, Indonesia, 2013.
15. Luttrell, C.; Resosudarmo, I.A.P.; Muharrom, E.; Brockhaus, M.; Seymour, F. *The Political Context of REDD+ in Indonesia: Constituencies for Change*; Elsevier Ltd.: Amsterdam, The Netherlands, 2014; Volume 35.
16. Poruschi, L. Progress towards national REDD-plus readiness in Vietnam. In *Developing National REDD-Plus Systems: Progress Challenges and Ways Forward—Indonesia and Viet Nam Country Studies*; Scheyvens, H., Ed.; Institute for Global Environmental Strategies (IGES): Hayama, Japan, 2010; pp. 53–76.
17. Scheyvens, H. *Papua New Guinea REDD+ Readiness—State of Play*; Institute for Global Environmental Strategies (IGES): Hayama, Japan, 2012.
18. Stewart, H.M.; Swan, S.; Noi, H. Final Evaluation of the UN-REDD Viet Nam Programme. Available online: https://g.zrj766.com/url?sa=t&rct=j&q=&esrc=s&source=web&cd=1&cad=rja&uact=8&ved=0ahUKEwi2o5nq2OXOAhUU-mMKHdfEA8AQFggaMAA&url=http%3A%2F%2Fwww.unredd.net%2Findex.php%3Foption%3Dcom_docman%26task%3Ddoc_download%26gid%3D10397%26Itemid%3D53&usg=AFQjCNGpl_I5_fe3rMasaYEGH_aaq1JNtQ (accessed on 20 April 2016).
19. Minang, P.A.; van Noordwijk, M.; Duguma, L.A.; Alemagi, D.; Do, T.H.; Bernard, F.; Agung, P.; Robiglio, V.; Catacutan, D.; Suyanto, S.; et al. REDD+ Readiness progress across countries: Time for reconsideration. *Clim. Policy* **2014**, *14*, 685–708. [CrossRef]
20. Cadman, T. *Quality and Legitimacy of Global Governance: Case Lessons from Forestry*; Palgrave Macmillan.: London, UK, 2011.
21. Lopez-Casero, F.; Cadman, T.; Maraseni, T. *Quality-of-Governance Standards for Carbon Emissions Trading*; Institute for Global Environmental Strategies (IGES): Hayama, Japan, 2012.
22. Leo, P.; Maria, B. When REDD+ goes national: A review of realities, opportunities and challenges. In *Realising REDD+ National Strategy and Policy Options*; Arild, A., Ed.; Center for International Forestry Research (CIFOR): Bogor, Indonesia, 2009; pp. 25–44.
23. Saunders, J.; Reeve, R. *Monitoring Governance for Implementation of REDD+*; Chatham House: London, UK, 2010.
24. Streck, C.; Gomez-Echeverri, L.; Gutman, P.; Loisel, C.; Werksman, J.; Gome-Echeverri, L. *REDD+ Institutional Options Assessment: Developing an Efficient, Effective, and Equitable Institutional Framework for REDD+ under the UNFCCC*; Meridian Institute: Washington, DC, USA, 2009.
25. Fujisaki, T. *Lao PDR REDD+ Readiness—State of Play, December 2012*; Institute for Global Environmental Strategies (IGES): Hayama, Japan, 2012.
26. Government of Indonesia. *Indonesia Readiness Preparation Proposal (R-PP)*; Forest Carbon Partnership Facility (FCPF), World Bank: Washington, DC, USA, 2009.
27. Government of Lao PDR. *Lao PDR Readiness Preparation Proposal (R-PP)*; Forest Carbon Partnership Facility (FCPF), World Bank: Washington, DC, USA, 2010.
28. Government of Papua New Guinea. *Readiness Preparation Proposal (R-PP): Papua New Guiena*; Forest Carbon Partnership Facility (FCPF), World Bank: Washington, DC, USA, 2013.
29. Government of Vietnam. *Readiness Preparation Proposal (R-PP): Social Republic of Vietnam*; Forest Carbon Partnership Facility (FCPF), World Bank: Washington, DC, USA, 2011.
30. Royal Government of Cambodia. *Cambodia Readiness Preparation Proposal (R-PP)-Revised*; Forest Carbon Partnership Facility (FCPF), World Bank: Washington, DC, USA, 2011.
31. Caldecott, J.; Indrawan, M.; Rinne, P.; Halonen, M. Indonesia-Norway REDD+ Partnership: First Evaluation of Deliverables. Availabel online: http://forestclimatecenter.org/files/2011-05-03%20Indonesia%20-Norway%20REDD%20Partnership%20-%20First%20Evaluation%20of%20Deliverables.pdf (accessed on 30 August 2016).
32. World Bank. Lao PDR Environment Monitor. Availabel online: http://siteresources.worldbank.org/NEWS/Resources/report-en.pdf (accessed on 30 August 2016).
33. REDD Vietnam: Institutional Arrangements for REDD in Vietnam. Available online: http://vietnam-redd.org/Upload/CMS/Content/Introduction/1-institutionalarrangementforREDDinVN_final.pdf (accessed on 10 April 2016).

34. REDD+ Cambodia: REDD+ Taskforce. Available online: http://www.cambodia-redd.org/category/national-redd-framework/redd-taskforce (accessed on 10 April 2016).

35. Presidential Decree of the President of Republic of Indonesia: Concerning the Task Force for Preparing the Establishment of REDD+ Agency (Unofficial Tlanslation). Available online: http://forestclimatecenter.org/files/2011%20Presidential%20Decree%20of%20The%20President%20No%2025%20Year%202011Task%20Force%20for%20Preparing%20The%20Establishment%20of%20REDD%20Agency.pdf (accessed on 20 April 2016).

36. Christy, L.C.; Di Leva, C.E.; Lindsay, J.M.; Takoukam, P.T. *Forest Law and Sustainable Development: Adressing Contemporary Challenges Through Legal Reform*; World Bank: Washington, DC, USA, 2007.

37. Geist, H.J.; Lambin, E.F. Proximate causes and underlying driving forces of tropical deforestation. *Bioscience* **2002**, *52*, 143–150. [CrossRef]

38. Hogl, K. *Reflection on International-Sectoral Co-Ordination in National Forest Programmes*; Tikkanen, I., Glück, P., Pajuoja, H., Eds.; European Forest Institute: Savonlinna, Finland, 2002.

39. Metcalfe, L. International Policy Co-Ordination and Public Management Reform. *Int. Rev. Adm. Sci.* **1994**, *60*, 271–290. [CrossRef]

40. Brockhaus, M.; Di Gregorio, M.; Mardiah, S. Governing the design of national REDD+: An analysis of the power of agency. *For. Policy Econ.* **2014**, *49*, 23–33. [CrossRef]

41. Shannon, M.A. *Mechanisms for Coordination*; Dube, Y.C., Schmithus, F., Eds.; The Food and Agriculture Organization Forestry Department: Rome, Italy, 2003.

42. Shannon, M.A.; Schmidt, C.H. *Theoretical Approaches to Understanding Intersectoral Policy Integration*; Tikkanen, I., Glück, P., Pajuoja, H., Eds.; European Forest Institute: Savonlinna, Finland, 2002.

43. Biggs, S.D. *Resource-Poor Farmer Participation in Research: A Synthesis of Experiences from Nine National Agricultural Research Systems*; International Service for National Agricultural Research (ISNAR): The Hague, The Netherlands, 1989.

44. Farrington, J. *Organisational Role in Farmer Participatory Research and Extension: Lessons from the Past Decade*; Overseas Development Institute (ODI): London, UK, 1998.

45. Pretty, J.N. Participatory Learning For Sustainable Agriculture. *World Dev.* **1995**, *23*, 1247–1263. [CrossRef]

46. Reed, M.S. Stakeholder participation for environmental management: A literature review. *Biol. Conserv.* **2008**, *141*, 2417–2431. [CrossRef]

47. Freeman, R.E. *Strategic Management: A Stakeholder Approach*; Pitman: Boston, MA, USA, 1984.

48. Sikor, T. Linking ecosystem services with environmental justice. In *The Justices and Injustices of Ecosystem Services*; Sikor, T., Ed.; Routledge: Oxton, UK, 2013; pp. 1–45.

49. Fisher, R.J.; Srimongontip, S.; Veer, C. *People and Forests in Asia and The Pacific: Situation and Prospects*; Food and Agriculture Organization Regional Office for Asia and the Pacific: Bangkok, Thailand, 1997.

50. Fischer, F. *Citizens, Experts and the Environment. The Politics of Local Knowledge*; Duke University Press: London, UK, 2000.

51. Beierle, T.C. *The Quality of Stakeholder-Based Decisions: Lessons from the Case Study Record*; Resources for the Future: Washington, DC, USA, 2000.

52. Reed, M.S.; Dougill, A.J.; Baker, T.R. Participatory indicator development: What can ecologists and local communities learn from each other? *Ecol. Appl.* **2008**, *18*, 1253–1269. [CrossRef] [PubMed]

53. Resosudarmo, B.P.; Ardiansyah, F.; Napitupulu, L. The dynamics of climate change governance in Indonesia. In *Climate Governance in the Developing World*; Held, D., Roger, C., Nag, E.M., Eds.; Polity Press: Cambridge, UK, 2013; pp. 72–90.

54. Michael, S. *Translating Vision Into Action: Indonesia's Delivery Unit, 2009-2012*; Princeton University: Princeton, NJ, USA, 2013.

55. International Tropical Timber Organisation (ITTO). *Status of Tropical Forest Management 2005*; ITTO: Yokohama, Japan, 2006.

56. Fox, J.C.; Yosi, C.K.; Keenan, R.J. Forest carbon and REDD+ in Papua New Guinea. In *Native Forest Management in Papua New Guinea: Advances in Assessment, Modelling and Decision-Making*; Fox, J.C., Yosi, C.K., Keenan, R.J., Eds.; Australian Centre for International Agricultural Research (ACIAR): Canberra, Australia, 2011; pp. 32–40.

57. REDD Vietnam. National REDD Network: Members. Available online: http://vietnam-redd.org/Web/Default.aspx?tab=member&zoneid=108&subzone=112&child=115&lang=en-US (accessed on 10 March 2016).

58. Am, P.; Cuccillato, E.; Nkem, J.; Chevillard, J. *Mainstreaming Climate Change Resilience into Development Planning in Cambodia*; IIED country report; IIED: London, UK, 2013.
59. Nomura, K. Selection Process for REDD + Consultation Group Representatives in Cambodia. Available online: www.unredd.net/documents/un-redd-partner-countries-181/asia-the-pacific-333/a-p-knowledge-management-a-resources/national-programme-documents/technical-reports-2065/safeguards-2071/cambodia-2206/11726-selection-process-for-redd-consultation-group-representatives-in-cambodia-11726/file.html (accessed on 20 April 2016).
60. REDD+ Cambodia. Terms of Reference: REDD+ Consultation Group. Availabel online: www.unredd.net/index.php?view=download&alias=9898-tor-of-counsultation-group-9898&category_slug=stakeholder-engagement-including-selection-processes-2969&option=com_docman&Itemid=134 (accessed on 20 April 2016).
61. Cambodia Forestry Administration (FA). Cambodia's National Forest Programme Background Document. Available online: http://accad.sean-cc.org/components/com_msearch/file_uploads/content_attachment/a30b91004deabbae6389ea13b2f04c2c.pdf (accessed on 10 April 2016).
62. Jeremy, C.-R. *Biodiversity Planning in Asia*; International Union for Conservation of Nature (IUCN): Gland, Switzerland; Cambridge, UK, 2002; pp. 111–128.
63. Pham, T.T.; Huynh, T.B.; Moeliono, M. Myth and reality: Security of forest rights in Vietnam. In *Analysing REDD+*; Angelsen, A., Brockhaus, M., Sunderlin, W.D., Verchot, L.V., Eds.; Center for International Forestry Research: Bogor, Indonesia, 2012; pp. 160–161.
64. McDougall, G. Mission to Viet Nam (2010). In *The First United Nations Mandate on Minority Issues*; Brill Nijhoff: Leiden, The Netherlands; Boston, MA, USA, 2015; pp. 222–252.
65. Royal Government of Cambodia. *National Strategy for Rural Water Supply, Sanitation and Hygiene 2011–2025*; Royal Government of Cambodia: Phnom Penh, Cambodia, 2011.
66. Friends of the Earth International. Availabel online: http://www.foei.org/ (accessed on 30 August 2016).
67. Mayers, J.; Bass, S. *Policy That Works for Forests and People*; Earthscan: London, UK, 2004.
68. Hutton, J.; Adams, W.M.; Murombedzic, J.C. Back to the Barriers? Changing Narratives in Biodiversity Conservation. *Forum Dev. Stud.* **2005**, *32*, 341–370. [CrossRef]
69. Sikor, T. REDD+: Justice effects of technical design. In *The Justices and Injustices of Ecosystem Services*; Sikor, T., Ed.; Routledge: Oxton, UK, 2014; pp. 46–68.
70. United Nations Framework Convention on Climate Change (UNFCCC). *Report of the Conference of the Parties on Its Sixteenth Session, Held in Cancun from 29 November to 10 December 2010*; UNFCCC: Bonn, Germany, 2011; pp. 1–31.
71. Forest Carbon Partnership Facility (FCPF). Guidelines on Stakeholder Engagement in REDD + Readiness. Available online: https://www.forestcarbonpartnership.org/sites/fcp/files/2013/May2013/Guidelines%20on%20Stakeholder%20Engagement%20April%2020,%202012%20(revision%20of%20March%2025th%20version).pdf (accessed on 20 April 2016).
72. Mayntz, R. Modernization and the logic of interorganizational networks. *Knowl. Policy* **1993**, *6*, 3–16. [CrossRef]
73. Keeley, J.; Scoones, I. *Understanding Environmental Policy Processes: Cases from Africa*; Earthscan: London, UK, 2003.

![forests logo] *forests*

MDPI

Article

Towards a Role-Oriented Governance Approach: Insights from Eight Forest Climate Initiatives

Mareike Blum * and Sabine Reinecke

Chair Group of Forest and Environmental Policy, University of Freiburg, Tennenbacher Str. 4, D-79106 Freiburg, Germany; sabine.reinecke@ifp.uni-freiburg.de
* Correspondence: mareike.blum@ifp.uni-freiburg.de; Tel.: +49-761-203-3722

Academic Editors: Esteve Corbera and Heike Schroeder
Received: 31 October 2016; Accepted: 25 February 2017; Published: 28 February 2017

Abstract: In forest climate governance processes such as REDD+, non-state actors take on various, more or less formal, but in fact potentially authoritative governance tasks when informing, financing, (co)deciding or implementing forest climate action. Drawing on the concept of social roles, we investigate eight different REDD+ governance processes and how a variety of practical authoritative roles are enacted in administration, finance, decision-making and knowledge production. By systematically revealing the distinct ways of how different roles were filled, we developed a first (potentially still incomplete) typology of role practices and underlying rationales within different governance settings. In this endeavor, the role concept offered a valuable and handy analytical tool for empirically operationalizing governance performance, which is principally compatible with both institutional and social constructivist approaches to legitimacy.

Keywords: forest climate governance; REDD+; roles; legitimacy

1. Introduction

Anthropogenic climate change is one of the most pressing contemporary societal challenges affecting human life and the environment across the globe [1–3] and induces high economic costs on human societies [4]. With a contribution of about 10% to 20% to global emissions of greenhouse gases (GHG) [5,6], the unsustainable use and conversion of forests mark a significant driver of climate change, especially in developing countries [7,8]. Considering forests' high potential as sinks for GHG emissions, the idea of establishing an international and results-based financial mechanism which addresses deforestation in developing countries was broadly welcomed by many Parties and stakeholders as early as 2005 at the 11th Conference of the Parties (COP) of the United Nations Framework Convention on Climate Change (UNFCCC) in Montreal, Canada.

However, it was not until COP 21 (2015) that negotiations were finalized for REDD+ ("Reducing emissions from deforestation and forest degradation and the role of conservation, sustainable management of forests and enhancement of forest carbon stocks in developing countries"). Formally, REDD+ is an international instrument that is implemented by nation-states, e.g., as part of their future Nationally Determined Contributions (NDCs). However, there exists a decisive national discretion in implementing NDCs; and, as the Paris Agreement highlights, there is indeed an increasingly important role for non-state actors in the design and implementation of concrete policies "to address and respond to climate change" [9]. Even before the actual REDD+ negotiations under UNFCCC were finalized, public and private actors were already actively engaged in activities to globally discuss or locally test REDD+, which de facto shaped the mechanism [10]. In addition, a number of preparatory steps were taken for capacity building for the implementing entities in ministries. Examples include the UN-REDD Program or the World Bank's Forest Carbon Partnership Facility (FCPF). These preparatory steps proved decisive during the REDD+ negotiation process [11], and rather than seeing their legacy disappear, it is likely that they will form the foundation for the future REDD+ system. This includes the many ties that already exist between public and private engagement.

The involvement of private actors in REDD+ governance offers great chances for experimenting with innovative solutions to collective action problems, e.g., finance [12,13]. For that reason, polycentric policy networks [14] are—in theory—often seen as normatively superior in complex cases such as climate change [15,16]. One prominent reason is that they help to combine different advantageous capacities of markets (efficiency) or communities (context-sensitive political support) [12]. In practice, however, governance across the state–society divide may similarly fail or perform poorly [17,18]. One concern is that polycentric governance involves several actors from business, civil society, government and science simultaneously, which disperses governance authority [19–21]. Even where processes appear genuinely state-led, non-state actors may in fact fulfill potentially authoritative governance tasks when *informing, financing, (co)deciding* or *implementing* climate action without proper legitimation which has critical repercussions for the provision of collective goods [17,22]. For instance, recent (Foucauldian) governmentality studies showed that even seemingly neutral and technical practices such as notation, accounting and auditing may be highly political and influential [23,24]. Accordingly, it would be too narrow to focus only at the more or less formally legitimized decisions made by official representatives in state-driven processes. Joint working of public and private actors may be misused for scapegoating or absolving from accountability [18,25,26]. In addition, the higher complexity of mixed or hybrid (state/non-state) policy networks at and across local and transnational levels makes it increasingly difficult to identify which actors are excluded in these processes [10,14]. More scrutiny is required regarding supposedly "neutral" actor roles such as administration or expertise in polycentric REDD+ governance [22] and how exactly they interplay with formal decision-making pathways.

When answering questions about the performance of distinctive governance settings, e.g., regarding their legitimacy, it appears essential to be able to identify and differentiate the specific roles that actors hold within polycentric settings and whether they actually carry authority—and thus whether they need to be held accountable. Revealing such authoritative roles, however, is no easy task when applied to the myriad hybrid cases that make up REDD+ reality. There are several cases that are concerned with technical information, capacity building, financing or implementation oriented actions, such as carbon standards in forests or reforestation projects—cases that often are not officially coined as REDD+ but have repercussions for how REDD+ performs. Each case simultaneously engages a number of actors from different societal spheres in a broad range of different roles or functions. However, the actual nature of actors does not automatically determine their respective governance strategy. Business is not per se always trying to maximize monetary profits. Likewise, public welfare is not guiding all actions of civil society representatives. It would be highly presumptive or even naïve to suggest that belonging to a certain actor group automatically disqualifies or preserves an actor from engaging in certain actions. Involving some government representatives will not always increase democratic legitimacy, nor will business involvement principally increase efficiency but undermine democratic accountability.

Drawing on a role-oriented account of governance in networks, we investigate eight distinct cases of polycentric governance in the broader context of inter- and transnational forest climate policies. Specifically, we want to elaborate on the disparate ways in which different actors (jointly) fill potentially authoritative roles. Covering a variety of different cases, we want to reveal the distinct patterns that supposedly similar roles—such as secretariat, expertise or financing—may take in polycentric forest climate governance. Based on the first empirical insights gained from different cases, we discuss how such typology may be usefully employed for evaluative considerations about democratic accountability and legitimacy and thus complement established institutional approaches to democratic legitimacy.

2. The Relevance of Roles for Legitimacy

Typically, and drawing on Scharpf's seminal work [27], legitimacy may most generally be conceptualized as having two analytical dimensions: input and output legitimacy. Input legitimacy generally refers to the inclusion of stakeholders in decision-making processes [28], and is concerned with the "right" procedures of participation, accountability and transparency [29,30]. Output legitimacy refers to the problem-solving effectiveness of a policy instrument and its result [31].

On the input side, representation is seen to be central. However, while official government delegates risk losing national mandates if their policies are not responsive to the voters' will, the input accountability of non-state actors in transnational governance is typically not as straightforward or formalized. From an institutionalist view, specific qualities of the process may be highlighted through which non-state actors are included in decisions. It is hoped that institutional arrangements or decision-making procedures, such as voting rules, participatory boards or grievance structures help to achieve these qualities.

Output legitimacy in Scharpf's understanding means effective problem-solving and, further, "that the policies adopted will generally represent effective solutions to common problems of the governed" [31]. As it is difficult to measure if the policy helped to address the original problem, possible criteria to capture output legitimacy are (1) the extent to which self-set goals were reached and (2) the level of compliance by the most important actors [28]. Moreover, output legitimacy depends, in large part, on expertise, where trusted actors with respective knowledge claims are entitled to define effective policies [32]. Where international organizations (IO) cannot deliver the needed expertise—and hence expert legitimacy—to public authorities in the transnational arena, other non-state actors may provide it [33]. However, the principal applicability of expert legitimacy in cases characterized by high uncertainty and stakes, as in cases such as climate change or natural resource use, may be questioned because expertise often excludes relevant stakeholders and their alternative ways of knowing from governance processes and their respective discourses [34,35]. For unstructured problems, there are no impartial evaluations of legitimate outcomes without considering the perspectives of those affected by decisions. Such "justification of actions" by referring "to those whom they affect according to reasons they can accept" [36] is a central theme in social constructivist notions of legitimacy. What matters in this view are the expectations that affected audiences have about an actor's involvement in governance and whether these expectations are eventually met [36,37]. Legitimacy necessarily varies as it is dependent on the context, issue area and specifically the different legitimacy understandings of distinct political communities [38] (p. 19). Rather than assuming that some sort of pre-defined mechanism or institution will ensure legitimate decisions, legitimacy is taken as a dynamic stance—as a two-sided process between different actors with reciprocal expectations. However, in research practice this assumption lacks sufficient specificity as to identify relevant processes that require attention because—in principle—any more or less formalized processes with state or non-state actor participation may be decisive. Rather than overly relativizing, we follow Bernstein and suggest that legitimacy depends on "historical understandings and the shared norms of the particular community or communities granting authority" [39]. Hence, in the forest climate governance communities around REDD+, we can assume that expectations are not just randomized, but structured and institutionalized. Such a reading of legitimacy has been used for analyzing both singular hybrid and private settings in environmental governance [40,41].

Analytically, the concept of actor roles offers a way of incorporating the advantages of both social constructivist and institutional schools of thought. Simply put, roles are the characteristic behavioral patterns that form part of social life [42] and describe the relationship between the individual and the social system [43]. On the one side, they are typically grounded and expressed in formal rules or informal norms (e.g., statutes, contracts or working task descriptions, or societal functions). At the same time, they rest in individual and subjective expectations of both the role holders and the affected communities toward the respective social position [42]. With this characteristic, roles not only reveal the congruence between a given norm (formal expectation toward role holder) and a practice, but also between different subjective expectations of the role holder and affected communities, which are tied to this social position. Covering, formal (de jure) and subjective (de facto) expectations, the role concept helps to understand legitimacy as a socially constructed and structurally institutionalized (i.e., patterned and hence empirically more easily detectable) phenomenon. If actors are able to fulfill the formal and subjective expectations tied to their roles, they create acceptance and credibility for their actions. Moreover, the concept is not limited to public or genuinely political actors but can be employed to all sorts of actors in governance settings. Such specification appears especially relevant regarding the mentioned concern that authoritativeness may not only be tied to official decision pathways and institutional settings (parliaments, boards, plenums etc.), but also to seemingly harmless functions related to administration, budget, auditing or expertise.

When investigating the plethora of novel polycentric cases in forest climate governance, it appears critical to draw on general analytical categories that help identifying relevant authoritative roles. Börzel and Risse (2005) offered a first attempt to analytically grasp different governance purposes (rule-setting, implementation, service provision) and functions (co-optation, delegation, co-regulation or self-regulation) that private actors may have in hybrid Public–Private Partnerships [17]. One notion was that a self-regulated setting which sets formal rules completely independent from government was more problematic than private governance having evolved just for the 'service' of government (i.e., delegation) [17]. While such typologies are very helpful for distinguishing different ways or degrees of exerting authority, they may still disregard apparently unproblematic functions which have potentially decisive impacts on the provision of collective goods [22]; particularly for such contested issues as climate change. Moreover, a functional typology provides no empirical insight as to what specific types of roles, tied to specific organizational strands or tasks of a setting, are typically relevant for analysis. We suggest that authoritative roles may principally be tied to four different organizational branches of governance that hold significance for the input or output legitimacy of this setting's operations: (1) administration, (2) funding, (3) internal decision-making and (4) knowledge making processes.

2.1. Administrative Roles

From the time Biermann and Siebenhühner conducted their seminal research about international bureaucracy, it has since become evident that administrative roles can be fairly relevant for questions about governance legitimacy [26]. The classical view in the public administration literature is that administrative roles encompass a "cohesive set of job-related values and attitudes" [44]. More specifically, attitudes that are assumed to ensure responsible behaviour. In practice, administrators evidently have to deal with contradictive role expectations that arise from simultaneously having discretion based on expertise while being obedient to external masters [43]. In principle, administrative roles may turn out to be fairly authoritative in operational decision-making, including agenda setting, book keeping, and communicating with stakeholders. However, rather than suggesting that an independent and strong secretariat is always necessary [45] or let alone legitimate, it is rather more important that such strength responds to valid expectations about the respective administrative role within a certain setting.

2.2. Financial Roles

While administration and finance appear as genuinely linked, because one and the same actor may be responsible for executive tasks, accountability relations from a role perspective may be completely different, not least because funding sources of governance setting are often external. The expectations tied to financial roles are typically linked to the dimension of input legitimacy and relate to concepts of transparency and accountability [29]. Considering that activities such as distributing and spending money always carry the risk of inefficiency or corruption, transparency becomes an important approach for ensuring input legitimacy of financial roles. Accountability refers to the "relationship between an actor and a forum, in which the actor has an obligation to explain and to justify his or her conduct. The forum can pose questions and pass judgment, and the actor may face consequences" [46]. Multilateral or bilateral finance is typically tied to a set of formal rules (conditions) and corresponding expectations about performance within or by the funded governance setting, a donor–recipient relationship well known from development cooperation [47]. Principally, recipients are accountable to providers and, for instance, they have to report regularly on the progress of their work, yet this accountability may be weaker if, for instance, the organization itself generates the necessary income (e.g., through selling services or products) or can draw on philanthropic (unconditional) donations.

2.3. Decision-Making Roles

On top of roles relating to financing and administration/organization [48] we need to consider roles that are more straightforwardly related to consequential decisions. Decisions are *consequential* if they relate substantially, rather than only procedurally, to a policy process. Such implementation and execution oriented roles determine fairly specific rules or methods for project definition, implementation or evaluation. A conceptual differentiation between more consequential, as compared to rather strategic or procedural decision-making modes is useful, because they may not be exerted by the same set of actors.

By contrast, strategic decision-making covers the overall goal orientation carrying expectations about what defines an effective outcome of the governance operations. These roles often relate to governance bodies such as Boards of Directors covering aspects of both input and output legitimacy. Evidently, the specific performance of such boards is not only defined by more or less formalized codes of conduct (statutes), but also depends on social–psychological processes regarding how actors participate or interact in such group and how exactly they exchange information or discuss controversial issues [49,50]. Regarding the input legitimacy of strategic decision bodies, adequate representativeness of those taking these decisions is important, whereas strategic roles may be filled by various actors, such as CEOs, owners, managers, shareholders or Boards of Directors.

2.4. Expert Roles

As mentioned earlier, expert legitimacy allows different non-state actors to be actively involved in public affairs. Experts might be invited to participate because they better understand the cultural context within a project area can provide lessons learned from pilot projects or bring in technical knowledge about legal or technological terms or scientific methods. The concept of epistemic communities [51] reflects this central role of expertise in (transnational) governance. Epistemic communities encompass "a network of professionals with recognized expertise and competence in a particular domain and an authoritative claim to policy-relevant knowledge within that domain or issue-area" [51]. Typically, policy processes draw on evidence-based science where technical, professional and bureaucratic expertise become decisive ingredients for decision-making [52]. In fact, expert knowledge carries high authoritative power because it influences consequential decisions (output dimension) fairly directly. The political power of knowledge is even more significant in unstructured environmental problems, where "both the nature of the 'problem' and the preferred "solution" are strongly contested" [53]. In such cases where ignorance and political opposition is strong, bureaucratic or scientific knowledge are hardly neutral. Rather, such knowledge is selective and partially supporting of certain positions while marginalizing others [54–56]. Revealing the different role related to knowledge and expertise and how they are institutionalized, including what expectations are tied to them, may be critical for evaluating the (output) legitimacy of governance in practice.

Given the aforementioned assumptions, seemingly neutral functions and their respective operational roles require analytical scrutiny. With the paper at hand, we attempt to systematically elaborate on the different rationales behind authoritative roles in forest climate governance such as REDD+ and, further, work to identify the different patterns that administrative, financial, decision-making and expert roles may take empirically. By exploring eight distinct cases of forest climate governance in a comparative fashion, we intend to provide a systematic typology of authoritative roles. In this way, we seek to generate valuable insights as to whether and how roles may serve as a helpful conceptual and methodological tool for operationalizing and, hence, evaluating the performance of governance settings, specifically concerning their legitimacy.

3. Methods

Our study of eight distinct global forest climate governance formats follows a comparative qualitative case study approach [44]. In order to carve out the variety of authoritative roles in polycentric governance, we follow a dual strategy. Our cases selected cover different hybrid governance settings that—first—range from more to less private actor intensive compositions and that—second—cover different fields of climate action (information, regulation, finance, project implementation). This way, we are able to provide an extensive informative basis that covers varied (potentially hidden) authoritative actor roles in governance across the whole policy cycle from problem-definition and agenda-setting to policy implementation and evaluation [45]. Broadly following the empirical variation in transnational policy processes that are more or less directly tied to the formal UN based REDD+ mechanism, this comparative study allows a nuanced discussion in different policy stages.

3.1. Case Selection

In an extensive online and literature review, we identified a set of 35 potential cases of transnational governance arrangements with relevance to the issue of forest climate policy and finance. This topical focus assured inter-case consistency and comparability. Finally, eight different settings were selected based on the criteria of fulfilling varied relevant REDD+ purposes (e.g., networking, implementing, standard-setting, evaluation and commitments). Four of these cases have—from a formal point of view—been strongly state-led activities, i.e., either informing global UNFCCC negotiations or national implementation: the REDD+ Partnership and the Standing Committee of Finance (both information exchange forums) as well as the Partnership for Market Readiness and the Association of Southeast Asian Nations (ASEAN) Forest Clearing House Mechanism (CHM) (two capacity building and exchange forums in support of national implementation of carbon markets and forest policy). Each time, however, some kind of private actor involvement was given—yet at different degrees and in different functions. In formal terms, the four other cases involve non-state or private actors in concrete implementation oriented activities more proactively, be it in the coordination and implementation of forest restoration projects in several countries, as in the case of the Bonn Challenge and the Green Belt Movement, or in developing and certifying global standards of sustainable land uses, as with the Roundtable on Sustainable Palm Oil and the Gold Standard.

Among the eight selected cases that engage in varied REDD+ related activities, four settings were rather closely tied to the UNFCCC process (UNFCCC Standing Committee on Finance, REDD+ Partnership, ASEAN Clearing House Mechanism and Partnership for Market Readiness), whereas the other four were relatively independent (Bonn Challenge, Green Belt Movement, Roundtable on Sustainable Palm Oil and the Gold Standard)—but still of relevance to the process. To gain in-depth insights into different performative and more or less authoritative roles in distinct governance settings, we look at different social roles of actors in internal operations. The idea is to not only list relevant actors in each setting, but to elaborate on their constellation and involvement in different decision-making procedures and contexts covering different operational branches, from administration and finance to strategic, procedural and consequential decisions as well as knowledge creation. Specifically, we asked which concrete actors are engaged in a certain role and with what expectation. By highlighting the major formal (institutionalized) and subjective (socially constructed) expectations, we are able to identify and reflect upon potential tensions and pitfalls tied to the specific roles within the respective setting, which would matter for considerations of the overall legitimacy of the cases.

3.2. Data Collection and Analysis

The eight case studies draw on extensive document analysis, participatory observation (where available) and expert interviews. Analyzed documents encompass, for instance, primary research about the setting, but also strategic documents and reports of the respective institution such as working programs or annual/evaluation reports. In addition, more popular public relations oriented documents (website, press releases, mission statement, self-description brochures) were used if they delivered insights into the internal procedures and the (expectations about the) roles of different actors. The insights from analyzing this material were iteratively enriched and validated with information gained from (typically) one additional qualitative semi-structured expert interview per case [48]. Most interviews were conducted face-to-face, by telephone or Skype mainly with experts at the strategic or operational level of the organization or with external actors (private or public stakeholders) that were considerably involved in or followed the governance operations at decisive points as to give an insight into the ways of how different actors work jointly together. Considering the diplomatic load of a number of the settings investigated and to encourage detailed and honest answers, interviewees are guaranteed full anonymity.

Table 1 provides an overview of the expert interviews and the year of the interview (besides the ASEAN CHM) all interviews were conducted between December 2015 and May 2016. In the case of the REDD+ Partnership, the Bonn Challenge and the Gold Standard, we build on our own observations collected during the participation in conferences, meetings and workshops. The respective codes are used to indicate the sources of information within the results.

Table 1. Interviewed experts.

Expert of . . .	Year	Code
Standing Committee on Finance (SCF) of the United Nations Framework Convention on Climate Change	2015	Expert 1
Reducing Emissions from Deforestation and Forest Degradation (REDD+) Partnership	2010–2014	Own observations
ASEAN Clearing House Mechanism (CHM)	2014	Expert 3
Partnership for Market Readiness (PMR)	2016	Expert 4
Bonn Challenge (BC)	2015	Expert 5
	2013–2017	Own observations
Green Belt Movement (GBM)	2016	Expert 6
Gold Standard (GS)	2015	Expert 7
	2015–2016	Own observations
Roundtable on Sustainable Palm oil (RSPO)	2016	Expert 8

All interviews were tape-recorded, transcribed and interpreted by means of qualitative methods of content analysis [48,49] along a set of predefined analytical governance dimensions (deductively) as well as on the basis of the empirical material itself (inductively). Four broader dimensions guided the analysis and comparison of the cases: administration, finance, decisions and expertise. To contextualize each case, a brief introduction is given about their development and set-up.

4. Introduction to the Cases

In the following, we briefly introduce each case (founding, objectives, general set-up) and its relevance for or relation to REDD+ before elaborating on the varied roles of different (esp. non-state) actors in administration, finance, strategic and consequential decisions as well as in knowledge making in a comparative fashion.

4.1. UNFCCC Standing Committee on Finance (SCF)

The SCF was established in 2010 to improve the coherence and coordination in the delivery of climate change financing by collecting and synthesizing relevant information to inform the ongoing negotiations under the UNFCCC. Accordingly, it is mainly composed of 20 delegates that are elected by members of the Conference of the Parties (COP) for two years. While (official) stakeholders may principally be admitted to attend regular meetings, there is also a forum for the "communication and continued exchange of information among bodies and entities". It is organized at least once a year and includes different actors from business, science or civil society [57]. REDD+ finance was a major topic at the 3rd forum on Enhancing Coherence and Coordination of Forest Financing which took place in Durban, South Africa, on 8–9 September 2015.

4.2. REDD+ Partnership

Another learning platform to inform the ongoing negotiations on REDD+ was the REDD+ Partnership, which was active between 2010 and 2014. It was composed of official party partners as well as more than 100 "institutional partners", i.e., stakeholders from science, business or civil society. It provided an informal setting for partners to discuss (but not formally decide on) activities, finance or technical and political issues related to REDD+ implementation. It held organizational meetings and technical workshops on selected topics of interest such as REDD+ safeguards, (forest) reference levels, measuring, reporting and verification (MRV) or REDD+ financing.

4.3. ASEAN Forest Clearing House Mechanism (CHM)

The ASEAN CHM was founded in 2004 and promotes networking and holistic knowledge management between ASEAN stakeholders, processes and institutions in the field of forest law enforcement and governance and REDD+. The knowledge and learning platform linked senior officials and regional experts of various ASEAN networks including state representatives, business actors and scientists. It is managed by the ASEAN Secretariat [58]. The network met twice a year and could rely on a broad digital infrastructure (interactive website, e-learning, webinar tools). The CHM served ASEAN stakeholders to agree on research agendas (e.g., about REDD+) and to learn about assessment tools related to forest law enforcement. A nominated network manager facilitated communications, but members were also expected to contribute actively [57,58].

4.4. Partnership for Market Readiness (PMR)

The PMR was initiated by the World Bank in 2010 with the objective to support middle income countries to prepare and implement financial climate change mitigation policies such as carbon pricing or carbon taxes. It provides capacity building through knowledge sharing as well as funding for technical assistance. The PMR consists of 13 contributing countries and 17 implementing countries, which together form the Partnership Assembly (PA). Decisions within the Partnership Assembly are made by consensus and may be informed by additional technical partners to provide expertise. The meetings are led by co-chairs, with one member from a contributing (high income) country and one from an implementing (middle income) country. The World Bank Group acts as secretariat and partner for the PMR. It organizes meetings, brings in experts, coaches the countries and monitors their progress based on implementation status reports. The technical assistance for the implementation of the proposals is done in partnership with the World Bank's operational units in each country [59].

4.5. Bonn Challenge (BC)

The Bonn Challenge was founded in 2011 and is an implementation platform for the three existing international commitments on forest restoration: Aichi Target 15 on Biodiversity, the "+" in REDD+ (esp. enhancement of carbon stocks), and the "zero net land degradation" goal of Rio+20. The Bonn Challenge is a declaration of intent and provides a platform for information exchange, networking and capacity building on reforestation for the "coalition of the willing". It promotes implementation processes by giving high level visibility to voluntary pledged commitments. Its aim is "to restore 150 million hectares of the world's deforested and degraded lands by 2020" and another 200 million by 2030 [60]. So far, it has received pledges from 30 countries, one private company and six sub-national regions. Together, these pledges account for 112 million hectares [60]. The civil society organization World Wildlife Fund (WWF), scientific advisors from International Union for Conservation of Nature and Natural Resources (IUCN) and the World Resource Institute contributed to the strategic development of the platform. The Global Partnership on Forest and Landscape Restoration (GPFLR) which is hosted by IUCN is providing the infrastructure of the Bonn Challenge (website, meetings, publications and capacity buildings). Repeatedly, financial actors have been invited to Bonn Challenge meetings to attract more private investment.

4.6. Green Belt Movement (GBM)

The Green Belt Movement, founded in 1977, has offices in Kenya, London and Washington. GBM is a Kenyan grassroots organization that encourages and financially supports women in Africa to jointly grow seedlings and plant trees. The GBM contributed to the development of the Kenyan REDD+ process and acted as a partner for the government. Since 1986, the GBM launches a Pan African Green Belt Network across Ethiopia, Tanzania, Uganda, Rwanda and other African countries. In this context, it offers trainings and practical capacity building for grassroots and development groups. Nobel-peace price awardee Wangari Maathai was the founder of the GBM and basically guided the organization until her death in 2011. Since then, Wanjira Maathai, the daughter of Wangari Maathai, is the new head of the organization.

4.7. Gold Standard (GS)

The Gold Standard is a foundation that offers a global standard to certify carbon emission reductions and sustainable development impacts for carbon offset projects both for the compliance market (Clean Development Mechanism under the Kyoto Protocol) and the voluntary market. It was founded in 2003 by WWF and is endorsed by more than 80 NGOs. Their role is to comment on new methodologies and project design documents. Moreover, 25 "market specialists" from consultancies, think-tanks, NGOs and research institutions constitute the GS Technical Advisory Board, which decides on new methods and oversees the whole certification process [61]. To date, the standard has been applied to 1100 projects in 70 countries [61].

4.8. Roundtable on Sustainable Palm Oil (RSPO)

The RSPO, founded in 2004, is a transnational multi-stakeholder roundtable that seeks to transform the palm oil production. It has developed a private standard for sustainable palm oil. The RSPO is a relatively large setting composed of 3100 members with participation rights and 1400 ordinary members with voting rights. The latter are mainly consumer good manufactures, processors and retailers. The remaining ordinary members are palm oil growers and financial actors. Civil society organizations make up only 3.5% [62]. In the annual general assembly, board representatives are elected and resolutions are made based on simple majority vote.

Table 2 provides an overview of the key features of each of the eight governance settings which are relevant for the analysis of the roles outlined before. The table illustrates the fact that non-state actors are essentially engaged in all settings. It also shows that the sources of funding vary, ranging from bi- and multilateral state-based financing (ODA) to (public or private) donations or self-financing through the provision of payable services. Yet, evidently, the simple fact that stakeholders are formally related to a setting as participants or funders does not tell us whether they actually take on decisive roles which require those actors to render accountability. Such evaluation requires empirical substantiation and a more refined look at who fills specific roles in administration, decision-making, finance and expertise. Additionally, understanding their expectations can help answer the aforementioned question for specific cases.

In order to get a more complete picture of the whole spectrum of possible authoritative roles within the settings, we specifically look for commonalities and differences across the cases as to how specific roles are enacted.

Table 2. Features of governance settings.

Features → / Governance Settings ↓	Stakeholders	Funding	Administration and Operational Decision-Making	Strategic Decision-Making / Consequential Decision-Making
UNFCCC SCF	SCF: States (20 delegates) SCF Forum: states, civil society, business, finance, science	Conference Of the Parties (COP) members (states)	**Administration:** UNFCCC Secretariat **Operational decision-making:** co-chairs	**Strategic:** State delegates **Consequential:** COP members (state delegates)
REDD+ P.	States, civil society, science, business	Conference Of the Parties (COP) members (states)	**Administration:** FMT/PT Facility Management Team (by FCPF, World Bank) and UN-REDD Program Team (by UNDP, UNEP, FAO) **Operational decision-making:** chairs	**Strategic:** States (ministerial level) **Consequential:** all members
PMR	States, financial institution	World Bank	World Bank	**Strategic:** World Bank and contributing countries **Consequential:** all members
ASEAN F. CHM	States, business, science	German Agency for Development Cooperation (GIZ)	ASEAN Secretary	**Strategic** GIZ, ASEAN **Consequential** all members
BC	(sub)states, business, science, civil society, financial institutions	Germany, IUCN	IUCN, GPFLR, WRI	**Strategic** Founders **Consequential** (implementing) states/sub-states/companies
GBM	Civil society	UN funds, donations, partly self-funded	Internal staff	**Strategic:** Founder **Consequential:** Founder
GS	Civil society, business, science	Two-thirds self-funded, one-third national, international and private funds	Internal staff	**Strategic** Advisory board **Consequential** Expert Committee
RSPO	Business, civil society, financial institutions	Self-funded (member fees)	Internal staff	**Strategic** Advisory board

5. Results

In the following section, we apply the role concept to the eight different cases. First, we examine administrative and operational decision-making roles that resemble each other regarding the more procedural pathways of influence, e.g., defining how operations are exercised or which actors are accepted to these operations and in what way. Second, we look at funding constellations and financing roles tied to different settings. Third, we analyze strategic and consequential decision-making roles, i.e., roles that are concerned with outcome-related decisions—both broadly and more specifically. Fourth, we aim to identify different expert roles in knowledge production and to reveal the respective rationale of and expectations about "expertise". For illustration, we provide a figure that covers the different subvariants for each role type. Where identifiable, outer circles describe the (common) rationales and corresponding expectations behind a specific (set of comparable) role(s). Specific expectations about and why a specific actor carries a distinct role (experience, reputation, capacity, professional aspects, etc.) are explicated within each bubble.

5.1. Administrative Roles in Forest Climate Governance

Assuming that administrations (in their *bureaucratic roles*) may be fairly authoritative [26] for governance performance, we investigated the eight cases regarding what formal requirements as well as inter-personal expectations are tied to this respective role. We identified three different types of administrative roles that seemed to follow different rationales for engaging specific actors. (1) The first type is that of *international bureaucracies*. In such cases, the secretariat is closely tied to or even embedded in the international UN processes. Two examples are the Standing Committee on Finance, which is administered by the UNFCCC secretariat, and the REDD+ Partnership which is administered by the secretariats of the two major running funding programs on REDD+ (UN REDD PT/FCPF FMT). (2) A second type of secretariat draws on *International Organizations* to provide the respective services, e.g., in the case of the Partnership for Market Readiness, which is administered by the World Bank, or the Bonn Challenge, which is administered by the ICUN secretary. (3) A third type is the approach of having *internal staff* that administer—often in private organizations. Examples here are the Gold Standard and the RSPO or NGOs such as the Green Belt Movement. Figure 1 gives an overview of the three different administrative roles.

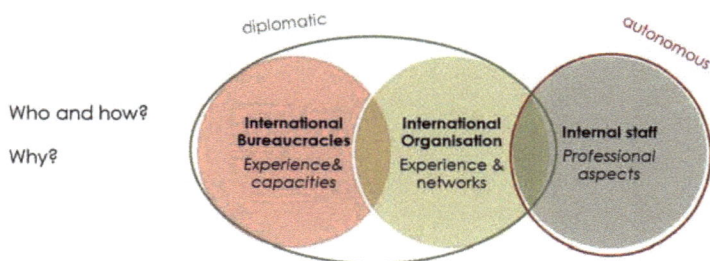

Figure 1. Administrative roles.

An important rationale for relying on (typically rather expensive and unwieldy) international bureaucracies (type 1) or international organizations (type 2) is that the respective actors are highly experienced in working professionally in diplomatic settings and possess specialized financial and organizational capacities or relevant networks for facilitating delicate policy processes with state representatives. Noteworthy to say, the cases of SCF and the REDD+ Partnership (type 1) are also more or less formally and directly oriented at and relating to ongoing international REDD+ negotiations. By contrast, such aspects appear to matter less in cases such as the Gold Standard or the Green Belt Movement, which mainly draw on their own (internal) staff. Overall, this administrative role seems to be more autonomous or independent from outside expectations and driven mainly by internal working requirements. The influence of internal staff secretariats (type 3) on relevant IO and UN networks or respective ongoing negotiations is formally much weaker than in cases of type 1 and 2 secretariats (such as the Bonn Challenge, PMR or SCF), where the same people physically take part in different settings. Figure 1 covers the spectrum of different administrative roles as identified in the eight forest climate governance cases. As mentioned earlier, each of the three smaller circles represents one role type (here: International Bureaucracies, International Organizations, Internal Staff) and the outer circles reflect the underlying (common) rationale (e.g., diplomatic vs. autonomous).

Secretariats with strong ties to external UN or IO organizations hold additional accountability relations, in part because of the need for reporting and information exchange with their home institutions (on working tasks, hours and respective funding). Such information however, is often made public, which increases the overall transparency of the setting. In a case such as the REDD+ Partnership, such an accountability relationship may become fairly complex. Its secretariat partly drew on staff from the UN-REDD Program, which itself is composed of three different international bureaucracies, UNDP, FAO and UNEP. Every UN organization has its own disclosure policy [58]. International organizations such as the World Bank, which administers the PMR, are also bound to formally codified transparency and reporting rules which assure accountability—at least internally. The UN is, in many cases, more accepted than the World Bank, especially among emerging economies which perceive the World Bank as an institution that serves primarily the interests of rich countries [63,64]. Even if the World Bank is just administering, it may affect the relationship between different constituencies in a setting and consequently hinder developing countries to participate at eye level (Expert 4, 2016).

Since organizations such as the Gold Standard or the Green Belt Movement are, by contrast, administered by their internal staff, such excessive external accountability relations are not a significant issue and typical employer–employee rules and legislation apply. For external communication, however, introducing the whole staff on their publicly accessible homepages is common sense and understandable if one considers that the administrative roles often include dealing with personal data or funding. By contrast, in type 1, where IO staff are concerned, such public presentation is less common; often to protect these authoritative people from external influence.

Operational decision-making appears to resemble administrative roles in terms of the day-to-day work they engage in, for instance as chairs of meetings that are concerned with the daily procedural decisions about how, when, and where to meet and what topics to set on the agenda. These seem like relatively harmless activities at first sight; but since such decisions, in fact, restrict certain (alternative) topics from being set on the agenda as much as they exclude certain stakeholders from contributing to decisions, this role is potentially highly authoritative. We identified four types of operational roles: (1) delegates, (2) administrative staff, (3) network facilitator and (4) stakeholders; these are visualized in Figure 2. Subtypes represented in the "bubbles" reflect the reasons for why a specific role was employed, in this case especially regarding expectations about professional skills. Moreover, we identified three common or distinct rationales tied to different operational decision subtypes: "closed shop", "nodal" or "buy in".

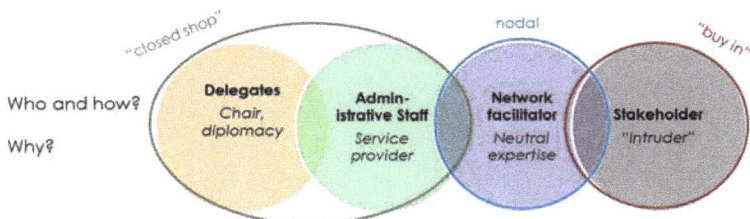

Figure 2. Operational roles.

The first subtype of an operational role is that of *official delegates* (1), i.e., officially entitled state representatives which hold (co)chair positions. This role is typically tied to a whole set of diplomatic rules of the game that were often lent from other official (UN) arenas for defining how to organize meetings etc. The co-chair role in these cases, as in that of the SCF or the REDD+ Partnership, is in fact fairly powerful and occasionally it happened that relevant stakeholders from within and outside the setting questioned the legitimacy of the process because co-chairs used their authority for promoting own political preferences (own observations, 2010–2014). At the forum on forest finance of SCF, stakeholder participants were given the impression that they are not equally welcome and that those present were the usual suspects which could actually not provide the new information hoped for. This is why a number of participants also showed disinterest and left the event earlier (Expert 1, 2015). The driving cause appeared to be the proximity to the politically charged UN process and that some co-chairs simply could not avoid the negotiation modus when setting the agenda and tried to shape the discussions while avoiding alternative views to be taken into consideration ("closed shop") (Expert 1, 2015).

On the other side, procedural roles of such settings that are close to international arenas are often also bound to make activities public on their website and to specify working procedures respectively, e.g., sharing work plans or progress reports—apart from exceptional cases of confidentiality. An example would again be the Standing Committee on Finance, which is administered by the UNFCCC secretary and co-chaired by two UNFCCC member state representatives. The meetings are videotaped and available online. A second type can be illustrated by the case of the Bonn Challenge, where administrative staff of a small team within the IUCN secretariat makes these day-to-day decisions. The role resembles that of a service-provider. This role also follows a "closed shop" rationale (though with less political charging), where only few people take potentially important operational decisions (see Figure 2). A third variant is when a neutral (3) *network facilitator* or mediator is employed to ensure a continuous communication process within a given network. This role appears to follow a "nodal" or mediating rationale as observed in the case of the ASEAN Clearing House Mechanism, where the idea was pursued that a neutral moderator with sufficient time resources and networking skills could be decisive for proper knowledge exchange in the rather political network (Expert 3, 2014). A fourth type of procedural role is that where all sorts of *stakeholders* (4) may actually "buy in" to procedural operations and contribute in administering the governance network if they wish to do so. An example of such a highly open setting that allows all more or less relevant actors to participate (intrude), is RSPO, where members can apply for such administrative roles given they are financed by their home organization, which may in fact favor members from large organizations while smaller ones do not have these capacities (Expert 8, 2016). We see the tendency for independent or self-administering governance settings to have as restrictive transparency rules as most of the international settings have. Dependent on what different audiences (civil society or transnational governance communities) expect and demand via diverse communication channels, they are obliged to justify their operations and make (most of) their decisions openly accessible [65,66]

Among the four types of roles, we find both rather hierarchical top-down roles (type 1 and 2) as well as more horizontally oriented roles of network facilitator or stakeholders settings. This mirrors the debate about the classical trade-off between inclusiveness and efficiency. However, the input legitimacy in settings with network facilitators and stakeholders involved might be significantly higher than in the case of delegates and administrative staff.

5.2. Financial Roles in REDD+ Governance

In the eight cases, we identified five different types of finance coming either from (1) multilateral funds, (2) international organizations, (3) nation states, (4) own funds, or (5) donations, each representing different accountability communities. Figure 3 gives an overview of the spectrum of different financial roles in terms of who provides the funding. The two wider circles illustrate how the funding relationship affects the work of the governance setting and how the role affects the possibilities to spend the money; either conditionally or autonomously. The words in italics below refer to the question of why these roles exist and what kind of relationship and what expectations are bound to them.

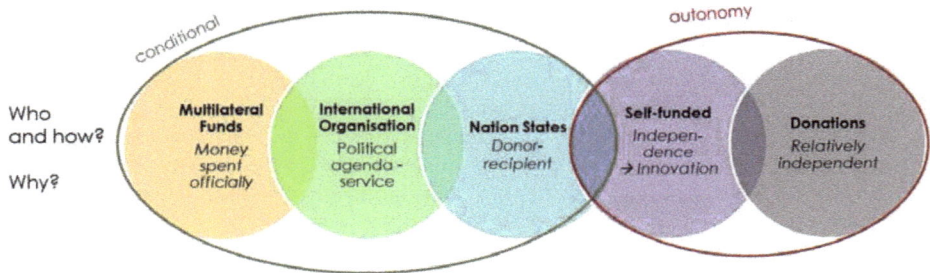

Figure 3. Financial roles.

Multilateral funds typically expect that money is spent according to official requirements and formal diplomatic standards. This accounts e.g., for the Standing Committee on Finance, which has to consider UN rates, expected professional standards for venues and technical support, as well as documentation requirements if it organizes a forum meeting [67].

If the funder is an international organization, such as the World Bank, and the governance setting is in the role of a service provider, it will often be urged to follow a specific political agenda (Expert 5, 2015). If states finance a setting, accountability constellations typically evolve from development cooperation between donor and recipient. An example may be the German Government in the Bonn Challenge, the German Development Agency GIZ within the ASEAN CHM or the contributing states within PMR. Contrarily, Bonn Challenge funds, that heavily draw on German support, are coined mainly for administrative staff, public relations or the regular ministerial round tables and have no link to actual implementation as pledges are defined and financed by partners. The contributing countries in the PMR had the possibility to effectuate the involvement of a preferred country, which they would like to support with the funds (Expert 4, 2016), showcasing that contributing countries typically expect something back for their money. In external funding arrangements such as this, freedom of decision may be sensitively constrained by the good will of funders. The two wider circles in Figure 3 with the catchwords "conditional" and "autonomous" illustrate this dynamic.

In contrast, self-funded settings such as the RSPO or the Gold Standard are principally more independent from such expectations, particularly if they offer their own products, ideas or services that are bought by customers. While this may principally mean a higher degree of freedom, it also means a higher dependency on market and customer demands (Expert 7, 2015). Besides the private market, international agreements such as the Paris Agreement or the Sustainable Development Goals also have the potential to create a demand for more sustainable and climate friendly products and services. To be responsive to both, the Gold Standard, for instance, has developed a third version of its standard and offers the certification of carbon offset projects that produce additional contributions to several Sustainable Development Goals related to biodiversity and poverty reduction [68]. In this way, they attract funding from multiple sources, such as the Luxemburgish Government, the World Bank or Goldman Sachs [68]. Governance settings that obtain donations, such as the Green Belt Movement, can act relatively independently, as long as they act according to their proclaimed vision, mission or values which typically serve as a source of motivation for donors to support them. Here, GBM is communicating its "love for environmental conservation, community empowerment, volunteerism and honesty" [69].

5.3. Strategic Roles in Forest Climate Governance

Strategic roles often defined the terms under which certain actors may take general, strategic decisions to guide the overall setting. They are also pointing at reasons why actors chose or are chosen to fill these roles. We identified four different strategic roles in decision-making among the eight cases: principals, experts, founders/individuals and stakeholders.

(1) Examples for the *principal* approach where authority derives from a ministerial mandate are the SCF or the PMR. Experienced state delegates involved in international forest climate negotiations come together in a smaller setting and try to synthesize and agree upon the strategic options (in line with ministerial agendas). Another way to take strategic decisions is to have (2) an *expert-based* advisory board such as in RSPO or the Gold Standard, where board members are elected or nominated based on their experience and the professional guidance they are able to provide. The Gold Standard's Board, for instance, does not rest in stakeholder membership but relies on experts from various sectors (civil society, business, science). The Board can consult the NGO supporter community if substantive change is desired; however, their decisions are instructed only by the Technical Advisory Committee [61]. A third type of strategic role rests in (3) the individual founders that initiated and spearhead the initiative. They gain authority from the charismatic leadership of the respective founder, such as Wangari Matthai in the Green Belt Movement or, to a smaller extent, the founding member of the Bonn Challenge, who was a renowned forest climate negotiator for many years. The Bonn Challenge also draws on celebrities, such as Bianca Jagger, to inspire further engagement. Particularly for the founding phases, individual leadership and charisma may be central in these cases to convince a critical amount of people about their ideas (Expert 5, 2015, Expert 6, 2016). Inspired by their commitment and visionary goals they may attract new members to follow. This "inspiring" spirit may also matter for decision-making processes beyond founding and enhance the overall acceptance of strategic top-down decisions (Expert 6, 2016). However, fulfilling this role requires consistency between decisions and commonly shared values about forest climate governance. A possible backside of such strategic role is that the setting depends on the permanent presence of the individual person. If they leave, instability of the setting might arise (Expert 6, 2016). Another possibility is that (4) *relevant stakeholders* exert strategic roles; possibly in conjunction with government representatives, as in the case of the REDD+ Partnership. Deriving legitimacy from the democratic inclusion of all stakeholders is also a concern for RSPO, which assures each of its seven stakeholder groups a proportional number of representatives within its Advisory Board. These Board members are democratically elected through the general assembly of the RSPO [70]. By contrast, the strategic role of stakeholders in the REDD+ Partnership was not fully achieved, mainly because they were excluded from the respective initial meetings where heated discussions among country partners about whether or not to include non-state actors at all or for specific sensitive topics (e.g., finance) were still ongoing [11,62]. Figure 4 visualizes the four role types including the main reason why a role is carried by an actor.

Figure 4. Strategic decision-making roles.

While the role of principals (type 1) is based on ministerial mandates, e.g., high-level public servant, experts (type 2) are elected or nominated because of the professional guidance they are expected to provide, e.g., in their role as technical advisory board members. Founders or individuals (type 3) carry their strategically important role due to their charismatic leadership. Lastly, stakeholders (type 4) themselves may take strategic decisions, e.g., in plenary discussion, mostly for the sake of ensuring that these decisions that concern everybody in the setting are democratic.

5.4. Consequential Roles in Forest Climate Governance

Decisions are consequential if they relate substantially, rather than only procedurally to a policy output, i.e., if they concern specific rules or methods for policy definition, implementation or evaluation. Such decisions may be a result of (1) a participatory process with *all members* of an organization (e.g., a General Assembly), such as in the cases of the REDD+ Partnership or RSPO, meant to increase the democratic legitimacy of these processes; (2) by scientific or professional *expert committees* such as in the case of the Technical Advisory Board of the Gold Standard (Expert 7, 2015). Moreover, such consequential decision-making roles may be filled (3) by *individuals* (esp. founders) such as in the case of the Green Belt Movement which is based on charismatic leadership. A fourth type is 4) decentralized decision-making that draws on a flexible, self-determined and hence potentially highly inclusive bottom-up approach such as in the case of the Bonn Challenge. Here, countries or regions autonomously decide on the ways of designing, implementing and evaluating *their* reforestation projects (own observations, 2015–2017). Figure 5 provides an overview of the four role types described above.

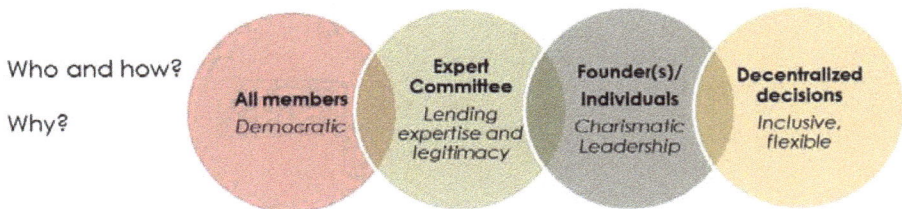

Figure 5. Consequential roles in decision-making.

When assessing the legitimacy of policies, assigning decision-oriented roles to a range of plural, potentially concerned stakeholders—as in type 1—can be critical for increasing the democratic legitimacy of decisions. Although decisions were of a rather information-based nature in the exchange platform REDD+ Partnership, meaningful stakeholder inclusion in meetings was a rule rather than the exception in various events and meetings. This was especially true in the last years of its existence and after the role of stakeholders has been formally accepted by all partners (own observations 2010–2014 [11,71]).

Also within the RSPO, consequential decisions are decided upon in a General Assembly (type 1). However, in terms of representation, economic actors significantly outnumber NGO representatives that make up only three per cent of the ordinary members [70]. Overall, the plurality of views present in the General Assembly (the central decision-making body) paralyzes rather than eases efficient decision-making. In the context of defining sustainability criteria, although principally supported and taken more seriously, the process become languid as discussions were shifted to smaller and potentially more efficient working groups. However, these have not come up with concrete solutions either, even after years (Expert 8, 2016). Because no other voting rules have ever been defined, consensus-based decisions are also the rule in the PMR, where all members jointly take operational decisions (type 1). In fact, in the PMR, a fairly homogenous group of public servants from different ministries meets in a comparably small setting moderated by World Bank staff. Lively discussions typically arise about how to distribute the money; such as whether to spend it on new members or to support existing members a second time (Expert 4, 2016). Despite the consensus rule, donor countries in fact hold a major share in authority, given that they can always bail out if they are not happy with the result (ibid.).

5.5. Expert Roles in Forest Climate Governance

As outlined above, expert roles matter for the output dimension of legitimacy. Despite the predominant model of neutral scientific and technical knowledge, expertise is hardly non-political and ways of selecting, framing or presenting knowledge is shaped by the beliefs and world views of so called experts. In our eight cases, we identified four types of roles concerned with knowledge delivery: (1) Delegates, (2) International Organizations and Bureaucracies, (3) Multiple Stakeholders and (4) Knowledge Partners which will be elaborated on in the following paragraph:

In the first sub-variant, (1) official *delegates* provide administrative and technical knowledge. Cases such as the REDD+ Partnership or the Standing Committee on Finance demonstrate that expectations about expertise in these settings can mean that only experienced diplomatic experts are eligible to assure political acceptance. Stakeholders within the REDD+ Partnership, for example, need an official accreditation to the adjacent official UNFCCC process as a prerequisite for participating in the discussions. That states' provision of official data about their territories and natural resources is seen as critical for ensuring political credibility (own observations 2010–2014, [11]). In other cases, expertise carried a more bureaucratic notion. (2) *International organizations* and *bureaucracies* which are experienced in diplomacy, such as experts from GPFLR/IUCN, may also be seen as trusted providers of knowledge, e.g., of spatial data for reforestation options or for organizing capacity building workshops as in the case of the Bonn Challenge (Expert 5, 2015) or technical assistance by the World Bank's operational country units in the PMR) [53].

Another possibility is to involve (3) *multiple stakeholders* purposefully for the sake that they would actually provide a valuable and diverse basis of knowledge which may allow for synergies and knowledge exchange in communication processes (Expert 3, 2014). Such a more practice oriented approach to expertise ("those who do, know best") that draws on different perspectives and experiences with a specific policy problem is the overall rationale, for instance, in the RSPO, the Gold Standard, the ASEAN CHM or the roundtable meetings of the Bonn Challenge. A fourth variant is more oriented toward pioneers where singular (4) *knowledge partners* acting as role models by providing technical advice on their leading example and respective experiences. The idea is to stipulate learning from actors that are seen as—presumably—very successfully working on relevant policies or actions. The Forest Stewardship Council is such a partner organization for the Gold Standard in the context of sustainable forestry projects that mitigate carbon emissions (Expert 7, 2015). The state of Quebec and the state of California play such a role within PMR as regions that are early movers in implementing carbon pricing mechanisms on a sub-national level [72]. Expectations behind such an exclusive role target policy diffusion and emulation of specific policy options rather than new (own) approaches or learning from a range of alternative approaches.

Overall, type 1 and 2 are closely related to expectations of receiving "official" and diplomatically valid input which may not be a surprise since, in our cases, the respective organizations also operated on a high political level where political experience or administrative, technical knowledge increased their credibility. By contrast, type 3 and 4 are meant to deliver practical personal or technical experience which is suggested to be rather related to implementation "on the ground" (e.g., from pilot projects or a leading region). Apart from bringing up best practices, this approach, in principle, may also allow critical voices to be heard more directly (e.g., discrimination of social groups or legal obstacles) as compared to type 1 or 2 where such concerns would have to attract the *official* expert's attention or "good" will.

Figure 6 displays the broader spectrum of different types of expert roles, visualizing (in bold) who is considered an expert and the respective qualitative features tied to these roles (outer circles). Accordingly, actors may be considered as experts based on their political experience (in italics, type 1), administrative (type 2) technical knowledge (type 2 and 4) or based on the different perspectives they bring (type 3).

Figure 6. Expert roles.

6. Discussion and Conclusions

Our paper departed from the observation that polycentrism in the global governance of problems—such as mitigating climate change in forests, where multiple actors take on varied governance functions—poses a severe challenge for evaluating the performance of the respective settings, due to increasingly blurred accountability relations in hybrid actor constellations [10,14,18,73–75]. Drawing on dominant institutional and social-constructivist accounts in the legitimacy literature that either stress the role of specific institutional arrangements for achieving input or output legitimacy (e.g., participatory structures) or interpersonal processes of expectations about what activity is acceptable to concerned communities, we proposed the sociological concept of social roles as an analytical vehicle for analyzing governance performance. We systematically investigated eight distinct forest climate governance settings regarding how different potentially authoritative (and hence accountable) roles related to governance purposes such as informing, financing, (co)deciding or helping implementing forest climate action, which appear harmless or neutral at first glance, but may have decisive repercussions on how things are done on the ground [17,22].

We identified several distinct variants within six broader types of relevant actor roles in governance—bureaucratic, operational, financial, strategic and consequential decision-making and expert roles—which have potential implications on the input or output dimension of legitimacy, or both. In our analysis, we compared which different actors filled a specific role and looked at related expectations regarding why exactly specific actors were preferred in a particular role or in the specific ways of engagement (how). Overall, distinct role patterns became visible when distinguishing different sub-variants of actors in the eight cases.

Most prominently, we found several roles across all dimensions that had a fairly formalized notion and respective diplomatic expectations as to what the right actor was and regarding the right way of engaging actors in a setting. This rationale was typical for settings with a high proximity to the politically charged UN process (such as SCF, REDD+ Partnership, PMR, and ASEAN CHM, not always across all dimensions). In the distinct set of *official* roles (e.g., international bureaucracies or delegates), concerns about diplomatic or technical/scientific correctness were central for the choice of specific actors and modes of engagement. Either internationally experienced actors, such as IOs, were preferred as they are experienced with diplomacy and deemed politically neutral, or officially legitimized actors such as country delegates filled the role, because they represent a country politically—as compared to some randomly assigned stakeholder. The respective approach to specific governance roles in administration, decision-making or knowledge making appeared to often result in an exclusively closed shop; inclusive only for the assorted set of official actors, as in the case of the SCF Forum.

In contrast, alternative rationales for the selection of specific actors or ways of their engagement were guiding the role approaches seen in cases that were not as close to or rather independent from the international climate negotiations under UNFCCC (such as GBM, Gold Standard, RSPO or Bonn Challenge). The respective "casting" rationales behind different roles were often related to ideas about facilitating the process in technical terms, allowing for more efficiency, autonomy, or effective implementation as well as lending a certain democratic quality to the overall operations. Here, one prominent concern was, for instance, to achieve a high level of *inclusivity*, e.g., by allowing stakeholders to "buy in" to operational tasks (if they could afford it) or by drawing on multiple stakeholders as experts or decision takers (e.g., RSPO, exceptionally also REDD+ Partnership). Giving relevant actors principally the right and opportunities to take on roles in governance settings can help to foster a sense of ownership for the respective decisions tied to that role and hence contribute to a higher actor's identification with the new setting. Autonomy, which means *political independence* and potentially allows experimenting with new approaches was another prominent motive for defining and filling roles, for instance, when drawing on rather unconditional schemes of funding (self-funding, donation) and internal administrative staff, or when engaging neutral mediators in the operations of the organization. At least in financial terms, accountability concerns are of minor importance in mainly self-financed settings such as the RSPO [70] or Gold Standard [68], which can be a blessing (for the organization) or curse (for those that wish to hold these actors accountable). In a few instances, these alternative cases drew on fairly exclusive leadership approaches when, for instance, charismatic founders of the organization made major strategic or consequential decisions (as in GBM or the Bonn Challenge) or when representatives from pioneering regions informed the knowledge creation process in the PMR—which from a legitimacy perspective, may require scrutiny.

An interesting deviant for legitimacy considerations is also the spatial distance and dispersion of authority in the case of the Bonn Challenge where implementation-related decisions were actually made outside of the setting and on the ground in the country context, either by national or sub-national entities or private actors. Authoritative discretion and with that respective legitimacy concerns appear deliberately externalized in this setting. The case also illustrates how the mix of different approaches to roles (charismatic leader, professional bureaucratic staff, decentralized decisions) across different dimensions can lead to less frustration and resistance, at least about what is done transnationally between the partners. Nonetheless, such approaches also require a certain level of trust in the implementation commitment of contributing partners.

Moreover, while knowledge production and expertise are theoretically often related to output oriented aspects of legitimacy, expertise rationales in our eight cases carried both input or output-related notions as well as a mix of both. The output dimension is highlighted whenever experts are pinpointed as providing a credible source of technical knowledge (e.g., IOs, pilot/pioneer knowledge). In contrast, an approach that draws on *on the ground* knowledge from multiple sources stresses the very ways of how alternative views of (local) stakeholders are represented (input). When involving project managers in the GBM or the Gold Standard, the intention appeared to be to satisfy both rationales.

In our smaller sample of eight cases, we could explore how formal (institutional) and interpersonal aspects (mutual expectations) form two complementary parts that are critical for governance roles and their performance. Specifically, our empirical data suggests that performance is at risk whenever the formal (institutions) and informal (expectations) components of roles become incongruent. For instance, in the case of the REDD+ Partnership where rules about working operations (esp. co-chairs roles) were hardly explicit and informal practices contradicted the ideas of the founding fathers (own observation 2010–2014, [11]).

Generally, the question of *who*, i.e., which concrete actor group, filled a role appeared relevant in several cases, but was far from deterministic. Although the nature of an actor does not a priori predefine performance, it mattered for relevant others who filled a specific role—at least in some cases. Secretariats that draw on staff from international organizations were expected (and evaluated) to better fit the diplomatic exercises and respective confidentiality expectations. This was a concern especially in settings with a high ratio of state representatives. With this focus on (external) actors that fill roles within a setting, the concept covers essential interpersonal linkages between distinct governance settings and allows—to a certain extent—to reconstruct pathways of mutual influence. These external ties were principally possible across all governance dimensions (administration, decisions, finance, expertise) because actors from other institutional realms could be engaged anywhere. Principally, however, our typology mainly focused on processes within a certain setting as on specific types of roles within particular governance dimensions, for which we identified different practical variants and underlying rationales. While we purposefully blinded out how different roles eventually play together, including whether trade-offs or mutual synergies exist or whether, for instance, it is advantageous or disadvantageous to have the same actor to fulfill different roles (e.g., checks and balances concerns), the role concept principally allows for such questions and may support respective analyses. These aspects, we believe, are central and should inform further research on social roles in forest climate governance and respective legitimacy considerations.

Overall, our analysis offers a more systematic account of the variation in different governance roles. Despite our narrow focus, we believe that different subtypes help to display and reflect different rationales or predominant expectations tied to these positions, while also covering institutional aspects of this position, e.g., predominant normative expectations about specific actor types. What the typology itself does not perfectly reflect are the contradictory expectations that are tied to a specific role, something that, however, is revealed empirically in the process of making use of and applying the typology to concrete cases. The same holds true for processes of social change, e.g., as regards the expectations that are tied to specific roles but may change as the actors learn by doing.

The described variety in types of actor roles—in only eight cases—vividly illustrates the many potentially contradictory expectations that governance settings may be confronted with internally and in relation with external actors. In practice, governance operations hardly ever please all expectations, but they may find better or worse strategies to outbalance and minimise the trade-offs between them. The Gold Standard is a telling example for how a setting tries to, and largely succeeded, in pleasing many sides; including expectations from civil society, specific states, markets and the UNFCCC. This case, together with the GBM, strongly deviated from the more state-led processes on almost all dimensions. Especially financial independence lends a fairly high *degree of autonomy* to such non-state actor based settings and with that power. The question then is, whether with this autonomy a respective demand for more democratic accountability arises or not [76].

In light of the insights deriving from this fairly exploratory study on roles in forest climate governance, which offers indeed no fully fleshed assessment of legitimacy for each governance setting, we believe that social roles may act as a useful analytical concept to draw on. Conceptually, we see it as highly appealing that roles have both institutional and social constructivist components, which allows linking and thus mediating between two valid claims in the legitimacy literature. Methodologically, the concept offers a handy analytical tool for empirically operationalizing governance performance for all sorts of mixed governance settings that integrate non-state actors to varying degrees and in varying positions. In this context, the role perspective has proven fairly compatible with the input and output dimensions of the legitimacy concept and hence with claims that stress the need for democratic legitimacy based on inclusive decision-making, accountability and deliberative processes [77] or that highlight the importance of problem-solving effectiveness [28,32,74] as tied to specific institutional settings and social interactions (expertise) within the setting.

Obviously, a role approach, as such, may not suffice for assessing legitimacy entirely, but the typology may still offer a useful analytical framework that any such legitimacy assessments could draw on and employ for deconstructing a variety of rather complex polycentric governance arrangements in a way that it allows such assessment to be more easily employed. A more nuanced typology, as developed in this paper, does not only illustrate the grand spectrum of possible roles that are more or less authoritative in transnational governance settings, but in research practice may in fact help identifying legitimacy-related issues, e.g., where roles appear harmless at first sight but are revealed as deserving further analytical attention. For instance, it can help to question the influence and the agendas of home organizations regarding ties to specific principals behind actors in administration, decisions or finance that might not be obvious at first. Also, in practical terms, our study results offer some reasonably useful insights. It illustrates that if new governance settings are founded, involved actors often bring in their previously developed concepts about roles (incl. working culture, values and administrative routines) which may not necessarily correspond with what the founders envisioned with the foundation in the first place. To avoid tensions at later phases, and for more clarity, mutual expectations and common conceptualizations of roles should ideally be iteratively discussed and communicated, especially during but also beyond the founding phase of the setting.

Acknowledgments: This study builds on the work in the research project "Integration of REDD? into a post-2020 climate agreement and linkages to the CBD". This project is financially supported by the German Federal Agency for Nature Conservation (BfN), with funds from the German Federal Ministry for the Environment, Nature Conservation and Nuclear Safety (BMUB). Moreover, Mareike Blum is financially supported by a scholarship of the German Federal Environmental Foundation (DBU). Special gratitude is dedicated to the three anonymous reviewers for very constructive and helpful feedback. The article processing charge was funded by the German Research Foundation (DFG) and the University of Freiburg in the funding programme Open Access Publishing.

Author Contributions: This paper is based on team work in which both authors contributed with fairly equal responsibilities esp. in the theoretical framework, research design and conclusions. Mareike Blum has a major share in drafting the case descriptions and conclusions and in collecting, analyzing and interpreting the empirical data. As corresponding author, she fulfilled major editing tasks. Dr. Sabine Reinecke was pursuing the overall research idea and was responsible for the introduction, respectively. During revision, she took a leading role for aligning theory, empirical insights and drawn conclusions.

Conflicts of Interest: The authors have declared no conflict of interest.

References

1. Adger, W.N.; Lorenzoni, I.; O'Brien, K.L. *Adapting to Climate Change: Thresholds, Values, Governance*; Cambridge University Press: Cambridge, UK, 2009.
2. IPCC. *Climate Change 2007. Synthesis Report. Contribution of Working Groups I, II and III to the Fourth Assessment Report of the Intergovernmental Panel on Climate Change*; IPCC: Geneva, Switzerland, 2007.
3. UNEP. *The Emissions Gap Report 2013*; United Nations Environment Programme (UNEP): Nairobi, Kenya, 2013.
4. Stern, N. The Economics of Climate Change. *Am. Econ. Rev.* **2008**, *98*, 1–37. [CrossRef]
5. Pan, Y.; Birdsey, R.A.; Fang, J.; Houghton, R.; Kauppi, P.E.; Kurz, W.A.; Ciais, P. A Large and Persistent Carbon Sink in the World's Forests. *Science* **2011**, *333*, 988–993. [CrossRef] [PubMed]
6. IPCC. *Climate Change 2014: Synthesis Report. Contribution of Working Groups I, II and III to the Fifth Assessment Report of the Intergovernmental Panel on Climate Change*; IPCC: Geneva, Switzerland, 2014.
7. IPCC. *Climate Change 2007: The Physical Science Basis. Summary for Policymakers. Contribution of Working Group I to the Fourth Assessment Report of the Intergovernmental Panel on Climate Change*; IPCC: Geneva, Switzerland, 2007.
8. IPCC. *Working Group III Contribution to the Intergovernmental Panel on Climate Change Fourth Assessment Report. Climate Change 2007: Mitigation of Climate Change. Summary for Policymakers*; IPCC: Geneva, Switzerland, 2007.
9. PARIS AGREEMENT, 2015. Available online: https://unfccc.int/resource/docs/2015/cop21/eng/l09r01.pdf (accessed on 15 July 2016).
10. Corbera, E.; Schroeder, H. Governing and implementing REDD+. *Environ. Sci. Policy* **2011**, *14*, 89–99. [CrossRef]
11. Reinecke, S.; Pistorius, T.; Pregernig, M. UNFCCC and the REDD+ Partnership from a networked governance perspective. *Environ. Sci. Policy* **2014**, *35*, 30–39. [CrossRef]
12. Lemos, C.M.; Agrawal, A. Environmental Governance. *Annu. Rev. Environ. Resour.* **2006**, *31*, 297–325. [CrossRef]
13. Metlay, D.; Sarewitz, D. Decision strategies for Addressing Complex, "Messy" Problems. *Bridge J. Nat. Acad. Eng.* **2012**, *42*, 6–16.
14. Kjaer, A.M. *Governance: Polity*; Cambridge University Press: Cambridge, UK, 2004.
15. Marks, G.; Hooghe, L. *Contrasting Visions of Multi-Level Governance in Multi-Level Governance*; Bache, I., Flinders, M.V., Eds.; Oxford University Press: Oxford, UK, 2004.

16. Ostrom, E. A Multi-Scale Approach to Coping with Climate Change and Other Collective Action Problems. *Solutions* **2010**, *1*, 27–36.

17. Börzel, T.; Risse, T. Public-private partnerships: Effective and legitimate tools of international governance? In *Complex Sovereignty. Reconstituting Political Authority in the Twenty-First Century*; Grande, E., Pauly, L.W., Eds.; University of Toronto Press: Toronto, ON, Canada, 2005.

18. Burris, S.; Kempa, M.; Shearing, C. Changes in Governance: A Cross-Disciplinary Review of Current Scholarship. *Akron Law Rev.* **2008**, *41*, 1–66.

19. Bell, S. Do we really need a new "Constructivist Institutionalism" to explain institutional change? *Br. J. Political Sci.* **2011**, *41*, 883–906. [CrossRef]

20. Held, D. Regulating Globalization? The Reinvention of Politics. *Int. Sociol.* **2000**, *15*, 394–408. [CrossRef]

21. Kooiman, J. *Modern Governance: New Government-Society Interactions*; Sage: London, UK, 1993.

22. Abbott, K.W. The Transnational Regime Complex for Climate Change. *Environ. Plan. C Gov. Policy* **2012**, *30*, 571–590. [CrossRef]

23. Lövbrand, E.; Stripple, J. Making climate change governable: Accounting for carbon as sinks, credits and personal budgets. *Crit. Policy Stud.* **2011**, *5*, 187–200. [CrossRef]

24. Miller, P.; Rose, N. *Governing the Present*; Polity Press: Cambridge, UK, 2008.

25. Stoker, G. Governance as theory: Five propositions. *Int. Soc. Sci. J.* **1998**, *50*, 17–28. [CrossRef]

26. Biermann, F.; Siebenhühner, B. *Managers of Global Change: The Influence of International Environmental Bureaucracies*; MIT Press: London, UK, 2009; pp. 8–9.

27. Scharpf, F.W. *Regieren in Europa: Effektiv und demokratisch?* Campus Verlag: Frankfurt, NY, USA, 1999.

28. Lederer, M. From CDM to REDD+—What do we know for setting up effective and legitimate carbon governance? *Ecol. Econ.* **2011**, *70*, 1900–1907. [CrossRef]

29. Gupta, A.; Mason, M. *Transparency in Global Environmental Governance: Critical Perspectives*; The MIT Press: Cambridge, MT, USA, 2014.

30. Keohane, R.O. Accountability in World Politics. *Scand. Political Stud.* **2006**, *29*, 75–87. [CrossRef]

31. Scharpf, F.W. *Problem Solving Effectiveness and Democratic Accountability in the EU*; Political Science Series of Max Planck Institute for the Study of Societies: Vienna, Austria, 2006; pp. 1–2.

32. Mayntz, R. Legitimacy and Compliance in Transnational Governance. MPIfG - Max-Planck-Institut für Gesellschaftsforschung. Available online: http://edoc.vifapol.de/opus/volltexte/2011/3011/ (accessed on 14 February 2017).

33. Green, J.F. *Rethinking Private Authority: Agents and Entrepreneurs in Global Environmental Governance*; Princeton University Press: Princeton, NJ, USA, 2013.

34. Lövbrand, E. Revisiting the politics of expertise in light of the Kyoto negotiations on land use change and forestry. *For. Policy Econ.* **2009**, *11*, 404–412. [CrossRef]

35. Steffek, J. Discursive legitimation in environmental governance. *For. Policy Econ.* **2009**, *11*, 313–318. [CrossRef]

36. Williams, M. Citizenship as Agency within Communities of Shared Fate. In *Unsettled Legitimacy: Political Community, Power, and Authority in a Global Era*; Bernstein, S., Coleman, W.D., Eds.; University of British Columbia Press: Vancouver, BC, Canada, 2010; pp. 33–52.

37. Bernstein, S. Legitimacy in Global Environmental Governance. *J. Int. L. Int. Relat.* **2005**, *1*, 139–166.

38. Bernstein, S. Legitimacy in intergovernmental and non-state global governance. *Rev. Int. Political Econ.* **2011**, *18*, 17–51. [CrossRef]

39. Bernstein, S.; Cashore, B. Can non-state global governance be legitimate? An analytical framework. *Regul. Gov.* **2007**, *1*, 347–371. [CrossRef]

40. Bernstein, S.; Cashore, B. Nonstate global governance: Is forest certification a legitimate alternative to a global forest convention? In *Hard Choices, Soft Law: Voluntary Standards in Global Trade, Environment and Social Governance*; Kirton, J.J., Trebilcock, M.J., Eds.; Routledge: London, UK, 2004; pp. 33–64.

41. Esteban de la Plaza, C.; Visseren-Hamakers, I.J.; de Jong, W. The legitimacy of certification standards in climate change governance. *Sustain. Dev.* **2014**, *22*, 420–432. [CrossRef]

42. Biddle, B.J. Recent development in role theory. *Ann. Rev. Sociol.* **1986**, *12*, 67–92. [CrossRef]

43. Stout, M. *Logics of Legitimacy: Three Traditions of Public Administration Praxis*; CRC Press: Boca Raton, FL, USA, 2013.

44. Selden, S.C.; Brewer, G.A.; Brudney, J.L. Reconciling competing values in public administration: Understanding the administrative role concept. *Adm. Soc.* **1999**, *31*, 171–204. [CrossRef]

45. Hale, T.N.; Mauzerall, D.L. Thinking globally and acting locally: Can the Johannesburg partnerships coordinate action on sustainable development? *J. Environ. Dev.* **2004**, *13*, 220–239. [CrossRef]
46. Bovens, M. Analysing and assessing accountability: A conceptual framework. *Eur. Law J.* **2007**, *13*, 447–468. [CrossRef]
47. Degnbol-Martinussen, J.; Engberg-Pedersen, P. *Aid: Understanding International Development Cooperation*; Zed Books: London, UK, 2003.
48. Goodstein, J.; Boeker, W. Turbulence at the Top: A New Perspective on Governance Structure Changes and Strategic Change. *Acad. Manag. J.* **1991**, *34*, 306–330. [CrossRef]
49. Forbes, D.P.; Milliken, F.J. Cognition and Corporate Governance: Understanding Boards of Directors as Strategic Decision-Making Groups. *Acad. Manag. Rev.* **1999**, *24*, 489–505.
50. Elbanna, S. Strategic decision-making: Process perspectives. *Int. J. Manag. Rev.* **2006**, *8*, 1–20. [CrossRef]
51. Haas, P.M. Introduction: Epistemic communities and international policy coordination. *Int. Org.* **1992**, *46*, 1–35. [CrossRef]
52. Moravcsik, A. Is there a "Democratic Deficit" in world politics? A framework for analysis. *Gov. Oppos.* **2004**, *39*, 336–363. [CrossRef]
53. Head, B.W. Wicked problems in public policy. *Public Policy* **2008**, *3*, 101–118.
54. Cash, D.; Clark, W. From Science to Policy: Assessing the Assessment Process. Faculty Research Working Papers Series (November 2001), John F. Kennedy School of Government, Harvard University. Available online: http://dx.doi.org/10.2139/ssrn.295570 (accessed on 14 February 2017).
55. Hunt, J.; Shackley, S. Reconceiving Science and Policy: Academic, Fiducial and Bureaucratic Knowledge. *Minerva* **1999**, *37*, 141–164. [CrossRef]
56. Van der Sluijs, J. Uncertainty as a monster in the science-policy interface: Four coping strategies. *Water Sci. Technol.* **2005**, *52*, 87–92.
57. Association of Southeast Asian Nations (ASEAN) Clearing House Meachnism. Available online: http://www.aseanforest-chm.org/asean-forest-clearing-house-mechanism-chm/ (accessed on 31 October 2016).
58. Goehler, D.; Schwaab, J. Managing Knowledge and Regional Policy Advice: ASEAN Forest Clearing House Mechanism. Available online: http://forest-chm.asean.org/document_center/afcc_fs/managing_knowledge_and_regional_policy_advice_asean_forest_clearing_house_mechanism.html (accessed on 14 February 2017).
59. PMR Website. Available online: https://www.thepmr.org/ (accessed on 30 January 2017).
60. Bonn Challenge. Available online: http://www.bonnchallenge.org/content/challenge (accessed on 31 October 2016).
61. Gold Standard Website. Available online: http://www.goldstandard.org/ (accessed on 31 January 2017).
62. RSPO Website. Available online: http://www.rspo.org/about (accessed on 31 October 2016).
63. Karsenty, A. The World Bank's endeavours to reform the forest concessions' regime in Central Africa: Lessons from 25 years of efforts. *Int. For. Rev.* **2016**, *18*, 1–16.
64. Ongolo, S.; Karsenty, A. The politics of forestland use in a cunning government: Lessons for contemporary forest governance reforms. *Int. For.* **2015**, *17*, 195–209. [CrossRef]
65. RSPO Website. Available online: http://www.rspo.org/key-documents/supplementary-materials/minutes-reports-of-rspo-ga-ega (accessed on 28 January 2017).
66. Gold Standard Website. Available online: http://www.goldstandard.org/our-work/innovations-consultations (accessed on 28 January 2017).
67. United Nations Website. What is Common System? Available online: http://www.un.org/Depts/OHRM/salaries_allowances/common.htm (accessed on 14 February 2017).
68. Gold Standard Website 2017. Available online: http://www.goldstandard.org/resources/faqs (accessed on 28 January 2017).
69. GBM Website 2016. Available online: http://www.greenbeltmovement.org (accessed on 31 October 2016).
70. RSPO Website 2017. Available online: http://www.rspo.org/about/who-we-are/board-of-governors (accessed on 29 January 2017).
71. Pistorius, T.; Reinecke, S. The interim REDD + Partnership: Boost for biodiversity safeguards? *For. Policy Econ.* **2013**, *36*, 80–86. [CrossRef]
72. PMR Website 2017. Available online: https://www.thepmr.org/pmrimplements/2 (accessed on 30 January 2017).

73. Bäckstrand, K. Multi-stakeholder partnerships for sustainable development: Rethinking legitimacy, accountability and effectiveness. *Eur. Environ.* **2006**, *16*, 290–306. [CrossRef]

74. Bäckstrand, K.; Khan, J.; Kronsell, A.; Lövbrand, E. *Environmental Politics and Deliberative Democracy: Examining the Promise of New Modes of Governance*; Edward Elgar Publishing: Cheltenham, UK, 2010.

75. Hajer, M.A. Policy without polity? Policy analysis and the institutional void. *Policy Sci.* **2003**, *36*, 175–195. [CrossRef]

76. Uhlin, A. Democratic Legitimacy of Transnational Actors: Mapping Out the Conceptual Terrain. In *Legitimacy Beyond the State? Re-examining the Democratic Credentials of Transnational Actors*; Erman, E., Uhlin, A., Eds.; Chippenham and Estbourne: Palgrave, Macmillan, 2010; pp. 16–37.

77. Dingwerth, K. Private Transnational Governance and Its Democratic Legitimacy (Transformations of the State). In *The New Transnationalism: Transnational Governance and Democratic Legitimacy*; Basingstoke: Palgrave, Macmillan, 2007.

Article

Using REDD+ Policy to Facilitate Climate Adaptation at the Local Level: Synergies and Challenges in Vietnam

Pamela McElwee [1,*], Van Hai Thi Nguyen [2], Dung Viet Nguyen [2], Nghi Huu Tran [3], Hue Van Thi Le [4], Tuyen Phuong Nghiem [4] and Huong Dieu Thi Vu [4]

[1] Department of Human Ecology, School of Environmental and Biological Sciences, Rutgers, The State University of New Jersey, New Brunswick, NJ 08901, USA
[2] People and Nature Reconciliation (PanNature), Hanoi 10000, Vietnam; van@nature.org.vn (V.H.T.N.); dungnv@nature.org.vn (D.V.N.)
[3] Tropenbos International Vietnam (TBI), Hue 530000, Vietnam; nghi@tropenbos.vn
[4] Centre for Natural Resources and Environmental Studies (CRES), Vietnam National University, Hanoi 10000, Vietnam; thivanhue@gmail.com (H.V.T.L.); tuyennghiem_cres@yahoo.com (T.P.N.); huongvudieu@yahoo.com (H.D.T.V.)
* Correspondence: pamela.mcelwee@rutgers.edu; Tel.: +1-480-252-0999

Academic Editors: Esteve Corbera and Heike Schroeder
Received: 1 November 2016; Accepted: 19 December 2016; Published: 24 December 2016

Abstract: Attention has recently been paid to how REDD+ mitigation policies are integrated into other sectoral policies, particularly those dealing with climate adaptation at the national level. But there is less understanding of how subnational policy and local projects are able to incorporate attention to adaptation; therefore, we use a case study in Vietnam to discuss how REDD+ projects and policies address both concerns of mitigation and adaptation together at subnational levels. Through stakeholder interviews, focus groups, and household surveys in three provinces of Vietnam with REDD+ activities, our research sought to understand if REDD+ policies and projects on the ground acknowledge that climate change is likely to impact forests and forest users; if this knowledge is built into REDD+ policy and activities; how households in forested areas subject to REDD+ policy are vulnerable to climate change; and how REDD+ activities can help or hinder needed adaptations. Our findings indicate that there continues to be a lack of coordination between mitigation and adaptation policies in Vietnam, particularly with regard to REDD+. Policies for forest-based climate mitigation at the national and subnational level, as well as site-based projects, have paid little attention to the adaptation needs of local communities, many of whom are already suffering from noticeable weather changes in their localities, and there is insufficient discussion of how REDD+ activities could facilitate increased resilience. While there were some implicit and coincidental adaptation benefits of some REDD+ activities, most studied projects and policies did not explicitly target their activities to focus on adaptation or resilience, and in at least one case, negative livelihood impacts that have increased household vulnerability to climate change were documented. Key barriers to integration were identified, such as sectoral specialization; a lack of attention in REDD+ projects to livelihoods; and inadequate support for ecosystem-based adaptation.

Keywords: REDD+; household livelihoods; climate adaptation; vulnerability; forest policy; land

1. Introduction

There has been increasing concern in recent years for the need to link climate mitigation and adaptation policies together, particularly with regard to forests [1,2]. Forest policies to respond to climate change often involve either mitigation actions, such as biological carbon sequestration, or adaptation actions, such as promoting resilience of ecosystems, but rarely are both considered together. Although the Intergovernmental Panel on Climate Change (IPCC) since 2007 has called for combined approaches, little has happened to facilitate mitigation and adaptation policies within the UN Framework Convention on Climate Change (UNFCCC), and individual country policies vary considerably in how much they integrate both approaches [3,4]. The IPCC defines adaptation as "adjustments in practices, processes, or structures" that can "moderate or offset the potential for damage or take advantage of opportunities created by a given change in climate" [4], and thus can encompass a wide range of potential policies for both forests and forest-using peoples.

Reducing Emissions from Deforestation and Degradation (REDD+) is the most well-known forest mitigation strategy to lower land-use generated greenhouse gas (GHG) emissions; the fundamental premise of REDD+ is that if households and governments are given payments and other types of rewards that equal or exceed what is earned from deforestation, then forests will be better protected, carbon emissions will be reduced, and these areas can serve as greater sinks for future GHG mitigation [5]. The rollout and implementation of REDD+ policies in various countries over the past decade has received much scholarly attention [6–10], although these has been less attention to how adaptation policies have been integrated into REDD+. Most existing studies of REDD+ and adaptation have been at the national level and have assessed how different ministries and sectors have coordinated through REDD+ projects to incorporate adaptation and mitigation concerns [11–17], while a smaller number of studies have looked at how voluntary forest carbon projects include adaptation measures [18–20]. However, we know less about how sub-national policymakers are treating adaptation in the development of forest carbon policies, and how households in areas with REDD+ projects, particularly those already vulnerable to climate impacts, are affected in terms of their adaptation options by REDD+ activities.

This article uses a case study in Vietnam to explore how REDD+ projects and policies link both concerns of mitigation and adaptation together at subnational levels in both policy and household impacts, and if not, what the barriers to doing so are. We build off a previous assessment for Vietnam that determined there was potential to address adaptation in REDD+ at national levels, as many stakeholders recognized the importance of integration of this sector [13]. Our project follows up at local levels to see if these potentials have been realized by exploring two main questions:

(1) To what degree do REDD+ policies and projects on the ground at subnational levels acknowledge that climate change is likely to impact forests and forest users, and how is this built into REDD+ policy and activities?

(2) How are households in forested areas subject to REDD+ policy also vulnerable to climate change, and how can REDD+ activities help or hinder needed adaptations?

Overall, we find that there continues to be a lack of coordination between mitigation and adaptation policies for forests in Vietnam, particularly with regard to REDD+. Policies at both the national and provincial level, and site-based projects, have paid little attention to the adaptation needs of local communities, and how REDD+ activities could facilitate increased resilience in livelihoods. While there were some implicit and coincidental adaptation benefits of REDD+ activities, most of the projects and local policies that we examine did not explicitly consider their activities to touch on adaptation or resilience, and in at least one case, negative livelihood impacts that have increased household vulnerability to climate change were documented. We conclude the article with insights into the barriers that continue to exist that keep REDD+ and adaptation from being considered more holistically.

2. Background: Intersections between REDD+ and Climate Adaptation in International and National Policy and Practice

REDD+ policies have been discussed as part of the UN Framework Convention on Climate Change (UNFCCC) since 2005, and has been on the agenda of all subsequent Conference of the Party

(COP) meetings, as technical subcommittees have worked out elements of how REDD+ might be implemented [21]. COP 19 in Warsaw in 2013 adopted a number of important technical decisions on REDD+, including on results-based finance, coordination of support, forest monitoring systems, safeguards, reference levels, measuring, reporting and verifying (MRV), and addressing the drivers of deforestation [22]. Pilot programs to prepare countries for "REDD+ readiness" have been underway in many nations, funded by bilateral and multilateral donors, and involving new institutions like the United Nations' UN-REDD+ program and the Forest Carbon Partnership Facility (FCPF) of the World Bank [23].

Very few of the decisions taken at various COPs have explicitly linked REDD+ and adaptation approaches. For example, the 2015 Paris Agreement that was negotiated at COP 21 entered into force in November 2016, following ratification by at least 55 parties accounting for 55% of total global emissions (Vietnam ratified the agreement in November 2016). REDD+ is explicitly mentioned in article 5 of the agreement, but it does not state how countries are to implement forest sinks, or how results-based payments will be made, and leaves such decisions up to individual countries. These actions will be clarified by states in their submissions of "nationally determined contributions" (NDCs) that the Paris Agreement regularly requires. Article 7 notes that adaptation actions are similarly to be decided at the country level, through National Adaptation Plans (NAPs) also to be regularly filed with the UNFCCC, and such actions should *"follow a country-driven, gender responsive, participatory and fully transparent approach, taking into consideration vulnerable groups, communities and ecosystems, and should be based on and guided by the best available science and, as appropriate, traditional knowledge, knowledge of indigenous peoples and local knowledge systems, with a view to integrating adaptation into relevant socioeconomic and environmental policies and actions, where appropriate"* [24].

Within the international literature, there is increasing reference to how forest mitigation and adaptation activities might intersect [1]. In some countries, the same actors are in charge of both REDD+ and adaptation plans [11]. (However, this is not the case in Vietnam where REDD+ activities and actors are centered in the Ministry of Agriculture and Rural Development (MARD), and adaptation activities are mainly driven by the Ministry of Natural Resources and Environment (MONRE)). Some privately financed forest carbon projects, such as those certified by the Climate, Community and Biodiversity Alliance (CCBA) standard, have required attention to social adaptation as part of their certifications [19]. Other REDD+ policies, at national and local levels, have had little to say about adaptation. Some authors have assumed that attention to safeguards under REDD+ (a requirement from the Warsaw COP in 2013) demonstrates a sufficient approach to livelihoods, which will result in adaptation benefits [11], while other authors are more skeptical that safeguards and local participation (which might lead to adaptation benefits) are actually happening in REDD+ projects on the ground [25,26]. Of the existing reports on combined mitigation and adaptation policy, very few assess the complicated relationships between livelihoods, climate impacts, and REDD+ through on-the-ground surveys or interviews, leaving most discussions at higher levels of policy. One of the few papers to tackle this problem did find that while many forest mitigation projects in a case study in Belize did not have specific adaptation actions embedded in them, nearly half did have adaptation-related outcomes, such as improved livelihoods, and 90% of adaptation-explicit projects also reported mitigation-related outcomes, such as enhanced carbon stocks [18].

This topic deserves further study, as a lack of integration of REDD+ and adaptation into subnational and local policy could have serious consequences. Several authors have noted that activities taken to increase mitigation of land-based GHGs under REDD+ might have unintended impacts on the livelihoods of forest-dependent people [27,28], and therefore also have impacts on the ability of these households to adapt to climate change. A hypothetical example might be a forest plantation created for carbon sequestration that reduced water availability for nearby households, who then might become more vulnerable to climate-change induced droughts [18]. Overall, communities' and individuals' ability to cope with many forecasted climate changes, like localized changes in rainfall timing and amounts, among other impacts, are likely to be strongly conditioned on their ability to access and mobilize resources like land, trees, water, fish and other means of livelihood [29,30].

If access rights to forest change under REDD+ projects, this could render communities and households more vulnerable to the effects of climate change at local levels if traditional assets like forests that are used for adaptation responses (e.g., as a source of quick cash or as food) become inaccessible [25,31]. Alternatively, REDD+ could potentially strengthen local access rights to forests through increased financing to ensure their protection from outside deforestation pressures, thus possibly increasing communities' resilience to climate change [32,33]. The aforementioned assessment of forest carbon mitigation projects in Belize show the positive benefits of attention to adaptation which resulted in diversified livelihoods, strengthened land tenure, and more robust local forest management [19].

Other reasons to combine mitigation and adaptation approaches include the need to maximize limited climate financing [11,18,34]; to harmonize sectoral policy and avoid institutional duplication and overlap in approaches [35]; and to potentially "climate-proof" mitigation and other development projects [36,37]. For example, the planting of biofuels (a mitigation policy) might be impacted by climate-induced changes in the future and would need to be planned for, otherwise projects' contributions to overall mitigation might decrease without adaptation measures [38].

However, numerous challenges face any attempt to integrate adaptation and mitigation together in policy and projects, at both national and local levels. For example, examination of existing REDD+ development at national scales has revealed major challenges in coordination across sectors already, with both duplications and gaps in how REDD+ works with other development policies [39–41]. Additionally, there are often mismatches between time scales for projects, with mitigation usually being more immediate and adaptation more longer term [19]. In cases where adaptation has explicitly been linked into a REDD+ mitigation project, there are often difficulties in financing and extended time spans for projects [42]. Further, future climate change impacts on both households and forests are variable and often depend on localized context, making generalizations about adaptation difficult to put into policy [43–46]. For example, one study that assessed climate change forecasts for the provisioning of ecosystem services from forests in Finland found a series of complicated impacts, some positive, some negative, with no clear direction for policy actions to increase adaptation [44].

On the positive side, one potentially promising new approach has been the concept of ecosystem-based adaptation (EBA), promoted for "the use of natural capital by people to adapt to climate change impacts, which can also have multiple co-benefits for mitigation, protection of livelihoods and poverty alleviation" [47]. EBA is often presented as a win-win for both mitigation and adaptation [48], and activities under this label include such activities as restoration or protection of coastal mangroves, which sequester carbon as well as helping coastal communities withstand the impacts of storms [49]. A recent review of EBA noted that forests can support human climate adaptation through: (1) provision of goods to vulnerable communities; (2) regulation of microclimates, especially for agriculture; (3) regulation of soil and water to buffer climate impacts; (4) coastal forest protection against storms; and (5) urban trees that regulate temperature and water [50,51]. However, there have been few assessments of the degree to which EBA approaches are integrated into either national or subnational REDD+ or other policies [52,53], and country experience shows that many challenges remain in operationalizing EBA [54].

Vietnam is a particularly appropriate country in which to look at both climate adaptation and forest-based mitigation, as it has been an early adopter of REDD+ activities, through the UN-REDD+ programme and the FCPF, as well as a number of voluntary projects. Slightly more than half of Vietnam's 2010 greenhouse gas emissions were from the agriculture, forestry, or land use sectors, indicating a high priority for investments in emissions reductions in this category [55]. Vietnam has also been identified as one of the top fifteen countries in the world vulnerable to natural hazards like drought and storms in terms of the number of people and scale of exposure [56], and forecasted temperature increases will exacerbate this condition to levels previously not experienced. The forecasted climate impacts to 2100 will likely be an increase in rainfall in wet seasons and decrease in dry of around 10% or more, increased intensity and frequency of storms and floods, and a likely sea level rise of at minimum one meter [57]. In order to minimize climate change impacts on Vietnam, adaptation projects to reduce vulnerability have been increasing in scale and importance in recent years, including in water management, health care provisioning, and land use planning, such as resettlement away from vulnerable zones [58–60]. Many of these actions have been combined with disaster-risk reduction strategies and aimed at increasing resilience of households to a multitude of climate related effects [61–64]. However, there have been relatively fewer adaptation actions directed at forest-dwelling and using communities, which tend to be located in mountainous areas of the country, while more adaptation attention and financing focuses on coastal and delta areas [57,65].

3. Materials and Methods

3.1. Fieldsites

This study was carried out in 3 provinces of Dien Bien, Kon Tum, and Kien Giang (see Figure 1a). These sites were selected as representative of the North, Center and South of the country and were sites in which preliminary REDD+ readiness projects were on-going, all sponsored by different donors, which gave us a range of project types to explore. In each study province, local communes were selected for in-depth study based on where existing projects for REDD+ or other forestry-focused projects have been operating (see Figure 1b–d).

Dien Bien is a mountainous area located in the Northwest along the border with Laos. The total natural area of this province is 956,290 ha, with 41.1% of the total area classified as forest. The total population is 547,785, and 47% of the province's households were considered under the government poverty line in 2016, the highest rate in the entire country. Livelihoods primarily consist of agricultural production, livestock husbandry, and forestry exploitation and development activities. Around 50,000 households, mostly consisting of ethnic minorities like Hmong, Thai, Dao, Kho Mu, and others, have participated in government payments for environmental services (PES) programs since 2011. A Japan International Cooperation Agency (JICA) REDD+ project was piloted in two districts from 2012 to 2013 and included activities such as raising awareness of REDD+, FPIC (Free, Prior and Informed Consent) agreements, agro-forestry extension, and development of the province's overall REDD+ policy.

Kon Tum is a mountainous area located in the Central Highlands, with a total forest area of 603,814 ha, 58.5% of the province. There is a large and vulnerable ethnic minority population (54% of total), dependent on both cash crop and subsistence agriculture, and 26% of the province's households were considered under the government poverty line in 2016. The major crops grown in this area are primarily cassava for subsistence, with only a little rice, corn and rubber for supplemental income. Households located in areas of the province with basalt soils have been able to transition into cash crop agriculture, particularly rubber but also coffee, tea, cashew and litsea in the past 10 years, but these activities have been faulted for deforestation and forest degradation. A REDD+ project has been piloted in Kon Plong district of Kon Tum in 11 villages of Hieu commune by Fauna and Flora International (FFI), Kon Tum Provincial People Committee's, Kon Tum's Agriculture & Rural Development Department (DARD), and PanNature (an NGO) in the period 2011–2014. Hieu commune has around 20,500 ha that is nearly 90% forest, and 660 households, mostly of the M'nam ethnicity, have participated in project activities, such as building capacity for local authorities and communities in order to directly implement REDD+ activities (including setting up a new local community-based institution which would have locally derived and implemented rules), with the hopes of providing financial benefits to forest-dependent local and indigenous people from selling carbon credits in the voluntary carbon market.

Kien Giang is located in the Mekong Delta area in the South. The total area of the province is 634,853 ha, in which forestry land is only 13.6% of the total area, mostly coastal mangroves. The total population is about 1.7 million people, of which ethnic minorities (primarily Khmer) are about 16% of the total, and the provincial poverty rate is 9.7%. Two-thirds of the population live in rural areas and have activities related to agriculture and aquaculture near mangrove forests. In 2008, with support from AUSAID and GTZ, Kien Giang began to implement a Conservation and Development of the Kien Giang Biosphere Reserve project. Within this project, activities have included surveys of mangrove species diversity; mangrove and coastline mapping via remote sensing and satellite image interpretation; studying biomass, carbon stocks, and biological diversity, including an assessment of forest regeneration needs and potential; and developing a REDD+ feasibility study. Local people have been involved in training courses and awareness raising activities held by the project and the local government agencies on the topic of PES and reducing emissions due to deforestation and agriculture production in the province.

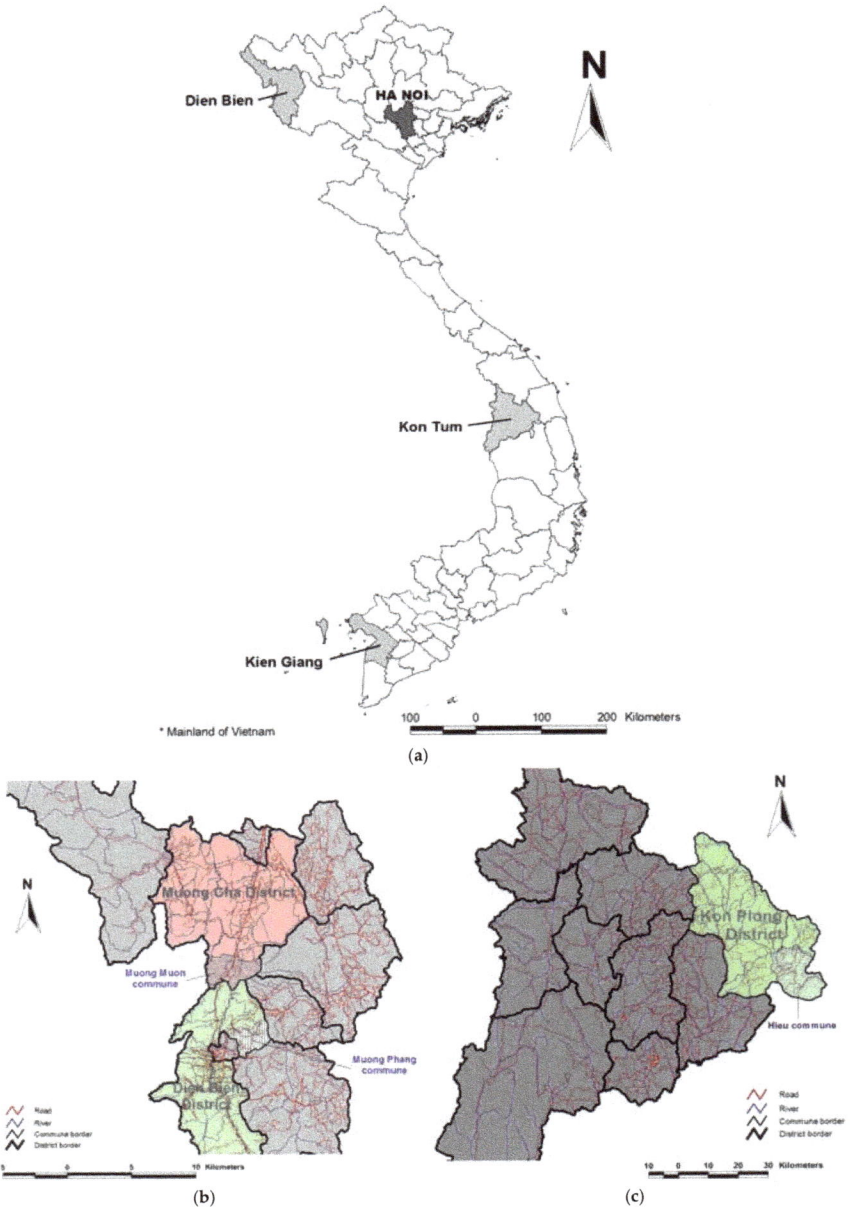

(a)

(b)

(c)

Figure 1. *Cont.*

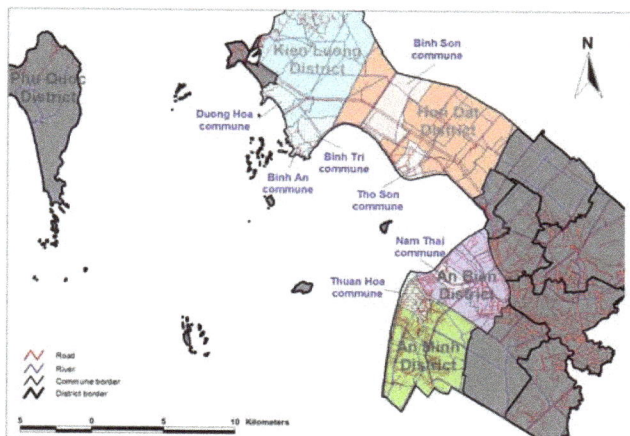

(d)

Figure 1. (**a**) Map of Vietnam showing provinces where research took place; (**b**) Dien Bien Province; (**c**) Kon Tum Province; (**d**) Kien Giang Province. Communes labelled in blue were sites of REDD+ projects and our research.

3.2. Data Collection

Within each province, we carried out a mixed methods approach to collecting social and environmental data, making several field trips to each province throughout 2012, 2013 and 2014, spending from two to three weeks collecting data in each site. In each province a standard questionnaire was administered to a sample of households (selected at random from a village census by choosing every *k*th household on list) proportionately spread across villages to generate 100 households per province, for a total sample size of 300 households. Households are usually the main units making land-use and livelihood decisions, and this project has used the standard Vietnamese government definition of households. The survey asked questions about livelihoods, income, assets, participation in forest projects, climate vulnerability and adaptation measures. The data from the surveys was entered into SPSS for analysis.

In addition, focus groups were carried out with small numbers of local residents in each study area to help us to build histories of resource use, determine how residents learned from one another and set up institutions for managing forests and reducing climate risk and vulnerability, how these institutions functioned in different situations, and how such institutions interacted with official forest policies like REDD+. In each research site, four focus groups (of approximately ten people per group) were run. A general focus group consisting of representatives of local civil society and government groups (such as the women's union, youth union, veteran's union, Communist Party, etc.) were asked to discuss the general issues of village such as main livelihoods, household economic status ranking, and seasonal calendars of the village, among other topics. A forest user focus group including families who were allocated forest land and who joined village forest patrolling groups discussed forest issues, changes of land and forest during the past ten to fifteen years, and participatory land-mapping exercises. For women's focus groups, representative women from different household types (poor, average, rich), and woman-headed households were selected to join for discussion. Finally, one focus group of officially designated "poor" households was run as well, and focused on risk-mapping, climate impacts on the poor, and other topics. In each case invitations were issued to attendees with the advisement of the village head and snowball sampling (e.g., asking invitees to bring along neighbors with related knowledge). We also targeted knowledgeable people in each community for lengthier, unstructured key informant interviews to collect life histories aimed at understanding social and climate vulnerability, as well as changes in resource-use patterns and access over time, among other topics.

Finally, we used stakeholder interviews with government officials and policymakers in each field site to gather information on the development of local forest policies. We asked how they were responsible for designing a forest policy incorporating social considerations; types of social data that policy makers use; and local input and participation to forest policy to their locality. We interviewed 15 policymakers at district and provincial levels in each fieldsite, and several national level stakeholders involved with REDD+ as well, for 60 policy interviews total.

4. Results

In the first part of this results section we discuss the national policy frameworks that have developed for REDD+ and for climate adaptation, and the degree to which they are integrated at subnational levels, while later we present data from a household survey undertaken in REDD+ project areas that aimed to understand forest-based livelihoods, climate vulnerabilities and adaptation, and the impacts of REDD+ projects on these.

4.1. National Policy Development for REDD+ and Adaptation

Vietnam developed a National Strategy on Climate Change in late 2011, which addresses both adaptation and mitigation, and REDD+ activities are a key element of the strategy [66]. It is estimated that through REDD+, 88.2 million tCO_2 emissions per year could be reduced [67]. Vietnam has recently submitted an intended "nationally determined contribution" (NDC) outlining plans to carry out the Paris Agreement, and the submission calls for an 8% reduction in greenhouse gas emissions by 2030 as compared to the business-as-usual scenario, which could be increased to a 25% reduction depending on international support. Specific activities in Vietnam's NDC related to forests and REDD+ include actions to:

- "Review and identify the areas and objects to apply sustainable forest management, afforestation and reforestation, biodiversity conservation, including special priority for regions with large forests that are important for forestry production and livelihoods of local communities of people;
- Develop and improve policies to promote sustainable forest management; mechanisms and policies to attract private sector investment for sustainable forest management, afforestation, reforestation, biodiversity conservation and livelihood development;
- Integrate and effectively use domestic and international resources for implementation of programmes and projects related to forest management and development, livelihoods and biodiversity conservation such as REDD+, the policy of payment for forest environmental services, etc.
- Strengthen and expand international cooperation for investment, technical assistance and capacity building, information and experience sharing on the sustainable forest management and development, biodiversity conservation and livelihood development" [68].

The development and governance structures for REDD+ in Vietnam have been reviewed by others, to which we refer readers for additional details [69–71]. We primarily focus on the adaptation linkages at national level in this paper. Most of the REDD+ activities in Vietnam have been carried out by various department of MARD, and have focused on traditional forestry activities like inventories and forest land use planning. However, one potential institutional linkage exists to facilitate adaptation into REDD+; a national steering network was set up in 2009 to coordinate REDD+ activities, and the Department of Meteorology and Climate Change of the Ministry of Natural Resources and Environment (MONRE) sits as a member the network, one of three non-MARD government departments (the other two are the Agro-economic Department of the Ministry of Planning and Investment and the Government Office, a prime-ministerial level coordinating office).

The steering network was charged with developing a National REDD+ Action Plan (NRAP), which was first completed in 2012 (and which is now under revision). The goals of the original 2012 NRAP were to (1) Build a national REDD+ Program and provincial REDD+ plans; (2) Enhance institutional capacity and coordination between ministries; (3) Raise awareness to stakeholders in forestry; (4) Improve technical capacity in reference levels, and monitoring; and (5) Develop benefit sharing systems and an information system on safeguards [72]. However, this initial NRAP made no mention of adaptation in any systematic way [13], although there are implied adaptation benefits from some of the REDD+ activities. For example, MARD is currently focused on a revision of forest criteria and classification for Vietnam to help clarify forest tenure agreements, which will likely have impacts on both REDD+ and adaptation projects in the future. The NRAP also affirms that by 2016–2020, there should be attention to "diversification and improvement of livelihoods of the forest owners and the people at large" [72], which would likely have positive adaptation benefits for households. Although previous researchers found limited policy support for adaptation in the forestry field at the national level in their analysis in 2014 [13], there has been some progress since then, particularly in research activities. For example, current climate adaptation focused activities in the forest sector include forest breeding of trees resistant to climate change; developing a national plan for adaptation to climate change in the forestry sector; and coordinating with the World Bank to conduct research on climate change adaptation in forestry [73].

Vietnam's NDC clearly states that adaptation will also be an important part of the country response. However, Vietnam has not yet filed a National Adaptation Plan (NAP) with the UNFCCC, though it intends to do so in the future. The NDC has a lack of specificity with regard to funding and priorities for adaptation beyond stressing its importance. Although the NDC states that costs of adaptation in Vietnam are estimated to exceed 3%–5% of GDP by 2030, how funding will be mobilized and for what specific activities is not yet clear, beyond a few priorities of: (1) responding pro-actively to disasters and improving climate monitoring; (2) ensuring social security, including EBA and community-based adaptation; and (3) responding to sea level rise and urban inundation [68].

While the ways adaptation may be carried out in the forest sector are not explicitly referenced in the NDC, the National Strategy on Climate Change refers more specifically to the ways that forestry can contribute to "preventing and coping with natural disasters, flash floods and landslides in mountain areas", and that policy will be needed to "improve quality of forests and afforestation, to turn bare lands and hills green, to effectively exploit different kinds of forest to secure and improve resistance against natural disasters, preventing desertification, land erosion and degradation; to enhance protection, management and development of mangrove forests and flooded ecosystems; to raise the forest coverage to 45% by 2020". There is also a stated goal to "preserve biodiversity, protect and develop ecosystems and species which can well resist climatic changes; to protect and preserve genes and species endangered by impacts of climate change". In addition, the National Strategy explicitly refers to "managing forest in a sustainable way, preserving and improving forests' absorption of carbon, and maintaining and diversifying local people's livelihood as well as helping them to adapt to climate change" [66].

The primary approach for livelihood improvements under REDD+ will likely be household payments, and it has been calculated that there will likely be a national-level payment rate from REDD+ of around 265,000 (12$US) VND/ha/year in the future. Together with the general financial support from the state budget for forest protection measures (100,000–400,000 VND (5–20$US) /ha/year), as well as the average payment for environmental services (PES) available for forest protection in some upland forests (on average 250,000 VND (11$US)/ha/year) [74], participation in forest protection and management under a combination of REDD+ and other programs can contribute to incomes of local people, especially for poor and vulnerable groups in remote areas. Existing PES payments to households in forested areas of Vietnam account for on average 6%–7% (and up to 30%) of household income in participating areas, and are often used to enable school fees and healthcare bills to be paid, to help ensure food security through purchases of rice or seedlings, and other forms of investment [75,76].

When payments can be made to community funds (done in some areas but not others), they can be used to build community infrastructure (e.g., roads, bridges, community houses, etc.); upgrading and buying common assets for the community (e.g., school supplies); paying for a village forest protection group; or even can setting up micro-loans for diversifying livelihood activities and other activities [77,78]. Through these ways, households and communities could be helped via REDD+ payments to adapt and recover following climatic shocks or disasters, although more research will need to be carried out once these payments begin to follow how these monies are actually being used (no areas of Vietnam have received national REDD+ payments as of 2016).

4.2. Subnational Policy on REDD+ and Adaptation and Interlinkages

The primary purpose of the NRAP is to set out key legal and institutional roles as well as priority interventions in REDD+ for the period 2011–2020 [72]. However, the document is mainly an enabling document rather than one providing detailed guidance to develop REDD+ interventions on the ground. Thus, the NRAP is to be supplemented by Provincial REDD+ Action Plans (PRAPs) and Site-based REDD+ Action Plans (SiRAPs). Depending on the particular context of sub-national levels, the PRAPs will help to develop mechanisms and set out suggested REDD+ prioritized interventions that are suitable for the local political, social and environmental conditions in order to support local actors to participate in REDD+ implementation more effectively and sustainability. Currently, fifteen provinces are developing PRAPs; ten are completed while five others are in the development process. (They include: Ca Mau, Lam Dong, Binh Thuan, Ha Tinh, Bac Kan and Lao Cai (supported by the UN-REDD+ program); Thanh Hoa, Ha Tinh, Nghe An, Quang Binh, Quang Tri and Thua Thien Hue (supported by the World Bank Forest Carbon Partnership Facility and the USAID-funded Vietnam Forests and Delta Program) and Dien Bien, Son La, and Hoa Binh provinces (supported under the aforementioned JICA project)). Of our three study provinces, only Dien Bien has an approved PRAP (finalized in May of 2014). In addition to the development of subnational PRAPs, pilot REDD+ activities for REDD+ readiness or voluntary carbon market accession have been implemented in many of Vietnam's provinces, funded by multilateral or bilateral donors or NGOs.

Many provinces have used these donor pilot projects to build off of for development of PRAPs. Because all provinces already were required to create Forest Protection and Development Master Plans to 2020 as required by MARD, REDD+ programs have piggybacked onto this process. This means that the PRAPs usually include both general forest protection and development activities, as well as specific activities for REDD+ pilot areas, divided into 5 different approaches: reducing deforestation, reducing degradation, sustainable forest management, conserving and enhancing forest carbon stocks. In addition, depending on the characteristic of the activities, PRAP activities also can be broken down into direct investment interventions (type I) and the supporting interventions (type II). Type I are defined as direct investment or activities for forest protection, management and development. Type II are supporting activities, which are mainly focused on providing incentives to forest owners, local communities and other relevant actors to carry out the type I activities. The supporting activities include, depending on the local demands, provision of community development funds, financial and technical support for sustainable local livelihoods, or capacity building on various aspects of forest or livelihoods activities. All these activities have been required to be considered in the development of PRAPs, as per government decision 5414/QD-BNN-TCLN dated on 25 December 2015. Although many of these activities likely have adaptation benefits, throughout the national guidance issued for PRAP development, the words "climate change", "mitigation" and "adaptation" are rarely used or not mentioned directly. Part of this inattention is likely due to the fact that PRAPs are primarily being developed by forestry departments at local levels, with less of the sector integration and multi-stakeholder engagement that has characterized the NRAP process at the national level.

We analysed the Dien Bien PRAP in some detail to understand the potential for synergies and linkages with climate adaptation policy. In addition, team members attended several workshops during the process of the development of the Dien Bien PRAP. Dien Bien was the first province in

Vietnam to have completed and launched a PRAP with the support from JICA under the project *"Sustainable Forest Management in the Northwest Watershed Areas (SUSFORM-NOW)"*. The PRAP was designed to cover the period 2013–2020 and was approved by the Dien Bien Provincial People's Committee in early 2014. Because the Dien Bien PRAP was developed and approved even before the national guidelines for PRAP development were issued, there are some now required activities that are not mentioned in this specific PRAP (see Table 1; the left hand column are activities now required to be addressed in most PRAPs). In particular, some interviewees noted that in fact Dien Bien's PRAP is more similar to a general Forest Protection and Development Plan rather than a specific REDD+ plan focused on forest carbon mitigation.

Key activities to be implemented in the future under the Dien Bien PRAP include forest patrolling and monitoring; forest land allocation; livelihood support; and forest plantations, but there are few details on how these will be accomplished yet. Similar to the NRAP, the Dien Bien PRAP does not mention adaptation directly, and is only implied indirectly through the livelihood development activities to support forest protection in implementing REDD+. For example, the Dien Bien PRAP has highlighted that forestland allocation to organizations, households, individuals and communities should be completed, as one of the supporting activities to reduce deforestation and forest degradation drivers at the local level. Such activities are considered to have the potential to indirectly improve the adaptive capacity of local people to tackle other climate impacts by providing more secure access rights to forests (see Table 1, far right column), although how this will play out in reality will need further research. Additionally, the PRAP does place emphasis on the need to incorporate climate-resilient forest species in future forest plantations, and has several other foci of potential adaptation benefits to forests themselves (Table 1, second column from right).

In addition to helping support the creation and implementation of PRAP for Dien Bien, JICA developed a guidebook for other provinces carrying out PRAP development as well [79]; however, the handbook makes no explicit mention of adaptation support or activities. The handbook outlines a 13-step process that is necessary to develop a PRAP, of which some of the key activities are forest and socio-economic surveys; calculations of reference emission levels; searches for potential funding of emissions reductions; surveys on forest monitoring; and policies on safeguards. The handbook does identify the global Adaptation Fund as a source of REDD+ financing, but does not identify what adaptation activities related to REDD+ should be incorporated into PRAPs. Additionally, the socio-economic surveys only indirectly touch on climate vulnerabilities, as primarily these surveys are to "identify forest status such as forest distribution & stock at present and past forest change, the driver of forest decrease and increase, socio-economic conditions such as demography and agriculture & forestry production, and to assess past programs & policies relating forest protection and development" [79]. Thus, there appears to be much more room for PRAPs and PRAP development guidance to focus explicitly on adaptation.

Table 1. Key sectors and activities in subnational Provincial REDD Action Plan (PRAP) in Dien Bien Province.

I. General Forest Sectors and Plans	In Dien Bien PRAP?	Potential Adaptation Benefit to Forests?	Potential Adaptation Benefit to Households?
1.1. Forest protection and natural forest regeneration and enrichment	x	Y	Y
1.2. Forest plantation	x	Unclear	Unclear
1.3. Forest rehabilitation and reforestation	x	Y	Y
1.4. Silvicultural-related construction projects			
II. Prioritized Activities carried out at the potential REDD+ areas			
2.1. Activities to reduce deforestation			
a. Reviewing and analyzing overall land-use planning and forest/forestland planning			
b. Reviewing and analyzing the overlap between forest/forestland planning and socio-economic development planning			Unclear
c. Strictly control the conversion of natural forests into other purposes	x	Y	Y
d. Supporting the forestland allocation, contracts and lease to households, individuals and local communities	x	Unclear	
e. Livelihood improvements			
f. Establishing small-scale local livelihood credit development funds			
g. Enhancing the forest protection and development law enforcement			
2.2. Activities to reduce forest degradation			
a. Reducing/ preventing illegal timber logging and utilization			
b. Establishing the administrative and technical monitoring system on timber legality assurance			
c. Promoting sustainable non-timber forest product (NTFP) models	x		Y
d. Market orientation to agro-forest products	x		Y
e. Developing village forest protection and development conventions	x		Y
f. Encouraging job creation in the REDD+ implementation sites			
2.3. Sustainable forest management			
a. Support to develop and implement the sustainable forest management plan and SFM certificates	x		Y
b. Improving the forest governance ability for forest owners			
2.4. Conserving carbon stocks			
a. Improving the quality of forests: forest enrichment, or diversification of crops to adapt impacts of climate change			
b. Biodiversity conservation in special-use forests and protection forest and payments for forest environmental services	x	Y	Y
2.5. Enhancing carbon stocks			
a. Technical support for forest plantation: climate change resilient seedlings, etc.	x	Y	Unclear
b. Forest enrichment in exhausted forest areas			
c. Afforestation by land-use conversion projects			

4.3. Project Based Interlinkages with Adaptation

The REDD+ projects in our three study sites differ from the three projects analysed previously by other reports on adaptation and REDD+ in Vietnam [13]. The REDD+ Community Carbon Pool Program (REDD+ CCP) is the first and only REDD+ project to be implemented in Kon Tum province and is focused at a site-based level in Hieu commune, Kon Plong District; a 3-year SiRAP had been developed and approved by the Kon Tum Provincial People's Committee. Under the REDD+ CCP project, 18,700 ha forests have been re-allocated with community land use rights, which is intended to lead to local access and control over forestland and forest resources. The development of local community forest management institutions that comprised equitable, easy to understand, locally devised and implemented rules was seen as a way to effectively reduce emissions and provide benefits to forest-dependent local people, and the project was hopeful to meet the requirements for CCBA certification, which explicitly includes projects that generate benefits for climate change adaptation.

In Dien Bien, the JICA SUSFORM-NOW project aimed to test REDD+ activities in two pilot communes; develop capacity for forest rangers and commune level field officers; and prepare the PRAP and related technical guidance documents as noted previously. Overall, the project aimed to build efficient models and develop capacity for provincial cadres to implement REDD+ themselves so that those models would be replicated to other areas. As such, the project was aimed to contribute to better forest management at the local level and respond to climate change in the forestry sector, although adaptation was not explicitly mentioned in project documents.

In Kien Giang, the GIZ/AusAid Conservation and Development program was originally designed as a climate change adaptation, mitigation and integrated coastal zone management project. In recent years it has also supported research into mangrove forests, awareness raising of forest protection, and training on REDD+ readiness in Kien Giang and Ca Mau provinces. One of the major activities funded by the project has supported local communities to develop "green fences" along the coastal area through mangrove plantations to protect against landslides and coastal erosion due to strong waves from the seas and future sea level rise. Local people were also supported by the project to establish mangrove nurseries. The project provided training courses for local people, especially women, on primary health care services and rural sanitation, for example building toilets and using clean water, among others, which have adaptation benefits. Kien Giang province also has had since 2005 an initiative of benefit sharing in mangrove forest protection that enables local people to combine forest protection and aquaculture development to secure local livelihoods. People are allocated two to three ha of protection mangrove forest, and they are allowed to use 30% of this area to raise shrimp, as long as they maintain 70% mangrove forest. By 2013, there were 1076 households participating who were allocated mangrove forest with "green books" (a long-term contract for forest protection) along 200 km of the coast of Kien Giang province. The program aims to give local people incentive to protect mangrove forest while developing shrimp production for their livelihood; however, according to GIZ/AusAid project officers' assessments, the expectation from local people is for a higher percentage of land devoted to shrimp production due to low profits and higher costs of maintaining mangrove forest as compared to non-participating shrimp farms.

4.4. Household Level Impacts from Climate Changes and REDD+ Projects

Our household survey provided a general summary of the characteristics of households in the three sites: around 31% of surveyed households were classified as poor by the government (based on guidelines of the Ministry of Labor, Invalids, and Social Affairs, as every household in Vietnam receives regular assessments of their economic status to determine some social benefits); 49% as average; and 11% better off (the rest did not know or did not answer). The average household generated 78 million (around 3580$US) per HH per year in subsistence and cash income (see Table 2), with great variation between the three sites, as Kien Giang households had over four times the income as those in Kon Tum. In each site, a different income source was the primary contributor to local livelihoods: business and trading in Dien Bien; livestock in Kon Tum; and aquaculture in Kien Giang.

The primary uses of forests by households were for fuelwood and forest product collection; as sites for aquaculture in the case of mangroves in Kien Giang (where households generated on average 4758$US per household from aquaculture); as sources of lands for fields in swidden cultivation systems in Dien Bien and Kon Tum; and as supplies of wood for house building, particularly for poorer or ethnic minority households. Forest-based subsistence and cash income was highest in Dien Bien, while it was surprisingly low in Kon Tum, despite having the most extensive forest cover of all three sites (and which we explain below was a result of restrictions on forest use instituted by the CCP project).

The living conditions of the populations in the research sites were generally low with poor infrastructure, making areas difficult to access in the rainy season, leading to limited access to information regarding markets and services. Local communities were all sensitive to changes in weather and had been very much impacted by natural disasters that appeared with higher frequency and more intensity in recent years. In general, local people at all fieldsites had concerns about climate changes and risk, although the specific type of impacts varied (see Table 3). In the Northwest mountainous area of Dien Bien, people were concerned about droughts, with 59% of surveyed households noting decreases in rainfall, and 53% noting longer dry seasons. In Kon Tum, people were more concerned with increases in storms and rainfall, leading to localized flooding. Households in the Mekong Delta area of Kien Giang were particularly concerned about salinity intrusions farther from the coast in the dry season. During group meetings, communities often linked these climate changes explicitly to forest policy and activities; for example, in Kon Tum, local people explained that the higher impacts of storms and flooding were due to deforestation in the last few decades in the area. In Dien Bien, local people noted that higher temperature and severe droughts had caused forest fires to increase as well (focus group data).

When asked to rank the most serious climate-related risk to their property and livelihoods, households choose a variety of answers (Figure 2a). Typhoons and storms were ranked as the most serious risk by nearly 30% of survey respondents, with landslides and drought chosen as the most serious risk by fewer than 15% and 10% of households respectively. Although households overall understood climate risks to be important and in most cases increasing, these risks were put in the context of other challenges that households had to face (Figure 1b). Health problems were considered the most serious and frequently encountered risk, with natural disasters and pests ranked second and third. Problems with access and quality and outcomes of education of their children were also considered risks, as people were worried they could not afford for their children go to school; these concerns are reflected as "children's schooling" in the Figure 2b. Other risks included labour shortages due to health and other problems (such as alcoholism or drug use). Poor infrastructure in the study areas, especially in health care, education, and public services, increased households' feeling of risky livelihoods. This findings confirmed that vulnerability to shocks, be it climate or health or unemployment, have long been one of the major challenges for the poor in Vietnam [57]. In particular, while these climate risks have the potential to impact all income groups, the poor tend to have less resilience, such as less access to insurance and less ability to rebuild or move away from affected areas [80,81].

Table 2. Average household livelihood sources and values in VND.

Province	Agriculture	Livestock	Aquaculture/Fishing	Forestry	Business	Total Average HH Subsistence and Cash Income
Dien Bien	14,493,982	24,124,279	4,464,852	12,008,551	30,811,538	79,772,542 (3626$US)
Kon Tum	5,352,315	20,136,242	0	3,754,609	10,415,938	30,078,040 (1367$US)
Kien Giang	55,395,454	19,965,853	104,679,297	5,951,666	35,293,787	126,419,650 (5746$US)
Average across all households	18,390,243	21,955,750	53,053,674	7,594,240	25,113,156	78,756,744 (3580$US)

* At the time of the survey, 1 USD = approximately 22,000 Vietnam Dong (VND).

Table 3. Household perceptions of weather and climate changes in recent years (% of households citing each reason).

Province	Increase in Rainfall	Decrease in Rainfall	Longer Rainy Season	Longer Dry Season	Forestry	Increase in Number of Storms	Increase in Storm Strength	Salinity Intrusion	Higher Temperatures
Dien Bien	14%	59%	5%	53%		3%	24%	n/a	30%
Kon Tum	37%	3%	26%	17%		74%	43%	n/a	9%
Kien Giang	48%	38%	70%	30%		23%	33%	100%	61%

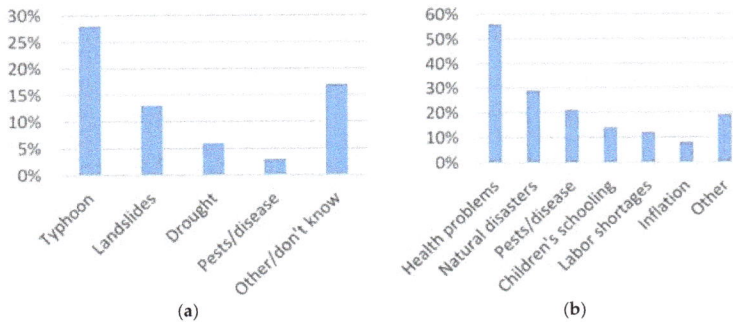

(a)

(b)

Figure 2. (**a**) The most serious climate/weather risk as identified by households; (**b**) The most serious of all risks as identified by households.

4.4.1. Impacts of REDD+ on Livelihoods and Vulnerability

In one of our studied projects, household level REDD+ activities had been confined to awareness raising of REDD+ and there were no significant livelihood or other activities directed at the household (Dien Bien). In Kon Tum however household activities had been taking place, and showed the potential negative impacts of REDD+ on both livelihoods and climate vulnerability. The commune where REDD+ activities were taking place is one of the 300 poorest communes in Vietnam [82]. Most of the local people are of the M'Nam ethnic group and have long directly depended on forestland and forest resources for livelihoods, such as rice or cassava cultivation on shifting cultivation plots in designated forestlands, firewood collection, and gathering and commercialization of diverse NTFPs. These forest activities are important because the local climate is highly variable, with rainy and cold weather affecting agricultural production. As a result, the growing season in Hieu commune is very short, around 6 months per year, from February until the end of September. Due to this weather, if households want to increase their rice or cassava production, they can only expand their cultivation areas (as they cannot diversify out of season), but this conflicts with REDD+ project priorities of conserving forest lands for emissions reductions [83].

In 2011 the initial stage of the REDD+ CCP project began, and the message that households could "protect forest for selling carbon" that the project delivered to local people through FPIC awareness-raising activities raised their hopes and expectation of gaining benefits and improving their livelihoods. The household survey and interviews conducted showed that the local people in Hieu commune perceived REDD+ as a new type of income source that would be used to replace existing practices. Therefore, the local people accepted the need to stop expanding their traditional swidden plots as well as restricting NTFP extraction of "*la kim cuong*" (*Anoectochilus setaceus*), a medicinal plant found in forests, as trade-offs in order to get income from forest carbon in the future. The project drew a lucrative picture about the benefits of REDD+ and forest carbon, but in reality, difficulties have already emerged from the project, which include scarcity of cultivable land as swiddening has stopped or been discouraged, and the loss of income from NTFP extraction. Data from the survey and interviews show that most people already changed their household livelihood strategies several years ago (household surveys showed lower forest income in Kon Tum than the other two sites, at only 170$US worth of timber, fuelwood and NTFPs extracted per household), but are still waiting for forest carbon benefits, which have not yet been paid. While the project finished at the end of 2014, it could not get carbon certificates to sell in the market as per the initial objective, and the project designer, FFI, has been trying to obtain further funding and access some carbon market to get emissions reductions certificates.

In the meantime, however, households with differences in cultivable land sources, capital and labor, have had to adjust their livelihood strategies in different ways. The poor and landless households are the most vulnerable group in this situation, and their way to adapt to the new context varies within this group. Some households have returned to make use of their old swidden plots or tried to find some small pieces of land near their paddy fields, or even have planted field crops like cassava in the garden surrounding their houses. Some other households, particularly those who do not have cultivable land, cannot open new swidden plots, and who also cannot wait two to three years for carbon credits, have decided to give up their rights to participate and get benefits from REDD+ in the future by leaving the villages or continuing to do restricted activities. For example, some husbands in these families have decided go to other communes or cities to find new jobs to compensate for the loss of income due to participating in REDD+, which has increased their families' overall vulnerability to risk and shocks. Further, "leakage" has been an on-going issue. Statistics by the local authority shows there has been no more forest clearance since the middle of 2013 in Hieu commune. However, there were some households who decided to leave and clear some forest of Bo E commune (a commune nearby but not in the REDD+ project) in order to establish new swidden plots illegally.

4.4.2. Household Adaptation Actions

At the household level in all research sites, some spontaneous adaptation actions are being taken to cope with climate risks that are already being felt. Households were most proactive in the agriculture sector, using adaptation strategies to adjust cropping patterns, harvesting time, selecting salt-resistant varieties, and so on, as to reduce damages to livelihoods. For example, 13% of households in Dien Bien and 12% of households in Kien Giang stated they had changed a crop variety, while only 3% of households in Kon Tum had done so. Changing crop calendars and harvesting crops early to avoid flood and disaster losses was another strategy, one that had been taken by 100% of households in Kien Giang, 90% of households in Kon Tum, and 60% of households in Dien Bien at least once. However, most households felt they were not doing enough to adapt to climate change, and needed more policy support. In particular, households rarely mentioned adaptation actions they were taking with regard to forestry, indicating that households felt less knowledge about this sector and needed guidance as to steps to take.

In group discussions, those residents who stated that they had not taken any adaptation actions explained that they knew that adaptation was necessary but they had a lack of resources. Poor households in particular stated they did not have enough human, physical or financial resources to protect their fields and homes from climate hazards, therefore they tended to lose relatively more when hit by floods and storms than wealthier households, and had a lower capacity to cope with and adapt to shocks due to lower access to savings. While acknowledging these challenges, local government officials interviewed stated that they lacked budgets to support households to carry out climate change strategies, plans and adaptation actions at community levels. These stakeholders at district and commune levels noted that national adaptation strategies and action plans were normally very ambitious but often not feasible due to lack of funding and low participation from local people, since there was very little funding to distribute.

5. Discussion and Conclusions

This study has confirmed that there continues to be a lack of coordination between mitigation and adaptation policies in Vietnam, particularly with regard to REDD+. While much lip service has been paid to combining approaches, in reality, policies at the national and provincial level, and site-based projects, have paid little attention to the adaptation needs of local communities, and how REDD+ activities could facilitate increased resilience in livelihoods. While there were some potential implicit adaptation benefits of REDD+ projected activities (such as promises of future activities to improve livelihoods, or to increase forest tenure security which might help local communities have better access to forest resources), local REDD+ planning through development of PRAPs has not systematically considered activities that focus on adaptation or resilience. This is despite the fact that many national policymakers, donors, and NGOs interviewed a few years ago expressed strong support for integrated attention to adaptation [13]. There was also little discussion in either PRAPs or site-specific REDD+

projects regarding EBA (with the exception of Kien Giang), again despite the fact that the idea has been enthusiastically discussed with regard to Vietnam [84,85].

These are missed opportunities, as our household data shows a great deal of need for assistance in reducing climate vulnerability (particularly in coping with drought and floods) and increasing adaptive capacity in many REDD+ project sites. There are potentials for integration with agricultural adaptation in particular. REDD+ activities in agroforestry or financial support for climate-smart agriculture could help households deal with the increasing climate risks they are facing, particularly those areas that are experiencing either drought or flooding as noted in household surveys. Tree shelterbelts that could be planted to enhance carbon stocks could help to reduce wind and drought pressures on nearby agricultural fields. Similarly, using trees to secure landslide and flood prone areas near fields can help reduce impacts on agricultural livelihoods from flooding [50]. Other examples of adaptation needs might include: a stronger focus on useful multipurpose tree species for reforestation and carbon stock enhancement that could provide for both carbon and increased livelihoods (e.g., food or products for sale); policies to increase value and marketing of forest goods that can be harvested sustainability (such as NTFPs); and policies to reduce woodfuel use but increase energy access (e.g., improved cookstoves or small-scale hydro-powered electrification projects), among others [19,86,87]. Poor households in particular showed needs for financial support to undertake adaptation actions in agriculture, which REDD+ payments could potentially be useful for, but households stated these needed to be coupled with training, education, and other forms of support so they would know what to invest in to increase resilience. Payments alone will not be sufficient.

Further, within both national and subnational REDD+ policy approaches, our analysis notes a lack of attention to the potential consequences of climate change on forest structure and composition, and the implications of this for REDD+ activities into the future. Several of the most serious climate impacts from the literature on forest vulnerability in Asia include forest fires; pest outbreaks; shifts in species distributions; higher tree mortality; changes in forest composition; or loss of wood volume [51,88–90]. Very few of these possible climate vulnerabilities of forests are discussed in the National Strategy on Climate Change (which only mentions the need to "increase capacity and efficiency of systems for evaluating, forecasting, preventing, monitoring, supervising and urgently responding to forest fires" [66]); in the NRAP; or the in the PRAP for Dien Bien. Future REDD+ projects to address forests' climate vulnerabilities might focus on reducing fire hazard and risk (e.g., community supported fire watches) or reforestation projects that prioritize drought and fire resistant native species (e.g., rather than introduced eucalyptus or acacia, which have been primary species in reforestation in the past but which are both drought and windfall-prone) [51].

Several pilot REDD+ projects examined also did not explicitly acknowledge how climate-induced changes might impact household livelihoods, and what role these vulnerabilities may play in REDD+ participation. In the one project we examined that did specifically have a focus on adaptation (Kien Giang), livelihood activities were developed that extended beyond typical REDD+ activities, such as addressing water scarcity and shoreline erosion, adaptation responses that were considered useful by households facing water and land erosion risks, and which may increase positive feelings and household participation in other parts of the project (e.g., tree planting). However, in another project site (Kon Tum), restrictions on livelihood activities had taken place under REDD+ and had caused negative consequences, particularly for the poorer households, due to restrictions on swidden agriculture and NTFP collection, the lack of suitable alternative livelihood plans, and delays in seeking carbon financing. Other studies in Vietnam have shown the importance of NTFPs as "insurance" for poor households, particularly for buffering unpredictable shocks like disasters or health problems [91]. Yet the Kon Tum project did not consider these roles of NTFPs, and how loss of access to these forests might create unforeseen negative impacts. Indeed, it appears that the project has made households more vulnerable to the risks of poverty and climate change impacts than before, particularly in the cases where REDD+ participation seems to have triggered male outmigration in some families.

Despite the negative outcomes of one of the examined projects, there appear to be opportunities to promote synergies with adaptation, where REDD+ could improve local actors' adaptive capacity. For example, in Kien Giang, the approach to forest management which allows some limited production activities in mangrove areas has proven moderately successful, and shows that co-management that allows for some livelihood activities, rather than complete abandonment of forest use as was the case in Kon Tum, can provide benefits for both people and forests. The fact that livelihoods are supported, and that mangroves provide useful protection against storms, landslides and river erosion, makes for a double adaptation benefit. Kien Giang could address concerns from households about low profitability by expanding the area of mangroves allowed under the 70/30 model from the current small pilot area to allow for each participating household to manage a larger area. Despite these successes, however, there have been shortfalls in incorporating mangroves into other REDD+ and PES policies. Although many reports within Vietnam have noted that mangroves are important for both mitigation and adaptation [92–94], there have been limited on-the-ground mangrove projects linked to REDD+, and there are currently no mangrove areas receiving national PES money due to an inability to determine who the buyers of mangrove ecosystem services are [95]. These are challenges that should be immediately prioritized in future REDD+ development.

All of these potential synergies rest on removing barriers to integration between REDD+ and adaptation. Key barriers to integration can be seen in both the PRAP development process and in individual site projects. These include:

(1)   Sectoral specialization: the PRAP process, as seen in Dien Bien, was led by forestry officials and primarily focused on narrow interpretations of forests, and did not involve much input from agricultural or climate adaptation offices. The minimal inputs of the Ministry of Environment and Natural Resources, the key climate adaptation ministry, in REDD+ policy development below the national level is further evidence of this disengagement.

(2)   A primary focus of REDD+ policy on technical measures rather than livelihoods: in the PRAP for Dien Bien and in national PRAP development guidance, there was far more attention devoted to reference levels of deforestation and carbon emissions equations than to outlining ways to involve local people in participatory forestry projects with livelihood benefits. The Kon Tum CCP project similarly spent time focused on meeting technical requirements for selling emissions reductions on the market, without interim livelihood activities.

(3)   A focus primarily on emissions rather than co-benefits or multipurpose trees: in the NRAP and in Dien Bien's PRAP, more attention was paid to maintaining large intact forest areas or plantations rather than support for small scattered tree plantings, such as in agroforestry or shelterbelts that would benefit farmers. A focus on maximizing emissions levels in REDD+ through extensive forestry thus might create a disincentive for more adaptive measures in local household-based forestry.

As we have shown, the potentials for integration of adaptation activities into REDD+ in Vietnam are there. Climate vulnerable households in our study sites, like many place in Vietnam, are already feeling the effects of some climate and weather changes now [57], and see natural disasters as serious risks threatening their families' wellbeing. REDD+ activities that provide ways to strengthen the adaptive capacity of these households would therefore be extremely useful and welcome. Future PRAP development in the remaining forested provinces of Vietnam would do well to consider better integration of adaptation considerations in future planning, and site-specific projects clearly need to learn from previous lessons with regard to the necessity of considering both livelihoods and forest outcomes simultaneously to avoid increasing the climate vulnerability of participating households.

Acknowledgments: This research was made possible by a grant from the National Science Foundation Geography and Regional Science Division for the project "Downscaling REDD+ policies in developing countries: Assessing the impact of carbon payments on household decision-making and vulnerability to climate change in Vietnam" (grant #11028793) to McElwee, support to McElwee from Hatch funding of the US National Institute for Food and Agriculture (NIFA), and a US Agency for International Development (USAID) Partnerships for Enhanced Engagement in Research grant to the CRES, Tropenbos and PanNature for the project: "Research and capacity building on REDD+, livelihoods, and vulnerability in Vietnam: developing tools for social analysis of development planning". Dao Minh Truong contributed the figures. Additional people who contributed to the collection of data include Dao Minh Truong, Le Trong Toan, Ha Thi Thu Hue, Ha Thi Tu Anh, and Nguyen Xuan Lam. Thanks are due to Lo Quang Chieu, Vice-Director of Department of Agriculture and Rural Development in Dien Bien, the Hieu Commune People's Committee in Kon Tum, and Doan Van Thanh, Le Thi Kim Anh, Ho Tuan Quang, Nguyen Minh Tan and Nguyen Van Thieu in Kien Giang who facilitated fieldwork in these provinces.

Author Contributions: All authors conceived and designed the research project; V.H.T.N., D.V.N. N.H.T., H.V.T.L., T.P.N., and H.D.T.V. conducted the fieldwork; all authors contributed to writing the paper.

Conflicts of Interest: The authors declare no conflict of interest. The funding sponsors had no role in the design of the study; in the collection, analyses, or interpretation of data; in the writing of the manuscript, and in the decision to publish the results.

References

1. Ravindranath, N.H. Mitigation and adaptation synergy in forest sector. *Mitig. Adapt. Strateg. Glob. Chang.* **2007**, *12*, 843–853. [CrossRef]
2. Matocha, J.; Schroth, G.; Hills, T.; Hole, D. Integrating climate change adaptation and mitigation through agroforestry and ecosystem conservation. In *Agroforestry—The Future of Global Land Use*; Advances in Agroforestry; Springer: Dordrecht, The Netherlands, 2012; pp. 105–126.
3. Füssel, H.; Klein, R. Climate change vulnerability assessments: An evolution of conceptual thinking. *Clim. Chang.* **2006**, *75*, 301–329. [CrossRef]
4. Intergovernmental Panel on Climate Change. *Climate Change 2007: Synthesis Report*; Intergovernmental Panel on Climate Change and Cambridge University Press: Cambridge, UK, 2008.
5. Pistorius, T. From RED to REDD+: The evolution of a forest-based mitigation approach for developing countries. *Curr. Opin. Environ. Sustain.* **2012**, *4*, 638–645. [CrossRef]
6. Olander, L.P.; Galik, C.S.; Kissinger, G.A. Operationalizing REDD+: Scope of reduced emissions from deforestation and forest degradation. *Curr. Opin. Environ. Sustain.* **2012**, *4*, 661–669. [CrossRef]
7. Gupta, J.; van der Grijp, N.; Kuik, O. *Climate Change, Forests and REDD: Lessons for Institutional Design*; Routledge: London, UK, 2013.
8. Minang, P.A.; van Noordwijk, M.; Duguma, L.A.; Alemagi, D.; Do, T.H.; Bernard, F.; Agung, P.; Robiglio, V.; Catacutan, D.; Suyanto, S.; et al. REDD+ Readiness progress across countries: Time for reconsideration. *Clim. Policy* **2014**, *14*, 685–708. [CrossRef]
9. Fischer, R.; Hargita, Y.; Günter, S. Insights from the ground level? A content analysis review of multi-national REDD+ studies since 2010. *For. Policy Econ.* **2016**, *66*, 47–58. [CrossRef]
10. Mbatu, R.S. REDD+ research: Reviewing the literature, limitations and ways forward. *For. Policy Econ.* **2016**, *73*, 140–152. [CrossRef]
11. Somorin, O.A.; Visseren-Hamakers, I.J. Integration through interaction? Synergy between adaptation and mitigation (REDD+) in Cameroon. *Environ. Plan. C Gov. Policy* **2016**, *34*, 415–432. [CrossRef]
12. Somorin, O.A.; Brown, H.C.P.; Visseren-Hamakers, I.J.; Sonwa, D.J.; Arts, B.; Nkem, J. The Congo Basin forests in a changing climate: Policy discourses on adaptation and mitigation (REDD+). *Glob. Environ. Chang. Part A* **2012**, *22*, 288–298. [CrossRef]
13. Pham, T.T.; Moeliono, M.; Locatelli, B.; Brockhaus, M.; Gregorio, M.; Mardiah, S. Integration of adaptation and mitigation in climate change and forest policies in Indonesia and Vietnam. *Forests* **2014**, *5*, 2016–2036.
14. Pramova, E.; Di Gregorio, M.; Locatelli, B. *Integrating Adaptation and Mitigation in Climate Change and Land-Use Policies in Peru*; CIFOR: Bogor, Indonesia, 2015.
15. Locatelli, B.; Evans, V.; Wardell, A.; Andrade, A.; Vignola, R. Forests and climate change in Latin America: Linking adaptation and mitigation. *Forests* **2011**, *2*, 431–450. [CrossRef]
16. West, S. *REDD+ and Adaptation in Nepal*; REDDNet Case Study; REDDNet: Bangkok, Thailand, 2012.
17. McFarland, W. *Synergies between REDD+ and Adaptive Capacity to Climate Change at the Local Level—A Ghana Case Study*; REDDNet Case Study; REDDNet: Bangkok, Thailand, 2012.
18. Kongsager, R.; Locatelli, B.; Chazarin, F. Addressing climate change mitigation and adaptation together: A global assessment of agriculture and forestry projects. *Environ. Manag.* **2015**, *57*, 271–282. [CrossRef] [PubMed]

19. Kongsager, R.; Corbera, E. Linking mitigation and adaptation in carbon forestry projects: Evidence from Belize. *World Dev.* **2015**, *76*, 132–146. [CrossRef]
20. Takacs, D. Carbon into gold: Forest carbon offsets, climate change adaptation, and international law. *Hastings West-Northwest J. Int. Environ. Law* **2009**, *15*, 39.
21. Fry, I. Reducing emissions from deforestation and forest degradation: Opportunities and pitfalls in developing a new legal regime. *Rev. Eur. Community Int. Environ. Law* **2008**, *17*, 166–182. [CrossRef]
22. Voigt, C.; Ferreira, F. The Warsaw Framework for REDD+: Implications for national implementation and access to results-based finance. *Carbon Clim. Law Rev. (CCLR)* **2015**, *2*, 113–129.
23. Cerbu, G.A.; Swallow, B.M.; Thompson, D.Y. Locating REDD: A global survey and analysis of REDD readiness and demonstration activities. *Environ. Sci. Policy* **2011**, *14*, 168–180. [CrossRef]
24. UNFCCC. The Paris Agreement. Available online: http://unfccc.int/files/essential_background/convention/application/pdf/english_paris_agreement.pdf (accessed on 31 October 2015).
25. Ingalls, M.L.; Dwyer, M.B. Missing the forest for the trees? Navigating the trade-offs between mitigation and adaptation under REDD. *Clim. Chang.* **2016**, *136*, 353–366. [CrossRef]
26. Poudyal, M.; Ramamonjisoa, B.S.; Hockley, N.; Rakotonarivo, O.S.; Gibbons, J.M.; Mandimbiniaina, R.; Rasoamanana, A.; Jones, J.P.G. Can REDD+ social safeguards reach the "right" people? Lessons from Madagascar. *Glob. Environ. Chang.* **2016**, *37*, 31–42. [CrossRef]
27. Jagger, P.; Sills, E.O.; Lawlor, K.; Sunderlin, W.D. *A Guide to Learning about Livelihood Impacts of REDD+*; Center for International Forestry Research (CIFOR) Occasional Paper 56; CIFOR: Bogor, Indonesia, 2010.
28. Corbera, E.; Brown, K. Offsetting benefits? Analyzing access to forest carbon. *Environ. Plan. A* **2010**, *42*, 1739–1761. [CrossRef]
29. Leach, M.; Mearns, R.; Scoones, I. Environmental entitlements: Dynamics and institutions in community-based natural resource management. *World Dev.* **1999**, *27*, 225–247. [CrossRef]
30. Ribot, J.C.; Peluso, N.L. A theory of access. *Rural Sociol.* **2003**, *68*, 153–181. [CrossRef]
31. Osborne, T. Tradeoffs in carbon commodification: A political ecology of common property forest governance. *Geoforum* **2015**, *67*, 64–77. [CrossRef]
32. Atela, J.O.; Minang, P.A.; Quinn, C.H.; Duguma, L.A. Implementing REDD+ at the local level: Assessing the key enablers for credible mitigation and sustainable livelihood outcomes. *J. Environ. Manag.* **2015**, *157*, 238–249. [CrossRef] [PubMed]
33. Kashwan, P. Forest policy, institutions, and REDD+ in India, Tanzania, and Mexico. *Glob. Environ. Polit.* **2015**, *15*, 95–117. [CrossRef]
34. Ayers, J.M.; Huq, S. The value of linking mitigation and adaptation: A case study of Bangladesh. *Environ. Manag.* **2008**, *43*, 753–764. [CrossRef] [PubMed]
35. Fujisaki, T.; Hyakumura, K.; Scheyvens, H.; Cadman, T. Does REDD+ ensure sectoral coordination and stakeholder participation? A comparative analysis of REDD+ national governance structures in countries of Asia-Pacific region. *Forests* **2016**, *7*, 195. [CrossRef]
36. Fankhauser, S.; Schmidt-Traub, G. From adaptation to climate-resilient development: The costs of climate-proofing the Millennium Development Goals in Africa. *Clim. Dev.* **2011**, *3*, 94–113. [CrossRef]
37. Thornbush, M.; Golubchikov, O.; Bouzarovski, S. Sustainable cities targeted by combined mitigation–adaptation efforts for future-proofing. *Sustain. Cities Soc.* **2013**, *9*, 1–9. [CrossRef]
38. Stromberg, P.M.; Esteban, M.; Gasparatos, A. Climate change effects on mitigation measures: The case of extreme wind events and Philippines' biofuel plan. *Environ. Sci. Policy* **2011**, *14*, 1079–1090. [CrossRef]
39. Murdiyarso, D.; Brockhaus, M.; Sunderlin, W.D.; Verchot, L.V. Some lessons learned from the first generation of REDD+ activities. *Curr. Opin. Environ. Sustain.* **2012**, *4*, 678–685. [CrossRef]
40. Korhonen-Kurki, K.; Brockhaus, M.; Bushley, B.; Babon, A.; Gebara, M.F.; Kengoum, F.; Pham, T.T.; Rantala, S.; Moeliono, M.; Dwisatrio, B.; et al. Coordination and cross-sectoral integration in REDD+: Experiences from seven countries. *Clim. Dev.* **2016**, *8*, 458–471. [CrossRef]
41. Atela, J.O.; Quinn, C.H.; Minang, P.A.; Duguma, L.A.; Houdet, J.A. Implementing REDD+ at the national level: Stakeholder engagement and policy coherences between REDD+ rules and Kenya's sectoral policies. *For. Policy Econ.* **2016**, *65*, 37–46. [CrossRef]
42. Pramova, E. *Integrating Adaptation into REDD+: Potential Impacts and Social Return on Investment in Setulang, Malinau District, Indonesia*; Center for International Forestry Research (CIFOR): Bogor, Indonesia, 2013.

43. Robledo, C.; Kanninen, M.; Pedroni, L. *Tropical Forests and Adaptation to Climate Change*; Center for International Forestry Research (CIFOR): Bogor, Indonesia, 2005.
44. Forsius, M.; Anttila, S.; Arvola, L.; Bergström, I.; Hakola, H.; Heikkinen, H.I.; Helenius, J.; Hyvärinen, M.; Jylhä, K.; Karjalainen, J.; et al. Impacts and adaptation options of climate change on ecosystem services in Finland: A model based study. *Curr. Opin. Environ. Sustain.* **2013**, *5*, 26–40. [CrossRef]
45. Schoene, D.H.F.; Bernier, P.Y. Adapting forestry and forests to climate change: A challenge to change the paradigm. *For. Policy Econ.* **2012**, *24*, 12–19. [CrossRef]
46. Kalame, F.B.; Nkem, J.; Idinoba, M.; Kanninen, M. Matching national forest policies and management practices for climate change adaptation in Burkina Faso and Ghana. *Mitig. Adapt. Strateg. Glob. Chang.* **2008**, *14*, 135–151. [CrossRef]
47. Munang, R.; Thiaw, I.; Alverson, K.; Mumba, M.; Liu, J.; Rivington, M. Climate change and ecosystem-based adaptation: A new pragmatic approach to buffering climate change impacts. *Curr. Opin. Environ. Sustain.* **2013**, *5*, 67–71. [CrossRef]
48. Munang, R.; Thiaw, I.; Alverson, K.; Liu, J.; Han, Z. The role of ecosystem services in climate change adaptation and disaster risk reduction. *Curr. Opin. Environ. Sustain.* **2013**, *5*, 47–52. [CrossRef]
49. Locatelli, T.; Binet, T.; Kairo, J.G.; King, L.; Madden, S.; Patenaude, G.; Upton, C.; Huxham, M. Turning the tide: How blue carbon and payments for ecosystem services (PES) might help save mangrove forests. *AMBIO J. Hum. Environ.* **2014**, *43*, 981–995. [CrossRef] [PubMed]
50. Pramova, E.; Locatelli, B.; Djoudi, H.; Somorin, O.A. Forests and trees for social adaptation to climate variability and change. *WIREs Clim. Chang.* **2012**, *3*, 581–596. [CrossRef]
51. Seppälä, R.; Buck, A.; Katila, P. *Adaptation of Forests and People to Climate Change—A Global Assessment Report*; International Union of Forest Research Organizations (IUFRO) World Series vol. 22; IUFRO: Vienna, Austria, 2009.
52. Vignola, R.; Locatelli, B.; Martinez, C.; Imbach, P. Ecosystem-based adaptation to climate change: What role for policy-makers, society and scientists? *Mitig. Adapt. Strateg. Glob. Chang.* **2009**, *14*, 691–696. [CrossRef]
53. Sikor, T. (Ed.) *The Justices and Injustices of Ecosystem Services*; Routledge: London, UK, 2013.
54. Chong, J. Ecosystem-based approaches to climate change adaptation: Progress and challenges. *Int. Environ. Agreem.* **2014**, *14*, 391–405. [CrossRef]
55. Hoa, N.T.; Hasegawa, T.; Matsuoka, Y. Climate change mitigation strategies in agriculture, forestry and other land use sectors in Vietnam. *Mitig. Adapt. Strateg. Glob. Chang.* **2012**, *19*, 15–32. [CrossRef]
56. Dilley, M.; Chen, R.S.; Deichmann, U. *Natural Disaster Hotspots: A Global Risk Analysis Synthesis Report*; International Bank for Reconstruction and Development/The World Bank and Columbia University: Washington, DC, USA, 2005.
57. McElwee, P.D.; Nghiem, P.T.; Van Hue, L.T.; Huong, V.; Be, N.V.; Tri, L.Q.; Trung, N.H.; Tuan, L.A.; Dung, L.C.; Duat, L.Q.; et al. *Social Dimensions of Climate Change in Vietnam*; World Bank Discussion Paper No 17; World Bank: Washington, DC, USA, 2010.
58. Dang, H.; Michaelowa, A.; Tuan, D. Synergy of adaptation and mitigation strategies in the context of sustainable development: The case of Vietnam. *Clim. Policy* **2003**, *3*, S81–S96. [CrossRef]
59. Beckman, M. Converging and conflicting interests in adaptation to environmental change in central Vietnam. *Clim. Dev.* **2011**, *3*, 32–41. [CrossRef]
60. Bruun, O. Sending the right bill to the right people: Climate change, environmental degradation, and social vulnerabilities in Central Vietnam. *Weather Clim. Soc.* **2012**, *4*, 250–262. [CrossRef]
61. Adger, W.N.; Kelly, P.M.; Nguyen, H.N. (Eds.) *Living with Environmental Change*; Routledge: London, UK, 2001.
62. Garschagen, M. Resilience and organisational institutionalism from a cross-cultural perspective: An exploration based on urban climate change adaptation in Vietnam. *Nat. Hazards* **2013**, *67*, 25–46. [CrossRef]
63. Tran, T.; Neefjes, K.; Ta, T.T.H.; Le, N.T. *Vietnam Special Report on Managing the Risks of Extreme Events and Disasters to Advance Climate Change Adaptation*; Vietnam Publishing House of Natural Resources, Environment and Cartography: Institute for Hydrology, Environment and Meteorology (IMHEM) and UN Development Programme (UNDP): Hanoi, Vietnam, 2015; pp. 1–456.
64. Huynh, P.; Resurreccion, B.P. Women's differentiated vulnerability and adaptations to climate-related agricultural water scarcity in rural Central Vietnam. *Clim. Dev.* **2014**, *6*, 226–237. [CrossRef]

65. Delisle, S.; Turner, S. "The weather is like the game we play": Coping and adaptation strategies for extreme weather events among ethnic minority groups in upland northern Vietnam. *Asia Pac. Viewpo.* **2016**. [CrossRef]

66. Socialist Republic of Vietnam. *National Strategy on Climate Change*; Socialist Republic of Vietnam, Prime Minister's Office: Hanoi, Vietnam, 2011.

67. Socialist Republic of Vietnam. *Vietnam's Submission on Reference Levels for REDD+ Results-Based Payments under the UNFCCC*; Submission to the UN Framework Convention on Climate Change by the Socialist Republic of Vietnam: Hanoi, Vietnam, 2016.

68. Socialist Republic of Vietnam. *Intended Nationally Determined Contribution of Viet Nam*; Submission to the UN Framework Convention on Climate Change by the Socialist Republic of Vietnam: Hanoi, Vietnam, 2015.

69. Pham, T.T.; Di Gregorio, M.; Carmenta, R.; Brockhaus, M.; Le, D.N. The REDD+ policy arena in Vietnam: Participation of policy actors. *Ecol. Soc.* **2014**, *19*, 22. [CrossRef]

70. Pham, T.T. *REDD+ Politics in the Media: A Case Study from Vietnam*; Center for International Forestry Research (CIFOR): Bogor, Indonesia, 2011; pp. 1–48.

71. Pham, T.T.; Moeliono, M.; Hien, N.T.; Tho, N.H.; Hien, V.T. *The Context of REDD+ in Vietnam: Drivers, Agents and Institutions*; Center for International Forestry Research (CIFOR): Bogor, Indonesia, 2012; pp. 1–98.

72. Socialist Republic of Vietnam. *National REDD+ Action Plan Vietnam*; Socialist Republic of Vietnam, Prime Minister's Office: Hanoi, Vietnam, 2012; pp. 1–18.

73. Ministry of Agriculture and Rural Development. *Action Plan Framework for Adaptation and Mitigation of Climate Change of the Agricultural and Rural Development Sector Period 2008–2020*; Ministry of Agriculture and Rural Development: Hanoi, Vietnam, 2009.

74. Dung, N.V.; Van, N.H. *Payment for Forest Environmental Services Policy in Vietnam and Its Implications on Sub-National Forest Management Institutions*; Policy Brief; People and Nature Reconciliation (PanNature): Hanoi, Vietnam, 2015.

75. Dung, N.V.; Van, N.H. *Payment for Forest Ecosystem Services Impact Assessment at Sub-National Level. Presentation at National Workshop on PFES and Multi-Stakeholder Participation in Vietnam*; People and Nature Reconciliation (PanNature) and Vietnam Fund for Forests: Hanoi, Vietnam, 2015.

76. Pham, T.T.; Dung, L.N.; Vũ, T.P.; Nguyen, H.T.; Nguyen, V.T. *Forest Land Allocation and Payments for Forest Environmental Services in Four Northwestern Provinces of Vietnam*; Center for International Forestry Research (CIFOR) Occasional Paper 155; CIFOR: Bogor, Indonesia, 2016.

77. Pham, T.T.; Moeliono, M.; Brockhaus, M.; Le, D.N.; Wong, G.Y.; Le, T.M. Local preferences and strategies for effective, efficient, and equitable distribution of PES revenues in Vietnam: Lessons for REDD. *Hum. Ecol.* **2014**, *42*, 885–899. [CrossRef]

78. Dung, L.N.; Loft, L.; Tjajadi, J.S.; Pham, T.T.; Wong, G. *Being Equitable is not Always Fair: An Assessment of PFES Implementation in Dien Bien, Vietnam*; Center for International Forestry Research (CIFOR) Working Paper 205; CIFOR: Bogor, Indonesia, 2016.

79. Ministry of Agriculture and Natural Resources and Japanese International Cooperation Agency. *PRAP Preparation Handbook*; Ministry of Agriculture and Natural Resources and Japanese International Cooperation Agency: Dien Bien, Vietnam, 2014.

80. Adger, W.N. Social vulnerability to climate change and extremes in coastal Vietnam. *World Dev.* **1999**, *27*, 249–269. [CrossRef]

81. Bruun, O.; Casse, T. *On the Frontiers of Climate and Environmental Change: Vulnerabilities and Adaptations in Central Vietnam*; Springer: Dordrecht, The Netherlands, 2013.

82. Socialist Republic of Vietnam. *Decision 204 QD-TTG on the Designation of Poor, Remote and Border Communes for Programme 135*; Socialist Republic of Vietnam: Hanoi, Vietnam, 2016.

83. Van, N.T.H. Embedding Forest Carbon in Vietnam's Forestland Property Relations. Master's Thesis, Forest and Nature Conservation Policy Group, Wageningen University, Wageningen, The Netherlands, 2014.

84. Asian Development Bank. *Ecosystem-Based Approaches to Address Climate Change Challenges in the Greater Mekong Subregion*; Asian Development Bank—Greater Mekong Subregion Environment Operations Center: Manila, Philippines, 2015; pp. 1–8.

85. Institute for Strategy and Planning on Natural Resources and Environment. *Mainstreaming Ecosystem-Based Adaptation in Vietnam*; Institute for Strategy and Planning on Natural Resources and Environment, Ministry of Natural Resources and Environment: Hanoi, Vietnam, 2013.

86. Kengoum, F. *Adaptation Policies and Synergies with REDD+ in Democratic Republic of Congo*; Center for International Forestry Research (CIFOR) Occasional Paper 135; CIFOR: Bogor, Indonesia, 2015.

87. Nkem, J.; Kalame, F.B.; Idinoba, M.; Somorin, O.A.; Ndoye, O.; Awono, A. Shaping forest safety nets with markets: Adaptation to climate change under changing roles of tropical forests in Congo Basin. *Environ. Sci. Policy* **2010**, *13*, 498–508. [CrossRef]

88. Innes, J.L.; Hickey, G.M. The importance of climate change when considering the role of forests in the alleviation of poverty. *Int. For. Rev.* **2006**, *8*, 406–416. [CrossRef]

89. Wang, G.; Wang, T.; Kang, H.; Mang, S.; Riehl, B.; Seely, B.; Liu, S.; Guo, F.; Li, Q.; Innes, J.L. Adaptation of Asia-Pacific forests to climate change. *J. For. Res.* **2016**, *27*, 469–488. [CrossRef]

90. Wang, G.; Mang, S.; Kryzanowski, J.; Guo, F.; Wang, T.; Riehl, B.; Kang, H.; Li, Q.; Innes, J.L. Climate change and forest adaptation in the Asia-Pacific. *J. Geogr. Res.* **2015**, *63*, 1–36. [CrossRef]

91. McElwee, P.D. Forest environmental income in Vietnam: Household socioeconomic factors influencing forest use. *Environ. Conserv.* **2008**, *35*, 147–159. [CrossRef]

92. Red Cross. *Mangrove Plantation in Viet Nam: Measuring Impact and Cost Benefit*; International Federation of Red Cross and Red Crescent Societies: Hanoi, Vietnam, 2012.

93. Schmitt, K. *Protection and Sustainable Use of Coastal Wetlands through Co-Management and Mangrove Rehabilitation with Emphasis on Resilience to Climate Change*; Gesellschaft fur Internationale Zusammenarbeit (GIZ): Soc Trang, Vietnam, 2015.

94. Nam, V.N.; Sasmito, S.D.; Murdiyarso, D.; Purbopuspito, J.; MacKenzie, R.A. Carbon stocks in artificially and naturally regenerated mangrove ecosystems in the Mekong Delta. *Wetl. Ecol. Manag.* **2016**, *24*, 231–244. [CrossRef]

95. McElwee, P.D.; Thanh, N.C. *Review of Three Years Policy of Payments for Forest Environmental Services in Vietnam*; Forest Protection and Development Fund, Ministry of Agriculture and Rural Development and Winrock International: Hanoi, Vietnam, 2015.

forests

MDPI

Article

Does the 'One Map Initiative' Represent a New Path for Forest Mapping in Indonesia? Assessing the Contribution of the REDD+ Initiative in Effecting Forest Governance Reform

Mari Mulyani [1,2,*] and Paul Jepson [1]

[1] School of Geography and the Environment, University of Oxford, Dyson Perrins Building,
     South Parks Road, Oxford OX4 1SW, UK; paul.jepson@ouce.ox.ac.uk
[2] School of Environmental Sciences, University of Indonesia, Gedung C. LtV-VI, Jl. Salemba Raya 4 Salemba,
     Jakarta 10430, Indonesia
*   Correspondence: mari.mulyani@gmail.com or mari.mulyani@ouce.ox.ac.uk;
     Tel.: +44-785-505-3892 or +62-81-197-7068

Academic Editors: Esteve Corbera and Heike Schroeder
Received: 9 October 2016; Accepted: 20 December 2016; Published: 27 December 2016

Abstract: This study investigates one notable result that the REDD+ ('Reducing Emissions from Deforestation and forest Degradation, and enhancing forest carbon stocks and conservation') initiative effected within Indonesia's forest institutions. It argues that during its interplay with existing National forest institutions REDD+ produced a significant benefit; namely, the 'one map initiative' (OMI) being the government's response to the call for greater transparency and enabling of REDD+ implementation. It asks: "Does the 'One Map Initiative' signify a switch to a new path of map-making, or is it just another innovation within an existing path dependence of forest governance?" Through eighty semi-structured interviews with 'REDD+ policy actors' and the deployment of 'path dependence' theory, this study seeks to determine the extent to which the REDD+ initiative created a 'critical juncture' (i.e., momentum for institutions to move to a new path). This study maps the institutional path dependence within forest-mapping as a means for the state to gain control of forest resources. In its development process the OMI has shown its ability to break the old path-dependence of map-making (e.g., lack of transparency, low level of public participation, and poor coordination amongst ministries). Moreover, this paper identified several historical events (i.e., 'critical junctures') that preceded the REDD+ initiative as contributing factors to the relative success of REDD+ in effecting forest governance reform.

Keywords: REDD+; One Map Initiative; Indonesia; forest governance reform; cross-sectoral coordination; participation; transparency; path dependence; critical juncture

---

1. Introduction

Maps, established via a variety of statistical and cartographic techniques, are fundamental tools in the construction of state and bureaucratic territories and the governance of natural resources, including forests [1]. As such, maps, and spatial data more generally, are deeply entwined in the construction of institutions governing forest resources. The quality of maps, and the extent to which they are assessable to other sectors and citizens, influences the political economy and democratic conduct of a nation.

Indonesia is the world's largest archipelago, fourth largest political unit, and third largest tropical forest nation. The mapping of forest resources has been central to the nation's political economy since the Dutch colonial era [2]. Since the country's independence in 1945, and in particular during the

Suharto New Order regime (1967–1998), forest resources were deployed to rapidly increase GDP, attract foreign investment, and consolidate and structure power across a vast archipelago [3,4]. More recently, the REDD+ ('Reducing Emissions from Deforestation and forest Degradation, and enhancing forest carbon stocks and conservation') initiative has created a new impetus for map-making practices, deploying both advanced technical techniques, but also norms of transparency and participation. This particular REDD+ contribution does not obscure the possibility of REDD+ being used by political and business elites to promote their own interests. For instance, concerns were expressed that REDD+ can provide validation for 'conservation-justified enclosures' by both national and district governments to reassert control over vast forest areas, and for oil palm companies to lobby decision-makers in ways that risk corrupting the forest land allocation process [5]. Moreover, Indonesia's map-making institutions have long been characterised by 'clientelism' ('clientelism' describes a long-standing form of social organisation, also known as 'patron–client relations', which involves complex chains of personal bonds between patrons and their clients founded on mutual advantage, whereby the patron offers resources and protection; in return their clients provide loyalty, services, and political support) and 'crony capitalism' ('crony capitalism' describes an economy which depends on close relationships between government officials and business people, elevating personal profit over public interest or the rule of law; it can be exhibited by favouritism in the distribution of legal permits, government grants, special tax breaks, or other forms of state intervention) [6] and have developed to further political interests: as a means for controlling territory and governing the access of political, business, and individual interests to land and its resources [7,8].

The 'One Map Initiative' (OMI), a national strategic policy instructed by President Yudhoyono in December 2010 [9], is a significant development with the potential to transform Indonesian forestry institutions. This is because the policy instructs the creation of a national spatial data infrastructure and the requirement that, (i) ministries, particularly those with land-based competencies such as Forestry, Agriculture, Public Works and Housing, work together to create a single authoritative state land-cover map; (ii) the public are provided with free access to digital land-cover maps and the opportunity to evaluate them; and (iii) indigenous customary lands are integrated into the state map [9]. This has the potential to address the problem of different land-cover maps being used by different ministries and at different levels of government, thereby addressing issues of overlapping concession areas, corruption, land-related conflicts, and social unrest [9,10].

The OMI was developed in the context of REDD+, a major international policy initiative that emerged from the United Nations Framework Convention on Climate Change (UNFCCC). However, this is not to suggest that the OMI is an inherent element of REDD+ institutions. Instead, as argued within this article, during its interplay with Indonesia's forest institutions the REDD+ initiative has prompted the development of the OMI. REDD+ is intended to provide developing countries with financial incentives to produce measurable reductions in carbon emissions beyond what would have occurred without it [11,12]. The REDD+ mechanism includes several activities that reduce emissions from forest deforestation and degradation, conserve and enhance forest carbon stocks, and manage forests sustainably [12].

Of particular importance in relation to the OMI is the performance-based approach within the REDD+ mechanism, whereby reductions in forest carbon emissions must be demonstrated before payments can be made [13]. This requires a transparent and accountable system for (a) measuring carbon stocks; and (b) monitoring, reporting, and verifying (MRV) reductions in carbon emissions [12,14]. This entails a significant investment in the development of accurate forest datasets and mapping across multiple spatial scales in REDD+ host countries [10,15]. In Indonesia, REDD+ has generated an 'interplay' between internationally-established REDD+ institutions that require the development of certain policies within national-level forest-related institutions.

Indonesia's forest governance institutions are complex [16], characterised by the uncertainty and ambiguity of the legal systems surrounding forest access and land ownership [17,18] and a rent-seeking environment that maintains deeply-embedded clientelist practices across government and society [3,19,20]. This paper examines the origin of the OMI and its implications for future forest governance and REDD+ implementation in Indonesia. It asks whether the One Map Initiative might signify a move to a new path of map-making that foregrounds transparency, coordinated cross-sectoral, and wider public participation, or whether it is just another innovation within Indonesia's existing forestry institutions. We employ the concept of 'path dependency' within 'historical institutionalism' and examine whether REDD+ has been instrumental in creating a 'critical juncture' in the form of momentum for Indonesian resource and land governance institutions to move to a new path.

By positioning REDD+ in this broader context of institutional interplay and political economy, this paper complements and extends literature on REDD+ governance that analyses (i) the challenges and opportunities for the REDD+ initiative to effect forest governance reform [21–30] and (ii) the potential social and economic impacts of REDD+ policies on local people and communities' participation on project implementations [31–36]. Moreover, concepts from political economy have been deployed to explore the challenge of REDD+ in creating incentives for forest conservation that are able to 'out-compete' other interests, such as timber, mining, and agricultural economics [10,37–39]. This paper addresses the need for more empirical evidence on the role of REDD+ in creating new institutional path-ways and the extent to which the REDD+ policy process is able to transform forest sector and cross sector regulations in host countries.

Indonesia is a key target country for REDD+ implementation for four important reasons: (a) it supports the third largest area of tropical rainforest with the fourth-largest forest carbon stock [40,41]; (b) an estimated 80% of the nation's carbon emissions result from forest clearance and degradation [42]; (c) it has an annual deforestation rate of around 1 million ha [42]; and (d) its forest governance has historically been regarded as weak [16,17]. The adoption of REDD+ policies in Indonesia is supported by nearly US\$2 billion of donor pledges including US\$1 billion from the Government of Norway [30,43,44]. Nonetheless, the potential of REDD+ to initiate a break from institutional path dependence is questioned since previous international policy initiatives have achieved minimal success in effecting forest governance reform [45,46].

2. Materials and Methods

We employ the concept of path dependency from within historical institutionalism to ascertain the role of REDD+ in the establishment of the One Map Initiative. Our analysis begins with a review of these conceptual frameworks and then applies them to Indonesia's forest institutions in order to highlight the strength of its institutional path dependence. The focus of our analysis is on identifying critical junctures relating to map-making institutions and particularly on the role of REDD+ as a driving force for the OMI movement. Finally, we discuss the implication of the OMI for future forest governance and REDD+ implementation.

2.1. Conceptual Framework

Historical Institutionalism views institutions as formal and informal procedures, norms and conventions, embedded in the organisational structure of a political economy that determine the behaviour of political and economic actors [47–49]. It asserts that history matters because the 'how and why' of policy choices made at one point in time affect choices at subsequent points [50,51]. Furthermore, historical institutionalism emphasises the asymmetry of power associated with the development and operation of institutions that produce forms of path dependency. Institutions tend to serve the interests of the powerful and give some groups greater access than others to the decision-making process [47]. For instance, how actors distribute and gain access to a decision-making process can be explained through an 'Actor-centred power' (ACP) approach which highlights three elements of power, namely coercion, (dis)incentives, and dominant information [52], which are summarised as follows.

Coercion builds upon the source of physical actions as the power to alter the behaviour of a subordinate, or the threat to use that force supported by laws [52]. This is particularly relevant when explaining how state actors use state laws and regulations to effect the obedience of a subordinate or relevant stakeholder. Laws can provide certain state actors with the authority to control and enforce sanctions [52]. Disincentives/incentives are defined as changing the behaviour of a subordinate through disadvantages or advantages [52]. In the realm of forest governance this is evident in the awarding of forest concession permits, or access to decision-making processes by political elites to their supporters. The opposite is effected through preventing access to policy-making or economic advantages (e.g., the revocation of permits) [30]. 'Dominant information' is particularly evident during policy-making processes, whereby powerful state actors (considered as the principal source of information) hold the power to influence a policy outcome, as they decide the type and verifiability of the information distributed, and to which actors the information is available [30,53].

This ACP framework provides the means to analyse how state agencies compete for resources (e.g., budget and staff), political authority, and mandate [30,53,54]. They seek to influence the development and operation of institutions, often by establishing coalitions with domestic and international actors; during this process their power may be reduced or increased [30,53,54]. As such, the state is not considered as a 'neutral broker' among competing interests, but as a complex of institutions capable of structuring the outcomes of group conflict to forwarding its own interests [47].

'Path dependency' theory posits that institutions are persistent because supporters believe deviation will compromise their political and economic interests [48,55,56]. Because institutions affect behaviour, over time they shape actors' preferences [51]. Moreover, the institutionalisation of policies leads to the formation of powerful interest groups that resist change [57]. This phenomena is termed 'institutional stickiness' whereby policy actors prefer a given institution and resist change, even when such change would lead to a better overall public policy outcome [58]. Whilst some believe that institutions remain essentially stable (at equilibrium) until they are faced with external or exogenous shock [48], others argue that institutional change ultimately results from a mixture of exogenous and endogenous pressures that lead to a series of path-changing initiatives [51,55]. Endogenous sources of change may include actions undertaken by actors within the institutional system as a response to the exogenous factors of change [59].

The concept of 'critical juncture' refers to a crucial founding moment that sets institutional development on a new path and that signifies a point when institutions move from a period of stability to change [48,60,61]. Critical junctures have been associated with periods of institutional 'fluidity' where structural institutions become less constrained and actors' decisions are freer, thereby enabling significant change to take place [59,62] During such periods contingency is enhanced, but whilst contingency implies that wide-ranging change is possible, and even likely, re-equilibrium to older institutional paths is not excluded [62,63]. Junctures become 'critical' when a particular decision is taken that sets institutions on new paths or trajectories that are subsequently difficult to alter [62,64].

Whilst there is no uniform idea as to what constitutes a critical juncture, three characteristics are often cited: (i) to constitute a 'juncture' the events must be brief relative to the duration of the relevant institutional path-dependent processes [62]; (ii) they are invariably initiated by an economic crisis, military conflict, and/or the presence of coalitions calling for governance reform [59,62]; and (iii) on account of structural institutions becoming less constrained, actors shape outcomes in a more voluntary fashion than normal circumstances permit [62,64]. These characteristics guide our assessment of the extent to which OMI signifies a critical juncture, or is just another innovation within the path-dependent institutions of Indonesia's forest governance, and the extent to which REDD+ can be considered as an initiating force.

2.2. Data Collection and Analysis

Our analysis draws on the findings of a broader body of research on the interplay between REDD+ and Indonesia's forest institutions across multiple levels of government [65,66]. This lead author conducted an interview survey of individuals engaged in the REDD+ policy process in Indonesia, reviewed literature on the history of Indonesian forest policy and contemporary policy documents, and participated in workshops convened by the REDD+ Task Force and local NGOs (non-governmental organisations). The latter included workshops on the topics of the forest moratorium and OMI conducted in Jakarta on 4/5 and 10/11 September 2012 respectively. The following policy documents, predominantly but not exclusively, were reviewed and used as the basis for analysis: (i) 'REDD+ Indonesia. A Catalyst for Change', a document released by the REDD+ Task-Force in 2012 [9]; (ii) Letter of Intent between the government of Indonesia and Norway on 'Cooperation on REDD+' in 2010 [44]; (iii) Presidential Instruction (PI) No. 10/2011 (renewed via Presidential Instruction No. 6/2013 issued in May 2013) on 'the suspension of licensing new concessions in the natural/primary forests and peatlands' [67]; (iv) Indonesia's Basic Forestry Law No. 41/1999 [68] and Geospatial Information Law No. 4/2011 [69]; and (v) The recommendations of the Corruption Eradication Commission after reviewing the forestry sector in 2010 [70,71].

Interviewees were selected to represent a wide range of perspectives on REDD+ and organisational types, comprising multi-level government officials (23), forest/carbon private actors (11), donor agencies (12), NGOs/civil society organisations (17), and academia/researchers (17). Interviews were conducted during the periods March–June 2011 (50 interviews) and April–September 2012 (20 interviews). A further ten OMI-specific interviews were conducted with key informants directly involved in the implementation of the OMI during March 2015 and repeated in March 2016 to seek the latest information on the OMI's progress. To assure anonymity while gaining meaningful insights, quotes from interviewees are referenced in a logical and consistent system (i.e., interviewees 1–80) and are referred to by their organisations.

Primarily, this research takes a qualitative approach for data analysis. Interviewees' responses on two main topics, (i) the origin of the OMI; and (ii) its implications for forest governance and REDD+ implementation, were transcribed, examined, and classified for common themes as well as those that were unique. This process resulted in six major categories of 'governance themes' (that addressed point (ii)) which emerged from the frequent and dominant responses, namely (i) coordination; (ii) transparency; (iii) participation; (iv) cost-efficiency; (v) improved data quality; and (vi) protection of indigenous peoples' land (discussed in Section 3.4). With regard to point (i), the origin of the OMI, interviewees' responses largely suggest that the REDD+ initiative played a major role in prompting the development of the OMI (discussed in Section 3.3).

3. Results

3.1. A Brief History of Indonesia Forest Law and Mapping Institutions

Present-day institutional pathways can be traced back to the Dutch colonial era and the 1870 Agrarian Law. This declared that all land without certified ownership was to be the domain (property) of the state [72,73]. This law reflected broader policies amongst colonial powers that sought to ensure the efficient exploitation of forests and in so doing framed forest resources as public goods that need to be managed in a strategic and rational manner [74].

In the early period of Indonesian independence, the Sukarno presidency (1945–1965) replaced the 1870 Agrarian Law with the 1960 Basic Agrarian Law (BAL). Whilst this introduced measures to improve justice in land rights and recognise customary law [73], the BAL retained the path of state sovereignty of forest resources by stating that state law is superior to customary law [75]. Importantly, the BAL introduced the right of peasants to settle, cultivate, and claim ownership of up two hectares of forest land and designated competency for issuing land titles to district level administrations [75,76]. Figure 1 outlines the Basic Agrarian Law (BAL) and Basic Forestry Law (BFL) in Indonesia's history.

This latter provision introduced a disjuncture between national level forest zoning and local forest conversion and created tension between local (district) government and higher levels of government. This was in part because the collection of rents associated with the issuance of land title permits provided local administrations with a valuable source of revenue, and, as a result, they had incentive to uphold forest zoning plans prepared by the national government [75].

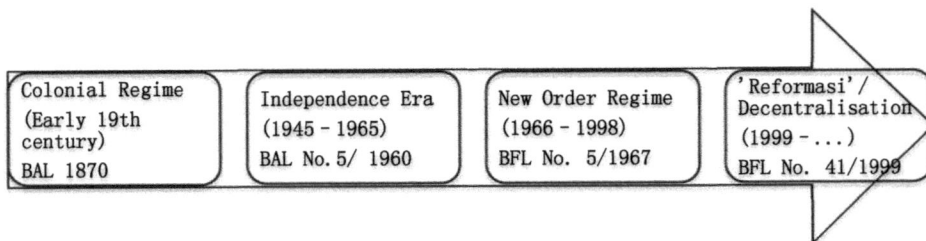

Figure 1. The Basic Agrarian Law (BAL) and Basic Forestry Law (BFL) in Indonesia's history.

The Suharto New Order regime (1966–1998) introduced separate laws governing forest resources and agriculture. The 1967 Basic Forestry Law (BFL) reasserted state control over the majority of Indonesia's 120–133 million ha of forest land, being 62%–69% of Indonesia's land area [10]. Importantly, the BFL introduced powers that enable the government to grant and manage forest concessions. This provision enabled the Suharto regime to consolidate and extend its power base through the awarding of lucrative forest concessions to supporters and led to the emergence of forestry institutions characterised by political cronyism and capital accumulation [3,76].

The imperative of the regime to control the allocation of lucrative forest concessions across a vast nation of 27 provinces (under the current administration of President Joko Widodo, there are now 35 provinces) and five major islands prompted three significant policies in the early 1980s: (i) elevating the department of forestry within the Ministry of Agriculture to a separate Ministry of Forestry (MoFor); (ii) establishing a state-level forest zoning and classification map known as the 'TGHK' (*Tata Guna Hutan Kesepakatan* or Forest Land Use Consensus); and (iii) establishing provincial representative offices of MoFor called *Kanwil Kehutanan* in 1983.

The TGHK mapping project (1981–1985) led by the MoFor further consolidated and stabilised state power over forest lands [37]. The resulting 1:500,000 scale TGHK 1983 was published in 1983 without meaningful input from either district government, civil society, or communities, and classified State forest into six use categories, namely: protection forest, conservation forest, limited production forest, production forest, conversion forest, and unclassified lands.

An important policy impetus for the TGHK was the continuation and acceleration of the transmigration programme by the New Order regime. Originally conceived by the Dutch colonial administration in 1905, this programme involved moving people from the densely populated islands of Java, Lombok, and Madura to the larger and less populated islands, principally Sumatra, Kalimantan (Indonesian Borneo), Sulawesi, and West Papua [77]. It was supported by a series of World Bank loans between 1976 and 1992 which received increasingly vigorous criticism from civil society groups for the lack of social and environmental safeguards. Among other things, the transmigration programme was criticised for accelerating deforestation through the issuance of concession licences, and through this, further embedded corrupt practices within the political economy of Indonesia [78].

To address some of these criticisms the office of Regional Physical Planning Programme for Transmigration ('RePPProt'), based within the National Coordination Agency on Survey and Mapping ('Bakosurtanal'), led a mapping project to establish a more reliable basis for transmigration. Funded by the UK Overseas Development Administration, this map-making project applied advanced techniques (for the time) to produce maps of land systems, land use, and land status and included a revision of the TGHK at the larger scale of 1:250,000 [9].

Notwithstanding this effort, the data on forests and land-cover maps still varied amongst ministries and was not well integrated [9]. Moreover, complexity and uncertainty in the governance of forest land and resources were exacerbated by government ministries and local government often using inaccurate and outdated maps, resulting in conflicting customary and statutory land tenure and overlapping forest-land concessions [10,30].

In 1992 the Spatial Planning Law No. 24/1992 was passed to reconcile the issue of different ministries working from different and/or their own land zoning maps. It introduced rules for spatial land use planning called 'RTRW' (*Rencana Tata Ruang Wilayah*/Regional Spatial Land Use Planning) under the authority of the National Development Planning Agency (Bappenas) and its provincial-level offices (Bappeda) [37]. In response, the MoFor engaged in the production of its own 'integrated maps' (*Peta Paduserasi*) published in 1997, but these were refused by some provinces who continued to refer to the earlier 'TGHK' map [37].

The above explains how institutions, via basic forestry and agrarian laws and policies related to map-making, were developed to ensure the strength of path dependency of the country's political economy. Through laws and regulations state actors, in particular the MoFor, exercise the 'coercive element' of power to ensure the obedience of citizens and the efficient exploitation of forest resources by awarding forest concession permits to supporters ('incentives'). At the same time, MoFor controlled the 'dominant information' on forest concession and map data by limiting public participation in the map-making process. Moreover, the above also reveals how state agencies (such as MoFor and Bappenas, amongst others) and district governments competed for the political mandate and resources to lead the map-making process.

3.2. The Fall of the Suharto Regime, 'Reformasi', and Forestry Institutions

The Suharto New Order regime was brought to an end in 1998 by the reform ('reformasi') movement, a response to the Asian financial crisis of 1997 which served as an 'exogenous shock' to the political system. ('Reformasi', a local term for 'Reformation', describes a public movement in Indonesia, driven by students and other intellectual actors, following the Asian financial crisis in 1997 which led to the fall of the Suharto regime in 1998. This movement demanded a governance system free of corruption, collusion, and nepotism). Viewed from a critical juncture perspective, 'reformasi' initiated a period of institutional fluidity including public demonstrations and media calling for an end to 'KKN' (*Korupsi, Kolusi, Nepotisme*) or corruption, collusion, and nepotism. This period was also marked by increasing civil society movements which challenged the long-standing collusion between the state and forest industrialists.

The 'reformasi' movement presented a major challenge to the unity of the Indonesian state and forestry institutions. In response, the Habibie government (1998–1999) acted to manage the political power of resource-rich provinces by delegating regional autonomy for resource planning, and devolving decision-making to the third (district) tier of government [3]. This policy was effected via Law No. 22/1999 on regional autonomy and Law No. 25/1999 on fiscal redistribution between the central and regional governments.

Under a new Basic Forestry Law (No. 41/1999) the state devolved management of all forest lands (other than national parks and reserves) to regional governments according to standards set by MoFor, with control over land tenure rights remaining with the National Land Bureau (NLB). This led to an increase in the overlapping authority and an intense power struggle between different levels of government and amongst ministries, representing a chaotic implementation of the decentralisation policy [16,17,46]. It also gave rise to the phenomenon of 'wild logging' ('wild logging', or *penebangan liar*, describes the illegal and uncontrolled logging of Indonesia's forests, particularly during 1999–2000; it is associated with the emergence of logging networks at the district level, creating a web of political, economic, and social exchanges amongst state and non-state actors that operate against State legal norms) involving government officials, community leaders, and 'local entrepreneurs' controlling logging networks [16,17].

In an effort to control this situation and effect forest governance reform, several new initiatives were implemented including the improvement of forest data and integrated forest-land maps (see Figure 2), most of which were supported by international donors. For instance, building on FAO support (1986–1991), in 2002 the World Bank financed a Forest Datasets Integration project which consolidated the plethora of forest maps and data sets. The resulting report emphasised the depleted state of Indonesia's forest resources [79] and, together with the chaos associated with the decentralisation policy, gave MoFor the pretext to gradually reassert control over forest resources [3,17].

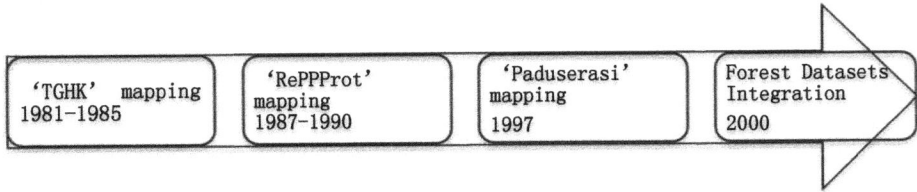

Figure 2. Previous efforts to integrate forest land maps and forest datasets. TGHK, *Tata Guna Hutan Kesepakatan* or Forest Land Use Consensus; RePPProt, Regional Physical Planning Programme for Transmigration.

Nonetheless, during this 'reformasi' period several other initiatives involving international donors and NGOs gained momentum. Together with REDD+ the European Union FLEGT (Forest Law Enforcement, Governance and Trade) is the latest in a series of international policy initiatives which aim to influence Indonesia's forest governance. In October 2013, the MoFor signed an agreement on FLEGT-VPAs (Voluntary Partnership Agreements) which aims to promote trade in legally-produced timber and has been praised for its design providing a means for stakeholder participation [80]. Effective April 2016 the European Commission confirmed Indonesia's status as the first country to satisfy the requirements for FLEGT licencing [81]. Furthermore, FLEGT is expected to provide another means for the state to institutionalise governance reform, including the recognition of community land and resource rights [81]. Prior to this, in 2009 the Indonesian government established its timber legality verification system, known as SVLK (*Sistem Verifikasi Legalitas Kayu*), which involved multi-stakeholder processes, though the state continues to hold final authority [81]. In relation to land and resource tenure the SVLK has shown limited reach [81,82]. The success of these two initiatives has yet to be determined, and will depend largely on the independent monitoring of activities on the ground and related anti-corruption measures [81,82]. These two policies were often cited by MoFor interviewees as their commitment to effect forest governance reform, given their aversion to REDD+ due to the mandate for its implementation being vested in another state agency (discussed in Section 3.3). In short, FLEGT and SVLK hold greater appeal to MoFor as both policies reassert their authority through legality verification [80,81].

The above-mentioned development is significant as it illustrates that, despite a series of internationally-supported initiatives to strengthen the capacity of other state agencies (e.g., state planning 'Bappenas' and mapping agencies 'Bakosurtanal') to integrate and standardise key datasets and land classification systems, MoFor continued to exercise unilateral control over forest lands [8]. In the case of REDD+, and as discussed in the following sections, initially there was strong resistance by Mofor (see also [65]), particularly with regard to their willingness to share information with the public on forest data and maps.

3.3. The Policy and Political Context of the OMI: The Role of REDD+ in Producing a Critical Juncture

Interviews revealed that the 'One Map Initiative' (OMI) emerged from the process of establishing REDD+ policy and measures at the national level. The stated objective of the OMI is to create an integrated map that will provide a single reference map of Indonesia for any decision-making related to land-based management [9]. The map is to be based on, (i) 'one reference', meaning a map-making process supported by one 'geodetic' control network; (ii) 'one standard' of thematic mapping, namely the Indonesia National Standard (SNI 7645:2000) for land-cover classification; (iii) 'one database', meaning one integrated database of spatial and non-spatial information compiled from across ministries and different levels of government; and (iv) 'one geo-portal' which stores all this information, a system intended to be fully transparent and enable public participation [9]. Interviewees from academia and research institutions identified the emphasis on agreed standards for land-use classification and the geospatial reference system as a significant development because the lack of these had previously compromised the production of integrated thematic maps.

Of particular importance was the timing of REDD+ entry into Indonesia's political and social arena in the wake of the 'reformasi' era. As mentioned earlier, 'reformasi' legitimised discourse on 'KKN' (*Korupsi, Kolusi, Nepotisme*) which led to the establishment of the Corruption Eradication Commission (*Komisi Pemberantasan Korupsi* or *KPK*) in 2002. This Commission gained wide public support in the midst of political uncertainty which followed the dismantling of an authoritarian regime and the introduction of democracy [83,84]. From a critical juncture viewpoint, this period of structural indeterminism or institutional fluidity provided levels of political freedom for policy actors and the public-at-large that had not previously been experienced. For instance, in 1999 Suharto's successor, President Habibie, revoked the 1963 Anti-subversion Law (a Law which can be used to criminalise people whose ideas are deemed by the government to be subversive). This meant that, and as confirmed by interviewees from NGOs and academia, reformists within the government became more open in their views and acted to form alliances with non-state actors for governance reform.

Significantly, hosting UNFCCC's Conference of the Parties (COP) 13 in Bali in 2007 accrued political capital to senior political actors internationally and domestically. In the context of the institutional fluidity following the end of the Suharto authoritarian regime, the confluence of populist support for anti-KKN policies, and the government occupying, albeit briefly, centre stage on international climate governance, enabled and empowered influential policy actors greater freedom with which to act. Most notably, President Yudhoyono's voluntary commitment in 2009 to reduce the country's greenhouse gas (GHG) emissions by 26% through its own effort, or 41% with international support by 2020, was cited by interviewees as an historic landmark in Indonesia's forest governance. This commitment led to the US$1 billion commitment from the Norway government to prepare Indonesia's REDD+ infrastructure [44].

Moreover, the appointment of Minister Kuntoro Mangunsubroto as the head of the REDD+ Task Force, a well-known reformist who was also the head of the Presidential Delivery Unit for Development Monitoring and Oversight (UKP4), created greater opportunity for the mobilisation of wider constituencies for reform. This demonstrated how endogenous and exogenous sources of institutional change become strongly interwoven when influential actors within government are involved. Figure 3 highlights the historical points relating to REDD+ which marked the institutional development of map-making towards the OMI.

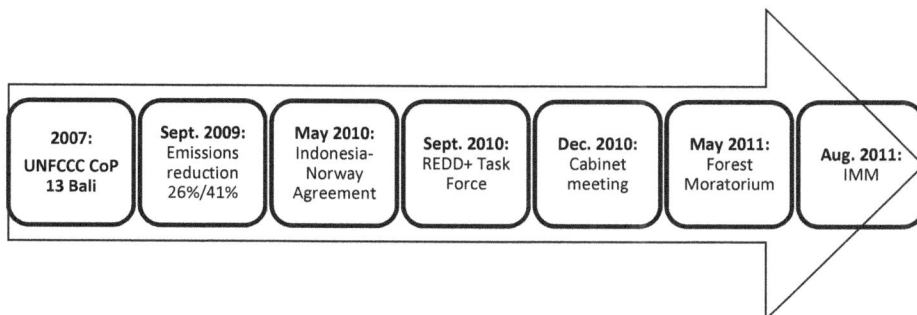

Figure 3. Historical events in Indonesia's REDD+ which lead to 'One Map Initiative'. UNFCCC COP, United Nations Framework Convention on Climate Change Conference of the Parties; IMM, Indicative Moratorium Map.

In accordance with the Indonesia-Norway partnership agreement in 2010, President Yudhoyono established a new institutional actor—the 'REDD+ Task Force'—outside the MoFor. The Task Force comprised representatives of ten relevant government ministries/agencies and was mandated with preparing the national REDD+ infrastructure and coordinating all activities related to REDD+ (Presidential Decree No. 19/2010, renewed by Decree No. 25/2011). Interviews revealed mixed opinions on the merits of this decision. The majority of interviewees from existing government ministries feared that widening membership of this important policy committee would exacerbate the problem of overlapping authority inherent in Indonesia's forest governance. On the other hand, the majority of interviewees from donor agencies, NGOs, and academia held that introducing new actors into forest policy is the only way to overcome institutional 'stickiness' within forest institutions and break the path-dependency of weak forest governance. One example often cited by interviewees from NGOs and civil society was that the recommendations of the 2010 Corruption Eradication Commission (KPK)'s review of the forestry sector had not been followed through by MoFor, suggesting 'institutional stickiness'.

The KPK identified significant regulatory uncertainty and inconsistency within MoFor that had facilitated mismanagement, misuse, and corruption [21]. Key recommendations were the review and improvement of current regulations for forest concession permits, and the establishment of a single map at 1:50,000 scale to be used as the only reference basis for issuing forest concessions [70,71]. Interviewees from the NGOs and academia revealed that the MoFor was reluctant to follow up these recommendations. The response of an interviewee from the private sector captures the wider view: *'I hold the strong belief that MoFor is more than capable of developing an accurate map with full transparency, but the question is whether they want to do it, as transparency would limit their income from rent-seeking'* (*Interviewee-41*). In sum, the REDD+ initiative, as viewed by many interviewees, has provided an external push for institutional change as initiated by domestic actors such as the KPK.

The REDD+ partnership agreement between Indonesia and Norway included a clause on the suspension of licensing new concessions in the natural/primary forests and peatlands for two years, known locally as the 'Forests Moratorium'. Through Presidential Instruction No. 10/2011 (extended for another two years via Presidential Instruction No. 6/2013 issued in May 2013) this moratorium policy included a clause on the production of an 'Indicative Moratorium Map' (IMM) with the MoFor assigned to lead the process. This policy, whilst receiving many criticisms for not providing adequate protection for 'secondary' type forests, has provided the means for efforts that improve coordination and communication, not only amongst government ministries/agencies but also between them and the indigenous people and civil society organisations (Table 1).

Table 1. Tasks given to several ministries to integrate data on forests (under Presidential Instruction No. 10/2011).

Ministries/Agency	Tasks
MoFor	to build an integrated database related to permits to manage, use and change the designation of land claimed by the indigenous/local communities
MoE	to integrate and build an e-application for an environmental permit for land utilisation
MoHA	to build an inventory system of data for the location of all permits, or recommendation of permits, issued by provincial and district governments
NLA	to build an integrated database for 'Business Usage Right', 'Transmigration Land Management Right', and land claims by indigenous people and local communities

Notes: MoFor, Ministry of Forestry; MoE, Ministry of Environment; MoHA, Ministry of Home Affairs; NLA, National Land Agency.

The Forests Moratorium and its map (scale 1:19,000,000) was released in May 2011 [85]. Interviews revealed that influential political actors within MoFor were initially reluctant to release the map. It was only after robust criticism from civil society and environmental groups of the map's small scale, format (absence of 'meta-data'—meta-data being the additional information associated with geospatial data that provides information about the data content, including for example when the image was created, who created it, and how the data was collected.), and lack of transparency on how it was generated that MoFor published the underlying land-cover data in 2009 and a series of larger scale (1:250,000) maps in June 2011. Notwithstanding these criticisms, international donor and academic interviewees held that the release of the IMM marked a significant step towards greater transparency, stressing the importance of the IMM as the embryo of the One Map Initiative. The bi-annual updating process of the IMM brings together key government ministries and agencies (see Table 2) and involves a stakeholder feedback mechanism.

Table 2. Ministries/Agencies tasked to oversee the implementation of the 'Indicative Moratorium Map' (IMM) updating process.

Ministries/Agency	Tasks
UKP4	to coordinate the whole process
BIG	as a technical coordinator and to provide topographic maps
MoFor	to provide forest cover and forest status maps
MoA	to provide peatland map
NLA	to provide land title map

Notes: UKP4, President's Delivery Unit for Development Monitoring and Oversight; BIG, Geospatial Information Agency; MoFor, Ministry of Forestry; MoA, Ministry of Agriculture; NLA, National Land Agency.

The IMM updating process involves ground verification by a monitoring team of the REDD+ Task Force and solicits input from government and private stakeholders across all levels of administration [9]. As such, this process can be seen as evidence of the persistence of a coalition of reformist governance actors within and outside government. One REDD+ Task Force official commented that: "*This is the first time in Indonesia's history of forest governance that various ministries joined force to update the state map which the public is able to view on-line and review*" (*Interviewee-59*).

Many interviewees cited President Yudhoyono's instruction to develop the OMI, delivered during a ministerial cabinet meeting (23 December 2010), as a significant moment in Indonesia's history of forest governance. This President's instruction was a response to reports from Minister Kuntoro Mangunsubroto, the head of REDD+ Task Force, concerning the challenges faced in publishing the IMM due to discrepancies in land-cover maps between the Ministries of Forestry and of Environment (MoEnv). In this regard, a MoFor official indicated that the older 1990 MoFor's map included data from a ground survey whereas the newer MoEnv's map was produced by a consultant. The same official said: "*I encouraged fellow officials at the MoFor to leave our 'ego-sectoral' position and support the One Map Initiative as it is good for our country*" (*Interviewee-5*). Several interviewees revealed that though the idea for 'one map' and initiatives to integrate forest data have been introduced before, it had never gained such momentum until it was attached to the REDD+ initiative. One interviewee said: "*The significance of REDD+ is that it emphasises public participation and transparency which are key elements of democracy. Once the people experience these elements it is very difficult to stop*" (*Interviewee-34*).

3.4. The Significance of the OMI for Forest Governance Reform

When asked about the benefits that the OMI brings to Indonesia's forest governance and the challenges faced with its implementation, interview responses fell into six general governance themes as stated within Table 3.

Many interviewees cited one of the OMI's most significant contributions as the contingency effect with respect to policy coordination amongst government ministries. The Geospatial Information Law No. 4/2011, which mandates the Geospatial Information Agency to lead the implementation of the OMI policy, including the preparation of system infrastructure and the standardisation of existing maps, was considered significant as it decouples mapping from 'ego-sectoral' strategies. Indeed, several interviewees expressed the hope that the OMI becomes a vehicle to improve sectoral policy integration by breaking forms of 'ego-sectoral' path-dependency that puts the interests of individual ministries before those of the nation. A government official stated: "*The OMI movement has certainly increased communication and the sharing of information amongst ministries as never before, and as the technical working group we have promised to guard the OMI to success*" (*Interviewee-61*). Improved data sharing and communication amongst Ministries (particularly Forestry, Agriculture, and Mining and Energy) was seen as a vehicle to improve efficiency in natural resource management. The inherently poor communication and coordination amongst them has resulted in major overlaps in resource concessions.

Table 3. Opportunities the OMI brings to Indonesia's forest governance according to interviews of five organisational categories (in 2011–2016) of REDD+ stakeholders in Indonesia *.

Governance Themes	Indication of Improved Governance	Interviewees * 'The Majority of Interviewees Cited Governance Themes'
(i) Coordination	Improved data sharing and communication amongst Ministries via the creation of a national spatial data infrastructure: ten government ministries/agencies are tasked with integrating data on forests, and working together to create a single state land-cover map (see Table 1); improved communication between government and NGOs/civil society	1, 3, 4, 5
(ii) Transparency	Greater transparency: provide public with access to digital land-cover maps and information about land-use licences/permits via 'one geo-portal'	1, 2, 3, 4, 5
(iii) Participation	Promote greater public participation in map-making: public can evaluate and provide input on state maps	1, 2, 3, 4, 5
(iv) Cost efficiency	'One point system': the process of acquiring and curating satellite data images is coordinated and provided by one government agency 'Lembaga Penerbangan dan Antariksa Nasional' (LAPAN)	3, 5
(v) Data quality	The procurement of technology and licences that enable the direct reception of satellite data at LAPAN's earth station, the use of one standard' of thematic mapping and 'one reference' of 'geodetic control network'	1, 3, 5
(vi) Protection of indigenous peoples' land	A total of 265 maps covering 2.4 million ha of indigenous peoples' land were registered for their inclusion within the State map	1, 3, 4, 5

Notes: LAPAN, The National Institute of Aeronautics and Space. * See Section 2.2 for the breakdown of interviewees and organisations: (1) multi-level government officials; (2) forest/carbon private actors; (3) donor agencies; (4) NGOs/civil society organisations; (5) and academia/researchers.

A REDD+ Task Force interviewee stated that the OMI is intended to promote greater transparency and provide all stakeholders with access to information about existing and new land-use licences and permits (*Interviewee-27*). In the process, a 'one geo-portal' (http://maps.ina-sdi.or.id) was established to provide the public with greater access to information about the state map as well as encourage them to provide input that improves it. For some this represents significant progress from previous institutional practice: in the words of a senior REDD+ Task Force official, "*The OMI is not merely a map or product, it is a movement towards greater transparency and public participation in map-making, and importantly transparency in the land-use licensing process, all of which represent a new paradigm in the governance of map-making*" (*Interviewee-14*). Whilst acknowledging these positives, interviewees from civil society organisations and environmental NGOs stated that it is still difficult to gain access to maps from MoFor, one of them saying: "*there is no deliberative action from MoFor to release data, and they will only do so when we ask*" (*Interviewee-78*). On this account, the 'one geo-portal' facility being established under the OMI policy represents a significant change from the old form of limited transparency towards a 'deliberative' form of transparency.

Interviewees from academia and donor agencies highlighted possible cost-savings resulting from the OMI, noting that this is the first time in Indonesia's history of map-making that the costly process of acquiring and curating satellite data images is coordinated and provided by one government agency. The National Institute of Aeronautics and Space (LAPAN) received a Presidential Instruction No. 6/2012 to act as the sole authority to acquire, process, and maintain satellite data on land-cover and land-use for distribution to other government ministries/agencies. The cost efficiency is estimated to be one fourth when compared with the previous prevailing method by which each ministry/agency bought satellite imaging data individually [9]. To this end, interviews revealed that the REDD+ Task Force has procured the technology and licences that enable direct reception of satellite data at LAPAN's earth station. This 'one point system' for acquiring and processing satellite data images is widely believed by interviewees to be much less expensive than has been practiced to date. Moreover, several interviewees again suggested that reformists within the REDD+ Task Force played a significant role in making this policy materialise.

Interviewees from organisations representing indigenous and forest-dependent communities view OMI as an important opportunity for indigenous people's land to be formally integrated into the State map. One said: "*The OMI has provided important momentum for us as our work focuses on community participatory mapping. We want to ensure that 'community-made maps' are used not only when there is land conflict, but for any decision-making process related to communities' development programmes*" (*Interviewee-76*). On 14 November 2012, the National Alliance for Indigenous People (AMAN) together with Participatory Mapping Network (JKPP) officially handed to the REDD+ Task Force 265 maps covering 2.4 million ha of indigenous peoples' land (*Interviewee-75*), an important event in the long running efforts within Indonesia's system of resource governance.

In summary, Figure 4 illustrates the historical points of institutional development that led to the OMI. As stated in the text accompanying the figure, it is evident that in the development process the OMI, as a policy initiative for map-making, has already shown significant signs of its potential to strengthen forest governance via the promotion of greater transparency and more inclusive stakeholder participation.

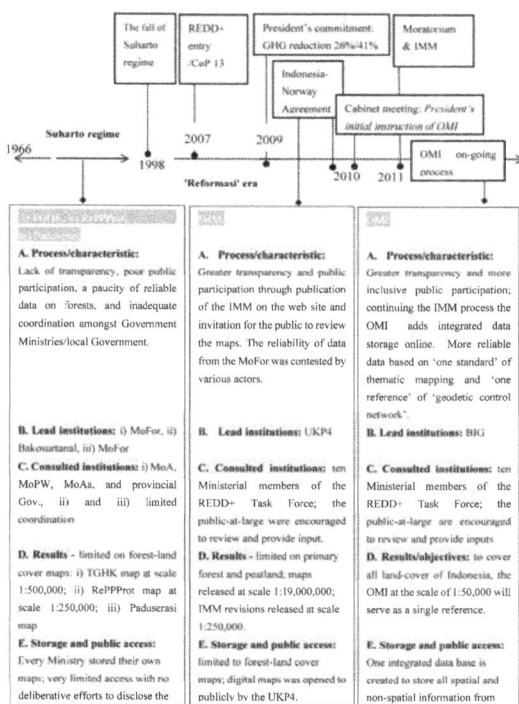

Figure 4. Summary of different map institutions in Indonesia (*Abbreviations are defined in the text*).

4. Discussion

This study asked whether the One Map Initiative might be considered a critical juncture that signifies a switch to a new pathway of map-making practice, or whether it is yet one more innovation within the existing path dependence of forestry institutions. The brief history of Indonesian forest governance presented in Sections 3.1 and 3.2 is consistent with themes in historical institutionalism that suggest strong path-dependency and state institutions that further strengthen the power of already powerful groups.

For instance, the successive revisions of agrarian and forest law and developments in forest resource mapping strengthened centralised state control of forest lands, and provided a means to develop and maintain the political capital vested in forest lands. The refusal of provinces to accept the 1997 'Peta Paduserasi' situation is consistent with the phenomenon termed 'institutional stickiness' [58] whereby policy actors prefer the status quo, even when new institutions may offer a better outcome; their behaviours have been shaped by the long-standing institutions they live within. With regard to the development and operation of institutions, this event also explained how state agencies competed to gain the source of power such as mandates and political responsibility [30,52–54]. In this regard, our account suggests that 'ego-sectoral' attitudes, where policy actors put the interest of their ministries first [45], both characterised and produced path dependency in Indonesia forestry institutions. A key driver of such attitudes is the fear that policy coordination amongst ministries, and across levels of government, will risk undermining the established rent-seeking practices used to finance departmental budget shortfalls [65].

Within the historical institutionalism literature it is recognised that, even when path dependence appears broken, re-equilibrium to the old institutions practices can occur. The institutional fluidity created by the 1998 'reformasi' movement, that created a political imperative to decentralise key competencies in forest governance from the state to the district level, appears to constitute a critical juncture. However, the 'Peta Paduserasi' process, a compilation of authoritative datasets and associated efforts on the part of MoFor to restore control over forest resources, confirms the findings of Gellert [46] and Nomura [86] that this act was insufficient to challenge the powerful domestic industrialist timber networks involving MoFor, and that until 2008 or even later the prevailing equilibrium within Indonesian forestry governance institutions remained. Moreover, several policy initiatives promoted by NGOs and foreign donors, that sought to influence forest governance reform, achieved limited success including the FLEGT-VPA (signed in 2013) whose success remains to be seen [82].

The forest moratorium suggests evidence of vertical institutional interplay whereby REDD+ (i.e., source institutions) prompted the development of new institutions within existing forest governance at the national level (i.e., target institutions). In this context, our interviews suggested that the OMI is a significant development within Indonesia's map-making due to its focus on more open and transparent access to spatial data. This exercise directly challenges the clientelist and rent-seeking practices built around MoFor's (or other departments) producing and curating practices, controlling access to spatial data (i.e., 'dominant information' [30,52]), and decision-making concerning territory and resources (i.e., '(dis)incentives' [52]).

The OMI signifies an intent to move towards a new path of map-making, one that embodies principles of open democracy relating to greater transparency and public participation. The commitment of the government to store, curate, and integrate all spatial and non-spatial information from across ministries and levels of government in one place, and to provide the public with free access to state maps and spatial meta-data via 'one geo-portal' named 'Geospasial untuk Negeri' or 'Indonesian Geospatial Portal' (although, nonetheless, the data provided may not be legible to all peoples including the indigenous communities—not only may they not have easy access to the internet, they may also lack the language fluency and ability to interpret and apply the data.) represents a major shift in established institutional practice. Previously, information could only be obtained by making an official request to the relevant ministry which took considerable time and often money. The adoption of transparency principles is particularly important because once the decision was taken to scale-up transparency and public participation, for example, the involvement of the public to review and improve the map, it became progressively more difficult to return to the initial point of path dependence. In this light, the OMI signifies a switch to a new pathway of map-making practice in Indonesia.

The president's decision to take the 'OMI path' and choose a new actor (BIG: *Badan Informasi Geospasial—Geospatial Information Agency*), which is now under the coordination of Bappenas (State Ministry for Development Planning/National Development Planning Agency) based on the latest Presidential Regulation No. 127/2015)) to lead its development, as opposed to continuing to use the MoFor (under the current administration of President Joko Widodo, 'MoFor'—the Ministry of Forestry is now the 'Ministry of Environment and Forestry')'s land-cover map, was critical in setting the development of map-making on a new path. A different decision could have plausibly led to a strengthening of the 'status quo' where would MoFor lead forest land-cover map-making. This finding resonates with that of earlier literature [30] which shows a clear decline in the relative power of the MoFor during the development process of the OMI, mainly because the policy task was given to other state agencies. Nonetheless, with regard to data on forests (e.g., concession permits, thematic maps) used for the OMI mapping exercise, MoFor remains the source of 'dominant information' [30]—a finding also revealed by our interviewees.

From the 'critical juncture' lens, the significance of this policy is also the contingency and downstream effects which appear to be unstoppable, such as the improvement of coordination amongst government ministries/agencies and between them and the indigenous and local communities. Whilst earlier literature [27,30] held a less optimistic view on the future implementation stage of 'improved coordination', this study maintains that the OMI has enabled significant progress when compared with the inherent coordination issues often cited within earlier literature on Indonesia's forest governance (e.g., [21,37,65,66]). These findings lead us to believe that the OMI does indeed signify a critical juncture. Further, to qualify as a critical juncture the event(s) should be brief relative to the period of institutional path dependence. This is the case with the OMI. It originated with President Yudhoyono's landmark 2009 commitment to reduce the country's GHG emissions by 26%/41%, and was formalised in his 2010 instruction to develop the OMI in December 2010.

Our study also sought to ascertain the role of REDD+ relating to the OMI and the construction of this critical juncture. The rise of REDD+ as a major international policy mechanism came shortly after the 'reformasi' movement and therefore entered Indonesia during a period of structural fluidity. During this period the president and other influential reformists within the government held a significant degree of political freedom given the high level of political support existing within the public-at-large. REDD+ was manifested in the international pressure and support to modernise natural resource (forest carbon) mapping and accounting, and the OMI created a vehicle to do this, at the same time further widening political reform in Indonesia. As part of a broader reform agenda it was supported by a broad constituency from both the domestic and international arena, a phenomenon that would not have occurred in 'normal times'. This study provided stronger evidence of the impact of an international policy (i.e., REDD+) on the resulting OMI, compared with earlier literature [30] which suggested only 'a reasonable extent' of international influence.

If we accept that REDD+ created the impetus for a critical juncture in the spatial data (mapping) component of Indonesian institutions, then it is important to analyse why the REDD+ mechanism has generated changes at a scale that has not been achieved by previous international forest governance initiatives. In our view the initial policy to generate funding from the carbon market via a performance-based system is of particular importance. This created a requirement for those countries wishing to receive payments to establish transparent systems capable of measuring and verifying the results of REDD+ programmes [13,27,87]. Key to this system is the development of accurate and consistent forest data sets and land-cover maps, but also the need to map, record, and resolve land ownership so that contractual arrangements relating to forest carbon can be upheld. This in turn requires coordination amongst ministries and across different levels of government.

REDD+ held the promise of new financial flows associated with carbon management policy, and the international donor community offered Indonesia substantial levels of 'REDD readiness' financial support. However, an expectation of this support was that Indonesia would introduce a moratorium on the issuance of forest concession licences. This stimulated the development and publication of the Indicative Moratorium Map (IMM) and in the process promoted transparency and public participation. The IMM then stimulated the development of the OMI. Moreover, because the REDD+ initiative embodies multiple agendas and objectives from climate change mitigation, to poverty reduction, and the protection of human rights and biodiversity, it created unparalleled levels of support from a wide range of international and domestic actors [24,30,54,88] who also lent their support to the OMI.

Whilst REDD+ policy has contributed to the development of the OMI at a critical juncture, this study finds that a range of exogenous and endogenous factors created a basis of support for REDD+ policy development in Indonesia. For instance, the 1997 Asian financial crisis served as an exogenous shock that generated the 'reformasi' movement which, together with an endogenous desire for change, led to the end of three decades of the Suharto regime. Given these circumstances REDD+ might be considered as a policy 'connector' that gave the issue of spatial data and transparency a political form and agency which manifested in the OMI (i.e., a modern spatial data infrastructure) as a tool for the reform of government institutions. The OMI was a product of the exogenous policy force of REDD+ interacting with an endogenous movement of high-ranking politicians and officials supported by domestic constituencies working for reform.

5. Conclusions

Deploying 'path dependency' within 'historical institutionalism' our study sought to determine whether the One Map Initiative signifies a switch to a new path of Indonesia's map-making practice. Research results show positive signs that the OMI, in its current stage, has already shown its ability to change the prevailing path-dependence of map-making which was characterised by a lack of transparency, a poor level of public participation, and inadequate coordination amongst ministries. During its map-making exercise this policy promotes several good governance norms, namely better coordination amongst government ministries, greater transparency and public participation, improved cost efficiency and forest data quality, and the protection of indigenous people's land (see Section 3.4). Notable is the establishment of 'one geo-portal' (named 'Geospasial untuk Negeri' or 'Indonesian Geospatial Portal') which shows government's commitment to provide the public with free access to state maps and spatial data, and encourage them to be involved in updating and providing input to state maps. This denotes a very significant shift in the established institutional practice of map-making.

Moreover, this article sought to determine whether REDD+ has been instrumental in creating a 'critical juncture' in the form of momentum for Indonesian resource and land governance institutions to adopt a new path. During the establishment process of the 'forest moratorium' to enable the implementation of REDD+, the embryo of the OMI (i.e., the 'Indicative Moratorium Map') was created (see Section 3.3). This 'forest moratorium' produces evidence of institutional interplay, with REDD+ prompting the development of new institutions within the forest governance prevailing at the national level. In this regard, the REDD+ 'performance-based system' has required host countries to establish transparent systems for measuring and verifying the results of REDD+ programmes, which in turn required the development of accurate and consistent forest data sets and land-cover maps. To this end, the REDD+ initiative has given spatial data and transparency a political form and agency which became manifested in the OMI. Viewed via a 'critical juncture' lens, the implementation of transparency principles is particularly important since the decision to scale-up transparency and public participation (e.g., the involvement of the public to review and improve the map) made it progressively more difficult to return to the initial point of path dependence. In short, REDD+ has contributed to the creation of a critical juncture via momentum for forest and land governance to move to a new path through greater transparency and public participation.

This study also finds that a number of exogenous and endogenous factors enabled a basis of support for REDD+ policy development in Indonesia. This includes the large presence of constituencies for governance reform, and the 'structural fluidity' resulting from the 'reformasi' movement following the Asian financial crisis of 1997 and the fall of the Soeharto regime in 1998 (see Sections 3.2 and 3.3). The time that the REDD+ initiative entered Indonesia's political area, shortly after the 'reformasi' movement, was critical to its relative success in effecting governance reform. REDD+ was promoted as an international policy initiative to reform the governance of forest resources and land via advanced forest carbon mapping and accounting. A 'critical juncture' lens reveals that this reform agenda was supported by a broad constituency from both the domestic and international arena during 'institutional fluidity', greater than would have occurred in normal times.

At the initial time of writing the future of the OMI was in question. However, the latest development, including a commitment by the current President Joko Widodo to accelerate the implementation of the OMI and position this policy high on the National Development Priority (based on Presidential Regulation No. 127/2015, the National Development Planning Agency (Bappenas) is given a mandate to coordinate activities falling under the Geospatial Information Agency) suggests another positive sign that its grounded implementation is viable. The implication of the OMI for the future implementation of REDD+ is threefold. Firstly, clarity on categories of forest land use produces certainty where REDD+ can take place. Secondly, transparency on forest data, particularly on the concessions already issued, means that there is no risk of REDD+ investment being undermined by overlapping concessions. Thirdly, and importantly, for the performance-based approach, and particularly for a market-based system that requires carbon emissions reduction to be assessed, the development of accurate forest data sets and land-cover maps is critical to the efficacy of REDD+ implementation.

Further research is necessary to follow the on-the-ground implementation of the OMI to ascertain whether, in the long run, this policy initiative can serve as a means to break the path dependency of map-making institutions, or whether there are signs that a 're-equilibrium' within map-making institutions will occur.

Author Contributions: The lead author conducted the field work research in Indonesia which included interviews of individuals engaged in the REDD+ policy process, participated in workshops convened by the REDD+ Task Force and local NGOs, conducted the initial data analysis, and reviewed contemporary policy documents as mentioned in Section 2.2. Working together closely, both authors reviewed literature on the history of Indonesian forest policy, deployed the academic conceptual framework as mentioned in Section 2.1, and wrote the article.

Conflicts of Interest: The authors declare no conflict of interest.

References

1. Demeritt, D. Scientific Forest Conservation and the statistical picturing of nature's limits in the progressive-era United States. *Environ. Plan. D Soc. Space* **2001**, *19*, 431–459. [CrossRef]
2. Fay, C.; Sirait, M.; Kusworo, A. *Getting the Boundaries Right: Indonesia's Urgent Need to Redefine Its Forest Estate*; World Agroforestry Center: Bogor, Indonesia, 2000.
3. Jepson, P.; Jarvie, J.K.; MacKinnon, K.; Monk, K.A. The End for Indonesia's Lowland Forests? *Science* **2001**, *292*, 859–861. [CrossRef] [PubMed]
4. Peluso, N.L. Coercing conservation? *Glob. Environ. Chang.* **1993**, *3*, 199–217. [CrossRef]
5. McCarthy, J.F.; Vel, J.A.C.; Afiff, S. Trajectories of land acquisition and enclosure: Development schemes, virtual land grabs, and green acquisitions in Indonesia's outer islands. *J. Peasant Stud.* **2012**, *39*, 521–549. [CrossRef]
6. Barber, C.V.; Schweithelm, J. *Trial by Fire: Forest Fires and Forestry Policy in Indonesia's Era of Crisis and Reform*; World Resources Institute, Forest Frontiers Initiative: Washington, DC (EUA), USA, 2000.
7. Fox, J. Mapping the Commons: The Social Context of Spatial Information Technologies. *Common Prop. Resour. Dig.* **1996**, *45*, 1–9.
8. Peluso, N.L. Whose woods are these? Counter-mapping forest territories in Kalimantan, Indonesia. *Antipode* **1995**, *27*, 383–406. [CrossRef]
9. REDD+ Task-Force. REDD+ Indonesia. A Catalyst for Change. REDD+ Task Force (Satgas Persiapan Kelembagaan REDD+). Available online: www.satgasreddplus.org (accessed on 16 October 2012).
10. Resosudarmo, I.A.P.; Atmadja, S.; Ekaputri, A.D.; Intarini, D.Y.; Indriatmoko, Y.; Astri, P. Does tenure security lead to REDD+ project effectiveness? Reflections from Five emerging sites in Indonesia. *World Dev.* **2014**, *55*, 68–83. [CrossRef]
11. UNFCCC (United Nations Framework Convention on Climate Change). Reducing emissions from deforestation in developing countries: Approaches to stimulate action. Submissions from Parties (Papua New

Guinea and Costa Rica). In *Conference of the Parties, Eleventh Session*; United Nations Framework Convention on Climate Change: Monreal, QC, Canada, 2005.

12. UNFCCC (United Nations Framework Convention on Climate Change). *The Cancun Agreements Decision 1/CP.16: Outcome of the Work of the Ad Hoc Working Group on Long-Term Cooperative Action under the Convention*; United Nations Framework Convention on Climate Change: Cancun, Mexico, 2011; pp. 1–31.

13. Wertz-Kanounnikoff, S.; Mcneil, D. Performance indicators and REDD+ implementation. In *Realising REDD+: National Strategy and Policy Options*; Angelsen, A., Brockhaus, M., Kanninen, M., Sills, E., Sunderlin, W.D., Wertz-Kanounnikoff, S., Eds.; Center for International Forestry Research: Bogor, Indonesia, 2012; pp. 233–246.

14. UNFCCC (United Nations Framework Convention on Climate Change). *UN Climate Change Conference in Warsaw Keeps Governments on a Track towards 2015 Climate Agreement*; United Nations Framework Convention on Climate Change: Warsaw, Poland, 2013; Available online: https://unfccc.int/files/press/news_room/press_releases_and_advisories/application/pdf/131123_pr_closing_cop19.pdf (accessed on 30 December 2013).

15. Boer, R. *Sustainable Forest Management, Forest Based Carbon, Carbon Stock, CO_2 Sequestration and Green Product in Order to Reduce Emission from Deforestation and Forest Degradation*; Indonesia Ministry of Forestry; International Tropical Timber Organization: Jakarta, Indonesia, 2012.

16. McCarthy, J.F. Changing to gray: Decentralization and the emergence of volatile Socio-Legal configurations in central Kalimantan, Indonesia. *World Dev.* **2004**, *32*, 1199–1223. [CrossRef]

17. Barr, C.; Resosudarmo, I.A.P.; Dermawan, A. Forests and Decentralization in Indonesia: An Overview. In *Decentralization of Forest Administration in Indonesia. Implications for Forest Sustainability, Economic Development and Community Livelihoods*; Barr, C., Resosudarmo, I.A.P., Dermawan, A., Mccarthy, J., Moeliono, M., Setiono, B., Eds.; Center for International Forestry Research: Bogor, Indonesia, 2006; pp. 1–15.

18. Marifa, I. Institutional Transformation for Better Policy Implementation and Enforcement. In *The Politics and Economics of Indonesia's Natural Resources*; Resosudarmo, B.P., Ed.; Institute of Southeast Asian Studies: Singapore, Singapore, 2005; pp. 248–258.

19. Larson, A.M.; Barry, D.; Dahal, G.R.; Colfer, C.J.P. *Forests for People: Community Rights and Forest Tenure Reform*; Center for International Forestry Research; Earthscan: London, UK, 2010; ISBN: 9781844079186, p. 263. Available online: http://www.cifor.org/library/2977/forests-for-people-community-rights-and-forest-tenure-reform/ (accessed on 2 October 2013).

20. Sunderlin, W.D.; Hatcher, J. *From Exclusion to Ownership? Challenges and Opportunities in Advancing Forest Tenure Reform*; Liddle, M., Ed.; AGRIS: International Information System for the Agricultural Science and Technology: Washington, DC, USA, 2008; Available online: http://agris.fao.org/agris-search/search.do?recordID=GB2013202125 (accessed on 9 October 2011).

21. Dermawan, A.; Petkova, E.; Sinaga, A.C.; Muhajir, M.; Indriatmoko, Y. *Preventing the Risks of Corruption in REDD+ in Indonesia*; United Nations Office on Drugs and Crime and Center for International Forestry Research: Jakarta and Bojor, Indonesia, 2011.

22. Corbera, E.; Schroeder, H. Governing and implementing REDD+. *Environ. Sci. Policy* **2011**, *14*, 89–99. [CrossRef]

23. Schroeder, H.; McDermott, C. Beyond carbon: Enabling justice and equity in REDD+ across levels of governance. *Ecol. Soc.* **2014**. [CrossRef]

24. McDermott, C.L. REDDuced: From sustainability to legality to units of carbon—The search for common interests in international forest governance. *Environ. Sci. Policy* **2014**, *35*, 12–19. [CrossRef]

25. Murdiyarso, D.; Dewi, S.; Lawrence, D.; Seymour, F. Indonesia's Forest Moratorium: A Stepping Stone to Better Forest Governance? Center for International Forestry Research, 2011. Available online: http://www.cifor.org/library/3561/indonesias-forest-moratorium-a-stepping-stone-to-better-forest-governance/ (accessed on 8 February 2012). [CrossRef]

26. Fujisaki, T. *Indonesia National REDD+ Readiness and Activities. State of Play: March 2012*; Institute for Global Environmental Strategies (IGES): Hayama, Kanagawa, Japan, 2012.

27. Astuti, R.; McGregor, A. Responding to the green economy: How REDD+ and the One map initiative are transforming forest governance in Indonesia. *Third World Q.* **2015**, *36*, 2273–2293. [CrossRef]

28. Kehbila, A.; Alemagi, D.; Minang, P. Comparative multi-criteria assessment of climate policies and sustainable development strategies in Cameroon: Towards a GIS decision-support tool for the design of an optimal REDD+ strategy. *Sustainability* **2014**, *6*, 6125–6140. [CrossRef]

29. Louman, B.; Cifuentes, M.; Chacón, M. REDD+, RFM, development, and carbon markets. *Forests* **2011**, *2*, 357–372. [CrossRef]
30. Wibowo, A.; Giessen, L. Absolute and relative power gains among state agencies in forest-related land use politics: The ministry of forestry and its competitors in the REDD+ Programme and the One map policy in Indonesia. *Land Use Policy* **2015**, *49*, 131–141. [CrossRef]
31. Pelletier, J.; Gélinas, N.; Skutsch, M. The place of community Forest Management in the REDD+ landscape. *Forests* **2016**, *7*, 170. [CrossRef]
32. Harada, K.; Prabowo, D.; Aliadi, A.; Ichihara, J.; Ma, H.-O. How can social safeguards of REDD+ function effectively conserve forests and improve local Livelihoods? A case from Meru Betiri national park, east java, Indonesia. *Land* **2015**, *4*, 119–139. [CrossRef]
33. Boissière, M.; Beaudoin, G.; Hofstee, C.; Rafanoharana, S. Participating in REDD+ measurement, reporting, and verification (PMRV): Opportunities for local people? *Forests* **2014**, *5*, 1855–1878. [CrossRef]
34. Barbier, E.B.; Tesfaw, A.T. Can REDD+ save the forest? The role of payments and tenure. *Forests* **2012**, *3*, 881–895. [CrossRef]
35. Brofeldt, S.; Theilade, I.; Burgess, N.; Danielsen, F.; Poulsen, M.; Adrian, T.; Bang, T.; Budiman, A.; Jensen, J.; Jensen, A.; et al. Community monitoring of carbon stocks for REDD+: Does accuracy and cost change over time? *Forests* **2014**, *5*, 1834–1854. [CrossRef]
36. Bellfield, H.; Sabogal, D.; Goodman, L.; Leggett, M. Case study report: Community-based monitoring systems for REDD+ in Guyana. *Forests* **2015**, *6*, 133–156. [CrossRef]
37. Brockhaus, M.; Angelsen, A. Seeing REDD+ through 4Is: A political economy framework. In *Analysing REDD+: Challenges and Choices*; Angelsen, A., Brockhaus, M., Sunderlin, W.D., Verchot, L.V., Eds.; Center for International Forestry Research (CIFOR): Bogor, Indonesia, 2012; pp. 15–30. Available online: http://www.cifor.org/library/3816/seeing-redd-through-4is-a-political-economy-framework/ (accessed on 8 January 2014).
38. Pacheco, P.; Aguilar-Støen, M.; Börner, J.; Etter, A.; Putzel, L.; Diaz, M.D.C.V. Landscape transformation in tropical Latin America: Assessing trends and policy implications for REDD+. *Forests* **2010**, *2*, 1–29. [CrossRef]
39. Pacheco, P.; Putzel, L.; Obidzinski, K.; Schoneveld, G. REDD+ and the global economy. Competing forces and policy options. In *Analysing REDD+: Challenges and Choices*; Angelsen, A., Brockhaus, M., Sunderlin, W.D., Verchot, L.V., Eds.; Center for International Forestry Research: Bogor, Indonesia, 2012; pp. 52–66.
40. The Food and Agriculture Organization of the United Nations (FAO). *The State of Forests in the Amazon Basin, Congo Basin and Southeast Asia*; The Food and Agriculture Organization of the United Nations (FAO); International Tropical Timber Organization (ITTO): Rome, Italy, 2011.
41. Strassburg, B.; Turner, K.; Fisher, B.; Schaeffer, R.; Lovett, A. An Empirically-Derived Mechanism of Combined Incentives to Reduce Emissions from Deforestation. Available online: https://ueaeprints.uea.ac.uk/24686/ (accessed on 8 February 2016).
42. Boer, R.; Sulistyowati, L.; Zed, F.; Masripatin, N.; Kartakusuma, D.A.; Hilman, D.; Mulyanto, H.S. *Summary for Policy Makers: Indonesia Second National Communication under the United Nations Framework Convention on Climate Change (UNFCCC)*; State Minister of the Environment, Government of Indonesia: Jakarta, Indonesia, 2009.
43. Fujisaki, T.; Hyakumura, K.; Scheyvens, H.; Cadman, T. Does REDD+ ensure Sectoral coordination and Stakeholder participation? A comparative analysis of REDD+ national governance structures in countries of Asia-Pacific region. *Forests* **2016**, *7*, 195. [CrossRef]
44. LoI. *Letter of Intent between the Government of the Kingdom of Norway and the Government of the Republic of Indonesia on "Cooperation on Reducing Greenhouse Gas Emissions from Deforestation And Forest Degradation"*; The Government of the Kingdom of Norway and the Government of the Republic of Indonesia: Oslo, Norway, 2010. Available online: http://www.regjeringen.no/upload/SMK/Vedlegg/2010/Indonesia_avtale.pdf (accessed on 10 December 2010).
45. Seymour, F.J.; Dubash, N.K.; Brunner, J.; Ekoko, F.; Filer, C.; Kartodihardjo, H.; Mugabe, J. *The Right Conditions: The World Bank, Structural Adjustment, and Forest Policy Reform*; Seymour, F.J., Dubash, N.K., Brunner, J., Eds.; Governance Program, World Resources Institute: Washington, DC, USA, 2000; pp. 8–155.
46. Gellert, P.K. Rival transnational networks, domestic politics and Indonesian timber. *J. Contemp. Asia* **2010**, *40*, 539–567. [CrossRef]

47. Hall, P.A.; Taylor, R.C.R. Political science and the three new Institutionalisms. *Political Stud.* **1996**, *44*, 936–957. [CrossRef]

48. Hall, P.A. Historical Institutionalism in Rationalist and Sociological Perspective. Available online: http://scholar.harvard.edu/hall/publications/historical-institutionalism-rationalist-and-sociological-perspective (accessed on 1 October 2013).

49. Vatn, A. *Institutions and the Environment*; Elgar, Edward Publishing: Cheltenham, UK, 2005.

50. North, D.C. *Institutions, Institutional Change and Economic Performance*; Cambridge University Press: Cambridge, UK, 1990.

51. Steinmo, S. 7 Historical institutionalism. In *Approaches and Methodologies in the Social Sciences*; Cambridge University Press: Cambridge, UK, 2000; Volume 118.

52. Krott, M.; Bader, A.; Schusser, C.; Devkota, R.; Maryudi, A.; Giessen, L.; Aurenhammer, H. Actor-centred power: The driving force in decentralised community based forest governance. *For. Policy Econ.* **2014**, *49*, 34–42. [CrossRef]

53. Giessen, L.; Sarker, P.K.; Rahman, M.S. International and domestic sustainable Forest Management policies: Distributive effects on power among state agencies in Bangladesh. *Sustainability* **2016**, *8*, 335. [CrossRef]

54. Rahman, M.S.; Giessen, L. The power of public bureaucracies: Forest-related climate change policies in Bangladesh (1992–2014). *Clim. Policy* **2016**. [CrossRef]

55. Crouch, C.; Farrell, H. Breaking the path of institutional development? Alternatives to the new determinism. *Ration. Soc.* **2004**, *16*, 5–43. [CrossRef]

56. Stubbs, R. The ASEAN alternative? Ideas, institutions and the challenge to "global" governance. *Pac. Rev.* **2008**, *21*, 451–468. [CrossRef]

57. Brockhaus, M.; Di Gregorio, M.; Mardiah, S. Governing the design of national REDD+: An analysis of the power of agency. *For. Policy Econ.* **2014**, *49*, 23–33. [CrossRef]

58. Boettke, P.J.; Coyne, C.J.; Leeson, P.T. Institutional stickiness and the new development economics. *Am. J Econ Sociol.* **2008**, *67*, 331–358. [CrossRef]

59. Deeg, R. Institutional change and the uses and limits of path dependence: The case of German Finance. *SSRN Electron. J.* **2001**. [CrossRef]

60. Collier, R.B.; Collier, D. *Shaping the Political Arena: Critical Junctures, the Labor Movement, and Regime Dynamics in Latin America*; Princeton University Press: Princeton, NJ, USA, 1991; ISBN: 9780691023137.

61. Thelen, K. Historical institutionalism in comparative politics. *Annu. Rev. Political Sci.* **1999**, *2*, 369–404. [CrossRef]

62. Capoccia, G.; Kelemen, R.D. The study of critical junctures: Theory, narrative, and Counterfactuals in historical Institutionalism. *World Politics* **2007**, *59*, 341–369. [CrossRef]

63. Mabee, B. Historical Institutionalism and foreign policy analysis: The origins of the national security council revisited. *Foreign Policy Anal.* **2010**, *7*, 27–44. [CrossRef]

64. Mahoney, J. Path-dependent explanations of regime change: Central America in comparative perspective. *Stud. Comp. Int. Dev.* **2001**, *36*, 111–141. [CrossRef]

65. Mulyani, M.; Jepson, P. REDD+ and forest governance in Indonesia: A Multistakeholder study of perceived challenges and opportunities. *J. Environ. Dev.* **2013**, *22*, 261–283. [CrossRef]

66. Mulyani, M.; Jepson, P. Social learning through a REDD+ "village agreement": Insights from the KFCP in Indonesia. *Asia Pac. Viewp.* **2015**, *56*, 79–95. [CrossRef]

67. Indonesian Government, 2011. Presidential Instruction No. 10, 2011: Moratorium on Granting of New Licenses and Improvement of Natural Primary Forest and Peatland Governance. Available online: http://blog.cifor.org/3003/indonesia-releases-presidential-instructions-for-logging-moratorium?fnl=en (accessed on 14 December 2012).

68. Indonesian Government, 1999. Law of the Republic of Indonesia Number 41 of 1999 Regarding Forestry by Mercy of the One Supreme God. Available online: http://theredddesk.org/sites/default/files/uu41_99_en.pdf (accessed on 14 December 2012).

69. Indonesian Government, 2011. The Law on Geospatial Information in Indonesia (Law No. 4 Year 2011). Available online: http://www.un-ggim-ap.org/article/Information/unggimap_meetings/plenary/LawNo.4Year2011GeoSpatialInformationofIndonesia-EnglishVersion.pdf (accessed on 14 December 2016).

70. KPK-a. KPK, 2010. Komisi Pemberantasan Korupsi (KPK): Matriks Temuan dan Saran Perbaikan Kajian Sistem Perencanaan dan Pengelolaan Kawasan Hutan pada Ditjen Planologi Kehutanan Kementerian

Kehutanan Republik Indonesia. 2010. Available online: http://www.kpk.go.d/uploads/PDdownloads/matriks_temuan_dan_saran_perbaikan_kajian_kehutanan.pdf (accessed on 19 October 2013).

71. KPK-b. KPK, 2010. Komisi Pemberantasan Korupsi (KPK): Paparan hasil Kajian KPK Tentang Kehutanan 2010. Available online: http://www.kpk.go.id/modules/news/article.php?storyid=726 (accessed on 19 October 2013).

72. Lev, D.S. Colonial law and the genesis of the Indonesian state. *Indonesia* **1985**, *40*, 57. [CrossRef]

73. McCarthy, J.F. Between Adat and state: Institutional arrangements on Sumatra's Forest Frontier. *Hum. Ecol.* **2005**, *33*, 57–82. [CrossRef]

74. Larson, A.M. Forest tenure reform in the age of climate change: Lessons for REDD+. *Glob. Environ. Chang.* **2011**, *21*, 540–549. [CrossRef]

75. Contreras, H.; Arnoldo, C.F. *Strengthening Forest Management in Indonesia through Land Tenure Reform: Issues and Framework for Action*; World Bank Group: Washington, DC, USA, 2005; pp. 1–55.

76. Resosudarmo, B.P. (Ed.) *The Politics and Economics of Indonesia's Natural Resources*; Institute of Southeast Asian Studies: Singapore, 2005; ISBN: 981230312X.

77. Fearnside, P.M. Transmigration in Indonesia: Lessons from its environmental and social impacts. *Environ. Manag.* **1997**, *21*, 553–570. [CrossRef]

78. Whitten, A.J.; Haeruman, H.; Alikodra, H.S.; Thohari, M. *Transmigration and the Environment in Indonesia: The Past, Present and Future*; International Union for Conservation of Nature (IUCN): Gland, Swizerland, 1987; ISBN: 2-88032-914-0.

79. Holmes, D.A. *"Where Have All the Forests Gone?" East Asia Environment and Social Development Departmental Discussion Paper*; World Bank: Washington, DC, USA, 2002.

80. Lesniewska, F.; McDermott, C.L. FLEGT VPAs: Laying a pathway to sustainability via legality lessons from Ghana and Indonesia. *For. Policy Econ.* **2014**, *48*, 16–23. [CrossRef]

81. Setyowati, A.; McDermott, C.L. Commodifying legality? Who and what counts as legal in the Indonesian wood trade. *Soc. Nat. Resour.* **2016**. [CrossRef]

82. Obidzinski, K.; Kusters, K. Formalizing the logging sector in Indonesia: Historical dynamics and lessons for current policy initiatives. *Soc. Nat. Resour.* **2015**, *28*, 530–542. [CrossRef]

83. Luttrell, C.; Resosudarmo, I.A.P.; Muharrom, E.; Brockhaus, M.; Seymour, F. The political context of REDD+ in Indonesia: Constituencies for change. *Environ. Sci. Policy* **2014**, *35*, 67–75. [CrossRef]

84. O'Rourke, K. *Reformasi: The Struggle for Power in Post-Soeharto Indonesia*; Allen & Unwin: Sydney, Australia, 2003; ISBN: 9781865087542.

85. Wells, P.L.; Paoli, G.D. *An Analysis of Presidential Instruction No. 10, 2011: Moratorium on Granting of New Licenses and Improvement of Natural Primary Forest and Peatland Governance*; Daemeter Consulting: Bogor, Indonesia, 2011.

86. Nomura, K. The politics of participation in Forest Management: A case from democratizing Indonesia. *J. Environ. Dev.* **2008**, *17*, 166–191. [CrossRef]

87. Jagger, P.; Brockhaus, M.; Duchelle, A.; Gebara, M.; Lawlor, K.; Resosudarmo, I.; Sunderlin, W. Multi-level policy dialogues, processes, and actions: Challenges and opportunities for national REDD+ safeguards measurement, reporting, and verification (MRV). *Forests* **2014**, *5*, 2136–2162. [CrossRef]

88. McDermott, C.L.; Coad, L.; Helfgott, A.; Schroeder, H. Operationalizing social safeguards in REDD+: Actors, interests and ideas. *Environ. Sci. Policy* **2012**, *21*, 63–72. [CrossRef]

Review

The Place of Community Forest Management in the REDD+ Landscape

Johanne Pelletier [1,*], Nancy Gélinas [2] and Margaret Skutsch [3]

[1] Woods Hole Research Center, 149 Woods Hole Rd, Falmouth, MA 02540-1644, USA
[2] Département des Sciences du Bois et de la Forêt, Faculté de Foresterie, Géographie et Géomatique, Université Laval, 2405, Rue de la Terrasse Pavillon Abitibi-Price, Québec, QC G1V 0A6, Canada; Nancy.Gelinas@sbf.ulaval.ca
[3] Centro de Investigaciones en Geografía Ambiental, Universidad Nacional Autónoma de México (CIGA-UNAM), Campus Morelia, Michoacán, 58190, Mexico; mskutsch@ciga.unam.mx
*   Correspondence: jpelletier@whrc.org; Tel.: +1-508-444-1537

Academic Editors: Esteve Corbera and Heike Schroeder
Received: 14 March 2016; Accepted: 28 July 2016; Published: 4 August 2016

Abstract: Community forest management (CFM) is identified by many actors as a core strategy for reducing emissions from deforestation and forest degradation in developing countries (REDD+). Others however see REDD+ as a danger to CFM. In response to these contrasting views, we carried out a systematic review of CFM case studies to look at CFM's potential role in achieving forest carbon benefits and social co-benefits for forest communities. We evaluated the potential impacts of REDD+ on CFM. Our review showed that there is strong evidence of CFM's role in reducing degradation and stabilizing forested landscapes; however, the review also showed less evidence about the role of CFM in reducing deforestation. For social benefits, we found that CFM contributes to livelihoods, but its effect on poverty reduction may be limited. Also, CFM may not deal adequately with the distribution of benefits within communities or user groups. These insights are important for CFM-based REDD+ intervention; measures should be adopted to overcome these gaps. Innovative incentive structures to existing CFM are discussed. The recognition of rights for forest communities is one first step identified in promoting CFM. We call for sound empirical impact evaluations that analyze CFM and CFM-based REDD+ interventions by looking at both biophysical and social outcomes.

Keywords: community forest management; reducing emissions from deforestation and forest degradation (REDD+); livelihoods; benefit sharing

1. Introduction

Recent estimates suggest that deforestation has been responsible for the emission of 2.9 Gigatonnes (Gt) CO_2 of carbon dioxide (CO_2) per year over the period of 2001–2015 and forest degradation for 1.0 Gt CO_2 per year over the period 1990–2015 [1]. Starting in Montreal and following a proposal from Costa Rica and Papua New Guinea, the parties to the United Nations Framework Convention on Climate Change (UNFCCC) initiated negotiations for the creation of a mechanism that would provide positive incentives to developing countries to address tropical deforestation. In Bali, at the 13th session of the conference of the parties (COP 13), countries decided to launch demonstration activities aiming at reducing emissions from deforestation and forest degradation in developing countries (REDD). A policy framework for the creation of a performance-based mechanism was adopted in Cancun (COP 16) that would compensate developing countries (Non-Annex 1 countries) on the basis of the measured success in reducing emissions from deforestation and forest degradation or increase carbon absorptions through carbon stock enhancement, sustainable management of forests and forest conservation (REDD+). The Warsaw Framework, including a set of seven decisions, consolidates

REDD+ as one of the key policy option in the climate change mitigation toolbox [2], and REDD+ is mentioned explicitly as a component of the 2015 Paris Agreement [3].

While REDD+ might bring an unprecedented level of funding to tackle deforestation, with the potential to significantly contribute to rural poverty reduction, one of the challenges that REDD+ faces in developing countries is to create a governance structure that can ensure genuine participation and equitable benefit-sharing of those who depend on forest for their livelihoods. The Cancun Agreement specifies safeguard measures that should be respected and monitored in unfolding REDD+ activities in developing countries, including respect for the knowledge and rights of indigenous people and full and effective participation of indigenous peoples and local communities and the enhancement of social benefits, while considering "the need for sustainable livelihoods of indigenous peoples and local communities and their interdependence on forests in most countries" [4].

Community Forest Management (CFM) has been identified as one interesting option to reduce emissions or increase removals from forests [5]. Various academic publications, policy reports, popular press and advocacy papers place CFM at the core of REDD+ implementation [5–11]. Further, the majority of the readiness proposals for REDD+ presented to the Forest Carbon Partnership Facility (FCPF) (http://www.forestcarbonpartnership.org/fcp/node/203) of the World Bank and most of those presented to the Collaborative Programme of the United Nations for REDD (UN-REDD) (http://www.un-redd.org) refer to community forest management. For some countries such as Ethiopia, Madagascar, Nepal, the United Republic of Tanzania and other countries in Central and South America, a community management program, organized at a national level, is central to the REDD+ national strategy proposed [12].

CFM at the heart of REDD+ is somewhat surprising given that it is well established that the major causes of deforestation and forest degradation in tropical countries are large scale conversion to commercial agriculture and commercial timber exploitation [13,14]. These activities are generally carried out by corporations with government approval, and not by rural communities, even if the local inhabitants are co-opted as labourers. The reason for CFM being placed at the centre of REDD+ in many countries probably relates to the fact that it is much more difficult, politically and economically, to tackle large scale enterprise-based deforestation than community forestry, and because support organizations for indigenous and local communities have been vociferous in the policy debate in demanding that these groups benefit from REDD+.

The popularity of CFM in the global REDD+ discourse at the international level can be partly explained by the fact that it is perceived as being a strategy to achieve dual objectives, forest protection and poverty alleviation. In contrast to protected areas where conservation policy may feature substantial social costs due to exclusion and/or restriction on access and use [15,16], CFM is seen as an option to protect the forest, as well as meeting social goals. Many supporters of CFM see REDD+ as a means of increasing finance for community management and securing new grounds for forest communities through tenure reform and decentralization. They have therefore promoted CFM as a win-win governance framework to be adopted in the planning of national REDD+ programs [7].

Not all of those who have traditionally promoted CFM as a sound policy approach for forest management however consider REDD+ as a good thing for community forestry [17], and some identify REDD+ as a potential threat to forest decentralization and devolution [18]. In the light of past forest tenure reforms, it has also been argued that there is little to expect in terms of benefits for forest communities from REDD+ national policies [19].

Several reviews exist of CFM impacts on forests, livelihoods or both, using more or less systematic approaches. Our systematic review is the first to focus directly on CFM in the REDD+ context in order to generate lessons from CFM experiences for REDD+ and evaluate how REDD+ may influence CFM. This study takes a new angle by examining the role of CFM in achieving REDD+ climate mitigation objectives and social co-benefits, and the potential of REDD+ in promoting and sustaining CFM through access to new income sources, among other co-benefits.

Through a systematic review of literature and in the context of an existing debate and contrasting views about the role of CFM in REDD+ and of REDD+ for CFM, we attempt to answer three main questions: (1) Does CFM appear to deliver carbon benefits? (2) Does CFM appear to deliver social benefits (co-benefits)? and (3) How can REDD+ affect CFM? For the first two questions, we look at reported evidence of positive biophysical and social outcomes in the CFM literature by analysing published CFM case studies. The first question is directly linked to the environmental effectiveness of REDD+ activities and assesses reported evidence of the success of CFM at reducing deforestation and forest degradation, conserving forests, sustainably manage forests and enhance forest carbon stocks (Section 3.1). The second question looks at reported evidence of the success of CFM at generating social benefits including on income, employment, security, empowerment and equity at the local level (Section 3.2). For biophysical and social outcomes, we also test for statistical relationships with potential contextual factors or modifiers that can have an influence on the outcomes of CFM. For the third question (Section 3.3), we use small-scale community carbon sequestration projects and early experiences in REDD+ demonstration projects to explore early reported evidence on the same outcomes as for the first and second questions, acknowledging that there are still limited quantitative cases studies. We also discuss different benefit-sharing proposals as well as opportunities and risks that are relevant for designing better interventions with REDD+ implementation in the CFM context or at the community level.

In the following section, we discuss the global trend and underlying rational for the promotion of CFM in developing countries. Then, we look at evidence that CFM is fulfilling REDD+ climate mitigation activities, after which we examine the reported social outcomes of CFM. We finally review experiences directly relevant to REDD+ benefit-sharing and participation in community forestry settings.

Community Forest Management: A Global Overview

CFM has various definitions and interpretations (Table 1) reflecting a variety of interventions, the specifics of which vary from place to place. It is a subset of the community-based management of natural resource practices that were initiated in the mid-1970s, designed specifically to promote forest management in a way that takes better account of the needs and interests of forest people. It was considered one of the most promising options for combining forest conservation with rural development, thus meeting poverty reduction objectives.

Table 1. Concepts of community forest management.

Definitions, Interpretations and Synonyms	Sources
An approach to forestry implying community or local control and management of forest resources	Glasmeier and Farrigan [20]
Any situation which closely involves local people in forestry activity	FAO [21]
The sustainable management of the forest for wood, non-timber forest products and other services with a social or environmental value, performed by forest-dependent families or smallholders, community groups and indigenous peoples	Growing Forest Partnerships [22]
Type of management in which communities have some degree of responsibility and authority in forest management that encompasses multiple uses involving subsistence and marketing with the goal to conduct an ecologically sustainable use of the forest	Charnley and Poe [23]
Associated terms: "community forestry", "community-based forest management", "social forestry", "participatory or collaborative forestry" or "agroforestry"	Arnold [24], Hajjar [25]

The rationale for promoting greater control by communities in forest protection lies in the following assumptions about the relationship between people and the forest [23]. First, people are more likely to take responsibility for forest resources if they have a sense of ownership and control over them. Secondly, because of their geographical proximity, local communities should be better able to provide effective protection of forests and enforce the rules of access and use. Third, local communities may be encouraged to defer current livelihood benefits and take future profits into account if they have more control over them [26]. Finally, there are numerous examples of communities managing forests sustainably under customary institutions [27].

In terms of social benefits, thanks to local control, a larger portion of forest rents and greater benefits should accrue to forest communities under CFM in comparison to other conservation approaches [28]. Indeed, it is often assumed that greater local control will produce more social and economic benefits for forest communities. The logic behind this assumption is that: (1) central governments are more likely to prioritize national interests and industries, while local communities will favor their own interests; (2) local institutions are able to respond to community needs more efficiently than central governments because of better information and accountability; and (3) local control provides more opportunities for marginalized groups to influence policy [23,29–31].

Since the mid-1970s, decentralization has become a major trend in global forest governance accompanied by reforms initiated by the governments of major forested countries [32]. This transition came about for multiple reasons. First, it was a response to mismanagement by central governments and international pressure from donors for better forest governance. Centralized control and management of public forests is increasingly regarded as untenable [28]. Public agencies have often been poor stewards of forests due *inter alia* to the difficulty of defending the forest against residents who have little interest in maintaining it if they do not have rights to it and also to abuse by political elites and corrupt interests [33]. Second, it was considered a way of reducing the financial burden of forest governance on governmental budgets. Third, this trend for decentralization has been in response to social demands, domestic and international, for the recognition of the rights of indigenous peoples and local communities over forest resources and for their greater role in managing local forests.

A concomitant forest tenure transition is also taking place. While public ownership of forests still predominates (Figure 1), the absolute area of public forest land administered by governments in 25 of the world's 30 most-forested countries has decreased from 2583 Mha in 2002 (80% of the global forest estate) to 2409 Mha in 2013 (73%) [33–35]. The percentage of forest in developing countries either owned or directly administered by indigenous peoples and other forest communities reached 27% in 2008 (This figure includes data from the 15 countries with most reliable data sets) (Figure 2), representing a significant share of forest area in developing countries. In reality, the forest area *de facto* managed by local people under customary tenure greatly exceeds the area of community and indigenous lands acknowledged by statuary law (see Table 2 for key concepts).

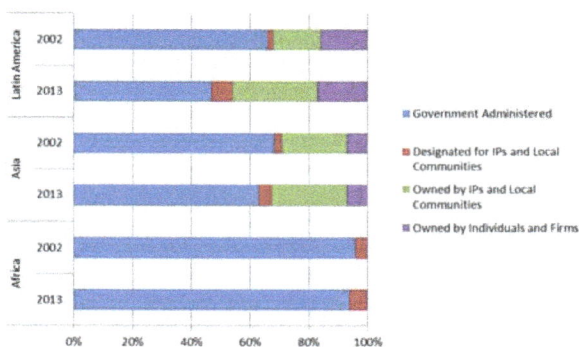

Figure 1. Changes in statutory forest tenure per region, from 2002 to 2013. Public forest ownership is predominant in South America, Asia and Africa. There has been a noticeable increase in forest area designated for indigenous peoples and local communities in Latin America and Asia. Forest owned by individuals and firms has increased in all three regions since 2002. Source: Rights and Resources Initiative—RRI's forest tenure database accessible through the Tenure Data Tool. Available at: http://rightsandresources.org/en/resources/tenure-data/.

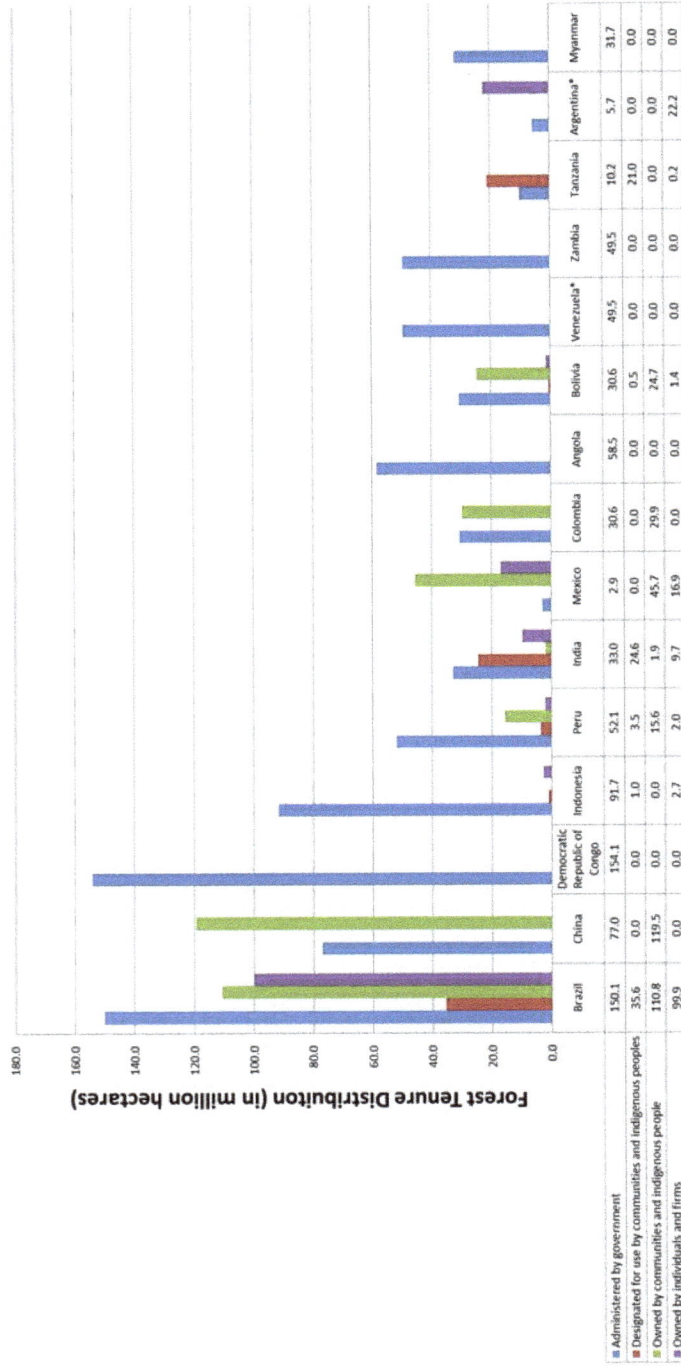

	Brazil	China	Democratic Republic of Congo	Indonesia	Peru	India	Mexico	Colombia	Angola	Bolivia	Venezuela*	Zambia	Tanzania	Argentina*	Myanmar
Administered by government	150.1	77.0	154.1	91.7	52.1	33.0	2.9	30.6	58.5	30.6	49.5	49.5	10.2	5.7	31.7
Designated for use by communities and indigenous peoples	35.6	0.0	0.0	1.0	3.5	24.6	0.0	0.0	0.0	0.5	0.0	0.0	21.0	0.0	0.0
Owned by communities and indigenous people	110.8	119.5	0.0	0.0	15.6	1.9	45.7	29.9	0.0	24.7	0.0	0.0	0.0	0.0	0.0
Owned by individuals and firms	99.9	0.0	0.0	2.7	2.0	9.7	16.9	0.0	0.0	1.4	0.0	0.0	0.2	22.2	0.0

Figure 2. Forest tenure distribution for the 15 most forested developing countries (millions of hectares) in 2013. Source: RRI's Forest tenure tool [35]; * No data were available for the year 2013.

Table 2. Key concepts for community forest management.

	Definition
Decentralization	The transfer of both decision-making authority and payment responsibility to lower branches of the government OR refers to a full or partial transfer of assets and power [a] from the central government to the lower branches of the government or local institutions. Decentralization is generally observed in forest management.
Devolution	The transfer of rights and responsibilities (of assets and power [a]) to non-state agents who are neither created nor controlled by the state including citizens or forest user organizations
Tenure systems	Rights that define ownership and resource specific user rights including duration and conditions
Customary tenure systems	Tenure systems established by custom or tradition and determined at the local level, rather than by law or contract, often based on oral agreements
Statutory tenure systems	Tenure systems applied by governments and codified by law.

[a] Charnley and Poe [23].

In CFM, the form of tenure and the institutional arrangements between local community or forest user groups and public agencies contribute to determining rights and responsibilities in the management of forests (with the involvement of non-governmental organization or not). There is a vast diversity of institutional arrangements and of realities that will have an impact on forests and livelihoods outcomes. Broadly, we can distinguish three different basic governance models: (1) collaborative/participatory forestry in which the land is formally held by the government and a contract or agreement is reached with local people regarding offtake rates and possibly silvicultural practices such as fire watching, in return for recognition of rights to certain forest products (e.g., Nepal, Tanzania, India, Kenya); (2) community-owned forest where the resources belong to the community and the government provides financial support for conservation through PES or for sustainable timber management (e.g., Mexico, Costa Rica) and (3) indigenous peoples' reserves in which granting of land rights (usufruct) is usually on traditional lands and which are usually more extensive.

2. Materials and Methods

In order to assess the role that CFM can play in REDD+, we examined the contribution of CFM to carbon and social benefits and generate lessons for CFM in the REDD+ context and for REDD+ implemented with forest communities. We performed a narrative systematic review, including also quantitative evaluation. Systematic review is a research methodology used to compile, critically appraise and assess results of primary research to build evidence that can be used to inform policy, highlight knowledge gaps and identify further research needs. This review is narrative as it uses textual and graphical descriptions of findings and key characteristics obtained from systematic review, and it is complemented by a quantitative synthesis with statistical testing.

Publications relevant to the issues were selected through search in the ISI web of knowledge, google scholar and internet. The keywords used in this search included: "Community forest management", "community forestry", "community forest", "community-based forest management", "social forestry", "participatory or collaborative forestry", or "agroforestry" and crossed with "carbon", "land cover", "deforestation", "degradation", "conservation", "carbon sequestration", "livelihoods", "development", "Clean Development Mechanism" and "Reducing emissions from deforestation and forest degradation". Individual terms were combined using appropriate Boolean operators. We also used available reviews on CFM to access original studies through their cited references and contacted expert authors to obtain their reference lists when not publicly available. The initial search generated >3000 results that were assessed for relevance using the title and abstract. The article screening was performed through an iterative process to examine the search results and select the studies included in this review (inclusion/exclusion criteria). We based our selection on the research questions developed to identify publications reporting on case studies. For the first and second research questions, we included studies of CFM-type of interventions that provided an evaluation of the biophysical outcome in terms of forest conditions, of social benefits including livelihoods and development outcomes or of both biophysical and social outcomes. For the third question, we considered experiences related to carbon mitigation including early REDD+ projects, Clean Development Mechanism for small-scale Afforestation and Reforestation project, and community-based carbon sequestration projects. We limited the geographical scope to developing country cases. The temporal scope chosen for this review ranges from 1992 to 2014. We excluded several publications that discussed the potential of CFM (many cases for small-holders agroforestry) for carbon mitigation (not empirically-based), studies evaluating the contribution without comparator that could be directly assigned to the CFM type of intervention, or cases providing accounts that were too specific (e.g., war context), too theoretical, or with a methodological focus, as well as studies that were not largely applicable and directly relevant to the questions at hand (Figure 3).

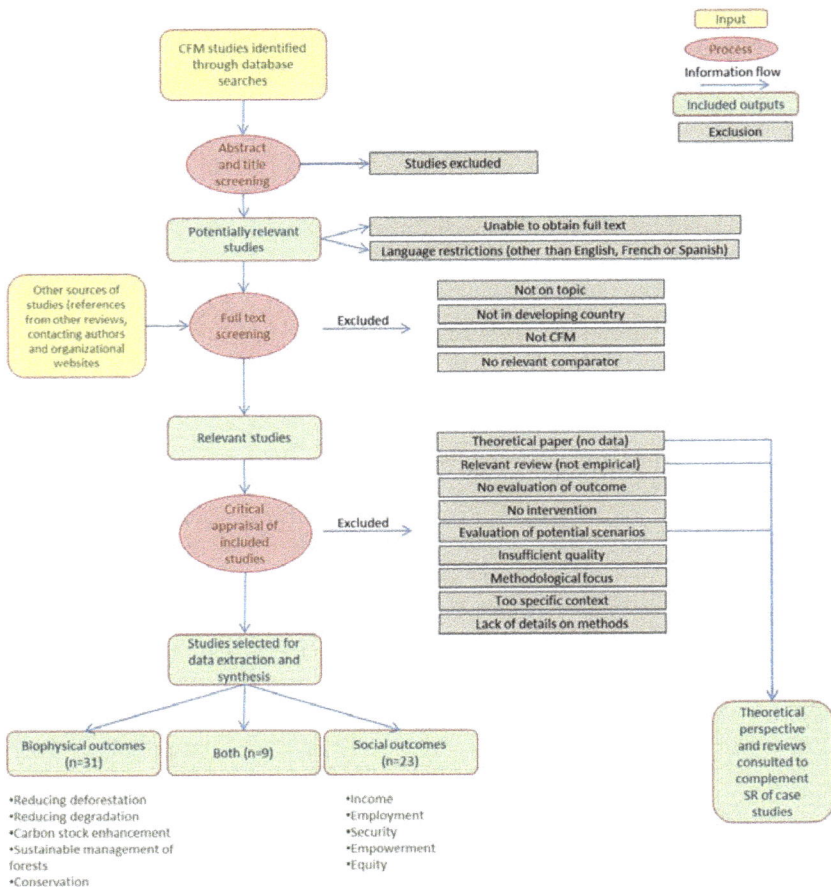

Adapted from CEE (2013) Guidelines for Systematic Review and Evidence Synthesis in Environmental Management. Version 4.2. template p.45.

Figure 3. Diagram of the selection and filtering process for the systematic review.

Three broad types of publications were used to answer our research questions: case studies (63/145), reviews (16/145) and papers providing theoretical perspectives (65/145). Most of these are from peer-reviewed publications (122/145) and from journal articles (119/145). We systematically reviewed empirical case studies to obtain a quantitative appraisal of the contribution in both social and biophysical outcomes of CFM (Questions 1 & 2), as well as of small-scale carbon sequestration and early REDD+ project cases (Question 3).

2.1. Critical Appraisal and Data Extraction

In the systematic review of case studies, we used the publication as the unit of analysis and we defined a coding approach to extract data that allows integrating opposite outcomes reported in the same publication. This approach was considered the most realizable despite its limitations since the pool of empirical studies varies greatly in terms of geographical scope and methods, which are often not clearly reported. Separating by specific area of CFM would have required much more information than what is currently published for most cases or relying on a very limited sample size. Our approach aimed to strike a balance between extractable information and sample size. For each study, we coded the outcome as: positive (1), neutral (0), or negative (-1). When multiple cases of CFM were compared in the same study with diverging outcomes or if the same case study was reporting both positive and negative outcomes on the same criterion (e.g., gain and loss of income), the outcome was coded as mixed ($1/-1$). When a case did not report on a criterion, it was considered a missing value (NA). Therefore, we provide the number of studies that reported on each aspect, or the sample size (n), and we report on the percentage of studies, based on the sample size.

We assessed the reported biophysical and social outcomes of CFM, with 31 studies addressing biophysical outcomes, 23 addressing social outcomes, and nine studies addressing both. For the biophysical outcomes, we assessed reported studies in relation to the five REDD+ activities, i.e., (1) reducing deforestation; (2) reducing forest degradation; (3) carbon stock enhancement; (4) sustainable management of forests; and (5) forest conservation. Positive outcomes at reducing deforestation were identified when the study showed a decline in deforestation rates over time under CFM or lower deforestation rates compared to an area that is not under CFM. Positive outcome at reducing degradation was identified when poor forest conditions were noted to have improved over time under CFM or in comparison to other non-CFM areas. A positive outcome in carbon enhancement was noted with quantified growing stocks or biomass increment. Positive outcomes in terms of sustainable management of forests were observed with no change in forest cover despite community extraction, and conservation impacts qualified as positive with improved or no change in forest conditions were assessed. For social outcomes, we adapted a framework developed by Lawlor, et al. [36] and looked at the reported results in terms of (1) income; (2) employment; (3) security; (4) empowerment and (5) equity. Positive outcomes in terms of security were identified if the case study results reported improvements in land ownership and management rights, access and use rights, carbon rights, health and education through infrastructure development or ecosystem services. Positive outcomes in terms of empowerment were identified if the case study results reported an increased participation in decision-making regarding local land-use and development, capacity-building and training for building social capital to participate more effectively. Positive outcomes for equity were identified if the case study results reported equitable or 'pro-poor' distribution of benefits among wealth groups. For both social and biophysical outcomes, we identified the approach of comparison used in the study for evaluating the impacts of intervention (Is the intervention being compared with no intervention or are alternative interventions being compared with each other? (See Pullin and Stewart [37])). We also included an assessment of the robustness of the methodological approach used by the case study based on the strength of counterfactuals used to demonstrate impacts of CFM interventions, with three tiers of quality (1 = Strong; 2 = Regular; 3 = Weak). The quality of the methodology was qualified as 'strong' when the success of CFM was evaluated with a control or counterfactual and/or over time (using before and after comparison), assessed with statistical methods and controlling for covariates, i.e., the heterogeneity of the context. 'Regular' quality was assigned for studies using comparators, but only when the study did not account explicitly for some relevant covariates. Study methodological approach was qualified as 'weak' when it relied on residents' perceptions and covariates are not accounted for. In evaluating the methodological robustness of case studies, we made a differentiation between studies assessing biophysical outcomes of CFM and those assessing social outcomes. The success of biophysical outcomes can be measured based on quantitative data with methods that are independent of the users, i.e., using a remote sensing approach and forest carbon inventories. For evaluating social outcomes, we acknowledge that measuring the success of the intervention, particularly for empowerment, security and equity criteria, is much harder to quantify and relies on inputs from the users. We took these differences into account when evaluating the methodological strength of case studies; however, for evaluating the outcomes of CFM interventions, we generally favored studies that apply rigorous counterfactuals, with a quantitative approach, including statistical testing to support evidence. The literature on CFM covering social outcomes includes a number of contributions based on qualitative research methods. These case studies were set in a separate class ('QR'), since they addressed different research goals and typically did not use formal counterfactuals.

Moreover, for each case study, we identify basic characteristics for each one including the country, the type of arrangement (as identified by its authors), the type of ownership (community ownership or designation for use by governmental authority), as well as the main type of extraction (subsistence, enterprise, or both). These characteristics (or modifiers) have been identified in the literature as influencing the outcome of CFM and were tested in this study. The full list of references and case studies evaluation is made available in Supplementary materials.

The reviews and theoretical perspectives publications were used to generate insights on key aspects of tenure, decentralization, participation, enforcement, equity and benefit-sharing, to obtain contextual information that typically characterizes the diverse array of CFM arrangements. Since REDD+ implementation has only just begun, there is a limited set of cases to compare. We have taken advantage of CFM experiences that are directly transferable to the REDD+ context, as well as papers with a more theoretical perspective to identify elements that are central to REDD+.

2.2. Data Synthesis

Data extracted was synthesized by determining the number of studies reporting on each of the ten outcomes. We also used Pearson's Chi-square test with simulated *p*-value for a relationship between the 10 outcomes and four CFM characteristics (the type of ownership, the type of extraction, the region and the type of arrangement), with a total of 40 contingency tables. For the type of arrangement, we reduce the number of categories from the authors' typology; we re-classified the 30 levels into seven categories: community forestry, indigenous reserve, joint forest management, community forest management, carbon project, REDD, and mixed. The null hypothesis for each Chi-square test is of independence between the two categorical variables. We use Freeman-Tukey (FT) deviates and/or the standardized residuals post hoc tests, to test for significant difference between the observed and the expected values in each cell of the contingency table [38]. The evaluation spreadsheet built for this review was imported in R statistical software [39] as a comma separated value (csv) file and summary statistics and statistical testing were performed based on selected criteria. We assessed the main cited criteria identified under CFM for obtaining positive forest condition, social condition and equity outcomes with relevance for REDD+ implementation at the community level by analyzing the content and recommendations of both empirical and reviews that aim to identify the factors promoting success in CFM. In the following sections, we provide the results from this review and summarize the main points identified in the case studies reviewed, as well as results from other reviews and theoretical work.

3. Results and Discussion

3.1. REDD+ Activities Realized by Community Forest Management

We sought to determine the contribution of CFM in slowing, halting and reverting forest cover loss and therefore its usefulness in terms of carbon benefits by looking to the five types of

REDD+ activities. This question is crucial, as the ultimate goal of REDD+ is to reduce emissions or increase removals of GHGs by forests.

Less than 23% of the studies reviewed report positive results at reducing deforestation (Figure 4). Although reducing deforestation contributes directly to forest conservation, the inverse is not true. Forest conservation does not necessarily imply a reduction in deforestation, a nuance that is important in the REDD+ context. To demonstrate its potential to reduce deforestation, CFM must be located in areas where the pressure of deforestation is strong; at the agricultural frontier, for example. This makes reducing deforestation harder to prove, since it has to be shown that without CFM intervention, forests would have been cut down. It is interesting to note that the studies reporting success at reducing deforestation are only for cases in Latin America, in indigenous reserves or CFM implemented in indigenous areas, with studies from Brazil, Mexico, Colombia, Panama, Guatemala, and Nicaragua. This relationship between success in reducing deforestation and region (Latin America) is statistically significant (FT deviate = 2.13; Critical value = 2.02, *df* = 6), see Table 3. For example, Nepstad, et al. [40] have shown the ability of indigenous reserves to control deforestation and fires in the Brazilian Amazon, where deforestation rates are very high. Aboriginal reserves also appear effective in preventing deforestation in the Guiana Shield in Colombia, despite their limited area [41]. Nelson and Chomitz [42] have demonstrated the effectiveness of Aboriginal reserves in Brazil in preventing forest fires compared to uninhabited protected areas. A recent study by Porter-Bolland, et al. [43] based on a statistical meta-analysis comparing 40 protected areas and 33 cases of community-managed forests found that forests managed by communities have slightly lower and less variable rates of deforestation than protected areas.

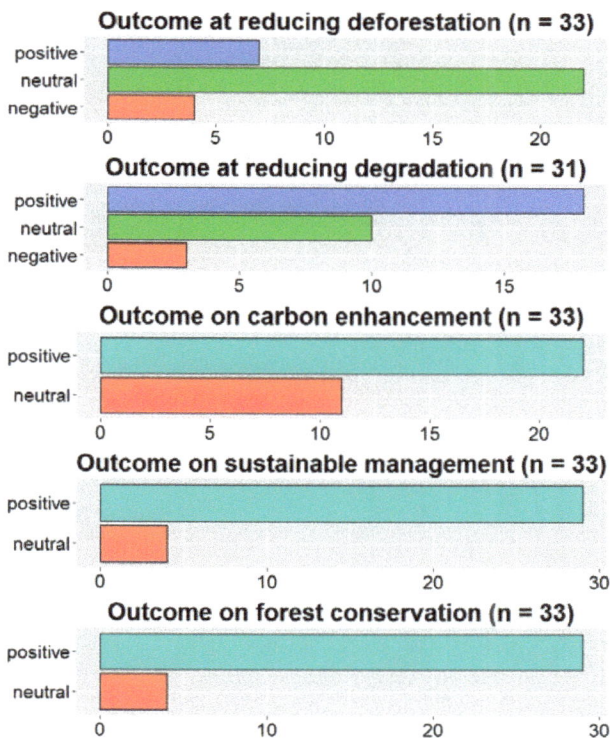

Figure 4. Frequency of the outcome identified for the five REDD+ activities evaluated in this systematic review. The number of studies reporting on each criterion is indicated as the sample size (*n*).

Table 3. Pearson's Chi-squared test with simulated p-value between the assessed outcomes and four CFM characteristics. The null hypothesis for each test is of independence between the two nominal variables.

Outcome	Characteristics							
	Type of Ownership		Type of Extraction		Region		Type of Arrangement	
	X-Squared	p-Value	X-Squared	p-Value	X-Squared	p-Value	X-Squared	p-Value
Reducing deforestation	7.851	0.116	2.890	0.639	15.630	0.013 *	11.924	0.500
Reducing degradation	6.905	0.139	5.692	0.229	2.827	0.900	13.620	0.359
Carbon stock enhancement	3.075	0.262	0.797	0.866	0.283	0.999	4.765	0.676
Sustainable management	1.381	0.658	0.556	0.999	4.315	0.202	5.181	0.626
Forest conservation	1.381	0.648	1.806	0.550	4.315	0.210	8.838	0.232
Income	3.560	0.775	9.936	0.035 *	9.973	0.384	22.080	0.087
Employment	6.530	0.248	7.736	0.088	12.450	0.069	10.401	0.453
Security	15.438	0.019 *	4.152	0.592	11.347	0.232	17.628	0.288
Empowerment	4.909	0.311	4.359	0.427	9.098	0.162	10.054	0.511
Equity	7.481	0.527	8.438	0.219	10.102	0.371	11.178	0.780

The asterix '*' shows a statistically significant result (at 95% confidence level).

378

The results of our systematic review of published case studies show that there is substantial evidence that CFM is effective at promoting sustainable landscapes through sustainable management of forests and forest conservation as this was reported in 87.5% of the cases [44–46]. CFM has also been shown to be effective at restoring forest cover and carbon density [47,48]. Sixty-nine percent of the studies indicated positive results for carbon stock enhancement and 60% for reducing degradation (Figure 4). For example, Hayes [49] showed the contribution of CFM to forest conservation and carbon stock enhancement for 163 forests located in 13 countries by reporting no significant difference between the forest conditions in protected areas and those managed by communities, with higher vegetation density in areas under local community control. In India, Somanathan, et al. [50], using state-managed forest as the comparison, showing no significant difference for broad-leaved forest and more percentage crown cover in pine forests under Village-Council forest management (VCFM), and also noted that VC forest management is seven times cheaper than management by the state. The positive contribution of CFM to the sustainable management of forests and forest carbon stocks enhancement was also measured directly through repeated forest carbon inventory by Karky and Skutsch [51] in Nepal.

Our results also show no statistical relationship between the five REDD+ activities and the types of ownership, extraction or arrangement (Table 3). We were unable to unveil clear patterns for these variables among the studies reviewed, but individual case studies identified that the size and type of forests, the quality of the resource, the type of utilization as well as a whole set of institutional and social factors can influence the forest outcomes. Some documented cases show persistent forest cover loss, indicating that CFM performance varies across communities and contexts. In Mexico, Dalle, et al. [52] show low rates of deforestation in *ejidos* of the Mayan zone of Quintana Roo. In Ecuador, indigenous reserves that do not overlap with land under conservation status display similar rates of deforestation to private lands. Furthermore, exogenous forces can influence the forest outcome, not just the activities of the communities themselves [12,53–56]. The success of CFM in protecting forest resources is more likely where population pressure is low, and less likely in the face of conflicts, market pressures, and rising population [57]. For example, while areas under CFM had lower rates of deforestation than protected areas under low colonization pressure in Mexico, in Guatemala both of these conservation strategies failed to maintain forest cover under high colonization pressure, an element symptomatic of weak governance [58].

Evaluation of the ecological outcomes of CFM has however been subject to criticism on the grounds of the methodology used [57,59,60]. In order to demonstrate the role of CFM in forest outcome, studies have to use data from comparable cases and/or counterfactuals. Most studies have compared CFM sites with other types of management (e.g., protected areas) in terms of forest cover change, while controlling for other confounding factors using appropriate statistical methods. Appropriate for quasi-experimental contexts, matching methods and propensity scores are especially interesting to avoid biased comparison and to control for heterogeneity across biophysical and community characteristics, and have been used in some of the studies we reviewed. Our analysis showed that 56% of the studies (*n* = 39) had a strong methodological approach. The most frequent approach to comparison combined before-and-after or time series of remote sensing images to assess forest cover change over time while controlling for confounding factors. When both social and biophysical outcomes are measured, the assessment of forest conditions often relied on local people's perceptions as ascertained through interviews, and was thus based on rather weak methodological underpinnings.

The lack of spatially-explicit national data on forests managed by communities has been and is still a major obstacle to a better understanding of the role of CFM in forest outcomes [61]. Information on the extent and location of CFM is highly fragmented, unavailable to the public or non-existent. Recent studies making use of newly available data sets for indigenous territory polygons in Panama [62] and Brazil [63] have been able to demonstrate the contribution of these reserves in conserving forests and the carbon they contain. In this sense, it is important to highlight the work of Rights and Resources Initiative for the creation of its Forest Tenure dataset (http://www.rightsandresources.org/

resources/tenure-data/) and the World Resources Institute's Status of Land Tenure and Property Rights map (http://www.wri.org/resource/status-land-tenure-and-property-rights-2005).

3.2. Livelihoods and Development Outcomes of Community Forest Management

We assessed the contribution of CFM in producing social co-benefits through improved livelihoods and development. Recent global comparative studies confirm the importance of forest resources in the livelihoods of the rural poor in developing countries, estimating that these represent on average 21.1% of total household income [64]. The importance of forest for subsistence purposes is particularly important where chronic poverty and forest cover overlap geographically [65]. This is due to the dependence of the poor on these environmental incomes, especially in remote areas, where often no substitute for forest products and services exists [66]. For communities living in poverty, restrictions in access to and use of forest resources can have a major impact on livelihoods, and such restrictions may conflict with the objective of poverty reduction [67].

The benefits derived from CFM depend on the quality of the forest resources, the access rights granted and the benefit-sharing mechanisms, which display high contextual variation. Almost all governments maintain certain rights of control over the use of land and resources, regardless of the formal property system [33,68]. Compared to the open access situation, CFM typically places new restrictive rules and regulations on extraction of forest-based resources. Even with statutory rights, communities do not automatically have rights to all resources (e.g., timber), and they are not always able to access or translate those rights into benefits [19].

We found that 44% of the studies reported beneficial impacts of CFM through an increase in forest incomes, and a further 44% reported neutral effects, neither positive nor negative (Figure 5). One case was reported as having negative impacts on income and two studies reported both negative and positive impacts. For the Pearson Chi-square test on income and the type of extraction (subsistence, enterprise or both), the null hypothesis was rejected at $p < 0.05$ significance level, meaning that the two variables are not independent; the fact of belonging to one category of the first variable influences the membership category of the second variable. However, none of the FT deviates or standardized residuals (Z statistic) post hoc tests showed a significant difference for all pairwise comparisons (FT critical value = 1.85 or Z critical value = 2.54, *df* = 4), indicating that the difference between the observed and the expected values is not large enough for individual cells to show significant results between income and the type of extraction. Benefits from employment were noted in 39% of the studies. However, the Chi-square test results were not significant for all the variables tested, meaning that the region, the type of extraction, arrangement or ownership are independent of the creation of employment benefits.

According to Mahanty, et al. [69], the potential for increases in the quality of life of rural populations through CFM depends largely on the type and size of the benefits created, how communities can make sure they get at least a part of these benefits, and how they are distributed locally. The importance of access rights was demonstrated in CFM in Ethiopia, as forest user group members had lower total income and assets than non-members where only subsistence use is allowed, while it was higher for members where timber harvesting is also permitted [70]. In Nepal, a study comparing the benefits and costs of CFM from eight user groups indicated that the impacts of this practice are highly variable within and between groups [71]. It was also shown that CFM tends to divert profits from individual households to the community level, with a decline in forestry-based income, but an increase in new sources of income (including grants, soft loans and other income-generating activities) that benefit poor households. Forest condition is also mentioned as a factor that will influence whether incomes generated through CFM are enough to cover the management costs and provide a direct benefit to communities [72]. In some cases, forests assigned by governments to communities can be of poor quality, as such forests have often been exploited and degraded by the logging industry and abandoned when the industry has no continued interest in exploiting them commercially [28,48]. It is also important to acknowledge that even if very small, the income generated by CFM can make a difference for very poor households, as exemplified in Malawi by Jumbe and Angelsen [73].

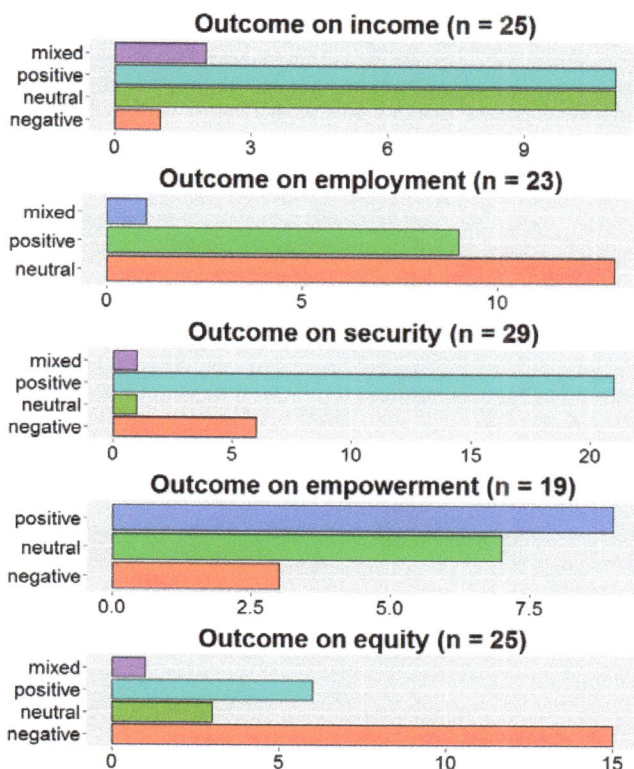

Figure 5. Frequency of the outcome identified for the five social criteria evaluated in this systematic review. The number of studies reporting on each criterion is indicated as the sample size (*n*).

In the studies reviewed, security was the most frequently mentioned benefit, in 72.4% of the cases (Figure 5). We also found a relationship between security and the type of ownership (Table 3), with positive outcomes on security with community-owned forests (FT deviate = 2.05; Critical value = 2.02, *df* = 6). This is not surprising since, generally, CFM involves a process that entails clarification and agreement on access and use rights. Benefits through empowerment, which is a more qualitative characteristic than variables such as income, were mentioned in 47% of the studies reviewed. In Mexico and Brazil, for example, Hajjar, et al. [74] showed that despite the fact that governments have maintained significant control over forest resources through heavy regulations on timber extraction, communities have effective decision-making power over the day-to-day planning and they derive considerable benefits from forest management.

One important finding is on the outcome observed on equity inside community-managed forest (Figure 5), and this is irrespective of the type of extraction, arrangement, region or ownership (Table 3). Sixty percent of the studies reported a decline in equity with respect to the distribution of local benefits for poorer and/or women-headed households under CFM. It is important to keep in mind that the notion of 'community' is a construct simplifying the heterogeneity of diverse actors found at the village-level who have diverse and sometimes competing interests (See also [75]). Indeed, processes and institutional arrangements that govern the implementation of CFM at the local level can easily be dominated by the wealthier or more powerful community members, producing results that reinforce and perpetuate social inequality, including gender inequality [68,76]. In Nepal, negative effects of devolution have been reported among the poorest households as a result of a reduced access to forest products necessary for their livelihood, due to more stringent harvesting regulations and more 'equitable' distribution of benefits from forest, without taking into account the fact that the poorer households generally need more forest resources [77,78]. Vyamana [72] also found that devolved management through CFM does not support an equitable local distribution of benefits and costs in Tanzania, and that arrangements exclude the poor from income-generating activities because of initial investment costs for participation.

Thirty-seven percent of the studies reviewed were classified as having strong methodological underpinnings for assessing evidence of social benefits. These studies evaluated the impacts of CFM using strong comparative indicators, and by removing rival explanations or confounding factors that were unrelated to this type of management. Propensity scores and covariate matching models have been used for controlling for these confounding factors and for achieving a better attribution of the outcomes of CFM programmes [70,73]. The importance of collecting baseline data or Before-After Control-Impact (BACI) methodology to evaluate the welfare outcomes is paramount [79,80]. As shown in these results, stratification by welfare groups needs to be done in order to evaluate local distribution of benefits. The contribution of qualitative research methods with rigorous approaches is also essential, especially for understanding power dynamics and the kind of benefits that are important to local people but hard to measure.

Only a few studies provide an analysis of both stated objectives of CFM, i.e., improved forest condition and livelihoods benefits [26,81]. Based on the International Forestry Resources and Institutions datasets (http://www.ifriresearch.net/resources/data/), factors associated with win-win outcomes for forest carbon and livelihoods are identified including: rulemaking autonomy, local enforcement rules, well-defined property rights, and the design of effective institutional arrangements. Factors promoting synergies in achieving positive forest and social outcomes are synthesized in Table 4, as well as factors promoting equity at the local level.

Table 4. Criteria identified under CFM for assessing successful outcomes with relevance for REDD+ implementation at the community level.

CFM Success Factors	Forest Carbon Benefits	Social Benefits	Equity	Sources
Poverty reduction as a stated objective		X		Schreckenberg and Luttrell [76]
Allowing for both subsistence and commercial use of forest products		X		Ibid.
Design of effective institutional arrangements	X	X	X	Ibid.
Transparent and equitable benefit-sharing mechanism at the local level			X	Ibid.
Sufficient support and training during establishment	X	X	X	Schreckenberg and Luttrell [76]; Hajjar, Kozak and Innes [74]
Well-defined property rights	X	X		Pagdee, et al. [82]; Robinson, et al. [83]
Community interests and incentives	X	X		Pagdee, Kim and Daugherty [82]
Fair representation and active participation of the poor and women/Pro-poor measures			X	Mahanty, Guernier and Yasmi [69]; Maharjan, Ram Dakal, SureshThapa, Schreckenberg and Luttrell [71]
Rulemaking autonomy	X	X		Chhatre and Agrawal [26]; Persha, Agrawal and Chhatre [81]; Ostrom [84]
Local rules enforcement	X	X		Chhatre and Agrawal [85]
Local power dynamics check and balance arrangements			X	Agrawal and Gibson [75]

3.3. REDD+ Benefits and the Incentive Structure for Community Forest Management

There are active debates around the potential of REDD+ in promoting CFM with some seeing it as a way of reinforcing existing CFM by promoting the generation of new income sources and community development, others as a model to be scaled-up for REDD+ national strategy [5] and yet others as a potential threat to CFM that could destabilize successful existing community forest governance through a top-down approach and re-centralization [17–19,86]. REDD+ success in each country is seen largely as contingent on how forest management rules and incentives in place shape local actions related to the use of forests and forest land conversion [87]. In order to reduce emissions or increase removals of GHG at the scale required, REDD+ would need to generate sufficient incentives and social acceptability at the local level to stimulate participation and sustain it through time. However, there are trade-offs between effectiveness in reducing GHGs, cost-efficiency of mitigation activities implemented, and equity between those who benefit and those who assume the costs. It is unrealistic to assume that ´win-win-win´ solutions can be found in all or indeed in the majority of cases, with existing tradeoffs [88]. It may be more realistic to aim for ´no-harm´ situations, by ensuring that this new global policy affecting land use and forest management does not exclude or damage existing CFM initiatives, for instance, by taking away existing rights through new restrictions or exclusion, by reducing local control and decision-making power or by capturing carbon payments that should belong to forest communities. Since REDD+ and CFM policy interventions do not share the same goals and mechanisms, analysis of available evidence of the potential contribution of CFM to REDD+ objectives and of potential benefits and risks for CFM is important [89].

3.3.1. Early REDD+ and Carbon Mitigation Projects in Low-Income Communities

The failure of the Clean Development Mechanism to promote widespread adoption of afforestation and reforestation (A/R) projects provides useful lessons on the distribution of costs and benefits and the generation of incentives in carbon-based projects. This mechanism was created as a way for developed countries to meet part of their emission reduction targets under the Kyoto Protocol, by purchasing credits from verified, sustainable and additional forest carbon sequestration activities carried out in developing countries. Only 68 out of 8705 CDM projects (http://cdmpipeline.org/Accessed on the 30 July 2014) (or 0.78%) are A/R projects, largely due to the complications in administration, finance and governance issues related to these projects [90,91]. The high transaction costs and long validation times associated with CDM have been proven prohibitive for small-scale projects [92,93]. Other issues include the modalities, financial and production risks, labour demands, liquidity/sunk costs, and perceived equity [94]. Mixed effects of carbon projects on local populations and on poverty alleviation have been found [95,96]. What we can learn from this is that the transaction costs, the administrative quagmire, the potential risks, and the modality arrangements that are negotiated for REDD+ are likely to have a significant impact on community incentives and participation in REDD+.

In this review, we found indications of negative impacts on security (four of nine cases), on empowerment (two of six cases) and on equity (four of eight cases) criteria of carbon mitigation projects. In Tanzania, Beymer-Farris and Bassett [97] provide a cautionary note for REDD+ projects that are modeled on decentralized forestry schemes that are not decentralized in practice. It appears that, in spite of extended policy discourses on devolved decision-making, justice and equity in terms of resource access and actual local-level decision-making are not always forthcoming. Early results from a REDD+ pilot project in Nepal show that it imposes direct additional costs through new restrictions on forest product harvesting and reduced grazing, and indirect costs in time and labour due to increased participation as well as forgone benefits [98]. These results suggest that REDD+ payments, if based only on the exchange value of the carbon saved, might not generate sufficient incentive in the long run [98]. In Mozambique, the carbon incomes generated were found to be small even with liberal carbon accounting [96]. Lack of equity between participants and non-participants in a CDM project is discussed in Vietnam, where new restrictions are imposed on non-participants who do not receive project revenues, threatening the long-term carbon mitigation outcomes of the project [99].

Concerns over additionality (Reduction of emissions or increased removals due to project activities compared to a business-as-usual scenario), leakage (displacement of emissions) and transaction costs have been voiced for community carbon and REDD+ projects [96,100]. It is also recognized that because most CFM projects are small, high transaction costs will be involved [100]. Since many CFM projects already have positive forest carbon outcomes, it is questionable whether they would be considered additional and thus deserving of REDD+ payments; although it would of course be possible to expand the areas under CFM and achieve additionality in this way. This option however implies paying the newcomers but not those who have protected their forests in the past, raising important questions regarding whether such an approach would be considered politically legitimate or fair. For ensuring fairness, existing CFM participants should also be provided with incentives.

3.3.2. Incentives Structures for REDD+ as CFM

Different REDD+ incentive structures have been identified for CFM under REDD+. Skutsch, et al. [101] proposed two possible REDD+ incentives structure approaches for CFM. In the first, payments are ex-post, based on performance measured in terms of the amount of avoided or sequestered carbon compared to a baseline (carbon outputs). In the second approach, payments are made to compensate for management inputs and formulated to incentivize specific norms of sustainable forest management, that is, they are related to management inputs. Management inputs could include forest monitoring where the community would be paid for measuring forest carbon stocks and changes [102].

For performance-based payments, REDD+ would likely entail new restrictions on use and access to land and forest resources. Recent research demonstrates that some non-trivial forest income flows could be at stake for the rural poor [103]. In this matter, it is relevant to make a distinction between communities that use the forest mainly for subsistence and family consumption and those who would use it commercially (e.g., timber, non-timber forest products) [56]. For subsistence use, it may not be possible to compensate a reduction in forest access and use; resources will have to come from elsewhere. In many cases, calculated opportunity costs underestimate the true value that forest has for communities [104]. Other alternatives must be created to compensate for a limit placed on the supply of forest resources. In Tanzania, Fisher, et al. [105] show that paying communities to reduce deforestation from fuelwood collection is not sufficient; alternatives should be implemented to avoid an increase in the value of firewood and a displacement of emissions. If there is no real alternative or if the supply is insufficient, forests will continue to be used by communities. In Nepal, Karky and Skutsch [51] reported that introducing forest communities to the carbon market involves high opportunity costs because forests provide many non-monetary benefits to the local population, and indeed these are the main reason they conserve and manage them; the carbon credits will not be sufficient to cover the costs engendered by not exploiting forest resources. In other cases, compensation to local communities in exchange for differing livelihood benefits could be attractive enough to strengthen carbon storage benefits [26]. In tropical dry forests of Guinea-Bissau, Mali and Senegal, Skutsch and Ba [106] showed that even if only 10% of the financial return on the carbon value from reduced degradation and carbon stock enhancement were to reach the community, this would be a significant incentive for their participation in REDD+. Therefore, depending on the context, different incentive structures for REDD+ can be made to ensure that REDD+ contributes to the sustainability of CFM interventions.

In cases where the primary focus of CFM is for commercial timber production, with benefits in terms of additional income or direct employment, it is possible that REDD+ can strengthen existing forest enterprises or stimulate the creation of new businesses. The economic viability of community forest enterprises dedicated to timber extraction, as well as incomes, may depend *inter alia* on the volumes of timber harvested [107]. For commercial CFM activities, Putz and Romero [108] proposed synergies between the forest management certification and forest product legality as established by forest auditors, by facilitating on-the-ground verification and allocation of additional carbon incentives to these reduced impact operations. Tomaselli and Hajjar [109] argue that REDD+ direct support should be oriented towards the development and sustainability of community forest enterprises by creating a conducive business environment and fostering the provision of business development and appropriate financial services.

Other authors suggest that the real way forward is to extend the coverage of forests under CFM [100], by using REDD+ to promote access and recognition of use rights, as well as defining and securing community forest tenure [7,110]. Effectively, as shown above, there are large forest areas already managed by communities under customary institutions and for which tenure rights recognition could contribute to both positive forest and social outcomes. The transfer of tenure to communities however might not be sufficient on its own for facilitating positive forest preservation outcomes. A recent review of the impacts of tenure form and tenure security on forest cover change concluded that it is tenure security that is associated with less deforestation, not the form of the tenure itself [111]. Effectively, tenure recognition is only a first step [19,112]. The implementation of those rights, their defence and ensuring access to the benefits by communities is a process through which REDD+ could help address fairness issues. To defend those rights in the context of REDD+ implementation, Robinson, Albers, Meshack and Lokina [83] argue that CFM REDD will face external forest change pressures similar to those of all previous enforcement programs aimed at preventing deforestation. Support from other institutions will be necessary for enforcement and for protecting those rights.

As we have noted, fears have been expressed that REDD+, which if carried out under UNFCCC rules will imply a coordinated national programme, may result in a return to re-centralization and top-down management [18]. The only way to avoid this trap is to design appropriate institutions at different levels which mediate between the social, economic, and environmental factors that cause tropical deforestation [87,113,114], without removing the management authority from local communities [18]. CDM experience illustrates the need for more effective institutions working at multiple levels to integrate local forest management into a national and global framework [115], with institutions providing grievance and redress mechanisms accessible at the local level. The development of multi-level or nested governance as a way to integrate REDD+ at multiple scale [116–118] remains a major challenge for policymakers and practitioners.

4. Conclusions

There is clear evidence of positive outcomes of CFM on forest conditions and terms of carbon benefits. However, our study indicates that CFM is more successful in forest conservation, sustainable management of forest which results in reduced rates of degradation, and enhancing carbon stock, than in reducing deforestation. The performance of CFM has been shown to be equivalent or better than that of protected areas in terms of maintaining forest cover. Reducing deforestation can be achieved through CFM, but several other factors exogenous to the governance and control of communities have to be taken into account. In terms of social benefits, there is some evidence of positive outcomes, but it would be prudent to say that, although CFM could provide a contribution to poverty alleviation, it is by no means a panacea to rural poverty. It is also clear that CFM does not deal very well with equity issues at the local level. We obtained very few significant statistical relationships when testing between forest or social outcomes and CFM characteristics (types of ownership, extraction and arrangement as well as region), perhaps because of the large heterogeneity of contexts found for CFM.

Revisiting assumptions about CFM experiences in order to derive realistic expectations based on strong evidence is paramount for designing better interventions in forest communities under the REDD+ context. Our results on equity in CFM indicate that other mechanisms or interventions would have to be put in place to ensure equitable distribution of benefits for the poor at the local level. REDD+ brings a whole new set of challenges and access to benefits in the form of carbon payments is no guarantee of equitable distribution, given what web have observed from CFM experiences. Deliberate action from government is needed to provide incentives as well as complementary poverty reduction interventions to CFM-type interventions. Recognition of rights for forest communities will be an important first step in promoting sustainable landscapes, but to address fairness in REDD+, a better evaluation of synergies and trade-offs between the different stated objectives is needed.

Further interventions will be necessary to avoid possible negative impacts on existing CFM cases and on the forest communities already involved in REDD+. Improving institutional coordination, an equitable benefit-sharing mechanism and capacity-building for community forest carbon monitoring are important areas requiring attention [89].

In order to keep learning about how to improve interventions for forest carbon and livelihood outcomes, reliable research methods using solid methodology are needed. Even if there have been recent improvements, the variation in methods still inhibits comparisons and meta-analyses of case studies that would provide the necessary quantitative evidence for policy recommendations [119]. Matching techniques and other contextual controls are essential in the selection of sites for comparison [103], and the construction of credible counterfactuals [79] must be a key element in the evaluation methods used. More emphasis on analyzing the human and natural aspects concurrently is much needed to be able to reach strong conclusions about these complex interactions, as well as to test for synergies and trade-offs between the two and with contextual factors [88]. The construction of global, spatially-explicit datasets of CFM will be crucial to evaluate the national and global outcomes of this approach, especially with the major changes happening in forest governance in the context of REDD+.

Supplementary Materials: The following are available online at www.mdpi.com/1999-4907/7/8/170/s1,: List of references and case studies evaluation.

Acknowledgments: This study followed an international workshop on Community Forestry and REDD+ organized at Laval University (Québec, Canada) in October 2011. We are grateful to the Environment, Development and Society Institute (Institute EDS) for providing funding to this research as well as FQRNT and NASA SERVIR Grant NNX12AL27G to J.P.

Author Contributions: J.P. and N.G. conceived and designed the study; J.P. performed the review and analyzed the data, with the contribution of N.G. and M.S.; J.P., N.G. and M.S. wrote the paper.

Conflicts of Interest: The authors declare no conflict of interest.

References

1. FAO. *Fao Assessment of Forests and Carbon Stocks, 1990–2015*; Food and Agriculture Organization: Roma, Italy, 2015.
2. UNFCCC. *Work Programme on Results-Based Finance to Progress the Full Implementation of the Activities Referred to in Decision 1/cp.16, Paragraph 70. Warsaw Framework for REDD-plus 2013, Decision 9/CP.19*; United Framework Convention on Climate Change: Bonn, Germany, 2013.
3. UNFCCC. Adoption of the paris agreement (annex-paris agreement), fccc/cp/2015/l.9. In Proceedings of the Conference of the Parties Twenty-First Session, Paris, France, 30 November–11 December 2015.
4. UNFCCC. *The Cancun Agreements: Outcome of the Work of the Ad Hoc Working Group on Long-Term Cooperative Action under the Convention*; United Framework Convention on Climate Change: Bonn, Germany, 2010.
5. Agrawal, A.; Angelsen, A. Using community forest management to achieve REDD+ goals. In *Realising REDD+: National Strategy and Policy Options*; Angelsen, A., Brockhaus, M., Kanninen, M., Sills, E.O., Sunderlin, W.D., Wertz-Kanounnikoff, S., Eds.; CIFOR: Bogor, Indonesia, 2009; pp. 201–211.
6. Bluffstone, R. Economics of REDD+ and community forestry. *J. For. Livelihood* **2013**, *11*, 69–74. [CrossRef]
7. Stevens, C.; Winterbottom, R.; Springer, J.; Reytar, K. *Securing Rights, Combating Climate Change: How Strengthening Community Forest Rights Mitigates Climate Change*; World Resources Institute: Washington, DC, USA, 2014.
8. Keeping it in the community. *The Economist*, 23 September 2010.
9. Hodgdon, B.D.; Hayward, J.; Samayoa, O. Putting the plus first: Community forest enterprise as the platform for REDD+ in the maya biosphere reserve, guatemala. *Trop. Conser. Sci.* **2013**, *6*, 365–383.
10. UN-REDD. *REDD+ and Community Foresty, Revisited*; UN-REDD Programme: Bonn, Germany, 2013.
11. UN-REDD. *Pondering the Role of Community Forestry in REDD+*; UN-REDD Programme: Bonn, Germany, 2013.
12. Skutsch, M.; McCall, M.K. The Role of Community Forest Management in REDD+. Available online: http://www.redd.ciga.unam.mx/files/FAOSkutschMcCall_workingPaper.pdf (accessed on 4 June 2016).

13. Geist, H.J.; Lambin, E.F. Proximate causes and underlying driving forces of tropical deforestation. *Bioscience* **2002**, *52*, 143–150. [CrossRef]

14. Rudel, T.R.; DeFries, R.; Asner, G.; Laurance, W.F. Changing drivers of deforestation and new opportunities for conservation. *Conserv. Biol.* **2009**, *23*, 1396–1405. [CrossRef] [PubMed]

15. Cernea, M.; Schmidt-Soltau, K. Poverty risks and national parks: Policy issues in conservation and resettlement. *World Dev.* **2006**, *34*, 1808–1830. [CrossRef]

16. West, P.; Igoe, J.; Brockington, D. Parks and peoples: The social impact of protected areas. *Annu. Rev. Anthropol.* **2006**, *35*, 251–277. [CrossRef]

17. Neupane, S.; Shrestla, K.K. Sustainable forest governance in a changing climate: Impacts of REDD program on the livelihoods of poor communities in nepalese community forestry. *OIDA Int. J. Sustain. Dev.* **2012**, *4*, 71–81.

18. Phelps, J.; Webb, E.L.; Agrawal, A. Does REDD plus threaten to recentralize forest governance? *Science* **2010**, *328*, 312–313. [CrossRef] [PubMed]

19. Larson, A.M. Forest tenure reform in the age of climate change: Lessons for REDD+. *Glob. Environ. Chang.* **2011**, *21*, 540–549. [CrossRef]

20. Glasmeier, A.K.; Farrigan, T. Understanding community forestry: A qualitative metastudy of the concept, the process, and its potential for poverty alleviation in The United States case. *Geogr. J.* **2005**, *171*, 56–69. [CrossRef]

21. FAO. *Forestry for Local Community Development*; Food and Agriculture Organization: Rome, Italy, 1978.

22. Growing Forest Partnerships. The Forests Dialogue: Investing in Locally Controlled Forestry (ilcf). Available online: http://www.growingforestpartnerships.org/ (accessed on 25 November 2015).

23. Charnley, S.; Poe, M. Community forestry in theory and practice: Where are we now? *Annu. Rev. Anthropol.* **2007**, *36*, 301–336. [CrossRef]

24. Arnold, J.E.M. *Forests and People: 25 Years of Community Forestry*; Food and Agriculture Organization of the United Nations: Rome, Italy, 2001.

25. Hajjar, R. Community Forests for Forest Communities: An Examination of Power Imbalances, Challenges and Goals in Brazil and Mexico. Ph.D. Thesis, University of British Columbia, Vancouver, BC, Canada, 2011.

26. Chhatre, A.; Agrawal, A. Trade-offs and synergies between carbon storage and livelihood benefits from forest commons. *Proc. Natl. Acad. Sci. USA* **2009**, *106*, 17667–17670. [CrossRef] [PubMed]

27. Ostrom, E. *Governing the Commons: The Evolution of Institutions for Collective Action*; Cambridge University Press: Cambridge, UK, 1990; p. 280.

28. Chomitz, K.M.; Buys, P.; de Luca, G.; Thomas, T.; Wertz-Kanounnikoff, S. *At Loggerheads? Agricultural Expansion, Poverty Reduction, and Environment in the Tropical Forests*; The International Bank for Reconstruction and Development, The World Bank: Washington, DC, USA, 2007; p. 284.

29. Ribot, J.C.; Agrawal, A.; Larson, A.M. Recentralizing while decentralizing: How national governments reappropriate forest resources. *World Dev.* **2006**, *34*, 1864–1886. [CrossRef]

30. Larson, A.M. Decentralisation and forest management in Latin America: Towards a working model. *Public Adm. Dev.* **2003**, *23*, 211–226. [CrossRef]

31. Brown, H.C.P. Gender, climate change and REDD plus in the congo basin forests of Central Africa. *Int. For. Rev.* **2011**, *13*, 163–176.

32. Agrawal, A.; Chhatre, A.; Hardin, R. Changing governance of the world's forests. *Science* **2008**, *320*, 1460–1462. [CrossRef] [PubMed]

33. White, A.; Martin, A. *Who Owns the World's Forests? Forest Tenure and Public Forests in Transition*; Forest Trends and Center for International Environmental Law: Washington, DC, USA, 2002.

34. Sunderlin, W.; Hatcher, J.; Liddle, M. *From Exclusion to Ownership? Challenges and Opportunities in Advancing Forest Tenure Reforms*; Rights and Resources Initiative: Washington, DC, USA, 2008.

35. RRI. Forest Tenure Data Tool. Available online: http://www.rightsandresources.org/en/resources/tenure-data/tenure-data-tool/ (accessed on 4 June 2016).

36. Lawlor, K.; Madeira, E.M.; Blockhus, J.; Ganz, D.J. Community participation and benefits in REDD+: A review of initial outcomes and lessons. *Forests* **2013**, *4*, 296–318. [CrossRef]

37. Pullin, A.S.; Stewart, G.B. Guidelines for systematic review in conservation and environmental management. *Conserv. Biol.* **2006**, *20*, 1647–1656. [CrossRef] [PubMed]

38. Legendre, P.; Legendre, L. *Numerical Ecology*, 3rd ed.; Elsevier Science BV: Amsterdam, The Netherlands, 2012.

39. R Core Team R. *A Language and Environment for Statistical Computing*; R Foundation for Statistical Computing: Vienna, Austria, 2015.

40. Nepstad, D.; Schwartzman, S.; Bamberger, B.; Santilli, M.; Ray, D.; Schlesinger, P.; Lefebvre, P.; Alencar, A.; Prinz, E.; Fiske, G.; et al. Inhibition of amazon deforestation and fire by parks and indigenous lands. *Conserv. Biol.* **2006**, *20*, 65–73. [CrossRef] [PubMed]

41. Armenteras, D.; Rodríguez, N.; Retana, J. Are conservation strategies effective in avoiding the deforestation of the colombian guyana shield? *Biol. Conser. Soc.* **2009**, *142*, 1411–1419. [CrossRef]

42. Nelson, A.; Chomitz, K.M. Effectiveness of strict vs. multiple use protected areas in reducing tropical forest fires: A global analysis using matching methods. *PLoS ONE* **2011**. [CrossRef] [PubMed]

43. Porter-Bolland, L.; Ellis, E.A.; Guariguata, M.R.; Ruiz-Mallén, I.; Negrete-Yankelevich, S.; Reyes-García, V. Community managed forests and forest protected areas: An assessment of their conservation effectiveness across the tropics. *For. Ecol. Manag.* **2012**, *268*, 6–17. [CrossRef]

44. Bray, D.B.; Antinori, C.; Torres-Rojo, J.M. The mexican model of community forest management: The role of agrarian policy, forest policy and entrepreneurial organization. *For. Policy Econ.* **2006**, *8*, 470–484. [CrossRef]

45. Bray, D.B.; Ellis, E.A.; Armijo-Canto, N.; Beck, C.T. The institutional drivers of sustainable landscapes: A case study of the "mayan zone" in quintana roo, Mexico. *Land Use Policy* **2004**, *21*, 333–346. [CrossRef]

46. Ellis, E.A.; Porter-Bolland, L. Is community-based forest management more effective than protected areas? A comparison of land use/land cover change in two neighboring study areas of the central yucatan peninsula, Mexico. *For. Ecol. Manag.* **2008**, *256*, 1971–1983. [CrossRef]

47. Skutsch, M.; Solis, S. How much carbon does community forest management save? The results of k:Tgal´s field measurements. *Commu. For. Monito. Carbon Market* **2011**, *16*. Available online: http://www.communitycarbonforestry.org/NewPublications/How%20much%20carbon%20does% 20community%20forest%20management%20save%20website%20version.pdf (accessed on 4 June 2016).

48. Klooster, D.; Masera, O. Community forest management in Mexico: Carbon mitigation and biodiversity conservation through rural development. *Glob. Environ. Chang.* **2000**, *10*, 259–272. [CrossRef]

49. Hayes, T. Parks, people, and forest protection: An institutional assessment of the effectiveness of protected areas. *World Dev.* **2006**, *34*, 2064–2075. [CrossRef]

50. Somanathan, E.; Prabhakar, R.; Singh, M.B. Decentralization for cost-effective conservation. *Proc. Natl. Acad. Sci. USA* **2009**, *106*, 4143–4147. [CrossRef] [PubMed]

51. Karky, B.S.; Skutsch, M. The cost of carbon abatement through community forest management in Nepal Himalaya. *Ecol. Econ.* **2010**, *69*, 666–672. [CrossRef]

52. Dalle, S.P.; de Blois, S.; Caballero, J.; Johns, T. Integrating analyses of local land-use regulations, cultural perceptions and land-use/land cover data for assessing the success of community-based conservation. *For. Ecol. Manag.* **2006**, *222*, 370–383. [CrossRef]

53. Geist, H.J.; Lambin, E. *What Drives Tropical Deforestation? A Meta-Analysis of Proximate and Underlying Causes of Deforestation Based on Subnational Case Study Evidence*; International Human Dimensions Programme on Global Environmental Change (IHDP) International Geosphere-Biosphere Programme (IGBP): Louvain-la-Neuve, Belgium, 2001; p. 136.

54. Tacconi, L. Decentralization, forests and livelihoods: Theory and narrative. *Glob. Environ. Chang.* **2007**, *17*, 338–348. [CrossRef]

55. Blaikie, P. Is small really beautiful? Community-based natural resource management in Malawi and Botswana. *World Dev.* **2006**, *34*, 1942–1957. [CrossRef]

56. Alix-Garcia, J.; de Janvry, A.; Sadoulet, E. A tale of two communities: Explaining deforestation in Mexico. *World Dev.* **2005**, *33*, 219–235. [CrossRef]

57. Baland, J.-M.; Bardhan, P.; Das, S.; Mookherjee, D. Forests to the people: Decentralization and forest degradation in the Indian Himalayas. *World Dev.* **2010**, *38*, 1642–1656. [CrossRef]

58. Bray, D.B.; Duran, E.; Ramos, V.H.; Mas, J.F.; Velazques, A.; McNab, R.B.; Barry, D.; Radachowsky, J. Tropical deforestation, community forests, and protected areas in the maya forest. *Ecol. Soc.* **2008**, *13*, 56.

59. Bowler, D.E.; Buyung-Ali, L.M.; Healey, J.R.; Jones, J.P.G.; Knight, T.M.; Pullin, A.S. Does community forest management provide global environmental benefits and improve local welfare? *Front. Ecol. Environ.* **2012**, *10*, 29–36. [CrossRef]

60. Lund, J.F.; Treue, T. Are we getting there? Evidence of decentralized forest management from the tanzanian miombo woodlands. *World Dev.* **2008**, *36*, 2780–2800. [CrossRef]

61. Poffenberger, M. People in the forest: Community forestry experiences from southeast Asia. *Int. J. Environ. Sustain. Dev.* **2006**, *5*, 57–69. [CrossRef]
62. Vergara-Asenjo, G.; Potvin, C. Forest protection and tenure status: The key role of Indigenous peoples and protected areas in Panama. *Glob. Environ. Chang.* **2014**, *28*, 205–215. [CrossRef]
63. Walker, W.; Baccini, A.; Schwartzman, S.; Ríos, S.; Oliveira-Miranda, M.; Augusto, C.; Romero Ruiz, M.; Arrasco, C.S.; Ricardo, B.; Smith, R.; et al. Forest carbon in Amazonia: The unrecognized contribution of indigenous territories and protected natural areas. *Carbon Manag.* **2014**, *5*, 479–485. [CrossRef]
64. Wunder, S.; Börner, J.; Shively, G.; Wyman, M. Safety nets, gap filling and forests: A global-comparative perspective. *World Dev.* **2014**, *5*, 479–485. [CrossRef]
65. Sunderlin, W.D.; Dewi, S.; Puntodewo, A.; Müller, D.; Angelsen, A.; Epprecht, M. Why forests are important for global poverty alleviation: A spatial explanation. *Ecol. Soc.* **2008**, *13*, 24.
66. Sunderlin, W.D.; Angelsen, A.; Belcher, B.; Burgers, P.; Nasi, R.; Santoso, L.; Wunder, S. Livelihoods, forests, and conservation in developing countries: An overview. *World Dev.* **2005**, *33*, 1383–1402. [CrossRef]
67. Wunder, S. Poverty alleviation and tropical forests-what scope for synergies? *World Dev.* **2001**, *29*, 1817–1833. [CrossRef]
68. Edmunds, D.; Wollenberg, E. *Local Forest Management: The Impacts of Devolution Policies*; Earthscan: London, UK, 2003.
69. Mahanty, S.; Guernier, J.; Yasmi, Y. A fair share? Sharing the benefits and costs of collaborative forest management. *Int. For. Rev.* **2009**, *11*, 268–280. [CrossRef]
70. Ameha, A.; Nielsen, O.J.; Larsen, H.O. Impacts of access and benefit sharing on livelihoods and forest: Case of participatory forest management in Ethiopia. *Ecol. Econ.* **2014**, *97*, 162–171. [CrossRef]
71. Maharjan, M.R.; Ram Dakal, T.; SureshThapa, K.; Schreckenberg, K.; Luttrell, C. Improving the benefits to the poor from community forestry in the churia region of Nepal. *Int. For. Rev.* **2009**, *11*, 254–267. [CrossRef]
72. Vyamana, V.G. Participatory forest management in the eastern arc mountains of Tanzania: Who benefits? *Int. For. Rev.* **2009**, *11*, 239–253. [CrossRef]
73. Jumbe, C.B.L.; Angelsen, A. Do the poor benefit from devolution policies? Evidence from Malawi's forest co-management program. *Land Econ.* **2006**, *82*, 562–581. [CrossRef]
74. Hajjar, R.; Kozak, R.; Innes, J. Is decentralization leading to "real" decision-making power for forest-dependent communities? Case studies from Mexico and Brazil. *Ecol. Soc.* **2012**. [CrossRef]
75. Agrawal, A.; Gibson, C.C. Enchantment and disenchantment: The role of community in natural resource conservation. *World Dev.* **1999**, *27*, 629–649. [CrossRef]
76. Schreckenberg, K.; Luttrell, C. Participatory forest management: A route to poverty reduction? *Int. For. Rev.* **2009**, *11*, 221–238. [CrossRef]
77. Malla, Y.B. Impact of community forestry policy on rural livelihoods and food security in Nepal. *Unasylva* **2000**, *51*, 37–45.
78. Malla, Y.B.; Neupane, H.R.; Branney, P.J. Why aren't poor people benefiting more from community forestry? *J. For. Livelihood* **2003**, *3*, 78–92.
79. Caplow, S.; Jagger, P.; Lawlor, K.; Sills, E. Evaluating land use and livelihood impacts of early forest carbon projects: Lessons for learning about REDD+. *Environ. Sci. Policy* **2011**, *14*, 152–167. [CrossRef]
80. Ferraro, P.J.; Pattanayak, S.K. Money for nothing? A call for empirical evaluation of biodiversity conservation investments. *PLoS Biol.* **2006**, *4*, 482–488. [CrossRef] [PubMed]
81. Persha, L.; Agrawal, A.; Chhatre, A. Social and ecological synergy: Local rulemaking, forest livelihoods, and biodiversity conservation. *Science* **2011**, *331*, 1606–1608. [CrossRef] [PubMed]
82. Pagdee, A.; Kim, Y.-S.; Daugherty, P.J. What makes community forest management successful: A meta-study from community forests throughout the world. *Society Nat. Res.* **2006**, *19*, 33–52. [CrossRef]
83. Robinson, E.J.Z.; Albers, H.J.; Meshack, C.; Lokina, R.B. Implementing REDD through community-based forest management: Lessons from Tanzania. *Nat. Res. Forum* **2013**, *37*, 141–152. [CrossRef]
84. Ostrom, E. *Self-Governance and Forest Resources*; Center for International Forestry Research: Bogor, Indonesia, 1999; pp. 1–13.
85. Chhatre, A.; Agrawal, A. Forest commons and local enforcement. *Proc. Natl. Acad. Sci. USA* **2008**, *105*, 13286–13291. [CrossRef] [PubMed]
86. Bluffstone, R.; Robinson, E.; Guthiga, P. Redd+ and community-controlled forests in low-income countries: Any hope for a linkage? *Ecol. Econ.* **2013**, *87*, 43–52. [CrossRef]

87. Hayes, T.; Persha, L. Nesting local forestry initiatives: Revisiting community forest management in a REDD plus world. *For. Policy Econ.* **2010**, *12*, 545–553. [CrossRef]
88. Newton, P.; Oldekop, J.A.; Brodnig, G.; Karna, B.K.; Agrawal, A. Carbon, biodiversity, and livelihoods in forest commons: Synergies, trade-offs, and implications for REDD+. *Environ. Res. Lett.* **2016**, *11*, 044017. [CrossRef]
89. Newton, P.; Schaap, B.; Fournier, M.; Cornwall, M.; Rosenbach, D.W.; DeBoer, J.; Whittemore, J.; Stock, R.; Yoders, M.; Brodnig, G. Community forest management and REDD+. *For. Policy Econ.* **2015**, *56*, 27–37. [CrossRef]
90. Thomas, S.; Dargusch, P.; Harrison, S.; Herbohn, J. Why are there so few afforestation and reforestation clean development mechanism projects? *Land Use Policy* **2010**, *27*, 880–887. [CrossRef]
91. Benessaiah, K. Carbon and livelihoods in post-kyoto: Assessing voluntary carbon markets. *Ecol. Econ.* **2012**, *77*, 1–6. [CrossRef]
92. Lasco, R.D.; Evangelista, R.S.; Pulhin, F.B. Potential of community-based forest management to mitigate climate change in the Philippines. *Small Scale For.* **2010**, *9*, 429–443. [CrossRef]
93. Boyd, E.; Gutierrez, M.; Chang, M. Small-scale forest carbon projects: Adapting cdm to low-income communities. *Glob. Environ. Chang.* **2007**, *17*, 250–259. [CrossRef]
94. Coomes, O.T.; Grimard, F.; Potvin, C.; Sima, P. The fate of tropical forest: Carbon or cattle? *Ecol. Econ.* **2008**, *65*, 207–212. [CrossRef]
95. Asquith, N.M.; Vargas Rios, M.T.; Smith, J. Can forest-protection carbon projects improve rural livelihoods? Analysis of the noel kempff mercado climate action project, bolivia. *Mitig. Adapt. Strateg. Glob. Chang.* **2002**, *7*, 323–337. [CrossRef]
96. Jindal, R.; Kerr, J.M.; Carter, S. Reducing poverty through carbon forestry? Impacts of the N'hambita community carbon project in mozambique. *World Dev.* **2012**, *40*, 2123–2135. [CrossRef]
97. Beymer-Farris, B.A.; Bassett, T.J. The REDD menace: Resurgent protectionism in Tanzania's mangrove forests. *Glob. Environ. Chang.* **2012**, *22*, 332–341. [CrossRef]
98. Maraseni, T.N.; Neupane, P.R.; Lopez-Caseroc, F.; Cadman, T. An assessment of the impacts of the REDD pilot project on community forests user groups (cfugs) and their community forests in Nepal. *J. Environ. Manag.* **2014**, *136*, 37–46. [CrossRef] [PubMed]
99. Yamanoshita, M.Y.; Amano, M. Capability development of local communities for project sustainability in afforestation/reforestation clean development mechanism. *Mitig. Adapt. Strateg. Glob. Chang.* **2012**, *17*, 425–440. [CrossRef]
100. Balooni, K.; Lund, J.F. Forest rights: The hard currency of REDD+. *Conser. Lett.* **2014**, *7*, 278–284. [CrossRef]
101. Skutsch, M.; Vickers, B.; Georgiadou, Y.; McCall, M. Alternative models for carbon payments to communities under REDD+: A comparison using the polis model of actor inducements. *Environ. Sci. Policy* **2011**, *14*, 140–151. [CrossRef]
102. Skutsch, M.; McCall, M.; Larrazábal, A. Balancing views on community monitoring: The case of REDD+. *Biodivers. Conser.* **2014**, *23*, 233–236. [CrossRef]
103. Wunder, S.; Angelsen, A.; Belcher, B. Forests, livelihoods, and conservation: Broadening the empirical base. *World Dev.* **2014**. [CrossRef]
104. Gregersen, H.; el Lakany, H.; Karsenty, A.; White, A. *Does the Opportunity Cost Approach Indicate the Real Cost of REDD+? Rights and Realities of Paying for REDD+*; Rights and Resources Initiative: Washington, DC, USA, 2010; p. 23.
105. Fisher, B.; Lewis, S.L.; Burgess, N.D.; Malimbwi, R.E.; Munishi, P.K.; Swetnam, R.D.; Turner, R.K.; Willcock, S.; Balmford, A. Implementation and opportunity costs of reducing deforestation and forest degradation in Tanzania. *Nat. Clim. Chang.* **2011**, *1*, 161–164. [CrossRef]
106. Skutsch, M.M.; Ba, L. Crediting carbon in dry forests: The potential for community forest management in West Africa. *For. Policy Econ.* **2010**, *12*, 264–270. [CrossRef]
107. Humphries, S.; Holmes, T.P.; Kainer, K.; Koury, C.G.G.A.; Cruz, E.; de Miranda Rocha, R. Are community-based forest enterprises in the tropics financially viable? Case studies from the Brazilian Amazon. *Ecol. Econ.* **2012**, *77*, 62–73. [CrossRef]
108. Putz, F.E.; Romero, C. Helping curb tropical forest degradation by linking REDD+ with other conservation interventions: A view from the forest. *Curr. Opin. Environ. Sustain.* **2012**, *4*, 670–677. [CrossRef]

109. Tomaselli, M.F.; Hajjar, R. Promoting community forestry enterprises in national REDD+ strategies: A business approach. *Forests* **2011**, *2*, 283–300. [CrossRef]
110. Seymour, F.; la Vina, T.; Hite, K. *Evidence Linking Community Level Tenure and Forest Condition: An Annotated Bibliography*; Climate and Land Use Alliance: San Francisco, CA, USA, 2014; p. 61.
111. Robinson, B.E.; Holland, M.B.; Naughton-Treves, L. Does secure land tenure save forests? A meta-analysis of the relationship between land tenure and tropical deforestation. *Glob. Environ. Chang.* **2013**, *29*, 281–293. [CrossRef]
112. Agrawal, A.; Nepstad, D.; Chhatre, A. Reducing emissions from deforestation and forest degradation. *Annu. Rev. Environ. Resour.* **2011**, *36*, 373–396. [CrossRef]
113. Karsenty, A. The architecture of proposed redd schemes after Bali: Facing critical choice. *Int. For. Rev.* **2008**, *10*, 443–457. [CrossRef]
114. Poffenberger, M. Cambodia's forests and climate change: Mitigating drivers of deforestation. *Nat. Resour. Forum* **2009**, *33*, 285–296. [CrossRef]
115. Boyd, E. Governing the clean development mechanism: Global rhetoric versus local realities in carbon sequestration projects. *Environ. Plan. A* **2009**, *41*, 2380–2395. [CrossRef]
116. Doherty, E.; Schroeder, H. Forest tenure and multi-level governance in avoiding deforestation under REDD+. *Glob. Environ. Politics* **2011**, *11*, 66–88. [CrossRef]
117. Lawlor, K.; Weinthal, E.; Lander, L. Institutions and policies to protect rural livelihoods in REDD plus regimes. *Glob. Environ. Politics* **2010**, *10*, 1–12. [CrossRef]
118. Pettenella, D.; Brotto, L. Governance features for successful REDD plus projects organization. *For. Policy Econ.* **2012**, *18*, 46–52. [CrossRef]
119. Lund, J.F.; Balooni, K.; Casse, T. Change we can believe in? Reviewing studies on the conservation impact of popular participation in forest management. *Conserv. Soc.* **2009**, *7*, 71–82. [CrossRef]

MDPI AG

St. Alban-Anlage 66

4052 Basel, Switzerland

Tel. +41 61 683 77 34

Fax +41 61 302 89 18

http://www.mdpi.com

Forests Editorial Office

E-mail: forests@mdpi.com

http://www.mdpi.com/journal/forests

www.ingramcontent.com/pod-product-compliance
Lightning Source LLC
Chambersburg PA
CBHW051706210326
41597CB00032B/5389